Biosystems Engineering

Ahindra Nag

McGraw Hill

New York Chicago San Francisco
Lisbon London Madrid Mexico City
Milan New Delhi San Juan
Seoul Singapore Sydney Toronto

The McGraw·Hill Companies

Cataloging-in-Publication Data is on file with the Library of Congress.

McGraw-Hill books are available at special quantity discounts to use as premiums and sales promotions, or for use in corporate training programs. To contact a representative please e-mail us at bulksales@mcgraw-hill.com.

Biosystems Engineering

Copyright ©2010 by The McGraw-Hill Companies, Inc. All rights reserved. Printed in the United States of America. Except as permitted under the United States Copyright Act of 1976, no part of this publication may be reproduced or distributed in any form or by any means, or stored in a data base or retrieval system, without the prior written permission of the publisher.

1 2 3 4 5 6 7 8 9 0   DOC/DOC   0 1 4 3 2 1 0 9

ISBN 978-0-07-160628-8
MHID 0-07-160628-9

The pages within this book were printed on acid-free paper.

Sponsoring Editor
Taisuke Soda

Acquisitions Coordinator
Michael Mulcahy

Editorial Supervisor
David E. Fogarty

Project Manager
Somya Rustagi, International Typesetting and Composition

Copy Editor
Susan Giniger

Proofreader
Priyanka Sinha, International Typesetting and Composition

Indexer
Arc Films, Inc.

Production Supervisor
Pamela A. Pelton

Composition
International Typesetting and Composition

Art Director, Cover
Jeff Weeks

Information contained in this work has been obtained by The McGraw-Hill Companies, Inc. ("McGraw-Hill") from sources believed to be reliable. However, neither McGraw-Hill nor its authors guarantee the accuracy or completeness of any information published herein, and neither McGraw-Hill nor its authors shall be responsible for any errors, omissions, or damages arising out of use of this information. This work is published with the understanding that McGraw-Hill and its authors are supplying information but are not attempting to render engineering or other professional services. If such services are required, the assistance of an appropriate professional should be sought.

Contents

Contributors xiii
Preface xv

1 **Microarray Data Analysis Using Machine Learning Methods** 1
   1.1 Introduction 1
   1.2 Machine Learning Methods 3
       1.2.1 Neural Networks 4
       1.2.2 Support Vector Machines 7
       1.2.3 Fuzzy Systems 7
       1.2.4 Genetic Algorithms 9
       1.2.5 Particle Swarm Optimization 9
       1.2.6 Steps in Developing Machine Learning Models 10
   1.3 Microarray Technology 13
       1.3.1 cDNA Microarray 13
       1.3.2 High-Density Oligonucleotide Array 15
   1.4 Low-Level Analysis 16
   1.5 High-Level Analysis 18
       1.5.1 Clustering 18
       1.5.2 Classification 20
       1.5.3 Genetic Network Modeling 21
   1.6 Summary 26
   References 27
   Glossary of Terms 31

2 **Biosystems Analysis and Optimization** 33
   2.1 Introduction 33
       2.1.1 Definitions 33
   2.2 System Representation 34
       2.2.1 Block Diagram Representation 34
       2.2.2 Mathematical Representation 36
       2.2.3 The Laplace Transform 38
       2.2.4 Nonlinearities 44
   2.3 System Analysis 47
       2.3.1 Response of the System to a Step Input 47
       2.3.2 Magnitude and Phase of a Transfer Function 48
       2.3.3 Bode Diagram of Transfer Functions 51

Contents

2.4	System Identification		57
	2.4.1 Excitation Experiments		59
	2.4.2 Evaluating Experimental Data		60
	2.4.3 Estimating Nonlinearities		62
	2.4.4 Parameter Estimation		64
2.5	Feedback Controller Design		67
	2.5.1 Feedback Control Structure		67
	2.5.2 Stability		68
	2.5.3 Loop-Shaping Controller Design		69
	2.5.4 Controller Design Example: Depth Control for Slurry Injection		73
2.6	Nonlinear Optimization Example: Cruise Control on a Combine Harvester		75
	2.6.1 Motivation		76
	2.6.2 System Description		77
	2.6.3 Control Objective		77
	2.6.4 System Modeling		78
	2.6.5 Nonlinear MPC		83
	2.6.6 Experimental Setup		86
	2.6.7 Results		86
References			88

3 Soil and Water Conservation **91**

3.1	Introduction	91
3.2	Hydrological Cycle	92
3.3	Precipitation	96
	3.3.1 Rainfall Measurement	97
	3.3.2 Mean Areal Rainfall Depth	98
3.4	Evaporation and Evapotranspiration	99
	3.4.1 Some Relevant Concepts and Terminology	100
	3.4.2 Energy Balance Method for Estimating Evaporation	101
	3.4.3 Dalton's Law	102
	3.4.4 Pan Evaporation	103
	3.4.5 Evapotranspiration	103
3.5	Infiltration and Runoff	104
	3.5.1 Infiltration	105
	3.5.2 Runoff	107
	3.5.3 SCS Curve Number Method	108
	3.5.4 Rational Method	109
3.6	Soil Erosion and Sediment Transport	110
	3.6.1 Mechanics of Soil Detachment and Deposition	110

		3.6.2	Factors Affecting Soil Erosion and	
			Sediment Yield	112
		3.6.3	Types of Soil Erosion	113
		3.6.4	Universal Soil Loss Equation	114
	3.7	Best Management Practices		117
		3.7.1	Commonly Used BMPs	118
	References			123

4 Models for Heat Transfer in Heated Substrates ... 125
- 4.1 Heat Transfer in Soils ... 125
  - 4.1.1 The Soil Thermal Regime ... 125
  - 4.1.2 Factors Affecting Soil Temperature ... 126
- 4.2 Soil Thermal Properties ... 128
  - 4.2.1 Specific Heat or Heat Capacity ... 128
  - 4.2.2 Effective Thermal Conductivity ... 128
  - 4.2.3 Effective Thermal Diffusivity ... 130
  - 4.2.4 Determination of Soil Thermal Properties ... 130
- 4.3 Models for Predicting Soil Temperature ... 139
  - 4.3.1 Applications of the Models for Predicting Soil Temperature ... 139
  - 4.3.2 Methods Used to Build Models for Predicting Soil Temperature ... 142
- 4.4 Greenhouse Substrate Heating ... 146
- 4.5 Models for Predicting Temperature in Heated Substrates ... 148
- 4.6 An Analysis of Electric Cable Heating Systems ... 154
- References ... 160

5 Geographic Information System-Based Watershed Modeling Systems ... 169
- 5.1 Introduction ... 169
- 5.2 Watershed Models ... 170
  - 5.2.1 Need for Watershed Models ... 170
  - 5.2.2 Origin of Watershed Models ... 171
  - 5.2.3 Characterization of Watershed Models ... 171
  - 5.2.4 Important Components of Watershed Models ... 172
  - 5.2.5 Examples of Commonly Used Watershed Models ... 175

	5.3	Geographic Information Systems (GIS)	181
	5.4	GIS and Watershed Models	182
		5.4.1 Approaches for Interfacing GIS with Watershed Models	184
		5.4.2 Challenges with Interfacing	185
		5.4.3 Recent State-of-the Art GIS–Integrated Watershed Modeling Systems	186
	5.5	Future of GIS-based Watershed Modeling Systems	190
	References		191
6	**Design of Sustainable Water Management Systems**		**195**
	6.1	Introduction	195
	6.2	Physical Properties of Soil	196
	6.3	Water Management	197
		6.3.1 Plant Water Relations	197
		6.3.2 Assessing Plant Water Needs	198
	6.4	Need for Irrigation	200
		6.4.1 Meeting the Crop Water Demand	200
		6.4.2 Other Benefits of Irrigation	201
	6.5	Irrigation Systems	202
		6.5.1 Introduction	202
		6.5.2 Types of Irrigation Systems	202
		6.5.3 Irrigation Efficiency	202
		6.5.4 Design of Irrigation Systems	202
	6.6	Drainage Systems	207
		6.6.1 Importance of Drainage	207
		6.6.2 Methods for Dealing with Excess Water	207
	6.7	Salinity Control	209
		6.7.1 Introduction to Salinization	210
		6.7.2 Methods to Control Salinization	210
		6.7.3 Dealing with Water Quality Issues	211
	6.8	Summary	211
	References		211
7	**Biomass Pyrolysis and Bio-Oil Refineries**		**213**
	7.1	Introduction	213
	7.2	Biomass Composition	215
		7.2.1 Cellulose	215
		7.2.2 Hemicellulose	216
		7.2.3 Lignin	216

		7.2.4	Extractives	217
		7.2.5	Ash	217
	7.3	Overview of Biomass Pyrolysis Reactions		217
		7.3.1	Mechanism of Primary Reactions	218
	7.4	Single-Particle Models		220
	7.5	Pyrolysis Technologies		222
		7.5.1	Slow Pyrolysis	222
		7.5.2	Fast Pyrolysis	223
	7.6	Crude Bio-Oils		226
		7.6.1	Chemical Composition of Crude Bio-Oils	226
		7.6.2	Fuel Applications of Crude Bio-Oils	231
		7.6.3	Producing Chemicals from Bio-Oils	234
		7.6.4	Production of Transportation Fuels	238
	7.7	Bio-Oil Refineries		240
	7.8	Summary and Conclusions		242
	References			242
8	**Performance and Emissions of Biodiesel and Ethanol in Engines**			**253**
	8.1	Introduction		253
		8.1.1	Transesterification of Vegetable Oils	256
		8.1.2	Analyzing the Products	257
		8.1.3	Property Measurement of Transesterified Oils	258
		8.1.4	Measurement of Performance and Emission	260
		8.1.5	Comparing the Effect of Loads on Biodiesel Performance	260
		8.1.6	Comparison of the Effect of Load on Biodiesel Emission	261
		8.1.7	Exhaust Gas Temperature	265
	8.2	Ethanol as an Alternative Fuel		266
		8.2.1	Production of Ethanol	267
		8.2.2	Effect of Brake Power on Brake Thermal Efficiency	271
		8.2.3	Effect of Brake Power on Brake-Specific Diesel Fuel Consumption	272

		8.2.4	Effect of Brake Power on Diesel Substitutions at Various Ethanol Fumigation Rates	273
		8.2.5	Effect of Brake Power on NO_x Emission Levels at Various Ethanol Fumigation Rates	274
		8.2.6	Effect of Brake Power on CO Emissions at Various Ethanol Fumigation Rates	275
		8.2.7	Effect of Brake Power on Smoke Number at Various Ethanol Fumigation Rates	276
		8.2.8	Effect of Brake Power on Exhaust Gas Temperatures at Various Ethanol Fumigation Rates	277
	References			278
9	**Bioseparation Processes**			281
	9.1	Introduction		281
	9.2	Different Stages of Bioseparation Process		282
	9.3	Unit Operations in Bioseparation		284
		9.3.1	Disruption of Cells	284
		9.3.2	Centrifugal Separation	288
		9.3.3	Thickening	290
		9.3.4	Flocculation	291
		9.3.5	Filtration	292
		9.3.6	Evaporation	296
		9.3.7	Drying and Crystallization	297
		9.3.8	Chromatographic Techniques	300
		9.3.9	Membrane Technology	305
	Further Reading			308
10	**Food Safety Management**			309
	10.1	Introduction		309
	10.2	Zoonotic Foodborne Hazards		310
	10.3	Food Safety Measures		311
		10.3.1	World Trade	311
		10.3.2	European Trade	312
		10.3.3	Food Safety Targets	314
	10.4	Risk Analysis		315
		10.4.1	Stages in Risk Assessment	318
		10.4.2	Methodologies	319
	10.5	Perspective		320
	10.6	Conclusions		320
	References			321

Contents

11 Food Package Engineering **325**
  11.1 Introduction 325
  11.2 Packaging Materials 326
    11.2.1 Packaging Polymers 326
    11.2.2 Glass Packaging 332
    11.2.3 Metal Packaging 334
    11.2.4 Paper and Paper-Based
            Packaging 337
  11.3 Physical Properties of Packaging
        Materials 338
    11.3.1 Mechanical Properties 339
    11.3.2 Thermal Properties 340
    11.3.3 Optical Properties 341
    11.3.4 Properties of Mass Transport ... 342
  11.4 Recent Advances in Packaging 344
  References 346

**12 Evaluation of Transgenic Wood for Wood
     Productivity and Quality** **347**
  12.1 Introduction 347
  12.2 Overview of Transgenic Woods 348
    12.2.1 Transgenic Trees 348
    12.2.2 Challenges for Evaluation
            and Safety 350
  12.3 Productivity and Quality of Transgenic
        Wood 351
    12.3.1 Growth Rate 351
    12.3.2 Physical and Mechanical
            Properties 352
    12.3.3 Chemical Composition 355
    12.3.4 Durability 358
    12.3.5 Advanced Analysis Tools 359
    12.3.6 Impacts on Process and
            Utilization 363
  12.4 Conclusions 364
  References 364

**13 Extraction, Refining, and Stabilization
     of Edible Oils** **367**
  13.1 Introduction 367
  13.2 Determining Oil Content in Oil Seeds
        and Its Extraction 369
    13.2.1 Extracting Oil from Seeds 370
  13.3 Refining Edible Oil 374
  13.4 Bleaching and Deodorization 375

		13.5 Stability of Edible Oils and Antioxidants	379
		13.5.1 Mechanism of Antioxidants	380
		13.5.2 Functions of Antioxidants	382
		13.5.3 Different Types of Antioxidants	383
	13.6	Measuring Antioxidant Activities	385
	References	....	389
14	**Phenolic Substances from Olives (*Olea europeae*) and Related Olive Mill Products**		**391**
	14.1	Introduction	391
	14.2	Olives and Relative Olive Mill Products	392
	14.3	Nature of Olive Polyphenols	394
	14.4	Antioxidant and Nutraceutical Properties ...	398
	14.5	Future Biotechnological Applications	400
	References	....	401
15	**Effect of Exogenous Bioregulators on the Mineral Composition and Storability of Fruits**		**405**
	15.1	Introduction	405
		15.1.1 Auxins and the Mineral Status of Fruits	406
		15.1.2 Exogenous Use of Gibberellins	408
		15.1.3 Retardants Are Also Bioregulators ..	409
		15.1.4 Effect of Retardants on the Mineral Composition of Fruits	410
		15.1.5 Effect of Retardants on Calcium Uptake	412
	15.2	Fruit Quality	413
	15.3	Influence of Retardant and Auxin Treatment of Apple Shoots and Fruits on Calcium Uptake and Distribution	423
	15.4	The Effect of Fruitlet Thinning Using Bioregulators on Mineral Content and Storage Quality of Fruits	429
	15.5	The Influence of Thinning and Treatment with Calcium Chloride on the Quality of McIntosh Apples Treated with SADH	437
		15.5.1 Apple Size	439
		15.5.2 Red Color of Apples	440
		15.5.3 Ground Color of Skin	440
		15.5.4 Apple Firmness	440
		15.5.5 Physiological Disorders of Apples during Storage	440
		15.5.6 Chemical Composition of Apples ...	442
	References	....	442

16	**Supercritical Fluid Extraction Applications for Biosystems Engineering**		**447**
	16.1	Introduction	447
	16.2	Supercritical Mass-Transfer Mechanisms	450
	16.3	Carbon Dioxide Explosion Processes	451
	16.4	Biological Systems Applications	453
		16.4.1 Fungal Production of Extractable Oils	454
		16.4.2 Extracting Oil from Rice Bran	460
	16.5	Summary and Future Trends	463
	References		463
17	**Agriculture Management**		**469**
	17.1	Introduction	469
	17.2	Modern Agriculture Systems	470
		17.2.1 Land-Use Planning and Irrigation Management	470
		17.2.2 Preparing and Testing the Soil	470
		17.2.3 Seed Testing	480
		17.2.4 Agricultural Mechanization	483
		17.2.5 Manure and Vermicomposting	485
		17.2.6 Biological Control	499
		17.2.7 Harvesting	501
	References		502
	Index		**505**

About the Editor

Ahindra Nag, Ph.D., is a senior assistant professor in the Department of Chemistry at the Indian Institute of Technology, Kharagpur. He has three years of industrial and 21 years of teaching experience and has published 70 research papers in major national and international journals. Dr. Nag has guided six Ph.D. students. He was invited as a visiting professor in Academia Sinica (Taiwan), University of Campobasso (Italy), and Universidad de Córdoba (Spain). Dr. Nag is the author of five other books: *Analytical Techniques in Agriculture, Biotechnology, and Environmental Engineering; Environmental Education and Solid Waste Management; Foundry Materials and Pollution; Textbook of Agriculture Biotechnology;* and *Biofuels Refining and Performance.*

Contributors

Alina Basak *Plant Bioregulators Department, Research Institute of Pomology and Floriculture, Skierniewice, Poland* (CHAP. 15)

Tom Coen *INDUCT bvba, Zemst, Belgium* (CHAP. 2)

Enda Cummins *UCD School of Agriculture, Food Science and Veterinary Medicine, Agriculture and Food Science Centre, Belfield, Dublin, Ireland* (CHAP. 10)

Meidui Dong *Abengoa Bioenergy, St. Louis, Mo.* (CHAP. 16)

Michel Eskin *Department of Human Nutritional Science, University of Manitoba, Manitoba, Canada* (CHAP. 13)

Maria Dolores Fernandez *Universidad de Santiago de Compostela, Canaria, Spain* (CHAP. 4)

Manuel Garcia-Pérez *Department of Biological Systems Engineering, Washington State University, Pullman, Wash.* (CHAP. 7)

Bijoy Chandra Ghosh *Food and Agriculture Engineering Department, Indian Institute of Technology, Kharagpur, India* (CHAP. 17)

Latif Kalin *School of Forestry and Wildlife Science, Auburn University, Auburn, Ala.* (CHAPS. 3 and 5)

Dilip Lakshman *USDA-ARS, U.S. National Arboretum, Beltsville, Md.* (CHAP. 1)

Antonella De Leonardis *Department of Agricultural, Food, Environmental and Microbiological Science, and Technologies (DiSTAAM), University of Molise, Campobasso, Italy* (CHAP. 14)

Vincenzo Macciola *Department of Agricultural, Food, Environmental and Microbiological Science, and Technologies (DiSTAAM), University of Molise, Campobasso, Italy* (CHAP. 14)

Ahindra Nag *Natural Product Laboratory, Department of Chemistry, Indian Institute of Technology, Karagpur, India* (CHAPS. 8, 9, 13, 14, and 17)

Paresh Patel *Clemson University, Clemson, SC Meidui Dong, Abengoa Bioenergy, St. Louis, Mo.* (CHAP. 16)

Saroj K. Pramanik *Department of Biology, Morgan State University, Baltimore, Md.* (CHAP. 1)

Herman Ramon *Department of Biosystems, Katholieke Universiteit Leuven, Heverlee, Belgium* (CHAP. 2)

R. Sri Ranjan *Professor, Department of Biosystems Engineering, University of Manitoba, Winnipeg, Manitoba, Canada* (CHAP. 6)

Habtom W. Ressom *Lombardi Comprehensive Cancer Center, Georgetown University Medical Center, Washington, D.C.* (CHAP. 1)

Benildo G. de los Reyes *Department of Biological Sciences, University of Maine, Orono, Maine* (CHAP. 1)

Manuel Ramiro Rodriguez *Universidad de Santiago de Compostela, Canaria, Spain* (CHAP. 4)

Shyam S. Sablani *Department of Biological Systems Engineering, Washington State University, Pullman, Wash.* (CHAP. 11)

Wouter Saeys *Department of Biosystems, Katholieke Universiteit Leuven, Heverlee, Belgium* (CHAP. 2)

Puneet Srivastava *Biosystems Engineering, Auburn University, Auburn, Ala.* (CHAPS. 3 and 5)

Erich Vareed Thomas *Food and Agriculture Engineering Department, Indian Institute of Technology, Kharagpur, India* (CHAP. 17)

Terry H. Walker *Clemson University, Clemson, SC Meidui Dong, Abengoa Bioenergy, St. Louis, Mo.* (CHAP. 16)

Siqun Wang *Department of Forestry, Wildlife & Fisheries, University of Tennessee, Knoxville, Tenn.* (CHAP. 12)

Song Joong Yun *Institute of Agricultural Science and Technology, Chonbuk National University, Chonju, Korea* (CHAP. 1)

Preface

Biosystem engineering is pioneering innovations in academics as well as in global industry. New technologies include precision systems for irrigation, production, and harvesting; a new system for bioenergy production; advanced packaging systems to maximize product quality; recycling of materials and prevention of emissions to protect the environment; and information technologies to optimize bioprocess strategies.

Chapter 1 has a discussion on microarray technology. Using this technology, we can now classify genes based on their expression profile, infer their physiological role, and construct databases of genes involved in specific functions.

Chapter 2 discusses application of systems analysis to biologically related problems. It will be helpful for model building, system analysis through computer simulations, system identification, and optimization methods.

As soil and water are in constant interaction, a proficient conservation strategy should consider their complex interaction. Those processes, relevant theories, and recent advancements are presented in Chapter 3.

Soil temperature affects water availability to the plant. Temperature affects plant growth and development to such an extent that plant growth has often been described by a linear approximation dependent on temperature and time (thermal time). Chapter 4 describes different factors for heat transfer that influence physical, chemical, and biological processes in soil and plants.

Agricultural, urban, forest, and mining nonpoint source (NPS) pollutants continue to impact and degrade surface and groundwater quality. Chapter 5 describes geographic information systems based on watershed modeling. This chapter will be helpful for designing soil conservation practices, water table management, prevention of chemical pollution of surface water bodies and groundwater, and protection of aquatic biota.

The design of sustainable water management systems is one of the major areas within biosystems engineering. Chapter 6 discusses about soil's physical and chemical properties, and plant physiology to understand the interaction of the plant roots with the soil. Irrigation systems are designed to deliver water for optimum plant growth whereas drainage systems remove excess water and salts from the soil and maintain optimum aeration of the plant root zone.

The pyrolysis of biomass results in the formation of crude bio-oil, char, and gases. The current status, opportunities, and challenges to convert biomass into second generation transportation fuels via pyrolysis and bio-oil refineries are discussed in Chapter 7.

Biodiesel is the methyl or other alkyl esters of vegetable oils, animal fats, or used cooking oils. Different techniques of preparation of biodiesel and ethanol and their engine performance are discussed in detail in Chapter 8.

Chapter 9 discusses bioseparation processes which involve recovery, isolation, purification, and polishing of products synthesized by biotechnological processes.

Food safety management is of prime importance in maintaining consumer confidence and is critical for a healthy population and economy. Chapter 10 is the discussion of food safety management which refers to the process of ensuring and controlling food safety through regulation or other policy mechanisms, for the health and well-being of consumers.

Chapter 11 presents selected aspects of food package engineering which gives an overview of the most commonly used materials for foodstuff wrapping including their processing and manufacturing, physical and barrier properties of packaging materials, and recent technological advances.

Productivity and quality of transgenic wood are discussed in Chapter 12.

Chapter 13 discusses different processes of extraction and refining of edible oils. Extraction of antioxidants and their functions for the stabilization of edible oils have been presented.

Olive oil is very popular and highly used oil in the world because it has nutraceutical values. Chapter 14 discusses about the phenolic substances present in the olives (*olea europeae*) and related olive-mill products.

Bioregulators are known to have an effect on a number of physiological processes taking place during the acquisition of mineral content and storage quality of fruits. Chapter 15 discusses the effect of exogenous bioregulators on the mineral composition and storability of fruits.

Carbon dioxide is a potential extraction solvent alternative to the traditional organic solvents. Supercritical carbon dioxide has the advantages of low cost, nontoxicity, high diffusivities with appreciable solubility, and low viscosity, and its application in the biosystem are discussed in Chapter 16.

Chapter 17 describes how scientifically managed agriculture land can achieve maximum sustained productivity.

The book is organized in such a way as to cater to the needs of students, researchers, and managerial organizations. We welcome any opinions, suggestions, and added information which will improve future editions and help readers in future. Benefits for readers will be the best reward for the authors.

A. Nag

CHAPTER 1

Microarray Data Analysis Using Machine Learning Methods

Habtom W. Ressom[*], Dilip Lakshman[†],
Song Joong Yun[‡], Saroj K. Pramanik[§],
and Benildo G. de los Reyes[¶]

1.1 Introduction

With the popularity of genome studies in microbes, animals, and plants, the potentiality of microarray technology is beginning to be realized. Microarrays are increasingly used to study genomewide changes in gene expression, genome organization, and chromatin structure. Using this technology, we can now classify genes based on their expression profiles, infer their physiological roles, and construct databases of genes involved in specific functions.

Microarrays are being used to study developmental and coordinated regulation of gene expression, gene activation, suppression

---

[*]Lombardi Comprehensive Cancer Center, Georgetown University Medical Center, Washington, DC.
[†]USDA-ARS, U.S. National Arboretum, Beltsville, MD.
[‡]Institute of Agricultural Science and Technology, Chonbuk National University, Chonju, Korea.
[§]Department of Biology, Morgan State University, Baltimore, MD.
[¶]Department of Biological Sciences, University of Maine, Orono, ME.

and detection of gene splicing variants and members of multigene families. In *Arabidopsis* spp., the microarray has been used to monitor stochastic and epigenetic changes in gene expression of natural and chemically induced polyploids. The technique also allowed genome-wide fingerprinting of closely related species with no prior DNA sequence information (Jaccoud et al. 2001; Lezar et al. 2004).

In the field of animal husbandry, the prospect of simultaneous, genomewide, high-throughput analysis of gene expression opens novel strategies to improve breeding (Ushizawa et al. 2004), to study nutrition and tissue physiology (Band et al. 2002; Rodriguez-Zas 2003), to identify disease-resistant genes (Diez-Tascon et al. 2005), and to locate treatment targets for diseases (Huang et al. 2003). In addition, microarray techniques have been used to evaluate gene expression in different tissues, developmental stages, genomic duplication events, viral infection, drug treatments, tumors, and aging in animals.

Microarray technology creates massive expression patterns of thousands of genes. In addition to the high-dimensionality and complexity of gene expression data, there are many unknown and undiscovered functional relations in the physical delivery system used for collecting the data itself. Also, current microarray technologies provide data that are associated with a substantial amount of noise. These nonlinearities and noises adversely affect the extraction of relevant information from the data. To address this challenge, the first step is to use experimental designs that will ensure quality and reliable data with appropriate statistical power for detecting differences between treatment and control groups with respect to gene expression levels. It is essential to carry out replicate studies for key aspects of experiments so that measures of variability are available for testing hypotheses. The next step toward addressing this challenge is to understand the nature and properties of the data structure generated by current microarray technologies. Understanding the data structure helps in selecting/designing appropriate tools for analysis and compensation for technical variability. The discussion in this chapter focuses mainly on analysis of microarray data, assuming that appropriate experimental design was carried out prior to data generation.

Microarray data can be organized as matrices where the rows represent genes (or clones) and columns represent various sample phenotypes or experimental conditions. Each entry in the matrix corresponds to the expression level of a gene for a given condition or sample. A set of entries in a row or a column forms an expression pattern. A gene expression data matrix may consist of 10,000s of rows (genes) and 10s to 100s of columns (samples). Although microarray technology provides researchers with such large amounts of gene expression data, analysis of the massive data has been one of the major bottlenecks in using the technology effectively. Computational tools that can mine and discover biologically meaningful knowledge in such large and complex multivariate data are needed.

Many times, "raw" microarray data are not the best data for discovering biological knowledge. This is due to the substantial amount of noise and the technical variability caused by various stages such as labeling, hybridization, and scanning. Low-level analysis methods are applied to the raw microarray data to reduce background noise and normalize and transform the data into a form acceptable to a selected analysis method. Once the data are properly preprocessed, high-level analysis methods are applied to elucidate biologically significant information such as identifying differentially expressed genes, clustering genes to identify a new transduction pathway or novel genes that may be coregulated through the same known pathway, discovering/predicting of unknown phenotypic class, selecting genes that may have a functional role in specific phenotypes, and deciphering gene regulatory networks.

Many computational methods have been proposed to perform low- and high-level analysis of microarray data. Machine learning methods have received impetus in recent years for use in high-level microarray data analysis. This is due to their unique performance in capturing nonlinear relationships and analyzing large-volume and high-dimensional data.

The chapter is organized as follows: Sec. 1.2 presents an overview of machine learning methods. Section 1.3 describes the two commonly used microarray technologies: cDNA and high-density oligonucleotide microarrays. Section 1.4 highlights low-level analysis methods for background adjustment and normalization of gene expression data generated by the two types of microarray technologies. Section 1.5 reviews high-level analysis methods for clustering, classification, and genetic network modeling. Section 1.6 summarizes and concludes the chapter.

1.2 Machine Learning Methods

Machine learning is the field of scientific study that concentrates on induction algorithms and on other algorithms that can be said to "learn." Induction algorithms take as input specific instances and produce a model that generalizes beyond these instances. In this section, we present an overview of machine learning methods such as artificial neural networks (NNs), fuzzy systems (FS), genetic algorithms (GAs), particle swarm optimization (PSO), and support vector machines (SVMs). Most of these methods have their conceptual origins in biological systems. For example, NNs model biological neural systems. FS originated from studies of how organisms interact with their environment. They can be used to transform human expert knowledge into a mathematical description. GAs model natural evolution. PSO is an adaptive algorithm based on the social metaphor of flocking birds (or schooling fish or swarming insects). SVMs are learning kernel-based systems that use a hypothesis space of linear functions in high-dimensional feature spaces. Hybrid systems use a mixture of two or more machine learning methods to take advantage of their collective features. Machine learning methods have been used in various engineering and scientific applications to extract knowledge from high-dimensional and complex data and to solve optimization problems.

1.2.1 Neural Networks

Neural networks generally consist of a number of interconnected processing elements known as *neurons*. The way neurons are interconnected or how interneuron connections are arranged determine the architecture of a neural network. The method by which the strengths of the connections (known as weights or synaptic weights) are adjusted or trained to achieve a desired overall behavior of the network is governed by the learning algorithm used. NNs are classified according to their architecture and learning algorithms.

The most popular architecture is a feedforward neural network, where the neurons are grouped into layers. All connections are feedforward; that is, they allow information transfer only from an earlier layer to the next consecutive layers. Neurons within a layer are not connected, and neurons in nonadjacent layers are not connected. Input signals are presented to the network via an "input layer." The nodes in the input layers do not process input signals but pass them to one or more "hidden layers" where the actual processing is done via a system of weighted "connections." The hidden layers then link to an "output layer" that provides the outputs of the network.

Figure 1.1 depicts a multilayer feedforward neural network that has four layers: an input layer, two hidden layers, and an output layer. As

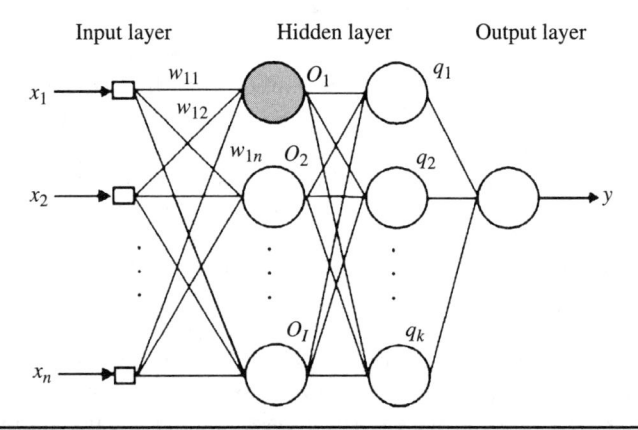

Figure 1.1 Multilayer feedforward neural network.

FIGURE 1.2 Details of a neuron.

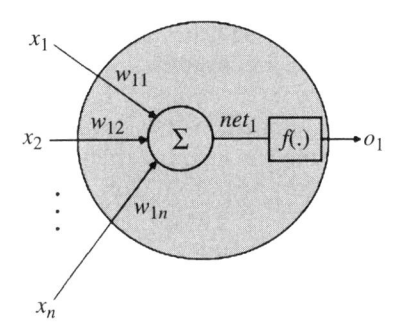

shown in the figure, the inputs to the network are $x_1, x_2, \ldots, x_n$, and the network output is y. Figure 1.2 illustrates the details of a single neuron where net_1, known as the activation level, is the sum of the weighted inputs to the neuron and $f(.)$ represents an activation function, which transforms the activation level of a neuron into an output signal. Typically, an activation function could be a threshold function, a sigmoid function (an S-shaped, symmetric function that is continuous and differentiable), a gaussian function, or a linear function. For example, the output of the neuron in Fig. 1.2 (i.e., the output of the first neuron in the first hidden layer of the network in Fig. 1.1) can be written as follows:

$$o_1 = f(net_1) = f\left(\sum_{j=1}^{n} w_{1j} x_j\right)$$

where the synaptic weights $w_{11}, w_{12}, \ldots, w_{1n}$ define the strength of connection between the neuron and its inputs. Such synaptic weights exist between all pairs of neurons in each successive layer of the network. They are adapted during learning to yield the desired outputs at the output layer of the network.

A multilayer feedforward network, whose neurons in the hidden layers have sigmoidal activation functions, is known as a multilayer perceptron (MLP) network. MLP networks are capable of learning complex input–output mapping. That is, given a set of inputs and desired outputs, an adequately chosen MLP network can emulate the mechanism that produces data through learning.

In a *supervised learning* paradigm, the network uses training examples (specifically, desired outputs for a given set of inputs) to determine how well it has learned and to guide adjustments of the synaptic weights in order to reduce its overall error. An example of a supervised learning rule is the back-propagation algorithm (Rumelhart and McClelland 1986), which is developed to train MLP networks based on the principle of steepest gradient method. The training of a neural network is complete when a prespecified stopping criterion is fulfilled. A typical stopping criterion is the performance of the network on a validation dataset, which is a portion of the training samples that was not used for updating the weights.

In contrast to supervised learning, *unsupervised learning* discovers patterns or features in the input data with no help from a teacher, essentially performing a clustering of the input space. Typical unsupervised learning rules include Hebbian learning, principal component learning, and Kohonen learning rules. These rules are applied in training self-organizing neural network architectures such as self-organizing maps (SOM). Although neural networks developed for both the supervised and unsupervised learning paradigms have performed very well in their respective application fields, improvements have been developed by combining the two paradigms. Examples include a radial basis function (RBF) network and learning vector quantization (LVQ) discussed as follows.

RBF networks are special feedforward networks that have a single hidden layer. The activation functions of the neurons in the hidden layer are radial basis functions, whereas the neurons in the output layer have simple linear activation functions. Radial basis functions are a set of predominantly nonlinear functions such as gaussian functions that are built up into one function. Each gaussian function responds only to a small region of the input space where the gaussian is centered. Thus, while an MLP network uses hyperplanes defined by weighted sums as arguments to sigmoidal functions, an RBF network uses hyperellipsoids to partition the input space in which measurements or observations are made. Thus, RBF networks find the input-to-output map using local approximators. The key to a successful implementation of RBF networks is to find suitable centers for the gaussian functions. Theoretically, the centers and width of the gaussian functions can be determined with supervised learning, for example, the steepest gradient method. They can also be determined through unsupervised learning by clustering the training data points. Once the centers and width of the gaussian functions are obtained, the weights for the connection between the hidden layer and the output layer are easily and quickly determined using methods such as linear least squares, as the output neurons are simple linear combiners.

LVQ selects training vectors with a known classification and presents them to the network to examine cases of misclassification. An LVQ network has a first competitive layer and a second linear layer. The competitive layer consists of reference vectors that learn to categorize input vectors. The linear layer transforms the competitive layer's classes into target classifications defined by the users. During training, a class labeled "input vector" (e.g., expression pattern) is picked at random and is compared with each reference vector. The input vector is assigned to a class that the most similar reference vector possesses. The reference vector is moved closer to the input vector if the assignment was correct; otherwise it is moved away. During operation, the unknown expression profiles are classified based on their similarity to the reference vectors.

Microarray Data Analysis Using Machine Learning Methods 7

1.2.2 Support Vector Machines

Support vector machines are suitable for classification problems that involve high dimensionality. They are learning kernel-based systems that use a hypothesis space of linear functions in high-dimensional feature spaces. Unlike artificial neural networks, which try to define complex functions in the input feature space, kernel methods perform a nonlinear mapping of complex data into high-dimensional feature spaces and then use simple linear functions to create linear decision boundaries. Thus, the problem of choosing network architecture is replaced here by the problem of choosing a suitable kernel for data projection.

The advantages of support vector machines over neural networks is that they are significantly faster to train, better suited to work with high-dimensional data, provide better generalization ability on an independent dataset, can be developed with few training examples, and allow for scaling the importance of outliers. SVM parameters are determined based on structural risk minimization. For example, in a classification problem involving two linearly separable classes (e.g., A and B in Fig. 1.3), an SVM search for one target is known as optimal hyperplane. Although various hyperplanes can separate the two groups correctly, the optimal hyperplane maximizes the margin of separation (ρ) between the hyperplane and the closest data points on both sides of the hyperplane.

1.2.3 Fuzzy Systems

Fuzzy logic is a superset of conventional two-valued (Boolean) logic that has been extended to handle the concept of partial truth. Thus, in fuzzy logic, the truth-value of a statement is defined in a continuous interval between 0 (completely false) and 1 (completely true). Fuzzy

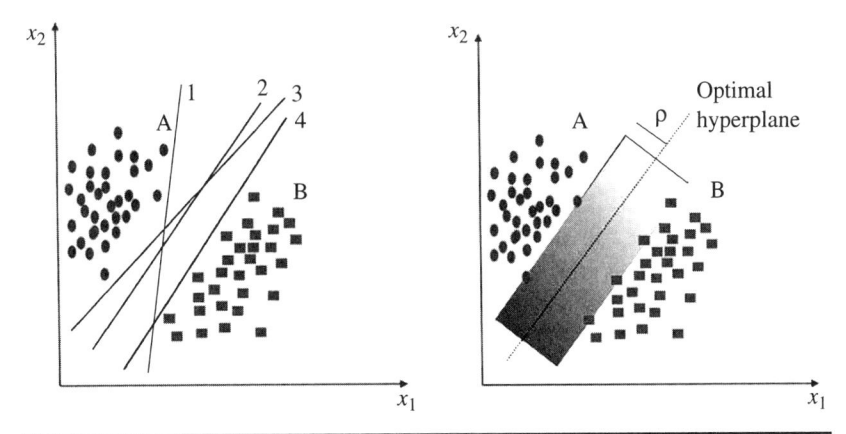

FIGURE 1.3 Different hyperplanes that separate the data points correctly (left figure) and optimal hyperplane (right figure).

8 Chapter One

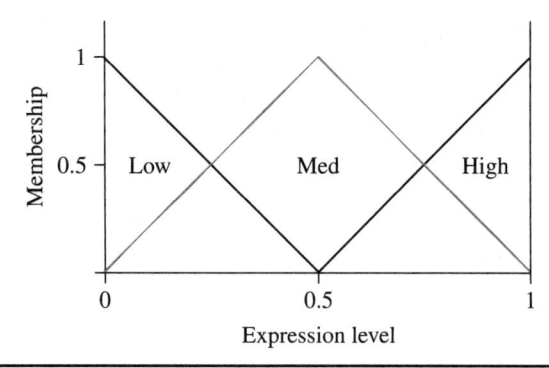

Figure 1.4 A membership function.

logic enables designers to simulate human thinking by quantifying concepts such as hot, cold, far, near, soon, high, and low. Thus, in contrast to traditional set theory that requires elements to be either part of a set or not, fuzzy logic allows an element to belong to a set to a certain degree of certainty. A membership function is used to associate a degree of membership of each of the elements of the domain to a fuzzy set. The degree of membership to a fuzzy set indicates the certainty that the element belongs to that set. For example, a gene expression level can be defined by the membership function in Fig. 1.4 as low, medium, and high.

Besides membership functions, a fuzzy system consists of a set of fuzzy rules. A fuzzy rule has two components, an *if* part (also referred to as premise) and a *then* part (also referred to as conclusion). Such rules can be used to represent knowledge and association, which are inexact and imprecise in nature, expressed in qualitative values that a human can easily understand. For example, one might say, "If gene x_1 is up-regulated and gene x_2 is down-regulated, then the probability of disease y is high."

Figure 1.5 depicts a fuzzy system that has four principal units: fuzzification, knowledge base, decision making (inference), and defuzzification. The fuzzy system accepts a set of inputs $(x_1, x_2, \ldots, x_n)$ as its information about the outside world (also referred to as crisp data).

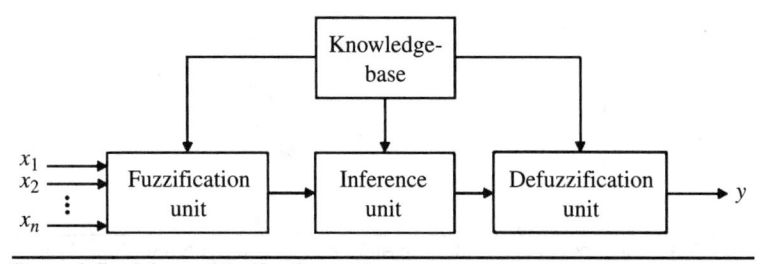

Figure 1.5 A fuzzy classification system.

The fuzzification unit converts these inputs into fuzzy sets based upon fuzzy values, such as "low," "medium," and "high." In this unit, membership functions defined on the input variables are applied to their actual values to determine the degree of truth for each rule premise. The knowledge base contains membership functions defining fuzzy values and a set of fuzzy rules. The inference unit executes these fuzzy rules. The truth value for the premise of each rule is computed and applied to the conclusion part of each rule. When all the rules are executed, a fuzzy region will be created for the output variable y. With the process of defuzzification, a crisp value of the output will be generated as a solution.

1.2.4 Genetic Algorithms

Genetic algorithms are global optimization algorithms that originated from mechanics of natural genetics and selection (Holland 1975; Goldberg 1989). They provide a method of problem solving that is based on genetic evolution. Based on probabilistic decisions, they exploit historic information to guide the search for better solutions in the problem space. Using a direct analogy of natural evolution, genetic algorithms work with a population of chromosomes where each chromosome encodes a possible solution to the problem as a string of bits (0100101010), a list of real values (0.2, 6.5, 4.1, 1.3), a string of characters (B2, A3, C1, A1), or some other representation. GAs start by randomly creating a population of chromosomes. The chromosomes in the population are evaluated by a fitness function indicating how good encoded solutions appear with respect to the problem under consideration. The algorithm takes the chromosomes with higher relative fitness (chosen as parents) and then creates a new generation of chromosomes using genetic operators such as crossover or mutation. In crossover, new offspring are created from two parents by swapping a portion of their strings. In mutation, offspring are identical to their parents, but they have random changes in portions of their strings. The algorithm repeats the foregoing steps until a predefined number of generations or fitness value is reached.

Genetic algorithms, therefore, begin with a random process and arrive at an optimized solution. They are thus well suited for those tasks that seek global optimization. They are highly effective in situations where many inputs interact to produce a large number of possible outputs or solutions. They are a robust search method requiring little information to search effectively in a large or poorly understood search space.

1.2.5 Particle Swarm Optimization

PSO is similar to GAs; each uses a population of potential solutions to explore the search space. Although GAs are based on survival-of-the-fittest approaches as in the theory of natural evolution, the PSO is an adaptive algorithm based on the social metaphor of flocking birds (or schooling fish or swarming insects). In PSO, a population of individuals adapt by a stochastic search of successful regions of the search space, influenced by their own success, as well as that of their neighbors. Individual particles move stochastically in the direction of their own previous best position, as well as the best position discovered by the entire swarm. Alternatively, a neighborhood approach can be used where instead of moving in the direction of the best position discovered by the entire swarm, each particle moves toward the best position discovered among a localized group of particles, termed the "neighborhood." Because the change in particle trajectory is based on the position of the particle's own best position, as well as the global (or neighborhood) best position, the essence of the PSO algorithm is that each particle will continuously focus and refocus the efforts of its search within these two regions. Each particle in the swarm represents a candidate solution to the optimization problem and is evaluated at each update by a performance function. PSO is a simple algorithm that has been shown to perform well for optimization of a wide range of functions, often locating optima in difficult multimodal search spaces faster than traditional optimization techniques. Detailed information on swarm intelligence and the PSO algorithm can be found in Clerc and Kennedy (2002) and Engelbrecht (2003).

1.2.6 Steps in Developing Machine Learning Models

Assuming that data preprocessing is performed, the steps for building a machine learning model (e.g., a classifier) involve model structure selection, learning, and model evaluation.

Model structure selection refers to the choice of a machine learning paradigm. It includes the choice of a neural network architecture, fuzzy rules, membership functions, fuzzy operators, genetic operators, coding scheme, and kernels. The selection of a neural network architecture includes choosing activation functions, number of layers, number of neurons in each layer, and interconnection of neurons and layers. It is known that too many neurons degrade the effectiveness of the model, as the number of connection weights in the model may cause overfitting and loss of a model's generalization. Too few hidden neurons may not capture the full complexity of the data. Many theories have been suggested for finding the optimal number of neurons in the hidden layer (Moody 1992; Amari 1995; Maass 1995). However, most users employ trial-and-error methods in which NN training starts with a small number of hidden neurons, and additional neurons are gradually added until some performance goal is satisfied.

In fuzzy logic–based modeling, model structure selection includes the choice of (1) shapes of the membership functions (trapezoidal, gaussian, etc.) of fuzzy sets; (2) a qualitative set of rules that can model the underlying process; and (3) fuzzy operators to handle rule conjunctions such as "and" in the following fuzzy rule: "if gene x_1 is up-regulated and gene x_2 is down-regulated, then the probability of disease y is high." Fuzzy operators include intersection, union, and complement. The fuzzy intersection operation is mathematically equivalent to the "and," the union to "or," and complement to "not" operation, respectively. The structure of the fuzzy rules could cause an exponential growth in the number of rules as the number of inputs increases, resulting in unwieldy rule bases. Thus, it is important to select an optimal set of fuzzy rules.

In SVM, parameters to select include cost of constrain violation, kernels (e.g., linear, polynomial, radial basis) and parameters related to kernels. Key selections in implementing genetic algorithms include the fitness function, the genetic representation or the chromosome coding scheme, and the genetic operators such as mutation and crossover. Similarly, parameters of PSO include number of particles, number of iterations, values for the stochastic component of the PSO algorithm, and measure of fitness.

Learning from data is the predominant feature of machine learning methods. Here, we define learning as the process of finding the free parameters of a machine learning model. Depending on the environment in which the model operates, two types of learning paradigms are commonly used, supervised and unsupervised. Supervised learning uses an external teacher to produce a desired output; the model learns from training examples. Conversely, unsupervised learning does not involve an external teacher; the network discovers collective properties of the inputs by self-organization. Also, two types of learning styles have been applied in neural network training. The first approach, referred to as pattern learning, where the weights of the network are adapted immediately after each pattern is fed in. The other approach, known as batch learning, takes the entire training dataset and updates the network parameters after the entire batch of data has been processed.

An example of a supervised learning paradigm is the back-propagation algorithm (Rumelhart and McClelland 1986), which is commonly used to train MLP networks. The back-propagation algorithm consists of two passes, forward and backward. During the forward pass, input signals are applied to the network and their effect propagates through the different layers and generates the network outputs at the output layer. Note that the synaptic weights of the network are all kept fixed during the forward pass. During the backward pass, synaptic weights are adjusted in accordance with the error correction rule. The network outputs are subtracted from the desired outputs to produce error signals. The error signals are propagated backward through the network against the direction of the synaptic connections to determine how the synaptic weights should be adjusted in order to decrease a predefined error function. A typical error function is the sum of square of the error signals between the network outputs and the desired outputs. Thus, during the backward pass, the back-propagation algorithm alters the synaptic weights in the direction of the steepest gradient of the error function to reduce the error function. The forward and backward passes are repeated until a prespecified stopping criterion is achieved or the error function is reduced significantly.

A number of methods are proposed to improve the performance of the steepest gradient method described earlier, which is based on the first derivatives of the error function with respect to the synaptic weights. Newton's method is used to speed up training by employing second derivatives of the error function with respect to the synaptic weights. The Gauss–Newton method is designed to approach second-order training speed without calculating the second derivatives. The Levenberg–Marquardt learning algorithm speeds up the learning process and produces enhanced learning performance by combining the standard gradient technique with the Gauss–Newton method.

An example of an unsupervised learning paradigm is the Kohonen learning rule (Kohonen 2001), which is used for training a self-organizing map (SOM). The design of an SOM starts with defining a geometric configuration for the partitions in a one- or two-dimensional grid. Then, random weight vectors are assigned to each partition. During training, an input pattern (input vector) is picked randomly. The weight vector closest to the profile is identified. The identified weight vector and its neighbors are adjusted to look similar to the input vector. This process is repeated until the weight vectors converge. During operation, SOM maps input patterns to the relevant partitions based on the reference vectors to which they are most similar.

Model evaluation is performed after learning is completed to prove the adequacy or to detect the inadequacy of the machine learning model. The latter could arise from an inappropriate selection of network topology, too small or too many neurons, or from insufficient training or overtraining. Incorrect input node assignments, noisy data, error in the program code, or several other effects may also cause a poor fit. The aim of model evaluation is to ensure that the model fit is correct, that the model satisfies the desired requirements, and that it serves as a general model. A general model is one whose input–output relationships (derived from the training dataset) apply equally well to new sets of data (previously unseen test data) from the same problem not included in the training set. The main goal of machine learning–based modeling is thus the generalization to new data of the relationships learned on the training set.

Various methods have been used to test the generalization capability of a model. These include the k-fold cross-validation, bootstrapping, and holdout methods. In k-fold cross-validation, we divide the training dataset into k subsets of (approximately) equal size. We train the model k times, each time leaving out one of the subsets from training, but using only the omitted subset to compute the prediction accuracy. If k equals the sample size, then the method is called "leave-one-out" cross-validation. In the leave-one-out method, one sample is selected as a validation sample and the model is trained using the remaining dataset.

The resulting model is tested on the validation sample. The process is repeated until all samples appear in the validation set. In the holdout method, only a single subset (also known as a validation set) is used to estimate the generalization error. Thus, the holdout method does not involve crossing. In bootstrapping, a subsample is randomly selected from the full training dataset with replacement at each iteration.

1.3 Microarray Technology

Construction of microarrays is generally dependent on information gained from genome sequencing or high-throughput expressed sequence tag (EST) sequencing projects that provide large sets of annotated clones and sequences. Based on methods of application of DNA substrates on to the slides, there are two major types of microarrays: predesigned chips and spotted microarrays. Predesigned chips synthesize 15 to 25 nucleotide oligomers directly on the chip using the photolithographic technique. In contrast, a spotting robot or an inkjet device is employed to place genomic DNA, cDNA, or 50- to 70-mer oligonucleotides. Of the two devices, the spotted microarray is cheaper to produce and offers more flexibility in terms of experimental design and data analysis. The sensitivities of spotted microarrays decrease from cDNA to long oligos and to short oligos; whereas the specificity of detection efficiency increases from cDNA to short oligos. Thus, short oligo microarrays have the potential to detect splicing variants and members of multigene families. In contrast, because of higher sensitivity, cDNA microarrays are more suitable for gene expression studies in related species.

1.3.1 cDNA Microarray

The construction of cDNA microarrays begins with the production of cDNA segments that represent each gene. Each segment is the complement to the actual DNA sequence of a gene and differs from the corresponding mRNA sequence only in that thymine in cDNA replaces uracil in mRNA. Each spot on the microarray is created by inserting copies of a gene's cDNA sequence on a glass slide or other substrate by a high-speed robotic process that physically binds the sequence to a small spot on the slide. A spot is created for each gene sequence to be used in the microarray. The substrate and the spots of DNA sequences are collectively known as the microarray. Each spot is referred to as a probe, while the hybridizing agent (cDNA or cRNA) is the target.

To measure gene expression for a cell population, mRNA is extracted from the cells and is reverse transcribed into cDNA. This cDNA sequence is identical to the DNA sequence for the gene found in the nucleus and is thus complementary to the cDNA probes on the microarray chip. The concentration of each sequence is multiplied proportionally through the chemical reactions. Chemical dyes (often green and red in microarray experiments [e.g., Cy3 and Cy5]) are bound to the sequences to allow for subsequent analysis of concentration. A solution of this dyed cDNA is created and exposed to the microarray. On the microarray, the cDNA sequences bind, or hybridize, to the probes that contain their complementary sequence. After a prescribed amount of time, the remaining cDNA solution is washed off the chip. What remains are the probes and the cDNA sequences that hybridized with them. The microarray is scanned with a laser set at the wavelength of the dye's color. The fluorescent intensity of each spot indicates approximately how many copies of the gene are bound to the spot, and thus, a relative perspective of the expression of that gene in the cell.

Unfortunately, fluorescence alone tells us very little when the gene expression from only one population is used; we cannot directly correlate the fluorescence of a probe to the copies of a gene on that probe. To alleviate the problem, we can add a second population whose cDNA sequences were treated with a different dye. This second population can be used as a control population; in the case of time series data, the second (control) population is often the cell population at a fixed point of time, whereas the first population is the same cell population at a later time. The two dyes should have colors of significantly different wavelengths to avoid "crosstalk" (i.e., a situation where one dye affects the measured fluorescence of the other). The relative difference in fluorescence of the two dyes on a particular spot should tell us how much a gene's expression differs between the two populations. Expression levels can be reported as some form of difference between the two fluorescences, such as a ratio. The appearance of a scanned microarray can be found in Fig. 1.6. The technology

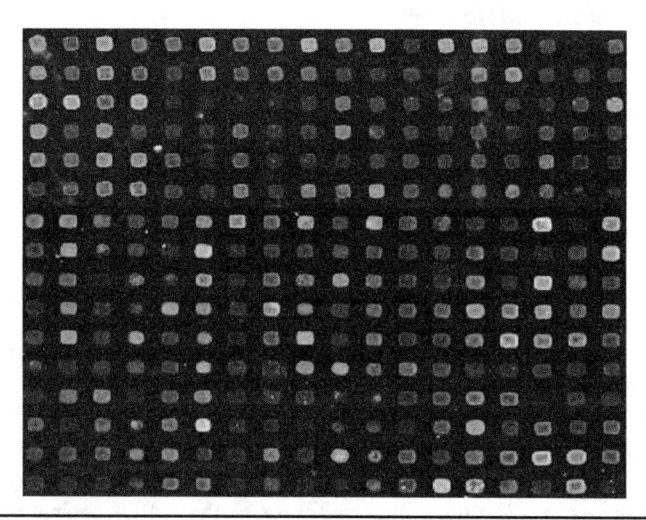

FIGURE 1.6 A scanned cDNA microarray.

is young and still has some problems. First, the fluorescence signal is unlikely to exactly match the level of expression of each gene. The target solution used is far from a free solution; the distribution of a certain cDNA sequence through the solution is not even. This problem may be partially alleviated by devoting several spots on the microarray to each gene and averaging the results, but it cannot guarantee the elimination of the problem. cDNA probes with similar, but not identical, sequences to a particular spot on the microarray may still hybridize to the spot with mixed results, exaggerating the expression of one gene, possibly at the expense of another. Kerr et al. (2001) named array effects, dye effects, populations, and genes as source of variation that have a significant effect on the relative expression of a gene from these microarray experiments. This variation can be viewed in terms of "noise" in a signal of gene expression for each gene.

1.3.2 High-Density Oligonucleotide Array

Oligonucleotide microarrays use a matrix of probes formed through a photolithographic printing. For example, Affymetrix GeneChip arrays use oligonucleotides with a length of 25 base pairs. These pairs are referred to as a perfect match (PM) probe and a mismatch (MM) probe. The MM probe is created by changing the thirteenth base of the PM probe with the intention of measuring nonspecific binding. Million copies of these base pairs are printed in each probe. Each gene is represented by 11 to 20 probe pairs that can uniquely identify a transcript and are referred to as probe sets (Fig. 1.7). By representing a gene with multiple probes, this technology is believed to provide reliable estimates of expression levels. Labeled RNA samples are hybridized with arrays. The arrays are stained, washed, and scanned. The scanned images are analyzed to obtain an intensity value for each probe. These intensities represent how much hybridization occurred for each probe. The expression value of a gene (probe set) is determined by combining its corresponding 11 to 20 probe pair intensities.

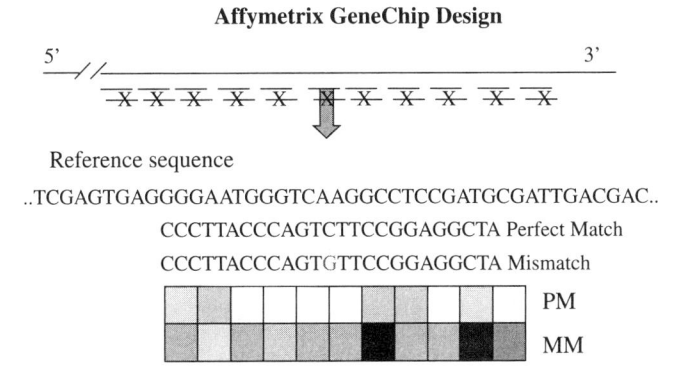

FIGURE 1.7 Affymetrix GeneChip design.

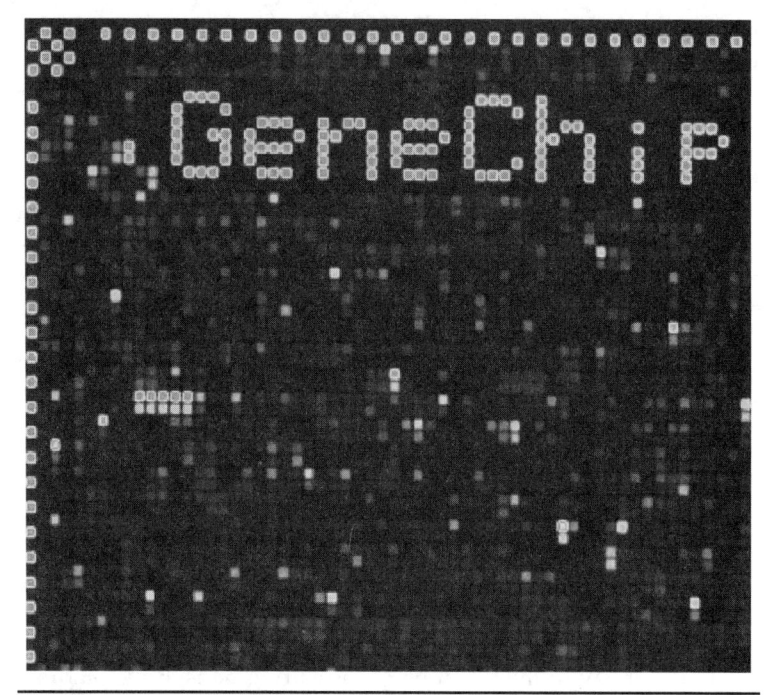

FIGURE 1.8 A scanned Affymetrix GeneChip.

Figure 1.7 illustrates Affymetrix GeneChip design. Figure 1.8 shows the appearance of a scanned Affymetrix GeneChip.

1.4 Low-Level Analysis

cDNA microarray data preprocessing involves image quantification, quality control, background correction, and normalization. Image quantification involves several steps: (1) finding where the printed spots are, (2) identifying the extent of each probe and separating foreground from background, (3) summarizing the varying brightness of the pixels in the foreground of each spot, (4) dealing with scanner saturation, and (5) dealing with variable backgrounds. Background correction is done by either subtracting the estimated local background or by subtracting values of negative controls nearby. Negative controls are probes designed for DNA sequences that should never occur in the sample. These controls should read low signals and should be fairly uniform. These signals give an estimate of the background due to nonspecific hybridization. Positive controls are replicate probes for sequences that should be abundant and are used to estimate the spatial variation in hybridization. Spike-in controls are probes for transcripts not expected in the sample but added to the sample with known concentration. These controls give an estimate of the accuracy and linearity of intensities and can be used to study and design background correction and normalization methods. The purpose of normalization is to ensure that the measurements from different arrays are comparable by compensating for systematic technical differences between arrays. The differences can be caused by variability in the labeling, hybridization, scanner setting, amounts of RNA, etc. By compensating for such technical differences, a better estimate of the real biological differences between samples can be made. The goal of most normalization approaches is thus to obtain the same data distribution across arrays. Normalization based on housekeeping genes—constitutively expressed genes—is commonly used (Wang et al. 2002). Min–max scaling preserves the relationships among the original data. Mean centering is more appropriate when the data contain no biases. Variance scaling is appropriate when training data are measured with different units. Z-score normalization is a combination of mean centering and variance scaling, and can be very useful when there are outliers present in the data. Linear and nonlinear fitting methods have also been proposed. The most common ones are locally weighted linear regression (lowess) and quantile normalization methods, whose goal is to obtain the same data distribution across arrays.

Affymetrix data preprocessing involves (1) image quantification, (2) quality control, (3) background adjustment to minimize the effect of nonspecific binding and optical noise, (4) normalization to ensure that the measurements from different arrays are comparable, and (5) summarization to obtain an expression value for a probe set by combining multiple probe intensities. The preprocessing methods used by Affymetrix's software, MicroArray Suite (MAS 5.0), have shown to be suboptimal. Li and Wong (2001) observed this and proposed an alternative model-based expression index (MBEI). Irizarry et al. (2003) proposed a robust multiarray analysis (RMA) method. These two methods are based on multichip models and are implemented in dChip and Bioconductor, respectively. RMA for background adjustment, quantile method for normalization, and robust multiarray average for summarization have provided better performance than MAS 5.0 and MBEI in detecting known levels of differential expression using spike-in Affymetrix data (Irizarry et al. 2003).

Other low-level analyses that need to be performed prior to undertaking a high-level analysis include handling missing values, screening outliers, data transformation, and dimensionality reduction.

It is common that microarray data have missing values. However, a value is required for each entry of a gene expression matrix. Although self-organizing models do not suffer under these problems, in supervised methods, missing values are a problem. Several options have been used to handle missing values such as removing the entire gene if there is a missing value, replacing a missing value with zero, and replacing with an average value. Troyanskaya et al. (2001) reported that methods based on weighted *k*-nearest neighbors and singular value decomposition surpass the commonly used row average method and filling missing values with zeros. Kim et al. (2005) proposed the local least squares imputation method that represents a missing value of a gene as a linear combination of similar genes.

An outlier is a data pattern that deviates substantially from data distribution. Outliers can have severe effects on accuracy. The problem can be addressed by removing outliers through statistical techniques such as multidimensional scaling or by using robust classification methods that are not influenced by outliers.

Transforming data into a form acceptable by an analysis method may be necessary. For example, for a statistical analysis (e.g., parametric t-test) that requires the data to be normally distributed, applying a logarithmic transformation of the data improves approximation to a normal distribution.

Dimensionality reduction is essential in exploratory data analysis, where the purpose is to map data onto a low-dimensional space for improved visualization. It also reduces the complexity of a problem and makes it easier to perform high-level data analysis such as building a classifier. Dimensionality reduction can be accomplished through feature extraction and selection where an optimum subset of features derived from the input variables (e.g., genes) is selected. Thus, feature selection methods keep only useful features and discard others. Note that feature selection is distinct from variable selection, because the former constructs new features out of the original variables and chooses the most relevant features. One well-known linear transformation used to reduce model dimensionality is principal component analysis (PCA). PCA transforms the input variables to a new set of variables (features). The new variables (also known as principal components) are computed as a linear combination of the original variables and are orthogonal to each other. PCA reduces input dimensionality by providing a subset of the principal components that captures most of the information in the original data. A classifier with the selected principal components as inputs may provide a better accuracy than a classifier with a large dimension of original variables consisting of coregulated genes.

1.5 High-Level Analysis

1.5.1 Clustering

Clustering is a useful exploratory technique for analysis of large-volume high-dimensional data when there is no a priori information about existing common properties. Clustering algorithms help discover common properties contained within the data and create groups of objects according to their properties. These groups could be objects clustered by samples or by gene. The former helps researchers identify which of the samples resulted in similar patterns or responses across all genes. The latter identifies genes with similar expression profiles across various experimental conditions. Genes with similar expression patterns might be transcriptionally regulated through a same transduction pathway or share some common function or regulatory elements.

Depending on how they cluster data, we can distinguish clustering algorithms into hierarchical and nonhierarchical clustering methods. Hierarchical clustering organizes the input patterns in a hierarchical tree structure, which allows detecting higher-order relationships between clusters of patterns. Nonhierarchical clustering begins from a predefined number of clusters and iteratively reallocates cluster members to minimize the overall within-cluster dispersion. We can also divide clustering algorithms into two categories, hard and fuzzy. Although a hard-clustering algorithm assigns each data point to only one of the clusters, fuzzy clustering assigns a certain degree of closeness or similarity to each object in a cluster.

A wide variety of clustering algorithms (hierarchical/nonhierarchical as well as hard/fuzzy) has been applied to group similar gene expression patterns together. In particular, hierarchical clustering (Eisen et al. 1998), self-organizing maps (SOM) (Tamayo et al. 1999), and k-means (Somogyi 1999) are widely used by the bioinformatics research community for clustering microarray data. Other clustering methods such as fuzzy c-means (Gasch and Eisen 2002) and adaptive resonance theory (Tomida et al. 2002) have provided useful results. However, detecting the number of clusters or selecting a "good" clustering algorithm remains a challenge. To address this challenge, several statistical methods have been proposed in the literature. For example, Yeung et al. (2001b) suggested a metric known as figure of merit (FOM) that they calculated by clustering the dataset while leaving out one experiment at a time. They used FOM to evaluate the performance of different clustering results as they vary the number of clusters. Tibshirani et al. (2000) proposed estimating the number of clusters in a dataset via the gap statistic. The gap statistic estimates the number of clusters by comparing within-cluster dispersion to that of a reference null distribution. Kerr and Churchill (2001) introduced a technique based on the application of a randomization technique (bootstrapping) for making statistical inference from clustering tools. They applied this technique to assess the stability of results from a cluster analysis of gene expression microarray data. Dudoit and Fridlyand (2002) developed a prediction-based resampling method to estimate the number of clusters in dataset. Fraley and Raftery (2002) provided functionality for visualizing cluster results in their model-based clustering technique. They characterized and compared various probability models in their clustering algorithm through a Bayesian information criterion (BIC) introduced by Yeung et al. (2001a). They calculated BIC scores for each model over a given range of number of clusters. A model with large BIC scores was selected, and the number of clusters with the highest BIC value for the chosen model was considered to be the best number of clusters. This model-based approach assumes that a finite mixture of underlying probability distributions such as multivariate normal distributions generates the data. With the underlying probability model, the problems of determining the number of clusters and of choosing an appropriate clustering method become statistical model choice problems. These and many other heuristic methods have provided useful results in gene expression clustering.

Su and Chang (2001) developed a technique known as double self-organizing maps (DSOM). Unlike SOM, DSOM nodes are represented not only by their weight vectors but also by two-dimensional *position vectors*. Weight vectors serve the same purpose as in SOM. Position vectors are projection of the weight vectors into a two-dimensional space and serve as a visualization tool for deciding how many clusters are needed, thus combining clustering and cluster visualization in one computational procedure. In other words, with the help of position vectors, DSOM adjusts its network structure during the learning phase so that neurons that respond to similar stimuli will not only have similar weight vectors but also move spatially nearer to each other. Ressom et al. (2003b) developed an adaptive double self-organizing map (ADSOM), which updates not only the weight vectors and position vectors but also all the free parameters involved in DSOM during the training process. After training, the final location of the positions vectors is used to detect the number of clusters by visually counting the clusters they form.

In fuzzy c-means, a cluster validity index is commonly used to estimate the best number of clusters. Several indices were proposed to help the detection of number of clusters. One of these is the partition coefficient introduced by Bezdek (1981), which ranges from 0 to 1. A partition coefficient of 1 indicates no membership sharing between clusters, whereas a low value indicates overlap between clusters. Thus, a high partition coefficient is desired. Although this is a good indicator, its calculation is solely based on membership values without involving the structure of the data. Other indices that are based on both membership values as well as the data structure include the Xie and Beni's index (Xie and Beni 1991), the Fukuyama and Sugeno's index (Pal and Bezdek 1995), the Gath and Geva's index (Gath and Geva 1989), and the Rezaee, Lelieveldt, and Reiber's index (Rezaee et al. 1998).

1.5.2 Classification

Classification involves the automated grouping of objects (e.g., samples or genes) into prespecified categories. Various statistical methods have been used to classify microarray data. These include discriminant analysis (linear, quadratic, and logistic), classification and regression trees (CART), generalized additive models, compound covariate predictor, weighted voting, *k*-nearest neighbor rule, and nearest centroid classifier.

Most classification algorithms perform suboptimally with thousands of genes and require the selection of the most relevant genes that are most predictive of a phenotype. Performing appropriate gene selection helps in achieving accurate classification. There are two objectives in gene selection: improving the prediction performance of the models and providing a better understanding of the underlying concepts that generated the data. Gene selection may start by filtering genes with no or significantly low fold change. A small subset of genes can be selected from the remaining genes using various techniques described in Sec. 1.4. Clustering methods can also be used to identify groups of coregulated genes; cluster centers of these groups can then be used as inputs to a classifier. Supervised methods identify the most informative genes using approaches such as (1) analysis of differential expression via a two-sample t-test, analysis of variance, etc., (2) selecting a gene's signal-to-noise ratio of above a prespecified cutoff, and (3) choosing genes that are correlated with an expected outcome (e.g., class labels). Optimizations methods can also be used in which a subset of genes is selected recursively (sequential or via "evolutionary" trial and error) and the best possible combination of genes is selected based on its classification performance.

Molecular classification based on machine learning algorithms have been shown to have statistical and clinical relevance for a variety of tumor types: leukemia (Golub et al. 1999), lymphoma (Shipp et al. 2002), brain cancer (Pomeroy et al. 2002), lung cancer (Bhattacharjee et al. 2001), and the classification of multiple primary tumors (Ramaswamy et al. 2001). The performance of machine learning methods in classifying microarray data can be enhanced if the most informative genes are used. For example, Guyon et al. (2002) applied a gene selection method that used SVM based on recursive feature elimination. They demonstrated experimentally that the selected genes yielded improved classification performance.

1.5.3 Genetic Network Modeling

With the help of global expression data—especially using time series microarray data—one can attempt to reverse engineer a network of gene interaction. The benefits of characterizing gene interaction are many; for example, the effects of drugs on a regulatory pathway can be characterized; tumor development in cells can be tracked, etc. Several methods have been proposed to develop maps of gene interaction, including linear equations (D'haeseleer et al. 1999; Weaver et al. 1999), differential equations (Chen et al. 1999), Boolean networks (Liang et al. 1998; Shmulevich et al. 2002), fuzzy logic–based methods (Woolf and Wang 2000; Ressom et al. 2003a), correlation-based approaches (Herrero et al. 2003; Schmitt et al. 2004), and Bayesian networks (Friedman et al. 2000).

Linear equations attempt to solve a weight matrix that represents a series of linear equations of the expression level of each gene as a function of the other genes. Unfortunately, there need to be as many time points as there are genes to develop a unique solution. When there is no unique solution, we cannot know if the model derived from a linear equation is correct. Differential equations model the expression level of genes as a function of other genes and their rates of change. The solution involves itself in solving for the constants in the differential equations. Unfortunately, this suffers from the same problem as linear equations. Boolean networks assume that genes are either "on" or "off" and attempt to solve the state transitions for the system. Assuming that genes are only in one of two states, however, is an oversimplification, although methods have been developed to get around the simplification.

Woolf and Wang (2000) introduced an approach based on fuzzy rules of a known activator/repressor model of gene interaction. Their algorithm transforms expression values into qualitative descriptors that can be evaluated by using a set of heuristic rules. Figure 1.9a shows the membership function used by the fuzzy logic–based model. The model finds triplets of activators, repressors, and targets in gene expression data by checking all possible triplets of genes if they fit to the fuzzy logic–based model governed by the rule-base decision matrix shown in Fig. 1.9b.

Woolf and Wang used data from the *Saccharomyces cerevisiae* cell cycle expression database (Cho et al. 1998) to test their model. The data consisted of 6321 time series gene expression profiles. Each gene expression profile represents expression levels of a gene at 17 time steps, thus forming a 6321 × 17 matrix. Using a normalized subset of this matrix, 1898 × 17, Cho et al. (1998) tested every possible combination of activators, repressors, and targets if they fit the fuzzy model. The model's output was compared to the expression

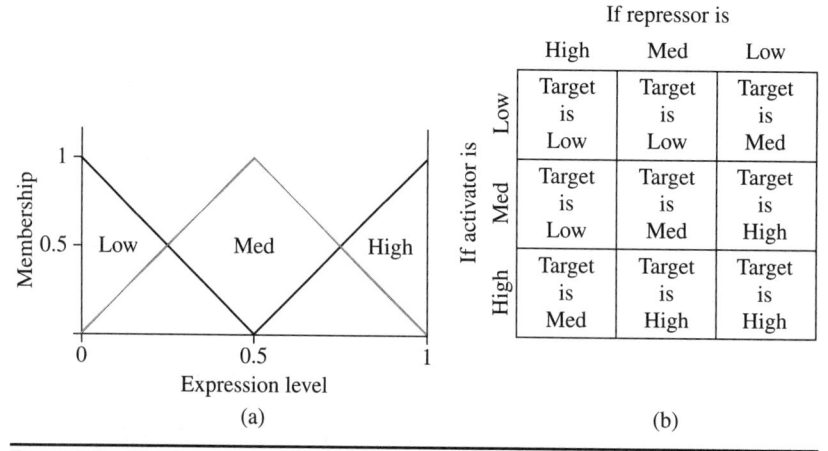

	If repressor is		
	High	Med	Low
If activator is Low	Target is Low	Target is Low	Target is Med
If activator is Med	Target is Low	Target is Med	Target is High
If activator is High	Target is Med	Target is High	Target is High

(a) Membership function plot with Low, Med, High triangular functions over Expression level 0 to 1.

(b)

FIGURE **1.9** Membership function and (a) rule base decision matrix (b).

level of the corresponding target gene. Gene combinations were ranked based on the residual between the model and the target gene and variance between the applications of the fuzzy rules over the given time series. Those combinations of genes that have a low error and cover most of the fuzzy rule base were inferred to exhibit an activator–repressor–target relationship. This method attempts to simulate what a human would do in comparing expression levels of genes to find the underlying relationships. Different fuzzy models can be developed for different models of interaction, including coactivators and corepressors as well as the presence of other factors in the cell, such as proteins or assorted compounds necessary for transcription. This method is intuitively pleasing, and the results are consistent with the literature of genetic networks of *S. cerevisiae*. The model itself is an interesting generalization of Boolean networks where genes are not either "on" or "off" but are often both "on" and "off" at the same time. This approach, although logical, is a brute force technique for finding gene relationships. It involves a significant computation time, which restricts its practical usefulness. It has an algorithmic complexity of N^3 where N is the number of genes analyzed. Furthermore, the model does not scale well to more complex gene interactions. Building a model that includes two activators and two repressors would increase the algorithmic complexity to N^5, making the analysis of 1898 genes not feasible. Also, Ressom et al. (2003a) showed that Woolf and Wang's model is susceptible to noise.

Ressom et al. (2003a) investigated the use of clustering as an interface to Woolf and Wang's method to improve its computational efficiency. This integrated approach significantly reduced the total number of gene combinations to be tested by first analyzing how well cluster centers fit the model. The algorithm ignores combinations of genes whose cluster centers are unlikely to fit, thereby gaining significant advantage over Woolf and Wang's approach in reducing computation time. To illustrate how clustering could reduce computation time in gene expression analysis, gene expression patterns representing cluster centers, grouped into two sets of triplets, are presented in Fig. 1.10. One can easily determine whether or not large groups of

Activator   Repressor   Target

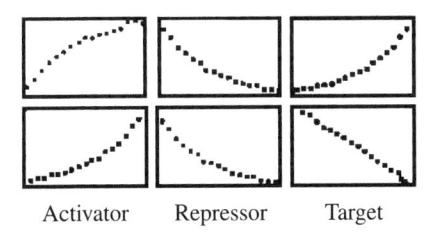

FIGURE 1.10 Cluster triplets that would fit the model well (top row) and cluster triplets that would not fit the model (bottom row).

genes, represented by these clusters centers, are likely to fit the fuzzy model described in Fig. 1.9. For example, from the top row in Fig. 1.10, we can see that an increasing activator and decreasing repressor would cause the target gene to increase quickly. These cluster triplets make sense intuitively and should be included in analysis. Figure 1.10 (bottom row) shows counterexamples to the combinations shown on the top. These cluster triplets do not make sense intuitively and should not be included in analysis.

There are many potential causes for noise, mostly originating from the stochastic nature of gene interactions and microarray technology. Investigating and using appropriate methods of conjunction and rule aggregation will increase resilience of the fuzzy logic–based gene regulatory model to noise. Ressom et al. (2003a) investigated the noise sensitivity of four models: Woolf and Wang's method, Mamdani's model, Kosko's standard additive model (SAM), and a hybrid model that attempts to take the best attributes of the Mamdani's model and SAM. It was observed that Mamdani's model has the best performance.

Sokhansanj et al. (2004) introduced an approach to gene network modeling based on a scalable linear variant of fuzzy logic, which was applied to real quantitative data where there were many inputs. The linear fuzzy logic model assigns distinct fuzzy rules for each individual input to a given output. After each individual rule is built, the intermediate evaluations of the fuzzy state of the output variables are aggregated by a fuzzy union operation (logic OR). This is different from traditional fuzzy logic, which defines all the rules related to the combinations of inputs. The advantage of the linear fuzzy model is that it can deal with complex multicomponent regulation because only rules for individual input and output are required.

Genetic algorithms have also been used to decipher genetic networks from microarray data (Kikuchi et al. 2003; Shin and Iba 2003). Ando and Iba (2001) developed an inference algorithm based on GAs and applied it to the optimization of the influence matrix of gene regulatory network. The GA inference method itself is not enough for the application of real data. It is suggested that the combination of GAs and clustering and perturbation analysis will make it easier and more accurate to infer gene networks from microarray expression data.

Herrero et al. (2003) applied time-lagged correlation to study the effect of genes at time t over them at $t + 1$. Noting that correlation is only an indication, but not a proof, of a causal relationship, they introduced a permutation test that takes into account the multiple testing nature of the results to check the reliability of their method. Also, to obtain a nonredundant dataset of gene expression profiles, they applied clustering as a preprocessing step to reduce the dataset's dimensionality. Through this approach Herrero et al. (2003) built a genetic network from time series expression profiles of yeast genes corresponding to 18 time points of alpha factor-arrested cells. Schmitt et al. (2004) adopted the time-series expression data using the *Synechocystis* microarray with the sampling interval (20 min) to study its physiological response to altering light conditions. Compared to earlier time-series experiments with DNA microarray, this smaller time interval enabled the time-lagged correlations to identify directional transcriptional relationships on a 20-min time scale. Khanin and Wit (2007) constructed the gene network of *Plasmodium falciparum* based on the method combining two types of correlations between each pair of genes: standard Pearson and partial correlations. The topology of the malaria gene expression network they obtained is consistent with scale-free behavior, similar to other biological networks.

Friedman et al. (2000) discussed how Bayesian networks describe interactions between genes and demonstrated their method on the *S. cerevisiae* cell-cycle measurements of Spellman et al. (1998). Based on the work of Friedman et al. (2000), Pe'er et al. (2001) extended this work by integrating a new discretization procedure and a principled way for learning with a mixture of observational and interventional data. To test Bayesian networks on gene expression data, several studies assessed the inference results on real (Zak et al. 2001) or simulated (Smith et al. 2002; Husmeier 2003) gene expression data, which allow estimating the proportion of spurious gene interactions incurred for a specified target proportion of recovered true interactions. Through these studies, it is demonstrated how network inference performance varies with the training set size, the degree of inadequacy of prior assumptions, the experimental sampling strategy, and the inclusion of further, sequence-based information.

Approaches that integrate gene expression data with other types of genomic information (e.g., promoter sequence) have also been proposed. Given known transcription factors (TFs), some researchers (Roulet et al. 1998; Krivan and Wasserman 2001; Grabe 2002; Halfon et al. 2002) have tried to find their binding motifs in the regions upstream of genes. Others have tried to predict gene targets of TFs using genomewide sequence searches of promoter regions for a TF whose target motif is known (Schuldiner et al. 1998; Zhu et al. 2002). A more powerful approach to determine targets of TFs whose binding motifs are unknown is the combined use of genomewide location and gene expression analyses. Genomewide location analysis (protein–DNA binding; ChIp–chip) combines techniques of chromatin immunoprecipitaion and microarray hybridization. In yeast, researchers have used this method to identify the targets of many TFs. On the other hand, gene expression data analysis uses a computational approach to identify the targets of TFs from time-course gene expression profiles. Bar-Joseph et al. (2003) and Hartemink et al. (2002) developed an algorithm that combines information from genomewide location and expression profiles to decipher regulatory networks. Qian et al. (2003) constructed a gene network using the data from two transcription databases: TRANSFAC (Wingender et al. 2001) and SCPD (Zhu and Zhang 1999). Qian et al. (2003) used SVM to predict the regulatory targets for 36 transcription factors in the *S. cerevisiae* genome based on the microarray expression data from many different physiological conditions. They assessed the performance of their regulatory network identifications by comparing them with the results from two recent genomewide ChIP–chip experiments. They found that the agreement between their results and these experiments was comparable to the agreement between the two experiments.

1.6 Summary

This chapter introduces computational methods for analysis of microarray data including gene clustering, marker gene selection, prediction of phenotypic classes, and modeling of genetic network. Because large-volume and high-dimensional data are being generated by the rapidly expanding microarray technology, the number of reported applications of machine learning methods is expected to increase. With increasing demand, however, comes the need for further improvements that can make implementation of machine learning algorithms in microarray data analysis more efficient. Key improvements include (1) enhanced computational power to handle high dimensionality and large-volume data; (2) improved microarray technology with a high-resolution scanner, low background noise, low technical variability, etc.; (3) enhanced quality control and protocol; (4) well-designed low-level analysis methods for background correction, cross-talk removal, normalization, outlier screening, and summary measures; (5) improved visualization tools to assess data quality and interpret results; (6) better data storage and retrieval mechanisms; and (7) advances in machine learning methods to enhance their speed and make them more accessible to the user.

Integration of gene expression data with genomic information (e.g., how transcriptional regulators bind to promoter sequences across the genome) and other prior biological knowledge is one of the future goals of bioinformatics. It is important, however, to ensure that existing biological knowledge is reliable. In particular, although supervised machine learning methods can take advantage of prior knowledge in constructing a model, their success is highly dependent on the quality of prior knowledge from previous experiments. For example, in constructing a classifier, if inaccurately labeled data are used for learning, the classification result will be impaired. Note that like other empirical models, machine learning models are only as good as the dataset to which they are applied; hence, the quality of the data collected is very important. Thus, we believe that the use of computational methods alone cannot provide a solution to the complex task of microarray data analysis. In addition to advanced computational methods that are capable of extracting knowledge from complex and high-dimensional data, this task requires careful experimental design (randomization and replication), sample collection, and preparation to control systematic bias. Ransohoff (2005) noted recently that bias presents the greatest difficulty at every step of design, conduct, and interpretation. Analysis results need to go through a thorough interlaboratory validation to ensure accurate biological interpretation.

References

Amari, S. 1995. Learning and statistical inference. In: *The Handbook of Brain Theory and Neural Networks*, ed. M. A. Arbib, 522–526. Cambridge, MA: MIT Press.

Ando, S. and Iba, H. 2001.Inference of gene regulatory model by genetic algorithms. 712–719. *Proceedings of the 2001 IEEE Congress on Evolutionary Computation*, Seoul, Korea.

Band, M. R., Olmstead, C., Everts, R. E., Liu, Z. L., and Lewin, H. A. 2002. A 3800 gene microarray for cattle functional genomics: comparison of gene expression in spleen, placenta, and brain. *Animal Biotechnology* 13:163–172.

Bar-Joseph, Z., Gerber, G. K., Lee, T. I., Rinaldi, N. J., Yoo, J. Y., Robert, F., Gordon, et al. 2003. Computational discovery of gene modules and regulatory networks. *Nature Biotechnology* 21(11):1337–1342.

Bezdek, J. C. 1981. *Pattern Recognition with Fuzzy Objective Function Algorithms*. New York: Plenum Press.

Bhattacharjee, A., Richards, W. G., Staunton, J., Li, C., Monti, S., Vasa, P., Ladd, C., et al. 2001. Classification of human lung carcinomas by mRNA expression profiling reveals distinct adenocarcinoma subclasses. *The Proceedings of the National Academy of Sciences, USA*, 98:13790–13795.

Chen, T., He, H. L., and Church, G. M. 1999. Modeling gene expression with differential equations. *Pacific Symposium on Biocomputing, Big Island, Hawaii* 4:29–40.

Cho, R. J., Campbell, M. J., Winzeler, E. A., Steinmetz, L., Conway, A., Wodicka, L., Wolfsberg, T. G., et al. 1998. A genomewide transcriptional analysis of the mitotic cell. *Molecular Cell* 2:65–73.

Clerc, M. and Kennedy, J. 2002. The particle swarm: explosion stability and convergence in a multi-dimensional complex space. *IEEE Transactions on Evolutionary Computing* 6(1):58–73.

D'haeseleer, P., Wen, X., Fuhrman, S., and Somogyi, R. 1999. Linear modeling of mRNA expression levels during CNS development and injury. *Pacific Symposium on Biocomputing* 4:41–52.

Diez-Tascon, C., Keane, O. M., Wilson, T., Zadissa, A., Hyndman, D. L., Baird, D. B., McEwan, J. C., et al. 2005. Microarray analysis of selection lines from outbred populations to identify genes involved with nematode parasite resistance in sheep. *Physiological Genomics* 21(1):59–69.

Dudoit, S. and Fridlyand, J. 2002. A prediction-based resampling method to estimate the number of clusters in a dataset. *Genome Biology* 3(7):0036.1–0036.21.

Eisen, M. B., Spellman, P. T., Brown, P. O., and Botstein, D. 1998. Cluster analysis and display of genomewide expression patterns. *Proceedings of the National Academy of Science, USA* 95:14863-8.

Engelbrecht, A. P. 2003. *Computational Intelligence: An Introduction*. Chichester, UK: Wiley.

Fraley, C. and Raftery, A. E. 2002. Model-based clustering, discriminant analysis, and density estimation. *Journal of the American Statistical Association* 97:611–631.

Friedman, N., Linial, M., Nachman, I., and Pe'er, D. 2000. Using Bayesian networks to analyze expression data. *Journal of Computational Biology* 7(3–4):601–620.

Gasch, A. P. and Eisen, M. B. 2002. Exploring the conditional coregulation through fuzzy k-means clustering. *Genome Biology* 3(11):research 0059.1–0059.22.

Gath, I. and Geva, A. B. 1989. Unsupervised optimal fuzzy clustering. *IEEE Transactions on Pattern Analysis and Machine Intelligence* 11(7):773–781.

Goldberg, D. E. 1989. *Genetic Algorithms in Search, Optimization, and Machine Learning*. Reading, MA: Addison-Wesley.

Golub, T. R., Slonim, D. K., Tamayo, P., Huard, C., Gaasenbeek, M., Mesirov, J. P., Coller, H., et al. 1999. Molecular classification of cancer: class discovery and class prediction by gene expression monitoring. *Science* 286:531–537.

Grabe, N. 2002. AliBaba2: Context specific identification of transcription factor binding sites. *In Silico Biology* 2:S1–15.

Guyon, I., Weston, J., Barnhill, S., and Vapnik, V. 2002. Gene selection for cancer classification using support vector machines. *Machine Learning* 46:389–422.

Halfon, M. S., Grad, Y., Church, G. M., and Michelson, A. M. 2002. Computation-based discovery of related transcriptional regulatory modules and motifs using an experimentally validated combinatorial model. *Genome Research* 12:1019–1028.

Hartemink, A. J., Gifford, D. K., Jaakkola, T. S., and Young, R. A. 2002. Combining location and expression data for principled discovery of genetic regulatory network models. *Pacific Symposium on Biocomputing* 437–449.

Herrero, J., Diaz-Uriarte, R., and Dopazo, J. (2003). An approach to inferring transcriptional regulation among genes from large-scale expression data. *Comparative and Functional Genomics* 4:148–154.

Holland, J. H. 1975. *Adaptation in Natural and Artificial Systems*. Ann Arbor, MI: University of Michigan Press.

Huang, D., Chen, W., He, R., Yu, F., Zhang, Z., and Qiu, W. 2003. Different cDNA microarray patterns of gene expression reflecting changes during metastatic progression in adenoid cystic carcinoma. *World Journal of Surgical Oncology* 1(1):28.

Husmeier, D. 2003. Sensitivity and specificity of inferring genetic regulatory interactions from microarray experiments with dynamic Bayesian networks, *Bioinformatics* 19(17):2271–2282.

Irizarry, R. A., Hobbs, B., Collin, F., Beazer-Barclay, Y. D., Antonellis, K. J., Scherf, U., and Speed. T. P. 2003. Exploration, normalization, and summaries of high density oligonucleotide array probe level data. *Biostatistics* 4(2):249–264.

Jaccoud, D., Peng, K., Feinstein, D., and Kilian, A. 2001. Diversity arrays: A solid state technology for sequence information independent genotyping. *Nucleic Acids Research* 29(4):e25.

Kerr, M. K., Martin, M., and Churchill, G. A. 2001. Analysis of variance for gene expression microarray data. *Journal of Comparative Biology* 7:819–837.

Kerr, M. K. and Churchill, G. A. 2001. Bootstrapping cluster analysis: Assessing the reliability of conclusions from microarray experiments. *Proceedings of the National Academy of Science, USA* 98(16):8961–8965.

Khanin, R. and Wit, E. 2007. Construction of malaria gene expression network using partial correlations. In: *Methods of Microarray Data Analysis*, eds. P. V. McConnell, S. M. Lin, and P. Hurban, 75–88. Springer.

Kikuchi, S., Tominaga, D., Arita, M., Takahashi, K., and Tomita, M. 2003. Dynamic modeling of genetic networks using genetic algorithm and S-system, *Bioinformatics* 19(5):643–650.

Kim, H., Golub, G. H., and Park, H. 2005. Missing value estimation for DNA microarray gene expression data: local least squares imputation. *Bioinformatics* 21:187–198.

Kohonen, T. 2001. *Self-Organizing Maps*. 3rd ed. New York, NY: Springer-Verlag.

Krivan, W. and Wasserman, W. W. 2001. A predictive model for regulatory sequences directing liver-specific transcription. *Genome Research* 11:1559–1566.

Lezar, S., Myburg, A. A., Wingfield, B. D., Wingfield, M. J., and Berger, D. K. 2004. Development of a microarray chip for fingerprinting *Eucalyptus grandis*. *Theoretical and Applied Genetics* 109:1329–1336.

Li, C. and Wong, W. H. 2001. Model-based analysis of oligonucleotide arrays: Expression index computation and outlier detection. *Proceedings of the National Academy of Science, USA* 98:31–36.

Liang, S., Fuhrman, S., and Somogyi, R. 1998. REVEAL, a general reverse engineering algorithm for inference of genetic network architectures. *Pacific Symposium on Biocomputing* 3:18–29.

Maass, W. 1995. Vapnik-Chervonenkis dimension of neural networks. In: *The Handbook of Brain Theory and Neural Networks*, ed. M. A. Arbib, 52226. Cambridge, MA: MIT Press.

Moody, J. 1992. The effective number of parameters: An analysis of generalization and regularization in nonlinear learning systems. In: *Advances in Neural Information Processing Systems*, eds. J. Moody, S. J. Hanson, and R. P. Lippmann, 847–854. San Mateo, CA: Morgan Kaufmann.

Pal, N. R. and Bezdek, J. C. 1995. On cluster validity for the fuzzy c-means model. *IEEE Transactions on Fuzzy Systems* 3(3):370–379.

Pe'er, D., Regev, A., Elidan, G., and Friedman, N. 2001. Inferring subnetworks from perturbed expression profiles. *Bioinformatics* 17:Suppl.:S215–S224.

Pomeroy, S. L., Tamayo, P., Gaasenbeek, M., Sturla, L. M., Angelo, M., McLaughlin, M. E., Kim J. Y. H., et al. 2002. Prediction of central nervous system embryonal tumour outcome based on gene expression. *Nature* (Lond.) 415:436–442.

Qian, J., Lin, J., Luscombe, N. M., Yu, H., and Gerstein, M. 2003. Prediction of regulatory networks: genomewide identification of transcription factor targets from gene expression data. *Bioinformatics* 19(15):1917–1926.

Ramaswamy, S., Tamayo, P., Rifkin, R., Mukherjee, S., Yeang, C. H., Angelo, M., Ladd, C., et al. 2001. Multiclass cancer diagnosis using tumor gene expression signatures. *Proceedings of the National Academy of Science, USA* 98:15149–1554.

Ransohoff, D. F. 2005. Bias as a threat to the validity of cancer molecular-marker research. *Nature Reviews Cancer* 5:142–149.

Ressom, H., Reynolds, R., and Varghese, R .S. 2003a. Increasing the efficiency of fuzzy logic-based gene expression data analysis. *Physiological Genomics* 13:107–117.

Ressom, H., Wang D., and Natarajan P. 2003b. Clustering gene expression data using adaptive double self-organizing map. *Physiological Genomics* 14:35–46.

Rezaee, M. R., Lelieveldt, B. P. F., and Reiber, J. H. C. 1998. A new cluster validity index for the fuzzy c-means. *Pattern Recognition Letters* 18:237–246.

Rodriguez-Zas, S. L., Band, M. R., Everts, R. E., Southey, B. R., Liu, Z. L., and Lewin, H. A. 2003. Analysis of gene expression patterns in the cattle digestive system. *Journal of Dairy Science* 86 (Suppl. 1):628.

Roulet, E., Fisch, I., Junier, T., Bucher, P., and Mermod, N. 1998. Evaluation of computer tools for the prediction of transcription factor binding sites on genomic DNA. *In Silico Biology* 1:21–28.

Rumelhart, D. E. and McClelland, J. L, eds. 1986. Parallel distributed processing: Explorations in the microstructure of cognition. *Foundations*, vol. 1. Cambridge, MA: MIT Press.

Schmitt, W. A. Jr., Raab, R. M., and Stephanopoulos, G. 2004. Elucidation of gene interaction networks through time-lagged correlation analysis of transcriptional data. *Genome Research* 14(8):1654–1663.

Schuldiner, O., Yanover, C., and Benvenisty, N. 1998. Computer analysis of the entire budding yeast genome for putative targets of the GCN4 transcription factor. *Currents in Genetics* 33:16–20.

Shin, A. and Iba, H. 2003. Construction of genetic network using evolutionary algorithm and combined fitness function, *Genome Informatics* 14:94–103.

Shipp, M. A., Ross, K. N, Tamayo, P., Weng A. P., Kutok, J. L., Aguiar, R. C., Gaasenbeek, M., et al. 2002. Diffuse large B-cell lymphoma outcome prediction by gene expression profiling and supervised machine learning. *Nature Medicine* 8:68–74.

Shmulevich, I., Dougherty, E. R., Kim, S., and Zhang, W. 2002. Probabilistic Boolean networks: a rule-based uncertainty model for gene regulatory networks. *Bioinformatics* 18(2):261–274.

Smith, V. A., Jarvis, E. D., and Hartemink, A. J. 2002. Evaluating functional network inference using simulations of complex biological systems. *Bioinformatics* 18: S216–S224.

Sokhansanj, B. A., Fitch, J. P., Quong, J. N., and Quong, A. A. 2004. Linear fuzzy gene network models obtained from microarray data by exhaustive search. *BMC Bioinformatics* 5:108.
Somogyi, R. 1999. Making sense of gene-expression data. *Trends in Biotechnology*, 17 (Supp.1):17–24.
Spellman, P. T., Sherlock, G., Zhang, M. Q., Iyer, V. R., Anders, K., Eisen, M. B., Brown, P. O., et al. 1998. Comprehensive identification of cell cycle-regulated genes of the yeast Saccharomyces cerevisiae by microarray hybridization. *Molecular Biology of the Cell* 9:3273–3297.
Su, M. and Chang, H. 2001. A new model of self-organizing neural networks and its application in data projection. *IEEE Transactions on Neural Networks* 12:153–158.
Tamayo, P., Slonim, D., Mesirov, J., Zhu, Q., Kitareewan, S., Dmitrovsky, E., Lander, E. S., and Golub, T. R. 1999. Interpreting patterns of gene expression with self-organizing maps: Methods and application to hematopoietic differentiation. *Proceedings of the National Academy of Science, USA* 96: 2907–2912.
Tibshirani, R., Walther, G., and Hastie, T. 2000. Estimating the number of clusters in a dataset via the gap statistic. *Technical Report 208*, Department of Statistics. Stanford, CA: Stanford University.
Tomida, S., Hanai, T., Honda, H., and Kobayashi, T. 2002. Analysis of expression profile using fuzzy adaptive resonance theory. *Bioinformatics* 18(8): 1073–1083.
Troyanskaya, O., Cantor, M., Sherlock, G., Brown, P., Hastie, T., Tibshirani, R., Botstein, D., et al. 2001. Missing value estimation methods for DNA microarrays. *Bioinformatics* 17:520–525.
Ushizawa, K., Herath, C. B., Kaneyama, K., Shiojima, S., Hirasawa, A., Takahashi, T., Imai, K., et al. 2004. cDNA microarray analysis of bovine embryo gene expression profiles during the pre-implantation period. *Reproductive Biology and Endocrinology* 2:77.
Wang, Y., Lu J., Lee, R., Gu, Z., and Clarke, R. 2002. Iterative normalization of cDNA microarray data. *IEEE Transactions on Information Technology in Biomedicine* 6(1):29–37.
Weaver, D. C., Workman, C. T., and Stormo, G. D. 1999. Modeling regulatory networks with weight matrices. *Pacific Symposium on Biocomputing* 3:112–123.
Wingender, E., Chen, X., Fricke, E., Geffers, R., Hehl, R., Liebich, I., Krull, M., et al. 2001. The TRANSFAC system on gene expression regulation. *Nucleic Acids Research* 29(1):281–283.
Woolf, P. J. and Wang, Y. 2000. A fuzzy logic approach to analyzing gene expression data. *Physiological Genomics* 3:9–15.
Xie, X. L. and Beni, G. 1991. A validity measure for fuzzy clustering. *IEEE Transactions on Pattern Analysis and Machine Intelligence* 13(8):841–847.
Yeung, K. Y., Fraley, C., Murua, A., Raftery, A. E., and Ruzzo, W. L. 2001a. Model-based clustering and data transformations for gene expression data. *Bioinformatics* 17:977–987.
Yeung, K. Y., Haynor, D. R., and Ruzzo, W. L. 2001b. Validating clustering for gene expression data. *Bioinformatics* 17:309–318.
Zak, E. D., Doyle, F. J. III, Gonye, G. E., and Schwaber, J. S. 2001. Simulation studies for the identification of genetic networks from cDNA array and regulatory activity data. *Proceedings of the Second International Conference on Systems Biology* California Institute of Technology, Pasadena, CA, USA. 231–238.
Zhu, J. and Zhang, M. Q. 1999. SCPD: A promoter database of the yeast Saccharomyces cerevisiae. *Bioinformatics* 15:607–611.
Zhu, Z., Pilpel, Y., and Church, G. M. 2002. Computational identification of transcription factor binding sites via a transcription-factor-centric clustering (TFCC) algorithm. *Journal of Molecular Biology* 318:71–81.

Glossary of Terms

Affymetrix GeneChips©   Microarrays in which the probes are oligonucleotides synthesized directly on the chip through a photolithography process.

cDNA arrays   Microarrays in which the probes are extracted from cDNA clones and are robotically printed onto a glass slide and subsequently hybridized to two differentially fluorescently labeled targets.

classification   Involves the automated grouping of objects into prespecified categories.

clustering   Helps discover common properties contained within the data and create groups of objects according to their properties.

defuzzification   Conversion of a fuzzy output value into to a crisp value.

DNA microarray   A tool for obtaining high-throughput global gene expression data.

feedforward neural network   A neural network with all connections are feedforward; that is, information transfer results only from an earlier layer to the next consecutive layers. Neurons within a layer are not connected, and neurons in nonadjacent layers are not connected.

fuzzification   Conversion of a crisp value of a variable into degrees of membership using membership functions assigned to the variable.

fuzzy logic   A superset of conventional two-valued (Boolean) logic that has been extended to handle the concept of partial truth.

gene expression   Rate at which a gene is used to produce functional RNA transcripts.

Gene expression data matrix   A matrix where each entry corresponds to the expression level of a gene for a given condition or sample.

genetic algorithms   Global optimization algorithms that originated from mechanics of natural genetics and selection.

genetic network   Conceptual network of interaction among gene expression events.

inference   Process of evaluating which and how fuzzy rules should be executed.

machine learning   Field of scientific study that concentrates on induction algorithms and on other algorithms that can be said to learn.

neural networks   Information processing systems that have certain performance characteristics in common with biological neural networks.

particle swarm optimization Adaptive algorithm based on the social metaphor of flocking birds (or schooling fish, or swarming insects).

principal component analysis Transforms the input variables to a new set of variables known as principal components that are orthogonal to each other. A subset of the principal components that captures most of the information in the original data is selected for dimensionality reduction.

supervised learning Uses an external teacher to produce a desired output; the model learns from training examples.

support vector machines Learning kernel-based systems that use a hypothesis space of linear functions in high-dimensional feature spaces.

unsupervised learning Does not involve an external teacher; the algorithm discovers collective properties of data by self-organization.

CHAPTER 2
Biosystems Analysis and Optimization

Wouter Saeys and Herman Ramon
Department of Biosystems
Katholieke Universiteit Leuven, Heverlee, Belgium

Tom Coen
INDUCT bvba, Zemst, Belgium

2.1 Introduction

Mathematical description of the dynamic behavior of technical systems has a long tradition, which first reached a climax in 1687 when Isaac Newton introduced his three fundamental laws to describe the motion of a point mass. Nevertheless, it was not until the nineteenth century before mathematical models were slowly introduced to describe biological systems (e.g., population growth model introduced by Pierre Verhulst in 1837), and it was only in the twentieth century that this discipline fully developed. It is expected that the twenty-first century will be the age of mathematical biology aiming at an accurate mathematical description of the most important biological processes (e.g., photosynthesis, respiration, metabolic pathways, tissue generation).

2.1.1 Definitions

- *Systems analysis* or systems theory—quantitative (mathematical) analysis of dynamic systems.
- *Dynamic system*—entity that can be isolated from its environment by a physical or conceptual border and whose state changes over time. Traditional examples of such dynamic systems are machinery (car, train, airplane, etc.), electronic circuits, and chemical processes (reactions, pathways, etc.). However,

all living organisms (human, animal, microorganisms, and even plants) are also dynamic systems because they change over time.

- A system can constitute different *subsystems* that interact with each other (e.g., organs in a human body) or can be part of a larger system (e.g., individual organisms in a population). Often, complex systems can be decomposed in subsystems on different levels or scales and are, therefore, called *multiscale systems*.

Example A combine harvester is a highly complex biologically related dynamic system, which can be split into different subsystems such as the header, the straw elevator, the threshing, the separation, the cleaning and the grain elevator. These subsystems themselves contain smaller subsystems. The cleaning section, for example, consists of a number of shaking sieves and a fan, which are dynamic systems themselves.

Goal The aim of biosystems analysis is to gain insight into the behavior of a system, such that we can predict the system behavior and eventually control it by actively adapting its dynamics.

This chapter is dedicated to the application of systems analysis to biologically related problems. The reader is introduced to the methods for model building, system analysis through computer simulations, system identification, and optimization methods.

The examples given in this chapter are limited to biotechnical systems (agricultural machinery). Some examples of biotechnical systems where we have applied the principles described in this chapter are control of slurry injectors (Saeys et al. 2007, 2008a, 2008b), cab suspension design (De Temmerman et al. 2004, 2005), spray boom suspension design (Ramon et al. 1998; Anthonis and Ramon 2003; Deprez et al. 2003), cruise control (Coen et al. 2008a, 2008b), and an autopilot (Coen et al. 2008c) for a combine harvester, and modeling and control of pasta drying (De Temmerman et al. 2007, 2009). However, the methods described can also be applied to purely biological systems.

2.2 System Representation

2.2.1 Block Diagram Representation

Throughout history, engineers have always sought to describe reality in a structured way. A common form of representation for dynamic systems is to use *block diagrams*. These diagrams are meant to represent the system as a block or a box where all signals entering (*inputs*) and exiting (*outputs*) the system are represented as directed lines.

Biosystems Analysis and Optimization 35

```
Multifunctional handle position  →  [ Combine harvester drive line ]  →  Driving speed
```

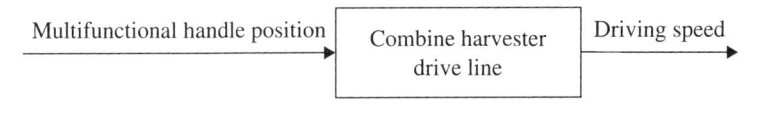

FIGURE 2.1 Simple block representation of the driving line of a combine harvester.

Figure 2.1 shows a simple block diagram for the driving line of a combine harvester. A change in the multifunctional handle (joystick) position causes the hydrostatic pump setting to change, which causes a change in the oil flow to the hydromotor, which in turn causes a change in the wheel speed and thus a change in driving speed.

By using the block diagram representation, we clearly define the system as an effect relationship between the *inputs* and the *outputs*. One way to classify systems is based on the number of inputs and outputs. The system in Fig. 2.1 has one input and one output and is therefore called a single-input single-output (SISO) system.

Of course, there also exist systems with multiple inputs and/or outputs. This leads to the definition of single-input multiple-output (SIMO), multiple-input single-output (MISO), and multiple-input multiple-output (MIMO) systems. For example, if we return to our combine harvester example, we should note that driving speed is not only influenced by the multifunctional handle position but also by the engine speed (rpm). Therefore, a better representation of this system, as a MISO system, is given in Fig. 2.2.

This block diagram is especially interesting when the system to be described consists of different subsystems that interact with each other. Each subsystem can then be represented by an individual block, and the interaction can be described graphically by drawing arrows between the different blocks indicating the signal flow from one subsystem to another. In Fig. 2.3, the block diagram for our combine

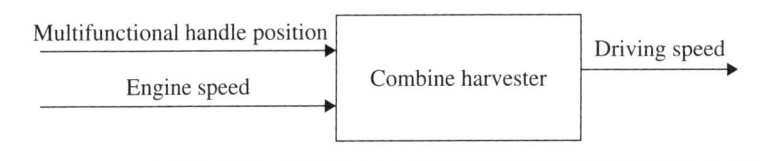

```
Multifunctional handle position ─┐
                                  ├─→ [ Combine harvester ] ─→ Driving speed
Engine speed ────────────────────┘
```

FIGURE 2.2 MISO block diagram for the driving line of a combine harvester.

```
[Multifunctional handle]  Steering current ─→ [Hydrostatic pump] ─Oil flow→ [Hydrostatic engine] ─Hydromotor speed→ [Mechanical gear box] ─Wheel speed→
[Diesel engine] ─Engine speed→
```

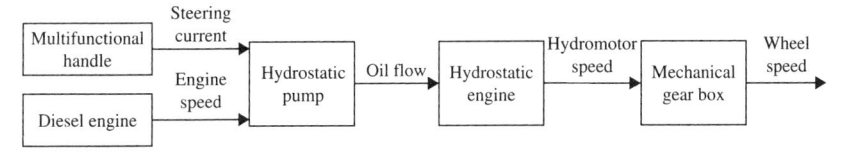

FIGURE 2.3 Block diagram for the driving line of a combine harvester including subsystem blocks.

harvester driving line (Fig. 2.2) is further detailed to include blocks for the individual subsystems.

2.2.2 Mathematical Representation

Although a qualitative block diagram representation, such as the one shown in Fig. 2.2, provides a structured insight in the connections between the different subprocesses of the system, for design purposes we usually need to describe its dynamic behavior quantitatively. For this purpose, we need a mathematical representation of each block.

Let us represent each signal by a time function. Then a system can be defined as a function G, which transfers the input signal $u(t)$ entering the system into an output signal $y(t)$ exiting the system:

$$y(t) = G(u(t)) \qquad (2.1)$$

Function G is referred to as an input–output transfer function, or simply *transfer function*.

For some simple linear systems, like basic electronic components, the dynamic behavior is well known and can be accurately described using mathematical models. A few examples of such simple systems and the corresponding models are given here.

The potential $v(t)$ over a *resistor* is equal to current $i(t)$ multiplied by the resistance R. When we consider the current $i(t)$ through the resistor to be the input $u(t)$ and the potential $v(t)$ over the resistor to be the output $y(t)$, the resistor transfer function can be written as follows:

$$v(t) = Ri(t) \quad or \quad y(t) = Ku(t) \qquad (2.2)$$

where the constant gain K is equal to the resistance R. The transfer function in Eq. (2.2) is called a constant gain multiplier, which is the simplest type of system. Other examples of constant gain multipliers are a linear mechanical spring, a tooth wheel connection, and conductive heat transfer through a material.

The voltage $v(t)$ over a *solenoid* is not proportional to the current $i(t)$ through it, but to its time derivative:

$$v(t) = L\frac{di(t)}{dt} \qquad (2.3)$$

where L is the inductance of the solenoid. If we again consider the current $i(t)$ to be the input $u(t)$ and the voltage $v(t)$ to be the output, Eq. (2.3) can be rewritten as follows to obtain the transfer function:

$$y(t) = K\frac{du(t)}{dt} \qquad (2.4)$$

This is a differentiator where the gain K equals the inductance L. Other examples of this differentiator action can be found: between the position of a damper shaft and the force exerted by the viscous damper, between the position of a conductor moving in a magnetic field, and the potential over this conductor, for example.

A *capacitor* or *condenser* can be seen as the counterpart of a solenoid, because the current $i(t)$ through it is proportional to the time derivative of the voltage $v(t)$ over it:

$$i(t) = C\frac{dv(t)}{dt} \qquad (2.5)$$

where C is the capacitance of the condenser. When the current $i(t)$ is again considered to be the input $u(t)$ and the voltage $v(t)$ to be the output, Eq. (2.5) can be rewritten as follows to obtain the transfer function:

$$v(t) = \frac{1}{C}\int_0^t i(t) \quad \text{or} \quad y(t) = K\int_0^t u(t) \qquad (2.6)$$

With this, the input–output relation is an integral function where the gain K is equal to the inverse of the condenser capacity C. Other examples of this integrator action can be found in the dynamic relation between the oil flow to a hydraulic actuator and the shaft position, between the water flow into a bath tub and the water level in the tub, and between the energy production in a room and the temperature in the room.

More complex systems can often be described as a combination of the higher mentioned building blocks by linking them in series or parallel.

Subsystems in Parallel

A popular mechanical example of such a combination of building blocks is the mass–spring–damper system. In this system, a mass (e.g., tractor) is linked to a fixed body (e.g., the ground) through a spring and a damper, which are mounted in parallel (Fig. 2.4).

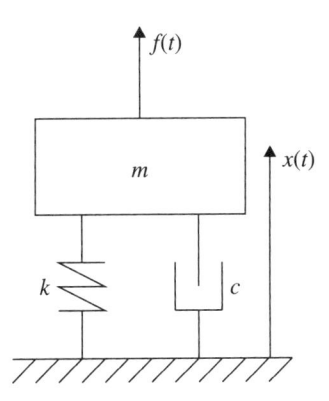

FIGURE 2.4 Schematic representation of a mass–spring–damper system; m is the mass, k is the spring constant, c is the damping constant, $f(t)$ is the time-dependent external force exerted on the mass, and $x(t)$ is the time-varying position of the mass with respect to the solid body.

The relation between the time-dependent force $f(t)$, generally called the forcing function, exerted on mass m and the position of the mass $x(t)$ with respect to a fixed body can be derived by applying Newton's second law:

$$f(t) + f_s(t) + f_d(t) = m\frac{d^2x(t)}{dt^2} \qquad (2.7)$$

where $f_s(t)$ is the force exerted by the spring with spring constant k when the mass has moved from the rest position x_0 to its current position $x(t)$:

$$f_s(t) = -k(x(t) - x_0) \qquad (2.8)$$

and $f_d(t)$ is the force exerted by the viscous damper with damping constant c:

$$f_d(t) = -c\frac{dx(t)}{dt} \qquad (2.9)$$

Combination of Eqs. (2.7) to (2.9), setting x_0 to zero, and moving all terms in $x(t)$ to the right gives us the equation of motion for the mass–spring–damper system:

$$f(t) = m\frac{d^2x(t)}{dt^2} + c\frac{dx(t)}{dt} + kx(t) \qquad (2.10)$$

This is a second-order differential equation describing the force $f(t)$ as a function of the position $x(t)$. However, we are interested in the effect of a force signal $f(t)$ on the position signal $x(t)$. For this purpose, it would be nice to have a mathematical description that allows us to consider the input signal, the output signal, and the system as different entities. Such a convenient description is available through the Laplace transform, which is described next.

2.2.3 The Laplace Transform

The Laplace transform $F(s)$ of a time signal $f(t)$ is defined as (Nise 2000)

$$\mathcal{L}[f(t)] = F(s) = \int_{0^-}^{\infty} f(t)e^{-st}\,dt \qquad (2.11)$$

where $s = \sigma + j\omega$.

The time t has been integrated out the time-variable function $f(t)$ to produce an algebraic function $F(s)$ of the complex variable s, which is also known as the Laplace variable. The real part of the Laplace variable $\text{Re}(s) = \sigma$ is related to the boundedness of the signal or the stability of the system, whereas the imaginary part $\text{Im}(s) = \omega$ is related to the frequency content. Note that to avoid confusion between signals and systems represented in the time domain and their Laplace transforms, the former are typically represented by lowercase letters and the latter by uppercase letters.

Thanks to this transformation, we can convert signals and systems from the time domain (function of time t) into the Laplace domain (function of the Laplace operator s). This transformation has some interesting properties, which are defined in the following theorems:

Linearity Theorem

$$\mathcal{L}[kf(t)] = kF(s) \qquad (2.12)$$

$$\mathcal{L}[f_1(t) + f_2(t)] = F_1(s) + F_2(s) \qquad (2.13)$$

Differentiation Theorem

$$\mathcal{L}\left[\frac{df(t)}{dt}\right] = sF(s) - f(0-) \qquad (2.14)$$

where $f(0-)$ is the value of the function $f(t)$ at $t = 0$, the minus sign indicates that when $f(t)$ is discontinuous at $t = 0$ we should take the left-side value.

Integration Theorem

$$\mathcal{L}\left[\int_{0-}^{t} f(\tau)d\tau\right] = \frac{F(s)}{s} \qquad (2.15)$$

Time-Delay Theorem

$$\mathcal{L}[f(t-T)] = e^{-sT}F(s) \qquad (2.16)$$

Time-Scaling Theorem

$$\mathcal{L}[f(kt)] = \frac{1}{k}F\left(\frac{s}{k}\right) \qquad (2.17)$$

Frequency-Shift Theorem

$$\mathcal{L}[e^{-\tau t}f(t)] = F(s+\tau) \qquad (2.18)$$

These properties of the Laplace transform are very interesting with respect to the description of dynamic systems. Instead of representing a dynamical system by (a set) of (coupled) differential equations in the time domain, we describe the system as a (set of) algebraic equation(s) on which well-established algebraic operations are applied. We illustrate this by converting the equation of motion of the mass–spring–damper system [Eq. (2.10)] from the time domain to the Laplace domain:

$$\mathcal{L}[f(t)] = \mathcal{L}\left[m\frac{d^2x(t)}{dt^2} + c\frac{dx(t)}{dt} + kx(t)\right] \qquad (2.19)$$

Instead of calculating the Laplace transform of the function on the left side of Eq. (2.19) by solving the integral from Eq. (2.11), we can use the interesting properties described in Eqs. (2.12) to (2.18). Thanks to the linearity theorem [Eqs. (2.12) and (2.13)], the Laplace transform of the equation of motion can be written as the sum of the Laplace transforms of the individual subsystems (inertia, damper, and spring). By introducing the Laplace transform $F(s)$ of the force signal $f(t)$ Eq. (2.19) can thus be rewritten as

$$F(s) = m\mathcal{L}\left[\frac{d^2x(t)}{dt^2}\right] + c\mathcal{L}\left[\frac{dx(t)}{dt}\right] + k\mathcal{L}[x(t)] \qquad (2.20)$$

Because the remaining functions to be transformed are just differentials of $x(t)$, we can now apply the differentiation theorem [Eq. (2.14)] and introduce the Laplace transform $X(s)$ of the position signal $x(t)$ to rewrite Eq. (2.20) as

$$F(s) = m[s^2X(s) - sx(0-) - \dot{x}(0-)] + c[sX(s) - x(0-)] + kX(s) \qquad (2.21)$$

where $x(0-)$ is the initial position x_0 of the mass and $\dot{x}(0-)$ is the initial speed v_0 of the mass. By separating the dynamic terms from the initial conditions [Eq. (2.21)] can be rewritten as

$$F(s) = [ms^2 + cs + k]X(s) - mv_0 - (ms + c)x_0 \qquad (2.22)$$

When we assume the mass to be initially at rest ($v_0 = 0$) and define the position x such that x_0 is zero, we can isolate the Laplace transform of the system from those of the signals:

$$\frac{F(s)}{X(s)} = ms^2 + cs + k \qquad (2.23)$$

This is a transfer function in the Laplace domain for the transfer of a position signal $X(s)$ into a force signal $F(s)$.

Remember that the external force applied to the mass–spring–damper system is actually the input, whereas the position of the mass is the output. In the time domain, the position $x(t)$ in Eq. (2.19) can only be written as an explicit function of the applied force $f(t)$ by solving the differential equation [Eq. (2.10)], which is often a tedious and time-consuming task. Thanks to the separation of signals and systems in the Laplace domain, the transfer function $G(s)$, which converts the applied force signal $F(s)$ into a position signal $X(s)$, can be obtained by simply inverting Eq. (2.23):

$$G(s) = \frac{X(s)}{F(s)} = \frac{1}{ms^2 + cs + k} \qquad (2.24)$$

The transfer function $G(s)$ in Eq. (2.24) is a rational polynomial in the complex variable s. This system is said to be of the second order, because the highest power of the Laplace variable s present in the denominator of the transfer function $G(s)$ is 2, which corresponds to the differential equation of the second order in the time domain [Eq. (2.10)].

Subsystems in Series

For a system consisting of two subsystems connected in series, the input signal of the second subsystem $u_2(t)$ equals the output signal of the first subsystem $y_1(t)$. In the time domain combination of the transfer functions of both subsystems would mean incorporating the first differential equation into the second differential equation. Thanks to the separation of signals and systems in the Laplace domain, combination of different subsystems placed in series also becomes a simple algebraic operation. For each subsystem the transfer function can be defined as a ratio of the output signal $Y_k(s)$ over the input signal $U_k(s)$. The transfer function of the system $G(s)$ consisting of two subsystems connected in series can then be defined as the product of the transfer functions of the subsystems $G_1(s)$ and $G_2(s)$:

$$G(s) = \frac{Y_2(s)}{U_1(s)} = \frac{Y_2(s)}{Y_1(s)} \frac{Y_1(s)}{U_1(s)} = G_2(s)G_1(s) \qquad (2.25)$$

For example, let us add a second mass m_2 to the previously described mass–spring–damper system (Fig. 2.4), which is connected to the first mass, now noted as m_1 for reason of clarity, by a hydraulic actuator oriented in the same direction as the spring and damper and with a time-dependent length $x(t)$ (Fig. 2.5). In this system, the oil flow to the actuator $q(t)$ can be seen as the input signal, and the position $x_2(t)$ of the second mass m_2 (e.g., height of the implement) can be seen as the output signal.

The hydraulic actuator subsystem converts the oil flow to the actuator $q(t)$ into an actuator length $x(t)$. Because the speed of the actuator shaft, and not the position, is proportional to the oil flow $q(t)$,

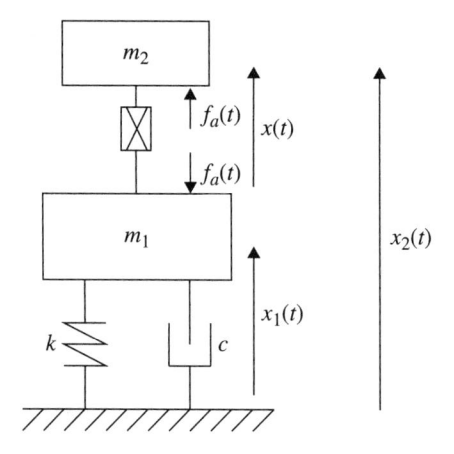

FIGURE 2.5 Schematic representation of a mass–spring–damper system with a second mass connected to it by an actuator; m_1 and m_2 are the masses, k is the spring constant, c is the damping constant, $f_a(t)$ and $x(t)$ are the time-dependent actuator force and length, respectively, and $x_1(t)$ and $x_2(t)$ are the time-varying positions of the masses with respect to the solid body.

the input variable $q(t)$ should be integrated over time to obtain a linear relation with the output variable $x(t)$ as follows:

$$x(t) = K \int_0^t q(t) \tag{2.26}$$

The resulting mechanical subsystem is still one-dimensional, but now has two translational degrees of freedom, giving rise to two position signals: positions $x_1(t)$ and $x_2(t)$ with respect to the solid body of the masses m_1 (e.g., tractor body) and m_2 (e.g., implement), respectively.

The equation of motion for the first mass has been derived previously [Eq. (2.10)]:

$$-f_a(t) = m_1 \frac{d^2 x_1(t)}{dt^2} + c \frac{d x_1(t)}{dt} + k x_1(t) \tag{2.27}$$

where $f_a(t)$ is the time-dependent actuator force acting on the first mass m_1.

The equation of motion for the second mass m_2 can be derived from Newton's second law:

$$f_a(t) = m_2 \frac{d^2 x_2(t)}{dt^2} \tag{2.28}$$

where $f_a(t)$ is the force exerted by the actuator on mass m_2. By eliminating the actuator force $f_a(t)$, we obtain a differential equation for the relation between the positions of both masses:

$$-m_2 \frac{d^2x_2(t)}{dt^2} = m_1 \frac{d^2x_1(t)}{dt^2} + c\frac{dx_1(t)}{dt} + kx_1(t) \qquad (2.29)$$

Because both masses are coupled by the actuator, the position $x_2(t)$ of the second mass m_2 (e.g., the implement) with respect to the solid body (e.g., the ground) is equal to the sum of the position of the first mass $x_1(t)$ and the length of the actuator $x(t)$:

$$x_2(t) = x_1(t) + x(t) \qquad (2.30)$$

In this mechanical subsystem, the length of the actuator is $x(t)$. Because we want to know the position $x_2(t)$ of the second mass (e.g., implement) with respect to the solid body (e.g., ground) and are not so much interested in the position $x_1(t)$ of the first mass m_1, we can eliminate the latter by combination of Eqs. (2.29) and (2.30):

$$(m_1 + m_2)\frac{d^2x_2(t)}{dt^2} + c\frac{dx_2(t)}{dt} + kx_2(t) = m_1 \frac{d^2x(t)}{dt^2} + c\frac{dx(t)}{dt} + kx(t) \qquad (2.31)$$

In the time domain, it is only possible to rewrite this equality of two differential equations as an equation giving the output signal $x_2(t)$ as a function of the input signal $x(t)$ by solving the differential equation [Eq. (2.31)], which is often a tedious task. Moreover, combining the mechanical subsystem model [Eq. (2.31)] with the hydraulic actuator subsystem [Eq. (2.26)] to eliminate the actuator length $x(t)$ further complicates the input–output relation. In the Laplace domain, isolation and combination of both subsystem transfer functions becomes incredibly easy. Let us transform the hydraulic actuator subsystem model [Eq. (2.26)] to the Laplace domain by applying the integration theorem [Eq. (2.15)] and rewrite it as a transfer function $G_1(s)$:

$$G_1(s) = \frac{X(s)}{Q(s)} = \frac{K}{s} \qquad (2.32)$$

where $X(s)$ and $Q(s)$ are, respectively, the Laplace transform of the actuator length signal $x(t)$ and the oil-flow signal $q(t)$.

Similarly, the mechanical subsystem model [Eq. (2.31)] can be transformed to the Laplace domain by applying the linearity and differentiation theorems [Eqs. (2.12) to (2.14)]. When we assume the initial conditions to be zero, we obtain the transfer function $G_2(s)$ for the mechanical subsystem as follows:

$$G_2(s) = \frac{X_2(s)}{X(s)} = \frac{m_1 s^2 + cs + k}{(m_1 + m_2)s^2 + cs + k} \qquad (2.33)$$

where $X(s)$ and $X_2(s)$ are, respectively, the Laplace transforms of the actuator length signal $x(t)$ and the position $x_2(t)$.

Combination of both subsystem transfer functions using Eq. (2.25) and elimination of $X(s)$ leads to the transfer function $G(s)$ for the total system:

$$G(s) = \frac{X_2(s)}{Q(s)} = \frac{m_1 s^2 + cs + k}{(m_1 + m_2)s^3 + cs^2 + ks} \quad (2.34)$$

2.2.4 Nonlinearities

Unfortunately, the dynamic behavior of many real-life systems cannot be accurately described by a combination of the nice linear building blocks mentioned earlier because they also exhibit some nonlinear behavior. Although modeling linear dynamic systems is fairly straightforward using the principles discussed in the previous sections, modeling nonlinear systems is more complex. It has, however, been shown in the literature that a wide range of nonlinear systems can be modeled as a cascade of linear dynamic systems and nonlinear static elements (Nelles 2001). When a system has a nonlinear element before the linear dynamic system, it is called a *Hammerstein model*, and a system with output nonlinearity is known as a *Wiener model*.

Two architectures are commonly used to describe nonlinear systems. The first consists of a linear dynamic system followed by a nonlinear static element and another linear dynamic system, known as the *Wiener–Hammerstein model*. The second architecture is composed of a linear dynamic system preceded and followed by nonlinear static elements, known as the *Hammerstein–Wiener model*.

In this section, we present an overview of some nonlinear elements that are common for biologically related systems, together with some examples of systems containing these nonlinear elements.

Dead Zone

A linear system model describes the output signal as a (frequency-dependent) multiple of the input. However, many real-life systems will only react when the input signal is sufficiently large. We can thus define a band or zone of input signals for which the system remains unaffected as the dead zone.

For example, in a tooth-wheel connection there typically is some slop, called backlash, which makes that the first wheel (input) has to rotate a little before touching the second wheel (output) and put it into rotation as well. Another example can be found in the dead zone of a hydraulic valve. Only when the current generates a sufficiently large magnetic force through its solenoid will the valve open and generate an oil flow. The static relation between the input and the output of a system with dead zone is illustrated in Fig. 2.6.

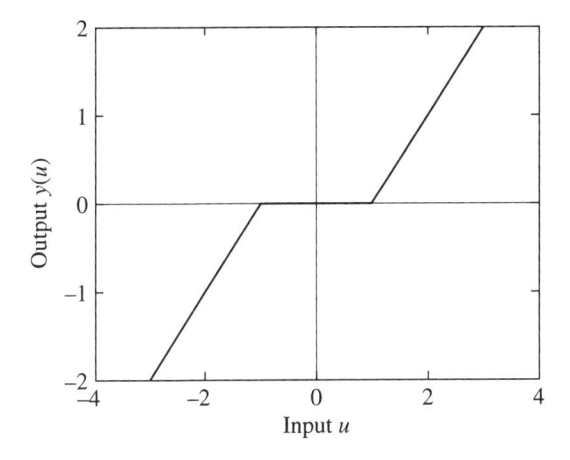

FIGURE 2.6 Dead zone in the static relation between input u and output $y(u)$.

Actuator Saturation

Apart from the minimal input needed for reaction, most real-life systems have physical boundaries, limiting the maximal output of the system. A typical example is actuator saturation where, in case of a hydraulic actuator, the output signal is limited by the stroke of the piston in the cylinder. Once the piston is fully extended or retracted, pumping more oil in the extended chamber will no longer result in a movement of the piston but will only lead to further compression of the oil in the chamber. A second application can be found in a fully open hydraulic valve. Further increasing the current applied to its solenoid will no longer increase the valve opening. The static relation between the input and the output of a system with actuator saturation is illustrated in Fig. 2.7.

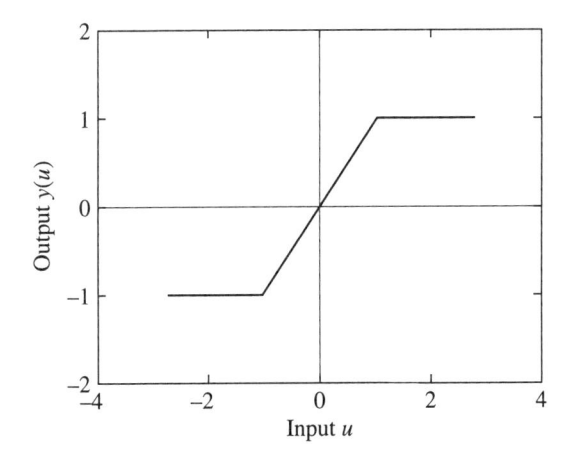

FIGURE 2.7 Illustration of actuator saturation in the static relation between input u and output $y(u)$.

Time Delay or Dead Time

In many biologically related systems, the response to the input signal can only be observed in the output signal some time after the input signal has been applied, due to transport phenomena. One can say that the system response is delayed or that the system does not respond for some time. If, for example, a combine harvester operator increases the ground speed of the combine during wheat harvesting, it takes approximately 1.2 s before the torque on the grain elevator belt rises due to increased crop flow into the machine. Because the crop has to move through the threshing and cleaning sections and has to be transported to the grain bin, it takes another 10 s before the grain mass flow into the grain bin starts to respond to the change in driving speed. These time delays are illustrated in Fig. 2.8 where the response of the feed rate signal and the grain mass flow signal to a step change in the driving speed of a combine harvester are shown.

These time delays play a critical role in controller design because they can easily lead to instability of the controlled system. Therefore, time delays have to be modeled accurately in the design model. In the Laplace domain, the transfer function of a pure time delay $G_{td}(s)$ is defined as

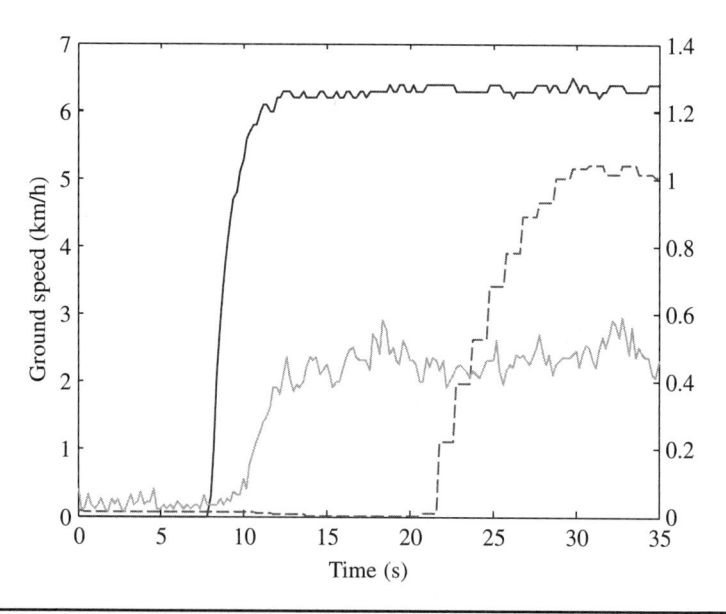

$$G_{td}(s) = e^{-s\tau} \tag{2.35}$$

Figure 2.8 Illustration of the effect of time delay in the response of feed rate (solid grey line) and grain mass flow (dashed line), both displayed on the Y-axis on the right on a step change in the driving speed (solid line and Y-axis on the left) of a combine harvester.

where τ is the time delay. Because this nonlinearity plays an important role with respect to the stability of a controlled system, it is often replaced by a linear approximation to be able to use linear stability analysis tools such as root locus (Franklin et al. 2006). A useful linear approximation is known as the second-order Padé approximant (Franklin et al. 2006), presented by the following expression:

$$G_{td}(s) = e^{-s\tau} \approx \frac{s^2 - \frac{6}{\tau}s + \frac{12}{\tau^2}}{s^2 + \frac{6}{\tau}s + \frac{12}{\tau^2}} \qquad (2.36)$$

2.3 System Analysis

Once a design model is built, we want to investigate and predict the system behavior under different conditions. For this purpose we can make use of computer simulation tools.

2.3.1 Response of the System to a Step Input

The description of a system in the time domain is typically given in terms of its response to a step input. The step response of a second-order system with natural frequency $\omega_n = 1$ rad/s and damping ratio $\zeta = 0.5$ is illustrated in Fig. 2.9. Several characteristics describing the dynamic behavior of the system can be derived from this step response:

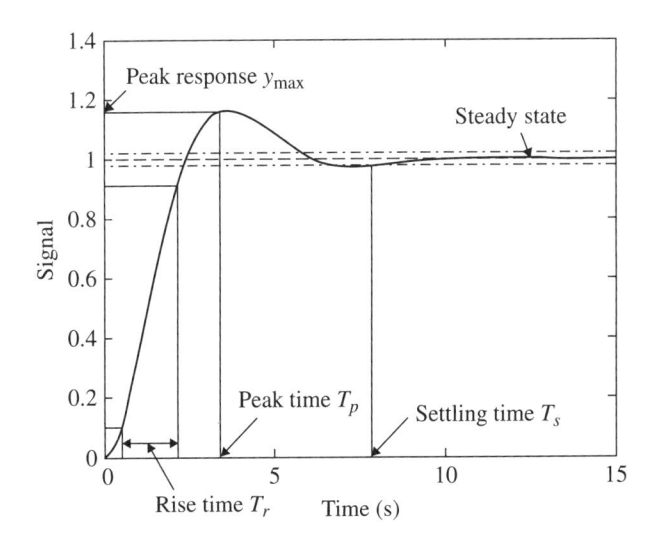

FIGURE 2.9 Time response of a second-order system (solid) to a unit step input (dash) with the 2 percent error band around the steady-state value (dash-dot) and the most important time response properties: rise time, peak response, settling time, and steady state.

1. *Steady state* is the time after which the transient behavior of a system in response to a step input has died out and the system behavior no longer changes in time.
2. *Settling time* T_s is the time required for the output signal to reach and remain within a given error band (e.g., 2 percent) around the steady-state value.
3. *Peak response* y_{max} is the maximal value of the step response, reached at the peak time T_p.
4. *Rise time* T_r is the time required for the output signal to change from a predefined low value to a predefined higher value, typically 10 and 90 percent, respectively, of the step height.

2.3.2 Magnitude and Phase of a Transfer Function

The description of a system in the frequency domain is given in terms of the response to a sinusoidal input signal after all initial transients have died out. Provided the system is linear, this steady-state output is a sinusoid of the same frequency (i.e., the forcing frequency) as the input, but with a shift of phase and a change of amplitude. The ratio of the amplitude of the output sine wave to the amplitude of the input sine wave is usually referred to as the magnitude (or sometimes as the magnitude ratio, amplitude ratio, or gain); the shift of phase of the output sine wave relative to the input is simply called the phase. Magnitude and phase are dependent both on the system transfer function and on the forcing frequency but not, for a linear system, on the amplitude (because magnitude represents an amplitude ratio). Variation of the magnitude and the phase with the frequency, traditionally known as the frequency response or the harmonic response information, can be found from an arbitrary transfer function $G(s)$ by replacing the Laplace variable s by $j\omega$. This variation can be presented graphically by a Bode diagram or Bode plot.

Note that the Laplace variable s is actually a complex number $a + j\omega$ with a real part a and a complex part ω in radians per second. Replacing s in $G(s)$ with $j\omega$ indicates that $G(s)$ is evaluated on an imaginary axis for $\text{Re}(s) = 0$.

We will show that the magnitude and phase (i.e., the ratio of the amplitude of the steady-state output to the amplitude of the input sine wave and the phase shift between the output and input sinusoids) are given by the modulus and argument of $G(j\omega)$, the transfer function where the Laplace variable s has been replaced by $j\omega$.

For a linear system, the transfer function $G(s)$ is the ratio of two polynomials in s, a polynomial of degree m in the numerator, and a polynomial of degree n in the denominator. Each polynomial can be given in factorized form:

$$G(s) = \frac{N(s)}{D(s)} = \frac{K(s-z_1)(s-z_2)\cdots(s-z_m)}{(s-p_1)(s-p_2)\cdots(s-p_n)} \quad (2.37)$$

where $z_1, z_2, ..., z_m$ are defined as the zeros of $G(s)$, values of s that annihilate the rational polynomial, and $p_1, p_2, ..., p_n$ are the poles of $G(s)$, values, making the rational polynomial infinite. Poles as well as zeros can be a real or a complex number. In the later case, each pole or zero has a complex conjugate.

If the input to the system is $u(t) = A\sin(\omega t)$, a sine wave of amplitude A and of frequency ω rad/s, then the Laplace transform of the input is

$$U(s) = \frac{A\omega}{s^2 + \omega^2} \tag{2.38}$$

The Laplace transform $Y(s)$ of the output $y(t)$ of the system is then calculated as

$$Y(s) = G(s)U(s) = A\frac{K\omega(s-z_1)(s-z_2)\cdots(s-z_m)}{(s^2+\omega^2)(s-p_1)(s-p_2)\cdots(s-p_n)} \tag{2.39}$$

This expression can be rewritten as the partial fraction expansion (Schwarzenbach and Gill 1978):

$$\frac{Y(s)}{A} = \frac{G(s)\omega}{s^2+\omega^2} = \frac{A_1}{s-j\omega} + \frac{A_2}{s+j\omega} + \frac{B_1}{s-p_1} + \cdots + \frac{B_n}{s-p_n} \tag{2.40}$$

where $A_1, A_2, B_1, B_2, ..., B_n$ are constants. Taking the inverse Laplace transform of Eq. (2.40) leads to the following time response:

$$\frac{y(t)}{A} = A_1 e^{j\omega t} + A_2 e^{-j\omega t} + B_1 e^{p_1 t} + \cdots + B_n e^{p_n t} \tag{2.41}$$

All individual terms in the right-hand side of Eq. (2.41) are called modes of the solution, which contain a system pole or an input signal pole in their exponent. Provided the system is stable and linear, poles $p_1, p_2, ..., p_n$ all have negative real parts and all of the modes in Eq. (2.41) decay to zero with increasing time t except the first two. The terms that decay to zero represent the transient response (or complementary function). The first two terms will not die out and thus represent the particular integral component of the solution (also called the steady-state response, particular solution, or stationary solution):

$$\left[\frac{y(t)}{A}\right]_{t\to\infty} = A_1 e^{j\omega t} + A_2 e^{-j\omega t} \tag{2.42}$$

To determine coefficient A_1, both sides of Eq. (2.40) are multiplied by $(s - j\omega)$ and s is replaced by $j\omega$:

$$A_1 = \left[\frac{(s-j\omega)G(s)\omega}{s^2+\omega^2}\right]_{s=j\omega} = \left[\frac{G(s)\omega}{s+j\omega}\right]_{s=j\omega} = \frac{G(j\omega)}{2j} \tag{2.43}$$

Sinusoidal steady-state transfer relation $G(j\omega)$ is a complex function of ω that can be expressed in terms of a real and an imaginary part; that is,

$$G(j\omega) = \text{Re}[G(j\omega)] + j\text{Im}[G(j\omega)] \tag{2.44}$$

Because a complex function can be represented in polar form, Eq. (2.44) can be rewritten as

$$G(j\omega) = |G(j\omega)| e^{j\angle G(j\omega)} \tag{2.45}$$

Using the expression in Eq. (2.45), Eq. (2.43) becomes

$$A_1 = \frac{1}{2j} |G(j\omega)| e^{j\angle G(j\omega)} \tag{2.46}$$

Similarly, to obtain A_2, both sides of Eq. (2.40) are multiplied by $(s + j\omega)$, and the Laplace variable s is replaced by $-j\omega$:

$$A_2 = \left[\frac{(s+j\omega)G(s)\omega}{s^2 + \omega^2} \right]_{s=-j\omega} = \left[\frac{G(s)\omega}{s - j\omega} \right]_{s=-j\omega}$$

$$= \frac{G(-j\omega)}{-2j} = -\frac{1}{2j} |G(j\omega)| e^{-j\angle G(j\omega)} \tag{2.47}$$

Using Eqs. (2.46) and (2.47), Eq. (2.6) is rewritten as

$$[y(t)]_{t\to\infty} = \frac{A|G(j\omega)|}{2j} \left\{ e^{j[\omega t + \angle G(j\omega)]} - e^{-j[\omega t + \angle G(j\omega)]} \right\}$$

$$= A|G(j\omega)| \sin\left(\omega t + \angle G(j\omega)\right) \tag{2.48}$$

Because $G(j\omega)$ is a complex function with a numerator and denominator, Eq. (2.44) is also rewritten as

$$G(j\omega) = \frac{N(j\omega)}{D(j\omega)} = \frac{\text{Re}[N(j\omega)] + j\text{Im}[N(j\omega)]}{\text{Re}[D(j\omega)] + j\text{Im}[D(j\omega)]} \tag{2.49}$$

Magnitude $|G(j\omega)|$ and phase $\angle G(j\omega)$ then become

$$mag[G(j\omega)] = |G(j\omega)| = \frac{\sqrt{\{\text{Re}[N(j\omega)]\}^2 + \{\text{Im}[N(j\omega)]\}^2}}{\sqrt{\{\text{Re}[D(j\omega)]\}^2 + \{\text{Im}[D(j\omega)]\}^2}} \tag{2.50}$$

$$phase[G(j\omega)] = \angle G(j\omega) = \arg[G(j\omega)]$$

$$= \arctan\left(\frac{\text{Im}[N(j\omega)]}{\text{Re}[N(j\omega)]}\right) - \arctan\left(\frac{\text{Im}[D(j\omega)]}{\text{Re}[D(j\omega)]}\right) \tag{2.51}$$

The magnitude and phase as functions of the frequency can thus be derived from transfer function $G(s)$ by replacing s with $j\omega$ and determining the modulus and argument of $G(j\omega)$, which, for any particular frequency, is a complex number.

2.3.3 Bode Diagram of Transfer Functions

A system's frequency response can be represented in several ways. One of the most popular graphical representations is in a Bode diagram that we treat in detail in this paragraph. A Bode diagram or Bode plot of a transfer function is composed of two curves, a magnitude curve or plot, $\log(|G(j\omega)|)$, drawn on a logarithmic scale and a phase curve or plot, $\underline{/G(j\omega)}$, drawn on a linear scale. Both curves are plotted against the frequency ω in radians per second, using a logarithmic scale. The magnitude is most commonly plotted in decibels, that is, $20\log_{10}(|G(j\omega)|)$.

We have shown that any transfer function $G(s)$ can be factorized into the form of Eq. (2.37). The factors in $G(j\omega)$ can be written as vectors in the complex s-plane in the following form:

$$s_{p_i} = j\omega - p_i \quad \text{for } i \in [1, n]$$
$$s_{z_k} = j\omega - z_k \quad \text{for } k \in [1, m]$$
(2.52)

These vectors are graphical representations of the contributions by the individual poles p_i and zeros z_k to the response of the system $G(s)$ on an input signal with amplitude 1 and frequency ω (Fig. 2.10).

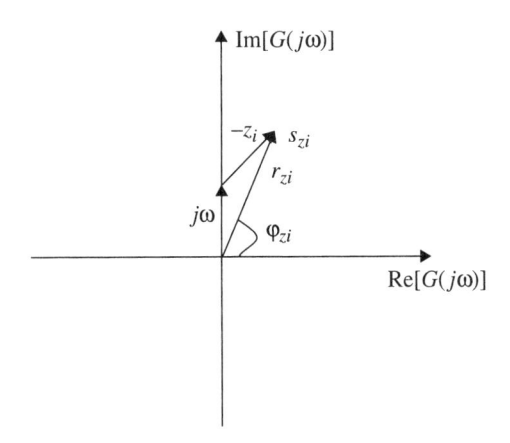

FIGURE 2.10 Graphical representation in the s-plane of the effect of a zero z_i, on the response of the transfer function $G(j\omega)$ as a vector in the s-plane.

If we define the magnitude and phase of these vectors s_p and s_z as follows:

$$\left|s_{p_i}\right| = r_{p_i} \quad \text{and} \quad \left|s_{z_i}\right| = r_{z_i}$$
$$\underline{\left|s_{p_i}\right.} = \varphi_{p_i} \quad \text{and} \quad \underline{\left|s_{z_i}\right.} = \varphi_{z_i} \tag{2.53}$$

Equation (2.37) can be rewritten as

$$G(j\omega) = \frac{s_{z_1} s_{z_2} \cdots s_{z_m}}{s_{p_1} s_{p_2} \cdots s_{p_n}} = \frac{r_{z_1} r_{z_2} \cdots r_{z_m}}{r_{p_1} r_{p_2} \cdots r_{p_n}} e^{j\left(\varphi_{z_1} + \varphi_{z_2} + \cdots + \varphi_{z_m} - \varphi_{p_1} - \varphi_{p_2} - \cdots - \varphi_{p_n}\right)} \tag{2.54}$$

From which the magnitude $|G(j\omega)|$ can be calculated as

$$|G(j\omega)| = \frac{r_{z_1} r_{z_2} \cdots r_{z_m}}{r_{p_1} r_{p_2} \cdots r_{p_n}} \tag{2.55}$$

and the phase $\underline{|G(j\omega)}$ is calculated as

$$\underline{|G(j\omega)} = (\varphi_{z_1} + \varphi_{z_2} + \cdots + \varphi_{z_m} - \varphi_{p_1} - \varphi_{p_2} - \cdots - \varphi_{p_n}) \tag{2.56}$$

Equations (2.55) and (2.56) demonstrate the advantage of using Bode diagrams in this form. Because the magnitude is plotted on a logarithmic scale, the overall magnitude and the phase information can both be obtained from the component parts by a graphical addition of the contributions by the different poles and zeros.

The zeros $z_1, z_2, \ldots, z_m$ and the poles $p_1, p_2, \ldots, p_n$ of the factorized transfer function $G(s)$ will each be real (including zero) or complex with a complex conjugate, and thus $G(s)$ can generally be considered to be composed entirely of terms of the four types appearing in the numerator or the denominator:

$$K; s; 1 + \tau s; \frac{s^2 + 2\zeta\omega_n s + \omega_n^2}{\omega_n^2} \tag{2.57}$$

Thus, $G(j\omega)$ is a composition of multiples or quotients of terms of the following form (Schwarzenbach and Gill 1978):

$$K; j\omega; 1 + j\omega\tau; \frac{\omega_n^2 - \omega^2 + 2\zeta j\omega\omega_n}{\omega_n^2} \tag{2.58}$$

Constant Term (Gain Term)

$$G(s) = K; G(j\omega) = K \tag{2.59}$$

Biosystems Analysis and Optimization   53

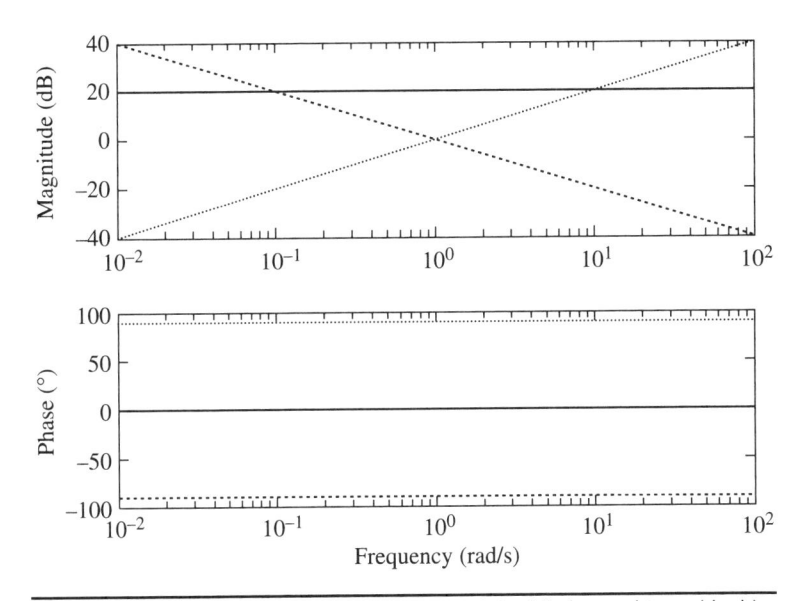

FIGURE 2.11 Bode plots for constant term K =10 (solid), integral term (dash) and derivative term (dot) transfer functions.

The magnitude is $|G(j\omega)| = 20\log_{10}(K)$ dB; the argument is zero. A gain term has a constant multiplying effect regardless of the frequency, and thus merely shifts the overall magnitude plot up or down by a certain number of decibels. There is no effect on the phase. The Bode plot of a constant term transfer function is illustrated in Fig. 2.11.

Integral Term (Pole at the Origin)

$$G(s) = \frac{1}{s}; G(j\omega) = \frac{1}{j\omega} = -\frac{j}{\omega} \qquad (2.60)$$

The magnitude can then be calculated as

$$|G(j\omega)| = \frac{1}{\omega} \; or \; |G(j\omega)| = -20\log_{10}(\omega) \, dB \qquad (2.61)$$

This means that the magnitude decreases by 20 dB for a tenfold increase in frequency. With the frequency ω plotted on a logarithmic scale, the magnitude is represented by a straight line of slope −20 dB per decade of frequency, which is passing through 0 dB for $\omega = 1$ rad/s.

The phase shift for this integral term is

$$\underline{|G(j\omega)} = \arg[G(j\omega)] = -\arctan\left[\frac{\omega}{0}\right] = -90° \qquad (2.62)$$

The phase shift is −90°, which means that the output lags behind on the input with 90°, regardless of the frequency. The Bode plot of an integral term transfer function is illustrated in Fig. 2.11.

Derivative Term (Zero at the Origin)

$$G(s) = s; G(j\omega) = j\omega \qquad (2.63)$$

The magnitude can then be calculated as

$$|G(j\omega)| = \omega \ or \ |G(j\omega)| = 20\log_{10}(\omega) \ dB \qquad (2.64)$$

This means that the magnitude increases by 20 dB for a tenfold increase in frequency. With the frequency ω plotted on a logarithmic scale, the magnitude is represented by a straight line of slope 20 dB per decade of frequency, which is passing through 0 dB for $\omega = 1$ rad/s.

The phase shift for this derivative term is

$$\underline{|G(j\omega)} = \arg[G(j\omega)] = \arctan\left[\frac{\omega}{0}\right] = +90° \qquad (2.65)$$

The phase shift is 90°, which means that the output would lead the input by 90° regardless of the frequency. The Bode plot of a derivative term transfer function is illustrated in Fig. 2.11.

Simple Lag (Real Pole First-Order System)

$$G(s) = \frac{1}{1+\tau s}; G(j\omega) = \frac{1}{1+j\omega\tau} \qquad (2.66)$$

The magnitude can then be calculated as

$$|G(j\omega)| = \frac{1}{\sqrt{1+\omega^2\tau^2}} \ or \ |G(j\omega)| = 20\log_{10}\sqrt{1+\omega^2\tau^2} \ dB \qquad (2.67)$$

This frequency dependency of the magnitude is often simplified by a linear asymptotic approximation, using the following asymptotes:

$$\begin{aligned} for \ \omega\tau \ll 1: \quad & |G(j\omega)| \approx 20\log_{10}(1) = 0 \ dB \\ for \ \omega\tau \gg 1: \quad & |G(j\omega)| \approx -20\log_{10}(\omega\tau) \ dB \end{aligned} \qquad (2.68)$$

The latter is a straight line of slope −20 dB per decade, which intersects the 0 dB line when $\omega\tau = 1$ (i.e., at $\omega = 1/\tau$). This frequency is termed the break-point, or corner frequency.

The phase shift for this simple lag is

$$\underline{|G(j\omega)} = \arg[G(j\omega)] = -\arctan\left[\frac{\omega\tau}{1}\right] = -\arctan(\omega\tau) \qquad (2.69)$$

which can also be approximated using the asymptotes:

$$\text{for } \omega\tau \ll 1: \quad \arg(G(j\omega)) \approx -\arctan\left(\frac{0}{1}\right) = 0°$$
$$\text{for } \omega\tau \gg 1: \quad \arg(G(j\omega)) \approx -\arctan\left(\frac{1}{0}\right) = -90°$$
(2.70)

A linear approximation can thus also be used for phase, that is, 0° for $\omega \leq 0.1/\tau$ and $-90°$ for $\omega \geq 10/\tau$ and a linear variation in between. The true curve is gently curving. The error between the real curve and the linear approximation is zero at the break-point frequency because the lag is exactly 45° when $\omega = 1/\tau$. The Bode plot of a simple lag transfer function is illustrated in Fig. 2.12.

Simple Lead (Real Zero First-Order System)

$$G(s) = 1 + \tau s; G(j\omega) = 1 + j\omega\tau \quad (2.71)$$

The expressions for magnitude and phase are identical to those for a real pole [Eqs. (2.67) to (2.70)], except that they have the opposite sign. The curves on the Bode plot are thus mirror images about the 0 dB and 0° lines. Magnitude and phase, therefore, both increase with frequency, the latter tending toward 90° (a phase lead) for frequencies in excess of $10/\tau$. The Bode plot of a simple lead transfer function is illustrated in Fig. 2.12.

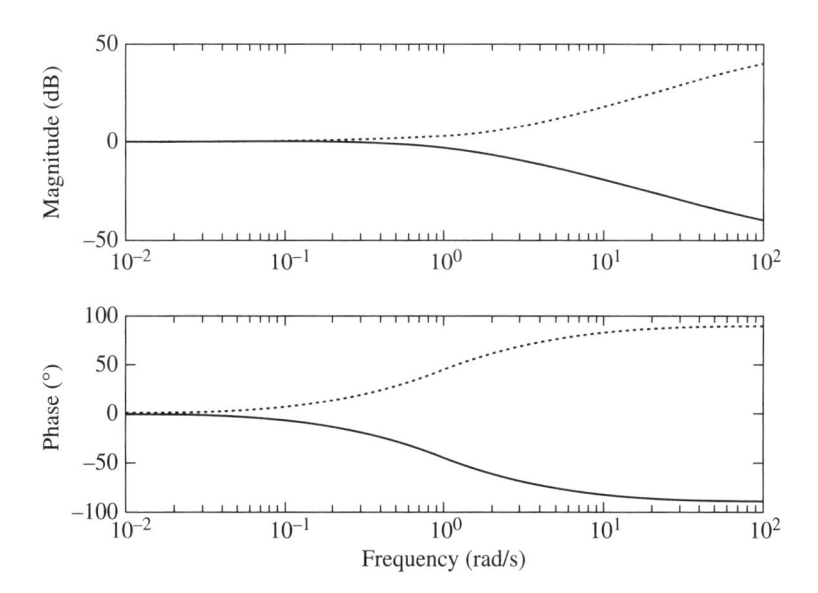

Figure 2.12 Bode plots for simple lag (solid) and simple lead (dash) transfer functions with time constant $\tau = 1$ rad/s.

Quadratic Lag (Second-Order System with a Pair of Complex Conjugate Poles)

$$G(s) = \frac{\omega_n^2}{s^2 + 2\zeta\omega_n s + \omega_n^2}; G(j\omega) = \frac{\omega_n^2}{\omega_n^2 - \omega^2 + 2\zeta\omega_n\omega j} \quad (2.72)$$

The magnitude for this transfer function is given by the following expression:

$$|G(j\omega)| = \frac{1}{\sqrt{\left(1 - \frac{\omega^2}{\omega_n^2}\right)^2 + \left(2\zeta\frac{\omega}{\omega_n}\right)^2}}$$

or (2.73)

$$|G(j\omega)| = -20\log_{10}\sqrt{\left(1 - \frac{\omega^2}{\omega_n^2}\right)^2 + \left(2\zeta\frac{\omega}{\omega_n}\right)^2} \, dB$$

This frequency dependency of the magnitude can again be simplified by a linear asymptotic approximation, using the following asymptotes:

$$\text{for } \frac{\omega}{\omega_n} \ll 1: \quad |G(j\omega)| \approx -20\log_{10}(1) = 0 \, dB$$

$$\text{for } \frac{\omega}{\omega_n} \gg 1: \quad |G(j\omega)| \approx -20\log_{10}\left(\frac{\omega^2}{\omega_n^2}\right) = -40\log_{10}\left(\frac{\omega}{\omega_n}\right) dB \quad (2.74)$$

The straight-line approximation for magnitude is thus a line at 0 dB for low frequencies, changing to a line of slope –40 dB/decade at the break point where the frequency ω equals the undamped natural frequency ω_n. The shape of the true curve depends on the value of the damping factor ζ. For $\omega = \omega_n$ and $\zeta = 0$, $|G(j\omega)| = \infty dB$. The slope of –40 dB/decade for frequencies higher than the undamped natural frequency is also called a roll-off rate of 40 dB/decade.

The phase shift for this quadratic lag system is given by the following expression:

$$\underline{|G(j\omega)} = \arg[G(j\omega)] = -\arctan\left[\frac{2\zeta\frac{\omega}{\omega_n}}{1 - \frac{\omega^2}{\omega_n^2}}\right] \quad (2.75)$$

which can also be approximated using the asymptotes:

$$\text{for } \frac{\omega}{\omega_n} \ll 1: \quad \arg(G(j\omega)) \approx -\arctan\left(\frac{0}{1}\right) = 0°$$

$$\text{for } \frac{\omega}{\omega_n} \gg 1: \quad \arg(G(j\omega)) \approx -\arctan\left(\frac{2\zeta}{\frac{\omega}{\omega_n}}\right) = -\arctan(0) = -180° \quad (2.76)$$

$$\text{for } \frac{\omega}{\omega_n} = 1: \quad \arg(G(j\omega)) \approx -\arctan\left(\frac{2\zeta\frac{\omega}{\omega_n}}{0}\right) = -\arctan(\infty) = -90°$$

The phase curve thus varies from 0 to –180° and passes through the –90° point at the undamped natural frequency ω_n. There is no convenient straight-line approximation for the phase shift, the transition again being a function of damping factor ζ and being most rapid for very small values of ζ. The Bode plot of a quadratic lag transfer function is illustrated in Fig. 2.13.

Quadratic Lead (Pairs of Complex Conjugate Zeros) (Second-Order System)

$$G(s) = \frac{s^2 + 2\zeta\omega_n s + \omega_n^2}{\omega_n^2}; G(j\omega) = \left(1 - \frac{\omega^2}{\omega_n^2}\right) + j\left(2\zeta\frac{\omega}{\omega_n}\right) \quad (2.77)$$

The magnitude and phase are numerically the same as for a pair of poles [Eqs. (2.73) to (2.76)] but of opposite sign and are thus represented on the Bode plot by families of curves that are the mirror images of those of the quadratic lag reflected about the 0 dB and the 0° lines. The Bode plot of a quadratic lead transfer function is also illustrated in Fig. 2.13.

2.4 System Identification

For most systems, many or all of the model parameter values are unknown and cannot be measured directly. Sometimes, even the model structure is not known and cannot be easily derived from first principles. Therefore, we want to estimate some or all of these model parameters based on experimental data for the input and output signals.

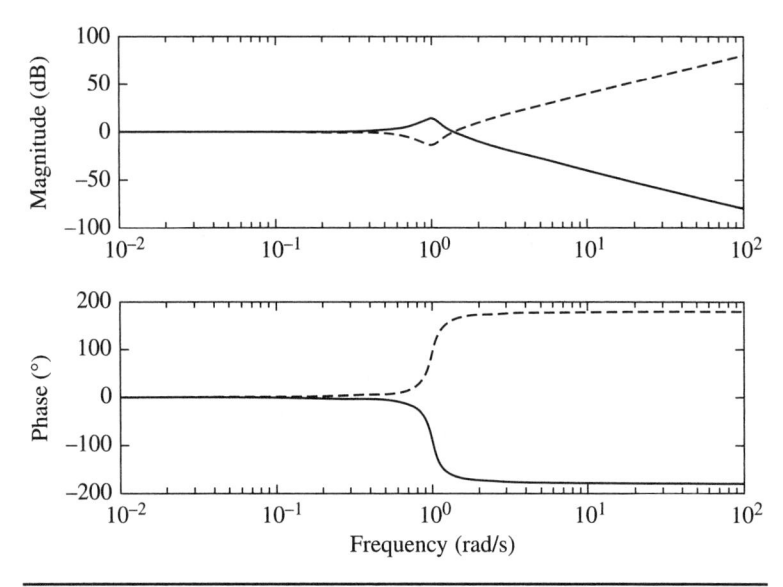

FIGURE 2.13 Bode plots for quadratic lag (solid) and quadratic lead (dash) transfer functions with natural frequency $\omega_n = 1$ rad/s and damping constant $\zeta = 0.1$.

The procedure of estimating the model or model parameters from experimental data is called system identification and consists of several steps:

- Gathering informative data by performing excitation experiments
- Evaluating experimental data
- Estimating system nonlinearities
- Estimating parameters

The type of experiments used and the procedure followed to estimate the model parameters depends on the purpose of the model to be estimated. If the control engineer is more interested in the frequency domain characteristics of the system to evaluate the stability of the designed control system, he or she wants the estimated model to fit the Bode plot well. Therefore, the control engineer desires good excitation of the dynamics in the frequency region of interest to obtain a good empirical estimate of the Bode plot and will estimate the model parameters as those that give the best fit to the experimental Bode plot. On the other hand, if the control engineer is more interested in prediction, he or she wants a model that fits the time domain signals (e.g., step response) very well. Therefore, the control engineer will use representative time domain signals to excite the system and will estimate the model parameters giving the best fit to these time domain responses.

2.4.1 Excitation Experiments

To obtain good estimates for the model parameters one needs experimental data that contains good information about the dynamics. Because we want to estimate the parameters of an input–output model, both the input and the output signals should have a signal-to-noise ratio as high as possible with respect to the dynamic relation between the input signal and the output signal. This can be obtained by applying an input signal to the system that has sufficient energy to excite the dynamics of interest. Therefore, this type of experiment is called an excitation experiment.

When we are mostly interested in modeling the frequency domain characteristics of the system, we will try to obtain a good empirical estimate of the Bode plot in the frequency region of interest. Because the bode plot is a plot of the amplification and phase shift by the system as a function of frequency, we could apply sine excitations at different frequencies of interest and measure the response of the system to these frequencies in terms of amplification and phase shift. Applying different sines one after another would, however, take quite some measurement time because we would have to measure several periods per sine to obtain a good estimation of the amplitude and phase in the output signal. Therefore, it is more efficient to excite the system with a multisine, which is a summation of sine waves of different frequencies within a certain spectrum. The sine waves have also been shifted in phase to obtain a periodic signal with a small crest factor resulting in high signal-to-noise ratios (Pintelon and Schoukens 2001). Obviously, the output signals will also be a multisine, but one where the amplitudes and phases have been changed. This multisine can then be split into its sine components by means of a Fourier transformation. When we have no idea about system dynamics, a multisine with a flat spectrum (equal amplitude for all frequency components) will be used. However, if we have a good idea about the expected shape of the Bode plot, we might apply a multisine with a different spectrum to obtain output signals with a nearly flat spectrum and thus have a better signal-to-noise ratio. Figure 2.14 is an example of a multisine signal that is given with a flat frequency spectrum.

When we are interested mostly in modeling the time domain characteristics of the system (model to be used for prediction), realistic input signals should be applied. If the model should be well capable to predict the response of the system to a step input, this is applied in the excitation experiments. This type of input has the advantage that it also excites many frequencies, but in a more realistic combination than the multisine. This is especially important for nonlinear systems that, differently from linear systems, exhibit dynamic behavior dependent on the input signal. Multiple upward and downward step inputs can be combined into an efficient excitation signal (e.g., block wave). Another commonly used input signal for time domain analysis is a ramp input, which can be combined into a sawtooth signal. Figure 2.15 shows some typical time domain excitation signals.

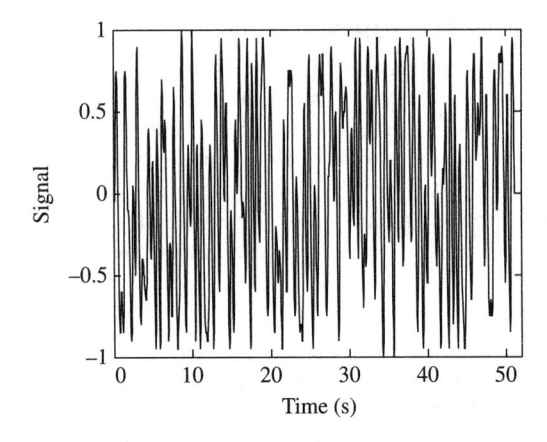

FIGURE 2.14 Example of one period of a multisine signal with amplitude 1 and a frequency range from 0.02 to 2 Hz sampled at 20 Hz.

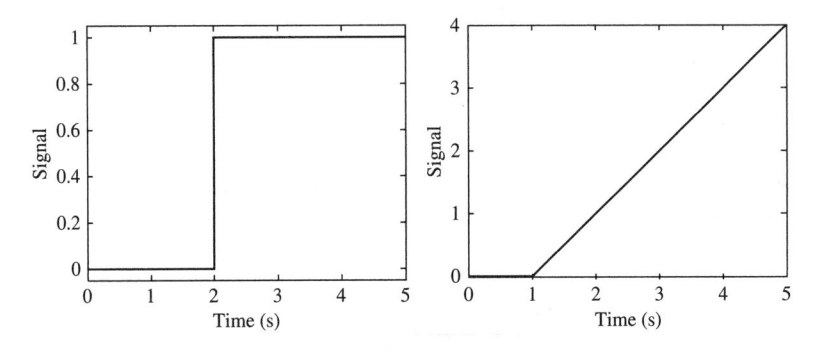

FIGURE 2.15 Examples of time domain excitation signals: step input (left), ramp input (right).

2.4.2 Evaluating Experimental Data

Once an excitation experiment has been performed, we have to investigate the experimental data. These data can again be looked at in the time domain or in the frequency domain, depending on the type of information desired.

In the frequency domain, we determine the frequency spectrum of both the input and the output by calculating the Fourier transform of the measured signals in the frequency region that has been excited. From these frequency spectra, an empirical, nonparametric, transfer function estimate (ETFE) can be calculated by dividing the Fourier components of the output spectrum by the corresponding components in the input spectrum for the excited frequencies. By plotting the magnitude and the phase shift for this empirical transfer function estimate, we obtain an empirical Bode plot. This procedure is illustrated in Fig. 2.16 for experimental data obtained of the transfer from

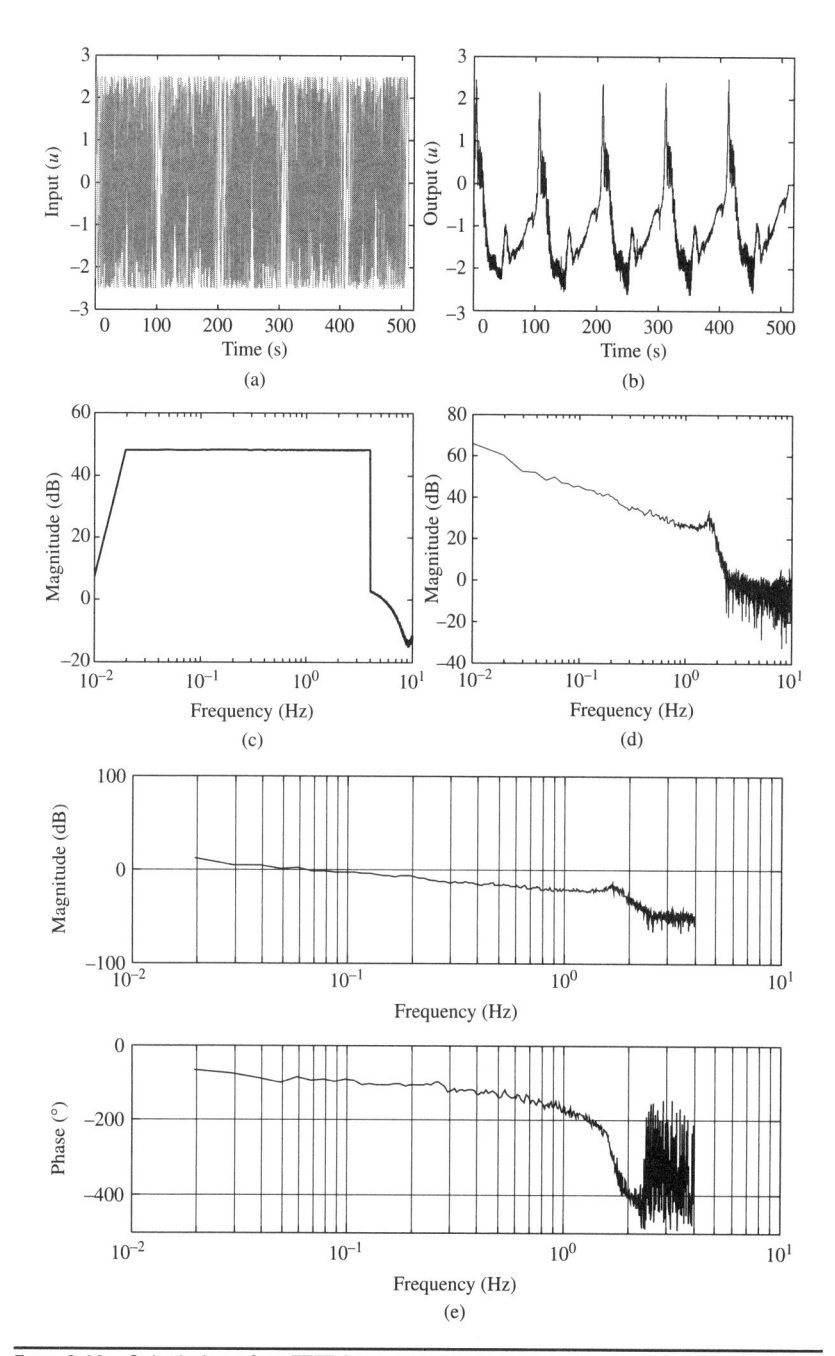

FIGURE 2.16 Calculation of an ETFE from experimental data of a depth control system for a slurry injector (Saeys et al. 2007). (*a*) Time signal for the input (five periods of the multisine). (*b*) Time signal for the output. (*c*) Frequency spectrum of the input signal. (*d*) Frequency spectrum for the output signal. (*e*) Empirical Bode plot.

voltage applied to a hydraulic valve into the height of a slurry injector above the soil (Saeys et al. 2007). Six periods of a multisine ranging from 0.02 to 4 Hz have been applied to the system at a sample frequency of 40 Hz. Because the output signal for the first period contains transient behavior of the system, this period has been removed from the data prior to calculating the Fourier transforms.

2.4.3 Estimating Nonlinearities

The larger part of our system-analysis tools is aimed at linear systems, whereas in practice most biologically related systems exhibit at least some nonlinear behavior. To investigate the contribution of system nonlinearities compared to linear system behavior and determine whether a linear model can be sufficient, we can further analyze the response of the system to a multisine.

For instance, apply a random odd multisine to the pump setting of a hydraulic pump and observe the response of the ground speed. This example is further elaborated later and can be seen in Fig. 2.3. A multisine is the sum of sines where the phase of each sine is random. Frequencies of the sines are multiples of the given base frequency, which is the lowest frequency contained in the multisine. For an odd multisine, only odd multiples of the base frequency are excited. Because the presence of harmonics is a good measure of nonlinearity, only exciting odd multiples of the base frequency enables us to quantify nonlinearities in the system. After all, the first (and most powerful) harmonic of an odd frequency is the double frequency, which, of course, is even and thus not excited directly. The magnitude of the response on even frequencies is a good measure for nonlinearity.

The response on the second harmonic is also very interesting. Because the second harmonic is again an odd multiple, it is very hard to distinguish the harmonic response from the excitation response. By removing one-fourth of the odd frequencies randomly from the excitation sequence, the system response on these nonexcited odd frequencies becomes a good indication of the second harmonic response.

The response on the nonexcited frequencies needs to be compared to a reference to be able to determine whether the effect is important or not. This reference is the noise level of the frequency response. The noise level can be estimated by applying multiple periods of the same multisine and calculating the variation of the response over the periods.

A multisine experiment has been performed on the propulsion system of the combine harvester with a sampling rate of 20 Hz. The length of the period was set to 1024 points (or approximately 50 s). This means that the base frequency is 20/1024 = 0.0195 Hz. Because the machine is not expected to respond to frequencies much higher than 1 Hz, the multisine only excites the frequency range between 0.0195 and 2 Hz. Four and one-half periods of the multisine are applied to the process. After the transition phase is removed, four full periods are available for analysis.

The response of the ground speed to the multisine excitation of the pump setting is split up in several parts:

- *Response at excited frequencies:* The frequency response of the process variable on the frequencies directly excited by the multisine is the linear process response on that frequency. Due to the structure of the excitation sequence, contamination of this response by harmonics is very low. Only second-order harmonics or higher can influence the response. The frequency response of the pump setting is also taken into account on the excited frequencies. The pump setting is the control variable to which the multisine is applied. In practice, the measured pump-setting spectrum is not identical to the applied spectrum because of timing and resolution issues. The frequency response on the pump setting is normalized so that it has a unit response on the first excited frequency. For each process variable, the response on the excited frequency is then divided by this normalized input spectrum on that frequency in order to obtain the transfer function from the pump setting to the process variable. The unit of this transfer function is the unit of the gain between the pump setting and the process variable. Because the input spectrum is more or less zero on the nonexcited frequencies, this rescaling can only be performed on the excited frequencies.

- *Response at even nonexcited frequencies:* The frequency response on these frequencies is most probably caused by the first harmonic of an excited frequency. If this response is clearly larger than the noise response, there are considerable nonlinearities present in the system.

- *Response at odd nonexcited frequencies:* This response shows the presence of second- or higher-order harmonics. A clear response on these frequencies indicates large nonlinearities.

- *Noise response:* Because multiple periods of the multisine are available—in this case, four full periods—the average process response can be calculated. Noise is defined as the variation around this average response (standard deviation per frequency over the four periods). This means that the system is assumed to be constant during excitation.

A multisine excitation is not the standard input for the pump setting. This means that some effects that are visible with this excitation may not need to be modeled in the actual controller. After all, only the effects that are encountered during normal controller operation need to be modeled. Due to nonlinear effects, the response to a multisine excitation may differ considerably from the response to a step input.

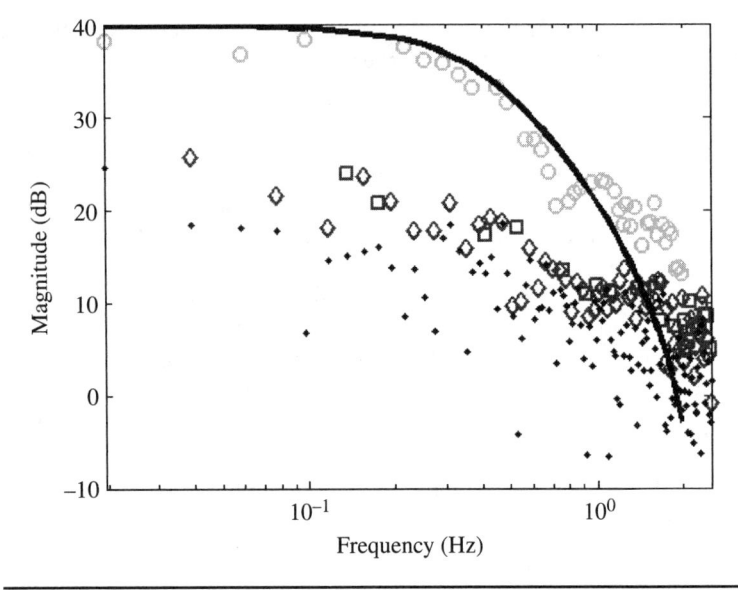

FIGURE 2.17 The frequency response averaged over four periods for ground speed. The response on the excited frequencies is divided by the normalized response on the pump setting. The solid diamonds (♦) indicate the noise level; the gray open circles (○) indicate the excited odd frequencies; the open squares (□) indicate the nonexcited odd frequencies; and the open diamonds (◊) indicate the (non excited) even frequencies. The solid line is a second-order system fitted to this data set.

The response of the ground speed is shown in Fig. 2.17. Several observations can be made from this response. At low frequencies (approximately 0.1 Hz), the response on the even frequencies, which are not excited, is considerable compared to the noise level. Even some of the odd nonexcited frequencies show a considerable response.

Also note that the response on the excited frequencies increases again at higher frequencies (more than 1 Hz). This is probably caused by propulsion system dynamics. The engine control unit determines the fuel injection of the diesel engine as a function of engine speed. There are no details about this factory-fitted controller available, but a differential action on this controller might result in such a bump on the ground speed spectrum.

2.4.4 Parameter Estimation

Now that we have empirical data for the system, we can use them to estimate the parameters of our system model. The objective is to find the model parameters for which the simulated model behavior corresponds best with the measured model behavior.

In case we select a time domain model (when good prediction is desired), parameters will be estimated by minimizing the deviation of the simulated time response from the measured response. For this purpose, the transfer function is redefined as a discrete time input–output model. Because this model has both in the numerator and denominator parameters to be estimated, a linear least squares optimization is not the best option. Therefore, the parameter estimation is typically done using the autoregressive moving average procedure where the equation error term is described as a moving average of white noise (Ljung 1987).

This procedure is illustrated in Fig. 2.18 for the driving line of a combine harvester where a second-order model has been fit to the measured response of the ground speed on a step change in the pump setting. The resulting transfer function has the following form:

$$G(s) = \frac{1.5}{s^2 + 1.8s + 1.5} \quad (2.78)$$

which corresponds to a resonance frequency ω of 1.23 rad/s and a damping ratio ζ of 0.73.

In a frequency domain model (for stability investigation and control), we search parameter values that minimize the difference

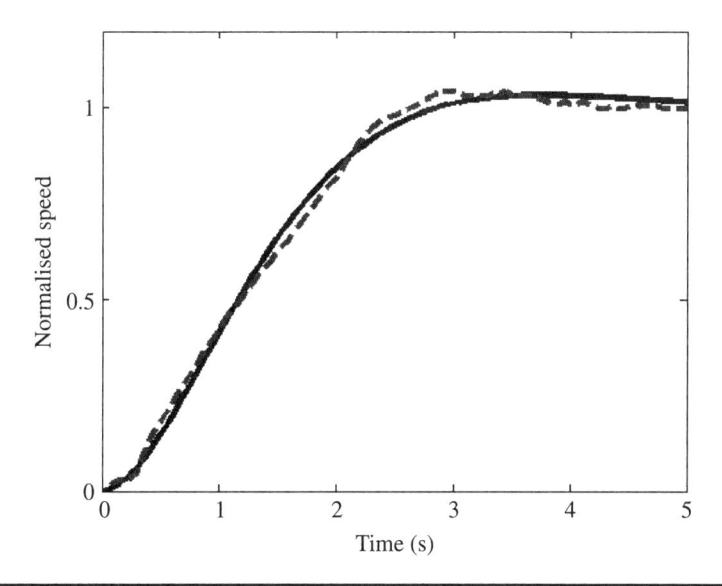

FIGURE 2.18 Normalized response of the second-order estimated model (solid line) and measured response of the real system (dashed line) on a step input.

between the measured and the simulated frequency response function. This can be done using the least squares fitting procedure but is more commonly performed using the nonlinear least squares procedure or a maximum likelihood procedure (Pintelon and Schoukens 2001). This procedure is illustrated for the depth control system of a slurry injector for which the determination of the empirical transfer function estimate has been described in this chapter (Sec. 2.3.2, Fig. 2.16). For this system, a second-order model structure in the numerator and a fourth-order model in the denominator has been derived from first principles (Saeys et al. 2007). The parameters of this model have been estimated by the nonlinear least squares fitting procedure. This resulted in a transfer function in the following form:

$$G(s) = \frac{Y(s)}{X(s)} = \frac{2.34s^2 + 6.18s + 663}{s^4 + 15.1s^3 + 139s^2 + 1560s} \quad (2.79)$$

The fit of the simulated Bode plot for this to the empirical Bode plot is illustrated in Fig. 2.19.

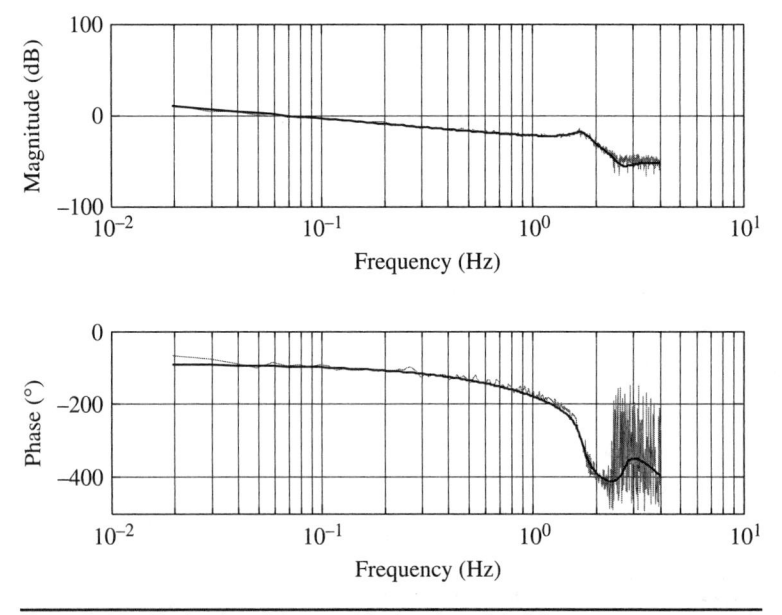

FIGURE 2.19 Bode plot of the empirical and modeled frequency response function for the transfer function of a depth control system for slurry injection (Saeys et al. 2007).

2.5 Feedback Controller Design

2.5.1 Feedback Control Structure

The most commonly used method to control the output of a dynamic system is by introducing a feedback loop. Figure 2.20 is a block diagram of such a feedback control system. The output $y(t)$ of the system that we want to control is fed back through sensor measurement H with measurement noise $v(t)$ to be compared with the reference value $r(t)$. The latter is the desired value, thus the difference between this reference value $r(t)$ and the measured output $y_m(t)$, called the error $e(t)$, has to be minimized. To reach this goal, controller C will translate the error signal $e(t)$ into an input signal $u(t)$ applied to system G under control. Because the loop between input and output is closed, this feedback controller is also known as a closed-loop controller. In practice, the output signal $y(t)$ will not only be determined by the input signal $u(t)$ applied to the system G but is also subject to disturbances $w(t)$.

So that we may be able to design a good controller C, we should be able to describe the transfer of a reference signal $r(t)$ into the output signal $y(t)$ using the closed-loop controlled system, the tracking performance. For this purpose, we define the closed-loop transfer function $G_c(s)$, describing the transfer from the Laplace transform of the reference signal $R(s)$ to the Laplace transform of the output signal $Y(s)$. When the effect of the disturbances $v(t)$ is not taken into account, the output signal $Y(s)$ is given by the following expression:

$$Y(s) = G(s)U(s) = G(s)C(s)E(s) = G(s)C(s)[R(s) - H(s)Y(s)] \quad (2.80)$$

By bringing all terms in $Y(s)$ to the left, and all transfer functions to the right, the closed-loop transfer function $G_c(s)$ can be derived as

$$G_c(s) = \frac{Y(s)}{R(s)} = \frac{G(s)C(s)}{1 + G(s)C(s)H(s)} = \frac{G(s)C(s)}{1 + L(s)} \quad (2.81)$$

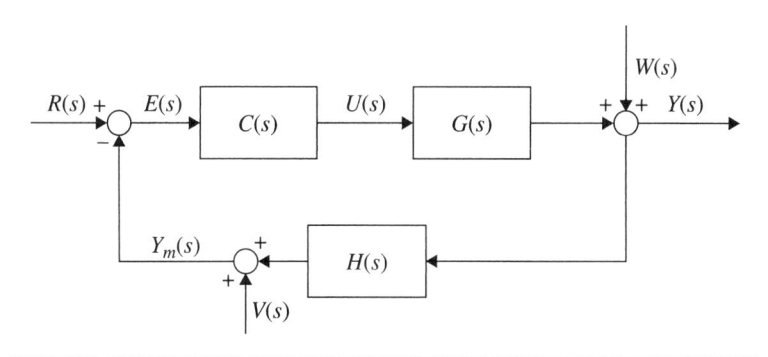

FIGURE 2.20 Block diagram of a feedback-controlled system.

where the return ratio $L(s)$ is defined as

$$L(s) = G(s)C(s)H(s) \tag{2.82}$$

The modulus of $L(s)$ [i.e., $|L(j\omega)|$] is called the loop gain. Besides tracking performance, the influence of disturbance signals $w(t)$ and measurement noise $v(t)$ on the output of the controlled system are also very important:

$$S(s) = \frac{Y(s)}{W(s)} = \frac{1}{1+G(s)C(s)H(s)} = \frac{1}{1+L(s)} \tag{2.83}$$

$$T(s) = \frac{Y(s)}{V(s)} = \frac{-G(s)C(s)}{1+G(s)C(s)H(s)} = \frac{-G(s)C(s)}{1+L(s)} \tag{2.84}$$

Finally, the transfer from the reference signal $R(s)$ into an actuator signal $U(s)$ applied to the system is given by the following expression:

$$\frac{U(s)}{R(s)} = \frac{G_c(s)}{G(s)} = \frac{C(s)}{1+G(s)C(s)H(s)} = \frac{C(s)}{1+L(s)} = C(s)S(s) \tag{2.85}$$

2.5.2 Stability

The simplified Nyquist stability criterion states that if an open-loop system is stable then the system with a closed loop is also stable provided that the magnitude of the loop gain $|L(j\omega)|$ does not exceed 1 when its phase shift $\arg[L(j\omega)]$ is $-180°$. Physically, the condition of instability can be visualized as follows. If the reference input to a closed-loop system is a sine wave, then the signal returning to the error detector will have a different amplitude and phase (which are known for each frequency of the sine wave from the Bode diagrams). If the phase lag is 180°, then the returning signal, when inverted and added to the reference input, will reinforce the signal. If the amplitude of the returning signal is less than that of the input signal at this phase shift $[|L(j\omega)| < 1]$ a steady-state condition will be reached, but if the amplitude is greater than 1, then the amplitude will build up continuously until the system is saturated. Even if the input signal is removed, the system will continue to oscillate.

Models of physical systems are only approximations. Some dynamical phenomena are not known or are neglected. These phenomena can cause extra gain variations and extra phase shifts. For example, in electrohydraulic control systems, where the dynamics of the hydraulic group as pump, hoses, and valves are not taken into account, these dynamic phenomena can cause extra phase shifts and gain variations at the input $u(t)$ of the feedback system that are unknown by the controller because they are not taken into account during controller design.

Therefore, the feedback control system must have some immunity with respect to these unmodeled dynamics. It is clear that the smaller $|L(j\omega)|$ will be ($\ll 1$) at the phase crossover (i.e., where $\arg[L(j\omega)] = -180°$), the more insensitive the feedback system will be to unmodeled dynamic phenomena. This means that these dynamic phenomena may add more gain to $|L(j\omega)|$ before instability occurs.

The *gain margin* (GM) is defined as the amount by which the system gain can be increased before instability occurs and is normally quoted in decibels. This can be calculated using the following equation:

$$\mathrm{GM} = 20\log_{10}\left(\frac{1}{|L(j\omega)|_c}\right) dB \qquad (2.86)$$

where $|L(j\omega)|_c$ is the magnitude of the loop gain $L(s)$ corresponding to a phase lag of 180° (i.e., at the phase crossover). The physical meaning of gain margin has already been explained. Gain margin is also known as amplitude margin.

The *phase margin* (PM) is defined as the amount by which the open-loop lag falls short of 180° at the frequency where the open-loop magnitude is unity (i.e., at the gain crossover). It is particularly significant when investigating the effect on the stability of system changes that primarily affect the phase shift $\arg[L(j\omega)]$. The phase margin also has a physical meaning. A sine wave with amplitude 1 and a frequency equal to the gain crossover frequency, which is applied to the closed-loop system, will reach the error detector with an unaltered amplitude of 1 but with a certain shift in phase. If this phase shift (i.e., phase lag) is 180°, then the returning signal, when inverted and added to the reference input sine wave, will just reinforce the signal and the system will be marginally stable. The smaller the phase margin (more negative), the larger the phase shift caused by unmodeled dynamic phenomena can be before instability occurs. This can be calculated using the following equation:

$$\mathrm{PM} = \arg[L(j\omega)] - 180° \qquad (2.87)$$

It should be noted here that the gain of a transfer function is always larger than or equal to zero.

2.5.3 Loop-Shaping Controller Design

In order to obtain the desired behavior of the feedback control system, we can design the frequency domain response (Bode plot) of the loop gain $L(s)$ to satisfy the control system requirements in terms of tracking performance, disturbance rejection, and stability. From this designed loop gain $L(s)$, the desired controller transfer function $C(s)$ can then be calculated. The reason for shaping the loop gain $L(s)$ is because the denominator of the closed-loop transfer function $G_c(s)$, also known as the characteristic equation, determines its tracking performance, disturbance rejection, and stability properties. Several design criteria can be defined for the loop gain $L(s)$:

1. *Disturbance rejection:* From Eq. (2.83) we can observe that improved suppression of the negative influence of disturbances $W(s)$ on the performance of the controlled system is obtained when

$$|L(j\omega)| \gg 1 \qquad (2.88)$$

The larger $|L(j\omega)|$, the better the influence of $W(s)$ will be attenuated.

2. *Noise filtering:* From Eq. (2.84) we can derive that the influence of sensor noise $V(s)$, which contaminates the measured output signal $Y_m(s)$, will be minimized when

$$|G(s)C(s)| \ll \frac{1}{|H(s)|} \quad or \quad |L(j\omega)| \ll 1 \qquad (2.89)$$

The better the sensor noise $V(s)$ is filtered out, the more accurate the measured signal $Y_m(s)$ becomes.

Apart from sensor noise, the measured signal $Y_m(s)$ can also be contaminated by unknown disturbance signals, generated by unmodeled high-frequency dynamics of the system. The better these are filtered out, the less influence these will have on the controlled output $Y(s)$. Badly filtered unmodeled high-frequency dynamics can distort $Y_m(s)$ so severely that the feedback system becomes unstable. This phenomenon is called *observation spillover*.

3. *Reference tracking:* From Eq. (2.81), the conditions can be calculated for which the output signal $Y(s)$ will follow the reference signal $R(s)$ accurately:

$$|G_c(j\omega)| \approx 1 \quad or \quad |L(j\omega)| \gg 1 \qquad (2.90)$$

4. *Input limitation:* The actuator signal, $U(s)$, is sometimes restricted (e.g., maximal available power, or maximal actuator speed); therefore, $U(s)$ has to be kept as small as possible. From Eq. (2.85) we can derive that this is obtained when

$$\left|\frac{C(j\omega)}{1+L(j\omega)}\right| = |C(j\omega)S(j\omega)| \ll 1 \qquad (2.91)$$

This can be reformulated to obtain design criteria for the controller $C(s)$:

$$|C(j\omega)| \ll |1 + G(j\omega)C(j\omega)H(j\omega)| \leq 1 + |G(j\omega)C(j\omega)H(j\omega)|$$

$$\leq 1 + |G(j\omega)||C(j\omega)||H(j\omega)| \qquad (2.92)$$

Equation (2.92) can be interpreted as follows. As long as $|G(j\omega)|$ is not much larger than 1, $|C(s)|$ must be as small as possible to keep $U(s)$ small in that frequency band. For most physical systems, this is at high frequencies, as they become insensitive to high-frequency disturbances (think about a tractor crossing a field, small ripples do not induce tractor vibrations any more, whereas large clods and field undulations do cause tractor vibrations and movements). At low frequencies, where $|G(j\omega)|$ is normally large, a high gain for $|C(s)|$ is allowed.

Thus, we may state that to avoid excitation of unmodeled high-frequency dynamics of the system, $U(s)$ must be kept as small as possible at high frequencies. The phenomenon where high-frequency modes are excited by the feedback control system so that they destabilize the feedback system is called *control spillover*.

Design rules (1) and (2) are conflicting so a trade-off has to be made between disturbance rejection [suppression of the negative influence of $W(s)$ on the performance] and noise filtering [suppression of sensor noise $V(s)$ and unmodeled high-frequency modes in order to keep the feedback system stable]. However, in most practical systems, $W(s)$ is dominated by low-frequency signals (low-frequency spectrum), whereas $V(s)$ is dominated by high-frequency signals (noise). Therefore, design rule (1) must be satisfied at low frequencies and design rule (2) at high frequencies.

Design rules (2) and (3) are conflicting too, so that there also exists a trade-off between suppression of sensor noise and unmodeled high-frequency modes and the improvement of reference input tracking. However, $R(s)$ typically only has a low-frequency spectrum so that (3) must only be satisfied at low frequencies.

Kopasakis (2007) defines a desired shape for the loop gain by combining these design rules, and making a trade-off between them. This desired return ratio starts with a pole at the origin for high-loop gain and zero steady-state error, followed by a zero to maintain a relatively high gain at the midfrequency range for disturbance attenuation and reference tracking, followed by a pole to attenuate the gain at higher frequencies to avoid exceeding the available actuator rate (input limitation) and to retain sufficient noise filtering. Such a design would also result in adequate stability margins with a phase margin of more than 90° and an infinite gain margin because the phase does not cross 180°. The trade-off between design criteria will then be made by the placement of the magnitude *crossover frequency*, which defines the *bandwidth* of the controlled system. For frequencies less than the crossover frequency, the reference signal is well tracked, and disturbance rejection is good. For frequencies higher than this crossover frequency, the controlled system can no longer follow the reference signal, and measurement noise is increasingly filtered out. The maximal bandwidth that can be reached with a given system is determined by the available hardware actuation speed or the actuator rate limit a_{rl} (in units of magnitude per second). Therefore, the maximal crossover frequency ω_{co} in radians per second can be derived from the actuator rate limit and the input step size r_m as

$$\omega_{co} = K \frac{a_{rl}}{r_m} \qquad (2.93)$$

where K is the gain factor between the actuator control rate and the system response rate.

The natural frequency of the closed-loop control system also depends on the lowest frequency zero of the return ratio, $L(s)$. Thus, both the magnitude crossover frequency and the time constant of the lowest frequency zero will influence the overall response time of the system. The magnitude crossover frequency of the system, which would be normally higher than the frequency of the lowest frequency zero, will therefore influence the initial response of the system. The lowest frequency zero will dictate the final response of the system before the system response settles. If it turns out that the return ratio has no zero less than the crossover frequency, the system time response will only depend on the magnitude crossover frequency. However, a zero placed before (actually, well before) the magnitude crossover frequency is a good design practice because it helps to boost the midfrequency gain for disturbance rejection and, increase the phase margin for stability, as well as increasing the control bandwidth.

A design procedure for the shaping of the return ratio would then consist of the following steps (Kopasakis 2007):

1. Choose the bandwidth of the controlled system based on the available hardware actuation speed (*input limitation*).
2. Choose the midfrequency gain based on the midfrequency *disturbance rejection* requirements.
3. Evaluate the *phase margin* at magnitude crossover. If the phase margin is not sufficiently large, the design should be adjusted (e.g., by lowering the midfrequency gain).
4. Compute the desired *lower zero frequency* by the settling time requirements of the response.
5. Calculate the gain of the closed-loop gain transfer function based on the *midfrequency gain* at the lower frequency zero.
6. Simulate the *Bode plots* of the return ratio and the closed-loop system response and make adjustments if necessary.

Once the return ratio $L(s)$ has been designed, the transfer function of the controller $C(s)$ can be derived from this.

2.5.4 Controller Design Example: Depth Control for Slurry Injection

The stability analysis and design principles that have been described in the previous sections will now be applied to the design of the depth control system for slurry injection for which the model has been identified in Sec. 2.3.4. The estimated transfer function was expressed as V/V, but we would like to know the measured depth in centimeters, therefore Eq. (2.79) is multiplied by a gain factor of 5 cm/V to get

$$G(s) = \frac{Y(s)}{X(s)} = \frac{11.7s^2 + 30.9s + 3315}{s^4 + 15.1s^3 + 139s^2 + 1560s} \quad (2.94)$$

The closed-loop gain $L(s)$ for this system can now be designed through the stepwise procedure described in the previous section. This can be done quite easily using the SISO design toolbox in MATLAB (The Mathworks, Natick, Massachusetts). The design requirements are the following:

1. The maximal actuator speed is determined by the maximal oil flow that can be generated through the proportional hydraulic valve and sent to the hydraulic actuators. The actuator rate limit a_{rl} for this system was found to be 5 cm/s. If we consider an input step size of 1 cm, the magnitude crossover frequency ω_{co} can be calculated using Eq. (2.93) to be 5 rad/s.

2. The desired disturbance attenuation rate at 0.5 rad/s is a factor of 10, which means that the desired magnitude of the closed-loop gain is $20 \log_{10}(10) = 20$ dB.

3. An integrator action is desired to avoid steady-state error.

4. High-frequency noise and unmodeled dynamics should be sufficiently filtered.

5. The phase margin should be at least 45° at crossover and the gain margin at least 10 dB.

Because of the integrator action, which is desired to avoid steady-state error, the bode plot decreases with 20 dB per decade. This makes that the disturbance attenuation requirement of 20 dB at 0.5 rad/s cannot be met without introducing a low-frequency zero when the crossover frequency is restricted to the actuator rate limit of 5 rad/s. Therefore, a single zero with a natural frequency of 0.1 rad/s is introduced to keep the midfrequency gain of the closed-loop gain more than 20 dB in the region around 0.5 rad/s. Introducing this zero causes the crossover frequency to move to a higher frequency, which violates design requirement (1). Because lowering the gain is not an option, where

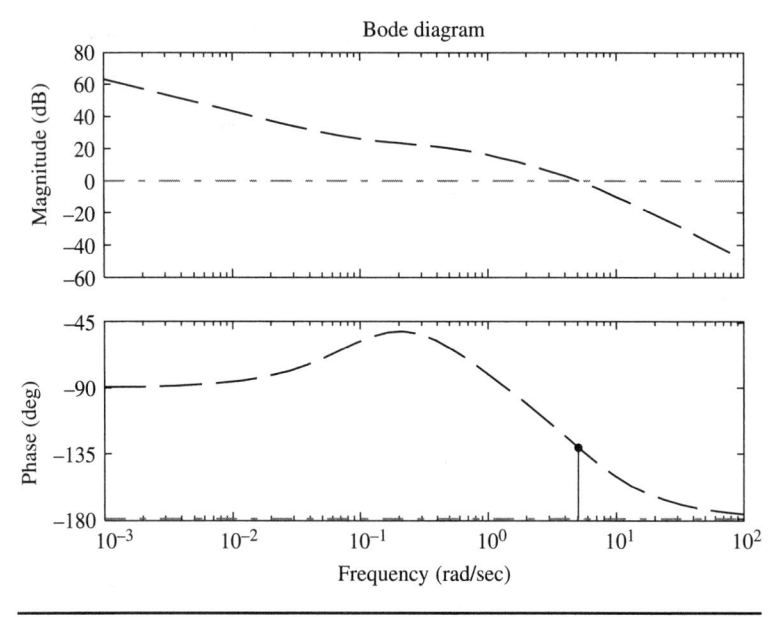

FIGURE 2.21 Bode plot for the designed return ratio L(s).

design requirement (2) would be violated, a single pole is added with a frequency of 0.5 rad/s. In order to obtain sufficient noise filtering, a second pole is added at the crossover frequency of 5 rad/s. Now that the return ratio $L(s)$ has been designed according to the first four design requirements, the phase and gain margin can be calculated to check whether this loop design is in agreement with the fifth design requirement. The designed return ratio is illustrated in Fig. 2.21. As we can see from the figure, the design also meets stability requirements with a phase margin of 49.1° and infinite gain margin. The designed return ratio can be described by the following transfer function $L(s)$:

$$L(s) = \frac{36(s+0.1)}{s(s+0.5)(s+5)} \quad (2.95)$$

If we assume that depth sensor H has no relevant dynamics, the desired controller transfer function $C(s)$ can be calculated by dividing the designed return ratio $L(s)$ by the open-loop system transfer function $G(s)$ [Eq. (2.95)]:

$$C(s) = \frac{L(s)}{G(s)} = \frac{\dfrac{36(s+0.1)}{s(s+0.5)(s+5)}}{\dfrac{11.7s^2 + 30.9s + 3315}{s^4 + 15.1s^3 + 139s^2 + 1560s}}$$

$$= \frac{3.1(s+0.1)(s+13.4)(s^2+1.69s+116)}{(s+0.5)(s+5)(s^2+2.64s+283)} \quad (2.96)$$

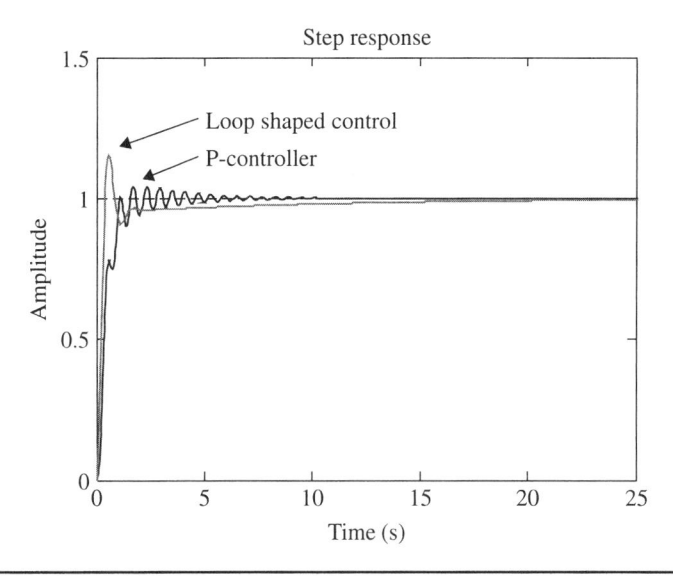

FIGURE 2.22 Step response for the closed-loop control system with designed return ratio $L(s)$ and with a proportional controller ($P = 1$).

The step response for the closed-loop control system is illustrated in Fig. 2.22, together with the step response of a closed-loop system with a simple proportional controller ($P = 1$).

Note that the system with the shaped return ratio moves more rapidly toward the desired value but needs a longer time to settle in the 2 percent region around the steady-state value. The reason for this high settling time can be found in the presence of the zero at 0.1 rad/s, which has been introduced to obtain sufficient disturbance attenuation in the midfrequency range. This attenuation is illustrated for both feedback-controlled systems in Fig. 2.23 for a sinusoidal disturbance with amplitude of 5 cm and frequency of 0.5 rad/s. The loop gain–shaped controller performs considerably better with an attenuation factor of 10, whereas the attenuation factor for the system with a proportional controller is only 1.3. This controller has thus traded some settling time performance (Fig. 2.22) for a considerable increase in disturbance attenuation performance (Fig. 2.23).

2.6 Nonlinear Optimization Example: Cruise Control on a Combine Harvester

In this section, a real-life optimization example will be elaborated, more specifically, a cruise control system on a combine harvester. This is a nonlinear optimization problem that will be solved by model-based predictive control. Because some of the concepts used in

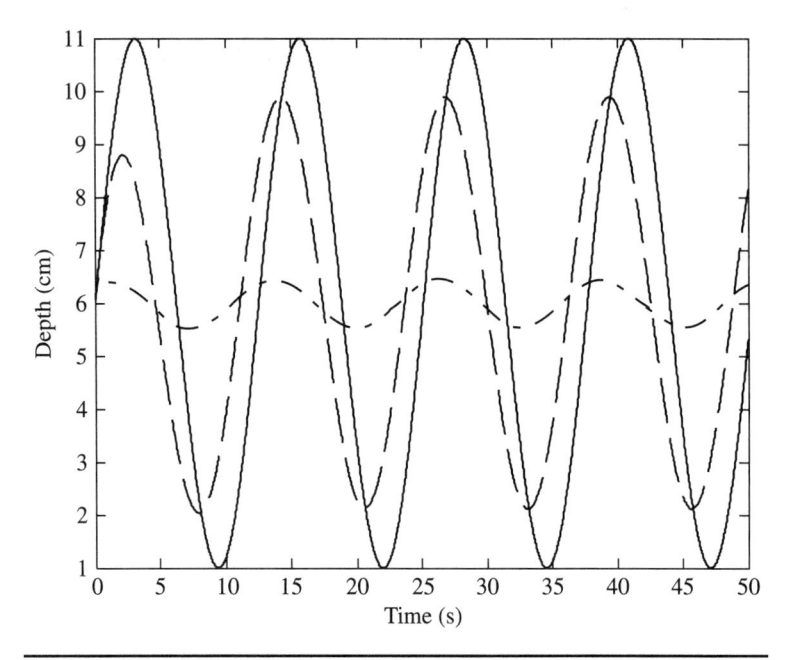

FIGURE 2.23 Influence of a sinusoidal disturbance with amplitude of 5 cm (solid) on the controlled response for the closed loop system with a P-controller (dash) and with designed return ratio $L(s)$ (dash-dot).

this example have not been explained in the theoretical part of this chapter, it should be seen as a showcase demonstrating to the interested reader what is possible with biosystems analysis and optimization rather than merely illustrating methods described here. We hope this example teases you, as a reader, to start reading more advanced books on the topic, like the ones cited throughout the text.

2.6.1 Motivation

Operator comfort is a hot topic in the world of farming machinery. Operators work long days, and good operators are increasingly hard to find. Ease of operation and operator comfort in general, are therefore important aspects in the competition between different manufacturers.

Environmental legislation is also becoming a more and more important catalyst for machine automation. Recent legislation has forced manufacturers to reduce the noise and the exhaust fumes of their machines during road transport (European Commission 2004, 2005, 2006). Thus, it makes sense to reduce engine speed whenever possible (i.e., in road conditions). A controller can now be designed to control the pump setting and the engine speed of the diesel engine such that the engine speed is minimized without losing acceleration performance (Coen et al. 2008a, 2008b).

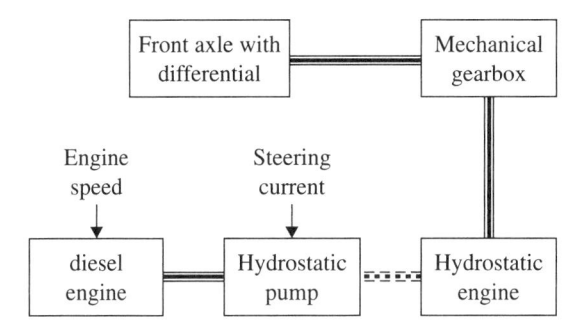

```
                    ┌─────────────────┐         ┌─────────────┐
                    │ Front axle with │═════════│ Mechanical  │
                    │  differential   │         │  gearbox    │
                    └─────────────────┘         └─────────────┘
                                                        ║
        Engine           Steering                       ║
        speed            current                        ║
          │                 │                           ║
          ▼                 ▼                           ║
      ┌─────────┐     ┌─────────────┐         ┌─────────────┐
      │ diesel  │─────│ Hydrostatic │- - - - -│ Hydrostatic │
      │ engine  │     │    pump     │         │   engine    │
      └─────────┘     └─────────────┘         └─────────────┘
```

FIGURE 2.24 A scheme of the propulsion system of a combine harvester.

2.6.2 System Description

The propulsion system of a combine harvester consists of a diesel engine, which drives a variable hydrostatic pump (Fig. 2.24). The hydrostatic flow thus depends on the speed of the diesel engine and on the pump setting. Hydrostatic flow, as well as engine speed, can be set independently by the operator. On our test machine (a New Holland CR combine harvester), engine speed can vary between a minimum (1300 rpm) and a maximum level (2100 rpm). Note however, that different combinations of engine speeds and pump settings result in the same machine speed.

During road transport, the operator can vary the pump setting as well as the speed of the diesel engine. This is not possible in field conditions. Because some components such as the cleaning system are mechanically connected to the diesel engine, the engine has to run at maximal speed while harvesting. Under road conditions, the speed of the machine can thus be controlled by two independent inputs: the pump setting and the engine speed of the diesel. It is clear that an operator does not have the time to use this extra degree of freedom and will always run the diesel engine at maximal engine speed while controlling the machine speed with the hydrostatic flow. However, the most straightforward way to reduce exhaust fumes and noise is to run the engine speed as low as possible.

2.6.3 Control Objective

The purpose of this controller is to run the combine harvester at the lowest possible engine speed without decreasing acceleration performance. In other words, the engine speed is allowed to increase during acceleration, but after the acceleration to the set point, the engine speed has to return to the lowest possible value without causing a set-point error. To reach this goal, the objective function is defined as follows:

Output error: The difference between the desired and the actual machine speed.

Input effort: Change of engine speed and change of pump setting. This is a regularization term to make the trade-off between both control inputs and between control inputs and control errors.

Engine speed: Engine speed minus minimal engine speed. This term leads to a minimization of the engine speed, and thus of engine noise and exhaust fumes.

2.6.4 System Modeling

Static Model

First, a static model is derived. This model statically links pump setting and engine speed to the machine speed. To derive this model, two test series have been performed:

- Slowly increase the pump setting, while the engine speed remains constant (for several engine-speed values).
- Slowly increase the engine speed, while keeping the pump setting constant (for several pump settings).

These experiments (Fig. 2.25) show that although the driving speed of the combine harvester is more or less linear in the engine speed, the relation between pump setting and machine speed contains a dead zone followed by a quadratic relation and, finally, saturation. Note that since these experiments are performed on an asphalt road, wheel slip is negligible.

Thus, machine speed (GS) can be written in terms of engine speed (ES) and pump setting (PS) as follows:

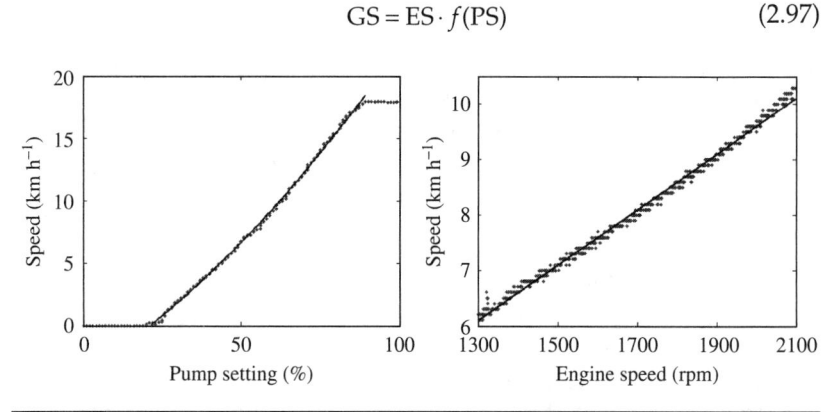

$$GS = ES \cdot f(PS) \tag{2.97}$$

Figure 2.25 The response of the speed to a slope input of (a) pump setting at 2100 rpm and (b) engine speed at a 60 percent pump setting. The measurements are indicated by the diamond markings. The solid line indicates the quadratic and the linear least squares approximation, respectively.

where f is a nonlinear function of the pump setting. This relation can thus also be written as

$$GS = \alpha(PS) \cdot ES \cdot PS \quad (2.98)$$

where α is the pump setting–dependent amplification. This amplification can be calculated by dividing the measured relation between pump setting and machine speed by the engine speed. In practice, however, a realigned model is often used. This model is independent of the actual plant response, because it is realigned to the measurements in every time step (Maciejowski 2002), predicting the change in output as a function of the change in input. Equation (2.98) then becomes

$$\Delta GS = \beta(PS) \cdot \Delta(ES \cdot PS) \quad (2.99)$$

where β is the local amplification of the system. Equation (2.99) is a local linear approximation (in the actual pump setting–engine speed point) of Eq. (2.98). Due to the dependency of the amplification factor on the pump setting, local and global amplification are not the same. Local amplification β is the amplification of a local linear approximation, or, in other words, β is the derivative of α (to the pump setting) in the working point. This local amplification is shown in Fig. 2.26.

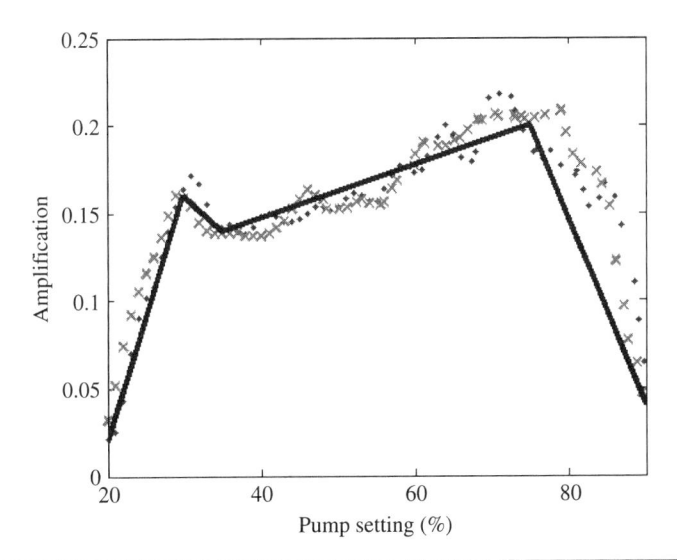

FIGURE 2.26 The local amplification between the pump setting–engine speed product and the machine speed as a function of the pump setting. Empirical amplification has been obtained for two engine speeds at 1300 rpm (♦) and at 2100 rpm (×).

Note that there are no significant differences between the local amplifications at two engine speed levels. The solid line represents a piecewise linear approximation. For the sake of clarity, amplification has already been rescaled by multiplying with the minimal engine speed (1300 rpm).

Dynamic Model

Because a change in travel speed requires acceleration and deceleration of the combine harvester subject to friction, it is clear that the system will only respond to relatively low frequencies due to inertial forces. Based on Sec. 2.6.4, in which the static model structure was shown, a dynamic state–space model structure can be proposed. Relying on physical knowledge, three states can be defined:

1. Actual pump setting to the hydrostatic valve
2. Actual engine speed of the diesel engine
3. Actual machine speed

The last state—machine speed—was shown to depend nonlinearly on the two other states through the earlier-derived static model.

Before we can determine the dynamic model, relationships in the system that may contain dynamic effects need to be identified. Assuming that there are no dynamic effects within the nonlinear relation (of the static model), three relationships remain in which dynamic effects can be present:

1. Between the pump setting set point and the actual pump setting
2. Between the engine speed set point and the actual engine speed
3. Between the static model and the actual machine speed

The general system structure, shown in Fig. 2.27, is called a Wiener–Hammerstein structure (Nesic 1999; Nelles 2001), because it is a linear dynamic system followed by a nonlinearity and a linear dynamic system.

Experiments have shown that the engine speed and the pump setting to the hydrostatic valve respond sufficiently fast so that those dynamics can be neglected. This means that only the relationship between the nonlinear transformation (of the pump setting and the engine speed) and the actual machine speed contains dynamics. Multisine excitations (Pintelon and Schoukens 2001) show that the dynamics form a low-pass filter, as could be expected.

The prediction performance of the model is essential for model-based predictive control (MPC). In addition, because the system is nonlinear, the system response depends on the type of input. Because the most typical input for the controller will be a stepwise increase of

```
Controlled
current
   │
   ▼
┌─────┐    ┌──────────────┐
│▨▨▨▨│───▶│Actual current│
└─────┘    │ to the pump  │
           └──────────────┘
                          ╲
                           ╲    ┌───┐    ┌─────┐    ┌──────────────┐
                            ───▶│ X │───▶│▨▨▨▨│───▶│Actual machine│
                           ╱    └───┘    └─────┘    │    speed     │
                          ╱                         └──────────────┘
           ┌──────────────┐
┌─────┐    │Actual engine │
│▨▨▨▨│───▶│    speed     │
└─────┘    └──────────────┘
   ▲
   │
Controlled
engine speed
```

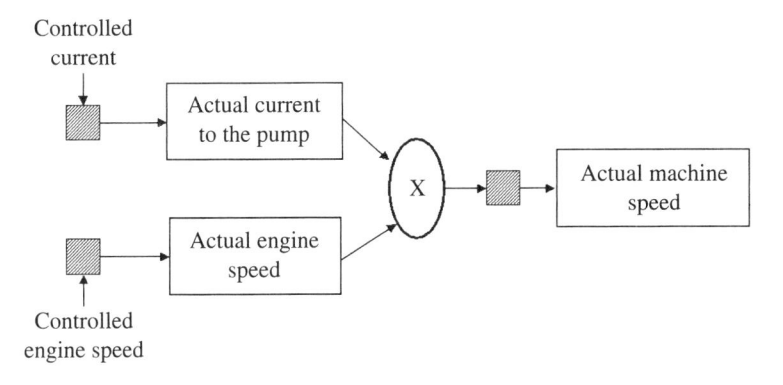

FIGURE 2.27 Overview of the model of the system. The actual pump setting, actual engine speed, and actual travel speed are states of the system. The oval (containing the "x") represents the nonlinear dependency of the travel speed on the engine speed and the pump setting. The squares represent the possible locations of the dynamics.

the reference speed, parameters of the second-order system model have been estimated on step responses (Sec. 2.4.4). The estimated transfer function is of the following form:

$$G(s) = \frac{1.5}{s^2 + 1.8s + 1.5} \qquad (2.100)$$

Linearization and Discretization

If possible, the best approach to nonlinear MPC is to linearize the model online, depending on the pump setting–working point. This way, the linearization error is minimal, and the optimization problem remains a quadratic programming problem. Working with a nonlinear model in the optimization requires more complex solution methods, which has a significant impact on the computation time.

First of all the system needs to be discretized. The sampling rate is chosen to be 20 Hz, which is the highest rate possible on the system. Discretization is performed with the Tustin discretization rule (Franklin et al. 2006). This leads to the following system:

$$\begin{cases} \mathbf{x}_{k+1} = \mathbf{A}\mathbf{x}_k + \mathbf{B}\mathbf{u}_k \\ \mathbf{y}_k = \mathbf{C}\mathbf{x}_k + \mathbf{D}\mathbf{u}_k \end{cases} \qquad (2.101)$$

where $\mathbf{u}_k$ is a function of the pump-setting–engine speed product [static model of Eqs. (2.98) and (2.99)], $\mathbf{y}_k$ is the ground speed, and $\mathbf{x}_k$ is the state vector of the system.

However, the actual inputs of the model should be a change of the pump setting ΔPS_k and a change of engine speed ΔES_k. Input $\mathbf{u}_k$

thus needs to be rewritten in terms of these variables, and then linearized. Using Eqs. (2.98) and (2.99), $\mathbf{u}_k$ can be written as follows:

$$\begin{aligned}\mathbf{u}_k &= \alpha(\mathrm{PS}_k) \cdot \mathrm{PS}_k \cdot \mathrm{ES}_k \\ &= \alpha(\mathrm{PS}_k) \cdot (\mathrm{PS}_{k-1} + \Delta \mathrm{PS}_k) \cdot (\mathrm{ES}_{k-1} + \Delta \mathrm{ES}_k) \\ &\approx \alpha(\mathrm{PS}_{k-1}) \cdot \mathrm{PS}_{k-1} \cdot \mathrm{ES}_{k-1} + \beta(\mathrm{PS}_{k-1}) \cdot \mathrm{PS}_{k-1} \cdot \Delta \mathrm{ES}_k \\ &\quad + \beta(\mathrm{PS}_{k-1}) \cdot \mathrm{ES}_{k-1} \cdot \Delta \mathrm{PS}_k + \beta(\mathrm{PS}_{k-1}) \cdot \Delta \mathrm{PS}_k \cdot \Delta \mathrm{ES}_k\end{aligned} \qquad (2.102)$$

where α is the global amplification and β is the local amplification. Note that after linearizing the system, the local amplification, as defined in Eq. (2.99) appears in the equations. By neglecting the second-order term and defining a new input vector $\mathbf{v}_k$ containing the change of the pump setting and the change of the engine speed at time k, $\mathbf{u}_k$ can be written as

$$\mathbf{u}_k = \alpha(\mathrm{PS}_{k-1}) \cdot \mathrm{PS}_{k-1} \cdot \mathrm{ES}_{k-1} + \mathbf{B}_u \mathbf{v}_k \qquad (2.103)$$

where

$$\mathbf{B}_u = \beta(\mathrm{PS}_{k-1}) \cdot [\mathrm{ES}_{k-1} \quad \mathrm{PS}_{k-1}]$$

$$\mathbf{v}_k = \begin{bmatrix} \Delta \mathrm{PS}_k \\ \Delta \mathrm{ES}_k \end{bmatrix} \qquad (2.104)$$

The system input thus becomes $\mathbf{v}_k$, and Eq. (2.103) becomes an extra system state with the state equation:

$$\mathbf{u}_k = \mathbf{u}_{k-1} + \mathbf{B}_u \mathbf{v}_k \qquad (2.105)$$

Note that $\mathbf{B}_u$ depends on the pump-setting state of the system. Using a constant $\mathbf{B}_u$ across the prediction horizon thus implies linearization. To keep the pump setting and the engine speed within their physical range, constraints can be introduced. However, constraints can only be placed on inputs, states, and outputs. Because the only inputs after linearization are $\Delta \mathrm{PS}_k$ and $\Delta \mathrm{ES}_k$, integrator states need to be added to calculate PS_k and ES_k. This way, constraints can be imposed on the calculated PS_k and ES_k. The extended state–space description then becomes

$$\begin{cases} \mathbf{x}_{k+1} = \mathbf{A}_L \mathbf{x}_k + \mathbf{B}_L \mathbf{v}_k \\ \mathbf{y}_k = \mathbf{C}_L \mathbf{x}_k + \mathbf{D}_L \mathbf{v}_k \end{cases} \qquad (2.106)$$

where system matrices $\mathbf{A}_L \in \mathbb{R}^{5\times 5}$, $\mathbf{B}_L \in \mathbb{R}^{5\times 2}$, $\mathbf{C}_L \in \mathbb{R}^{1\times 5}$, and $\mathbf{D}_L \in \mathbb{R}^{1\times 2}$. State vector $\mathbf{x}_k$ now contains five elements: the two model states of Eq. (2.100), the input state of Eq. (2.105) and two integrator states for PS_k and ES_k.

Three of these five system states cannot be measured: the two model states of Eq. (2.100) and the input state of Eq. (2.105). These states thus need to be estimated using a state estimator to compensate for model inaccuracy.

Disturbances

Several experiments have been performed to investigate the sensitivity of the model to disturbances as road slope and machine mass. The mass of the combine harvester may vary considerably as a function of the grain in the bin. The effect of this varying machine mass can be explained by the physical construction of the propulsion system.

The current to the electric coil of the hydraulic pump is called the pump setting. This electric coil is coupled to a servovalve through a flipper, which on its turn controls the swash plate position of the hydraulic pump. Note that there is mechanical feedback present between the swash plate and the flipper. A current change causes the flipper to move to the left or to the right. The servovalve then creates a pressure difference between both cylinders connected to the swash plate, causing the swash plate to tilt. The mechanical feedback places the flipper back in its central position. A current thus creates a force on the flipper, which is counteracted by the force of the mechanical feedback. The pump itself is driven by the diesel engine. Because of its construction, the flow only depends on its configuration (pump setting). As long as the diesel engine can deliver the necessary power, machine speed will only vary if the pump setting is changed. Thus, the mass of the grain bin and the terrain slope theoretically do not influence the propulsion system. However, because of the limitations of the diesel engine, the mass in the grain bin and the terrain slope may affect on the machine speed under field conditions (threshing mechanism on). In road conditions, however, it is very rare that the diesel engine cannot provide the required power.

2.6.5 Nonlinear MPC

The operator has two degrees of freedom to control the machine speed while driving on the road: pump setting and engine speed. The objective of the cruise control system is to use this extra degree of freedom to minimize the exhaust fumes and the noise production of the combine harvester without losing performance. The lower the engine speed of the diesel engine, the less exhaust fumes and noise.

Model-based Predictive Control

MPC is a control technique that calculates an optimal input sequence by minimizing an objective function subject to a plant model and other constraints over a fixed prediction horizon (Maciejowski 2002; Camacho and Bordons 2004). Most MPC approaches are based on time-domain techniques, which is beneficial for its interpretability by a wider public.

If the model is nonlinear, it can be approximated by a linear model. To minimize the approximation error, the linear model is calculated in every time step.

Objective Definition

As indicated previously, the purpose of the control system is to reduce engine speed without losing acceleration performance. The purpose of this controller is that the engine speed increases during acceleration, and then decreases again to the minimal possible engine speed needed to reach the requested speed set point.

The objective of the controller is to minimize the set-point error (between actual and desired machine speeds) with minimal input cost. The input cost is the change of engine speed and pump setting. This cost is chosen because it is related to the operator's comfort. To minimize engine speed, an extra penalty term in the engine-speed state is added to the objective function. This is the mathematical translation of our objective to operate the machine at minimal engine speed. Equality constraints are formed by the dynamic model of the system of Eq. (2.106). Inequality constraints are input constraints (thus on change of pump setting and engine speed) and state constraints (on the pump-setting and engine-speed state). Input constraints are imposed to guarantee the operator's comfort; state constraints confine the pump setting and the engine speed to their physical region. The squared set-point error, engine-speed cost, and input cost are summed to a quadratic objective function. The optimization problem at every time step thus becomes

$$\min_{\mathbf{x}_k, \mathbf{y}_k, \mathbf{u}_k} \sum_{k=1}^{N}(\mathbf{y}_k - \mathbf{r}_{y,k})^T \mathbf{S}(\mathbf{y}_k - \mathbf{r}_{y,k}) + \sum_{k=1}^{N}(\mathbf{x}_k - \mathbf{r}_{x,k})^T \mathbf{Q}(\mathbf{x}_k - \mathbf{r}_{x,k})$$

$$+ \sum_{k=0}^{M-1} \mathbf{v}_k^T \mathbf{R} \mathbf{v}_k \qquad (2.107)$$

subject to

$$\begin{cases} \mathbf{x}_{k+1} = \mathbf{A}_L \mathbf{x}_k + \mathbf{B}_L \mathbf{u}_k & k = 0 \ldots N-1 \\ \mathbf{y}_k = \mathbf{C}_L \mathbf{x}_k + \mathbf{D}_L \mathbf{v}_k & \\ \mathbf{x}_k \in \mathcal{X} & k = 1 \ldots N \\ \mathbf{y}_k \in \mathcal{Y} & k = 0 \ldots N-1 \\ \mathbf{v}_k \in \mathcal{V} & k = 0 \ldots M-1 \end{cases} \qquad (2.108)$$

where N is the prediction horizon, M is the control horizon, $\mathbf{r}_{y,k}$ is the desired machine speed, $\mathbf{S} \in \mathbb{R}^{1 \times 1}$ is the weight on the set-point error, $\mathbf{r}_{x,k}$ contains the desired state values, $\mathbf{Q} \in \mathbb{R}^{5 \times 5}$ is the weight on the

Biosystems Analysis and Optimization 85

state errors, and $\mathbf{R} \in \mathbb{R}^{2 \times 2}$ is the input cost. Matrices $\mathbf{Q}$, $\mathbf{R}$, and $\mathbf{S}$ have the following structure:

$$Q = \begin{bmatrix} 0 & 0 & 0 & 0 & 0 \\ 0 & w_{ES} & 0 & 0 & 0 \\ 0 & 0 & 0 & 0 & 0 \\ 0 & 0 & 0 & 0 & 0 \\ 0 & 0 & 0 & 0 & 0 \end{bmatrix} \quad R = \begin{bmatrix} w_{\Delta PS} & 0 \\ 0 & w_{\Delta ES} \end{bmatrix} \quad (2.109)$$

$$S = \begin{bmatrix} w_{error} \end{bmatrix}$$

where w_{ES} is the weight accorded to the engine-speed cost, $w_{\Delta PS}$ is the penalty for a unit change of the pump setting, $w_{\Delta ES}$ is the penalty for a unit change of the engine speed, and w_{error} is the penalty for a unit difference between the desired and the actual machine speed. The cost of the engine speed is defined as the (squared) difference between actual engine speed and minimal engine speed (i.e., 1300 rpm). $\mathcal{X}$, $\mathcal{Y}$, and $\mathcal{V}$ represent the feasible region for the vectors $\mathbf{x}_k$, $\mathbf{y}_k$, and $\mathbf{v}_k$, respectively. These regions are defined by the following inequality constraints:

$$\mathcal{X} = \left\{ \mathbf{x} \in \mathbb{R}^{5 \times 1} \middle\| \begin{bmatrix} 1 & 0 & 0 & 0 & 0 \\ 0 & 1 & 0 & 0 & 0 \\ -1 & 0 & 0 & 0 & 0 \\ 0 & -1 & 0 & 0 & 0 \end{bmatrix} \mathbf{x} \geq \begin{bmatrix} PS_{min} \\ ES_{min} \\ -PS_{max} \\ -ES_{max} \end{bmatrix} \right\}$$

$$\mathcal{Y} = \left\{ \mathbf{y} \in \mathbb{R}^{1 \times 1} \middle\| \begin{bmatrix} 1 \\ -1 \end{bmatrix} \mathbf{y} \geq \begin{bmatrix} speed_{min} \\ -speed_{max} \end{bmatrix} \right\} \quad (2.110)$$

$$\mathcal{V} = \left\{ \mathbf{z} \in \mathbb{R}^{2 \times 1} \middle\| \begin{bmatrix} 1 & 0 \\ 0 & 1 \\ -1 & 0 \\ 0 & -1 \end{bmatrix} \mathbf{v} \geq \begin{bmatrix} \Delta PS_{min} \\ \Delta ES_{min} \\ -\Delta PS_{max} \\ -\Delta ES_{max} \end{bmatrix} \right\}$$

where PS_{min} is the minimal pump setting (expressed in percent), PS_{max} is the maximal pump setting, ES_{min} is the minimal engine speed (expressed in rpm), ES_{max} is the maximal engine speed, $speed_{min}$ is the minimal machine speed, $speed_{max}$ is the maximal machine speed, ΔPS_{min} is the minimal change of pump setting (expressed in percent), ΔPS_{max} is the maximal change of pump setting, ΔES_{min} is the minimal change of engine speed (expressed in rpm), and ΔES_{max} is the maximal change of engine speed. This description assumes that the feasible region is constant over the horizon, but this formulation can easily be extended to time-varying constraints if necessary.

A logical way to define the trade-off between both control inputs is to base it on the squared ratio of their ranges. The maximal pump-setting change (ΔPS_{min}, ΔPS_{max}) is limited to ±5 per time step, whereas the engine speed change (ΔES_{min}, ΔES_{max}) is limited to ±100 per time step. Assume that the weight of the change of pump setting is 1, the weight of the change of engine speed then becomes $0.0025 = (5/100)^2$. The weight of the control error is tuned experimentally. The best results (trade-off between performance and comfort) are obtained with a control error weight equal to 1. If the control error weight is 1, a 1 km/h error leads to the same cost as a 1 percent change of the pump setting. Note that this trade-off is influenced by the control rate. For this application, the control rate is 20 Hz.

This formulation of the objective means that the system will always try to lower the engine speed to the minimum. Because high speeds can only be maintained if the engine speed is greater than the minimal engine speed, a trade-off will occur between the control error and the minimization of the engine speed. In other words, the engine speed will not return to its minimal value, but there will remain a control error in the steady state.

This issue can be solved by including an extra integrator state I in the system (enlarge the system to six states). This extra state will integrate the difference between the set point and the actual speed. In order not to interfere with fast acceleration dynamics, integrator I will be very slow. The objective function is then extended with a penalty term on this integrator state. As long as there is a control error present, the integrator term will increase, allowing for a larger engine speed.

Integrator I always needs to remain larger than zero and has to integrate the sum of the control error and the difference between the pump setting and the maximal pump setting. This way, I will increase if there is a control error and decrease if the pump setting is not at the maximal value. The combination of these two terms de facto disables the integrator during transients as long as the pump setting is increasing. The result of this approach is illustrated in the following section.

2.6.6 Experimental Setup

For the experiments, a New Holland CR combine harvester was used. This machine was equipped with a LabView PXI system (National Instruments). In order to test the controller, step changes were applied to the reference input of the controller. The experiments were performed on a private parking lot (asphalt road).

2.6.7 Results

The aforementioned controller is tested on the road for accelerations up until 12 km/h. The controller is tested with and without the engine-speed minimization term. The effect of adding this engine-speed minimization term is shown in Fig. 2.28. Because the engine-speed

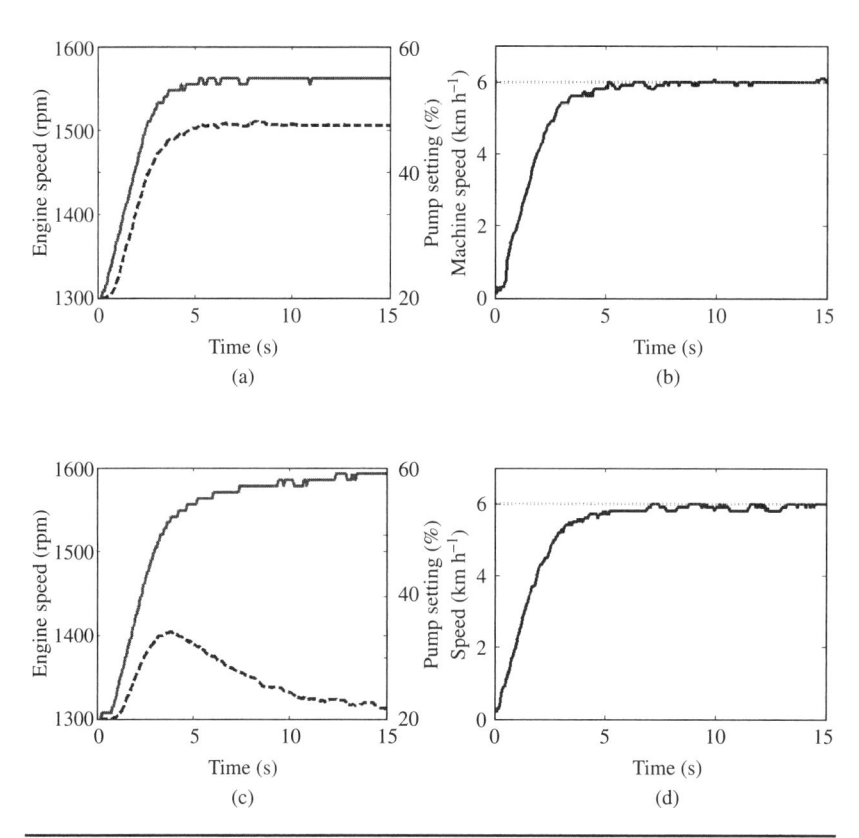

FIGURE 2.28 Response to a step change in the desired travel speed for an MPC controller without engine-speed minimization [(a) and (b)] and with engine-speed minimization [(c) and (d)]; (b) and (d) show the speed response where the dashed line represents the travel speed and the solid line represents the reference; (a) and (c) show the input signals where the solid line represents the pump setting (right axis), and the dashed line represents the engine speed (left axis).

minimization only kicks in after a few seconds, the operator does not feel a difference in acceleration speed. Moreover, because the noise is reduced at the reference speed, the cruise control with engine-speed minimization is experienced as being more comfortable by the operator.

Both MPCs start from minimal engine speed. In practice, the operator would always run the combine at maximal engine speed. Although acceleration is much slower when accelerating at minimal engine speed than when accelerating at maximal speed, the operator does not sense a difference between accelerating with the engine at maximal speed and the engine speed controlled by the MPC. Acceleration feels equally smooth.

As we can see in Fig. 2.28, the decreasing engine speed is compensated by the increasing pump setting. As previously described, this

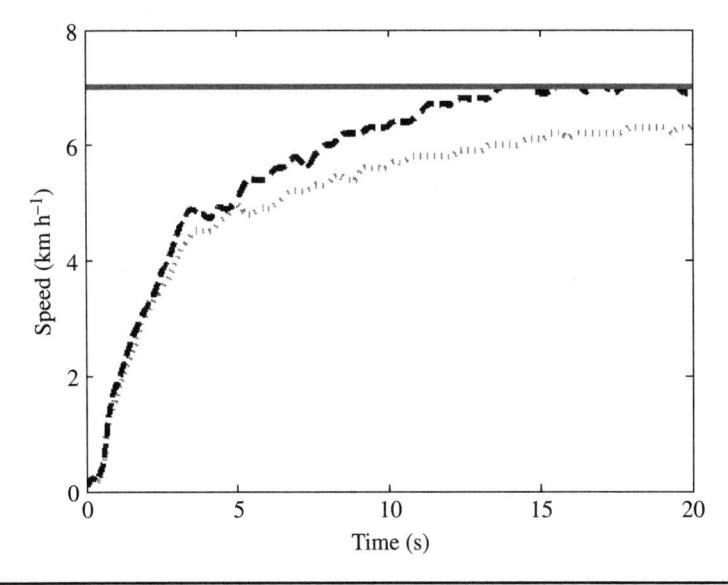

Figure 2.29 Response to a step change in the desired speed for an MPC controller with engine-speed minimization. The pump setting is limited to 50 percent to provoke a steady-state error at low speeds. The solid line represents the reference speed at 7 km/h, the dotted line represents the standard controller, and the dashed line represents the same controller with an extra integrator to avoid a steady-state error.

will cause problems at high speeds because the pump setting cannot become sufficiently large to compensate for the decreasing engine speed. Due to limitations at the test site, upper limit on the pump setting was lowered to 50 percent to be able to show this effect at lower speeds. The maximal speed at minimum engine speed was approximately 5.5 km/h. In Fig. 2.29, a step response to 7 km/h of the controller is shown with and without the integrator term. Adding the extra integrator state thus removes the steady-state error while keeping the engine speed as low as possible.

References

Anthonis, J. and Ramon, H. 2003. Design of an active suspension to suppress the horizontal vibrations of a spray boom, *Journal of Sound and Vibration* 266(3): 573–583.

Camacho, E. F. and Bordons C. 2004. *Model Predictive Control*. 2nd ed. London, UK: Springer.

Coen, T., Anthonis, J., and De Baerdemaeker, J. 2008a. Cruise control on a combine harvester using model-based predictive control with constraints, *Computers and Electronics in Agriculture* 63(2):227–236.

Coen, T., Saeys, W., Missotten, B., and De Baerdemaeker, J. 2008b. Cruise control on a combine harvester using model-based predictive control. *Biosystems Engineering* 99(1):47–55.

Coen, T., Vanrenterghem, A., Saeys, W., and De Baerdemaeker, J. 2008c. Autopilot for a combine harvester. *Computers and Electronics in Agriculture* 63(1):57–64.

De Temmerman, J., Deprez, K., Anthonis, J., and Ramon, H. 2004. Conceptual Cab suspension system for a self-propelled agricultural machine, part 1: Development of a linear mathematical model. *Biosystems Engineering* 89(4):409–416.

De Temmerman, J., Deprez, K., Hostens, I., Anthonis, J., and Ramon, H. 2005. conceptual cab suspension system for a self-propelled agricultural machine, part 2: Operator comfort optimisation *Biosystems Engineering* 90(3):271–278.

De Temmerman, J., Verboven, P., Nicolaï, B., and Ramon, H. 2007. Modelling of transient moisture concentration of semolina pasta during air drying. *Journal of Food Engineering* 80(3):892–903.

De Temmerman, J., Dufour, P., Nicolaï, B., and Ramon, H. 2009. MPC as control strategy for pasta drying processes *Computers and Chemical Engineering* 33(1):50–57.

Deprez, K., Anthonis, J., and Ramon, H. 2003. System for vertical boom corrections on hilly fields. *Journal of Sound and Vibration* 266(3):613–624.

European Commission. 2004. Emissions for non-road mobile machinery. Directive 2004/26/EC.

European Commission. 2005. Noise emissions for outdoor equipment. Directive 2005/88/EC.

European Commission, 2006. Machinery. Directive 2006/42/EC.

Franklin, G. F., Powell J. D., and Emami-Naeini, A. 2006. *Feedback Control of Dynamic Systems*. 5th ed. Upper Saddle River, NJ: Prentice Hall.

Kopasakis, G. 2007. Feedback control systems loop shaping design with practical considerations, NASA/TM—2007-215007 (available electronically at http://gltrs.grc.nasa.gov).

Ljung, L. 1987. *System Identification, Theory for the User*. Englewood Cliffs, NJ: Prentice Hall, Inc.

Maciejowski, J. M. 2002. *Predictive Control with Constraints*. London: Pearson Education.

Nelles, O. 2001. *Nonlinear System Identification*. Berlin, Germany: Springer-Verlag.

Nesic, D. 1999. Controllability for a class of simple Wiener-Hammerstein systems. *System and Control Letters* 36(1):51–59.

Nise, N. S. 2000. *Control Systems Engineering*. 3rd ed. Hoboken, NJ: Wiley.

Pintelon, R. and Schoukens, J. 2001. *System Identification: A Frequency Domain Approach*. Hoboken, NJ: Wiley-IEEE Press.

Ramon, H., Anthonis, J., Moshou, D., and De Baerdemaeker, J. 1998. Evaluation of a cascade compensator for horizontal vibrations of a flexible spray boom. *Journal of Agricultural Engineering Research* 71(1):81–92.

Saeys, W., Engelen K., Ramon H., and Anthonis J. 2007. An automatic depth control system for shallow manure injection, part 1: Modelling of the depth control system. *Biosystems Engineering* 98(2):146–154.

Saeys, W., Deblander, J., Ramon, H., and Anthonis, J. 2008a. High-performance flow control for site-specific application of liquid manure. *Biosystems Engineering* 99(1):22–34.

Saeys, W., Wallays, C., Engelen, K., Ramon, H., and Anthonis, J. 2008b. An automatic depth control system for shallow slurry injection, Part 2: Control design and field validation. *Biosystems Engineering* 99(2):161–170.

Schwarzenbach, J. and Gill, K.F. 1978. *System Modelling and Control*. London: Edward Arnold.

CHAPTER 3
Soil and Water Conservation

Latif Kalin
School of Forestry and Wildlife Science
Auburn University, Auburn, Alabama

Puneet Srivastava
Biosystems Engineering, Auburn, Alabama

3.1 Introduction

Soil and water are among the most precious resources human beings have. Not only can we not exist without those two invaluable resources, humans and all other living organisms will suffer if they are not conserved properly. Some past civilizations, not recognizing the importance of soil and water conservation or know-how to mitigate the adverse impacts of their agricultural or other practices on these resources, had to put up with its consequences, and some perished. In some instances, decision makers had to observe dramatic events taking place in front of them, which normally takes place over a long time, to move forward. A great example is the famous "Dust Bowl" in 1935 in the United States. In the spring of 1935, Hugh H. Bennett was testifying before a congressional committee on the bill that would create the Soil Conservation Service (SCS) when dust clouds from the Dust Bowl moved over Washington, DC. He knew that a dust storm was coming and used it to dramatically demonstrate the need for soil conservation. During his testimony, Bennett asked the legislators to look out the window, and then said, "This, gentlemen, is what I have been talking about." Congress passed the Soil Conservation Act, and SCS became a permanent federal agency in 1935 (Gilley and Flanagan 2007). It is hard to say that Congress would not have passed the bill if they had not observed the Dust Bowl, but it is fair to claim that the Dust Bowl had a huge effect in convincing Congress, as they were able to witness firsthand the effects of soil erosion. Unfortunately, not everyone was fortunate to witness such dramatic events to recognize the need and the significance of conservation.

The word "conservation" means preserving or protecting something from decaying, destruction, or deterioration. In this chapter we use it under a much wider framework. In addition to its customary meaning, we also consider mitigation and abatement as a part of conservation. In that sense, the word "conservation" will encompass any efforts trying to protect soil and water resources from deteriorating from a quality perspective, insufficiency (e.g., drought), or surplus (e.g., flooding) from a quantity standpoint, as well as any practices in trying to mitigate these undesirable conditions.

Conservation of water and soil could/should be carried out at two stages. The first stage is protecting the source, which is protecting the soil from erosion and water from being becoming polluted. In reality, even the most virgin lands are prone to erosion and contribute to pollutant loadings to some degree. Therefore, the goal for conservation cannot be total elimination but does include reducing rates of soil erosion and pollutant loadings. This will not only help alleviate the quality of downstream receiving water bodies but also preserve or even increase soil fertility. This first stage is referred to as *conservation at the source* or *source scale conservation*. The second stage involves various practices or mitigation strategies to improve the water quality of the water bodies and/or to minimize the potential adverse impacts of alleviated water quality and quantity problems on humans and the ecosystem. This stage is referred to as *conservation at transport stage* or *transport scale conservation*. In upcoming sections we will be discussing both of these stages/scales for soil and water conservation.

To develop and understand effective and efficient conservation strategies, we must understand the governing physical processes that take place in soil, water, and at their interfaces. Because soil and water are in constant interaction, a proficient conservation strategy should consider their complex interaction. In the following sections, we describe those processes and present relevant theories and recent advancements. We start with the hydrologic cycle. After an overall description of the hydrologic cycle, we delve only in the components of the hydrologic cycle that are most relevant to soil and water conservation. Then we talk about the physics of soil erosion and sediment transport processes. The last part of this chapter is dedicated to various best management practices, commonly referred to as BMP, that are useful at the source and transport scales.

3.2 Hydrological Cycle

Water on the earth and in the atmosphere is in constant movement and changes its form and its medium. The hydrologic cycle basically describes this circulation and could be defined as the continuous movement of

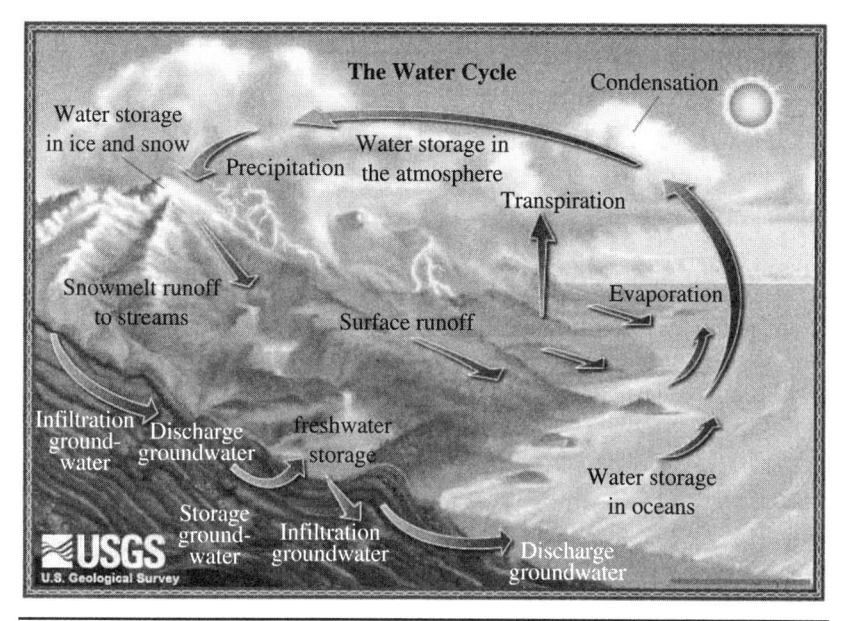

FIGURE 3.1 Schematic representation of the hydrologic cycle. [*Adapted from the United States Geological Survey (USGS).*]

water molecules in different forms (vapor, ice, liquid) on, above, and below the earth's surface (Fig. 3.1). The energy from the sun (i.e., solar radiation) is the main driver of the hydrologic cycle. Approximately 50 percent of the sun's energy reaching the earth's surface is used to vaporize water. The average solar radiation reaching the United States is about 400 cal/cm^2/day. Thus, on average, roughly 200 cal/cm^2 of energy is consumed per day by the earth's surface and by plants to vaporize water either by direct evaporation or by transpiration.

The hydrologic cycle includes many components and processes: evaporation, transpiration, condensation, precipitation in all forms (rain, snow, hail, sleet, fog, etc.), interception, infiltration, exfiltration, percolation, runoff, groundwater flow, groundwater recharge, and snow melt. These components and processes are tightly connected. However, it is not necessary to have all these processes for a complete water cycle. Often, some processes are bypassed or the hydrologic cycle is short circuited. Brief summaries of these components and processes are given as follows.

Evaporation This is the process of the formation of water vapor molecules from a liquid or gaseous (sublimation) form. This process requires solar energy. Factors affecting evaporation are temperature, wind speed, relative humidity, and solar radiation. There are two main limiting factors for evaporation to occur: water and energy. When there is plenty of water available, the temperature is the limiting factor for evaporation. The evaporation rate in this case is mainly driven by temperature.

Conversely, when there is not enough water, water is the limiting factor. Two important terms are *actual* and *potential evaporation*. Potential evaporation is the rate of evaporation when there is plenty of water, which corresponds to the temperature-limited case. When there is not enough moisture in the soil (i.e., water-limited case), the *actual evaporation* rate is smaller than the potential evaporation. Actual evaporation is always smaller than or equal to potential evaporation.

Transpiration This is the process of water molecules escaping from tree leaves and other vegetation into the atmosphere. This process is often combined with evaporation and is called *evapotranspiration*. Similar to evaporation, actual evapotranspiration is mostly less than potential evapotranspiration. In addition to the factors affecting evaporation rates, physiologic characteristics of the plants, their leaf shapes, and their leaf densities also affect transpiration rates.

Condensation This is the process of water vapors changing into liquid form as a result of cooling. Formation of clouds and fog is a result of condensation of water vapors.

Precipitation This is the process of the falling of condensed water onto the earth's surface (land or water bodies). Precipitation occurs mainly in the form of rain. However, snow, sleet, hail, dew, frost, and icy rain are all forms of precipitation.

Interception This is the process of the interception of a portion of the precipitation by vegetation or by trees. Intercepted water eventually evaporates into the atmosphere. In other words, it never reaches the soil and, therefore, does not contribute to surface runoff or to infiltration processes.

Infiltration This is the process of the movement of water from the soil surface into the soil as a result of gravity and capillary forces. Infiltration takes place at the soil–air interface, or in case water is ponded at the soil–water interface. In addition to soil texture and soil characteristics, available soil moisture content also plays a major role in determining the infiltration rate. The term *infiltration capacity* is used to define the rate of infiltration when there is plenty of water or water is ponded on the soil surface. This happens when the rainfall intensity is larger than the infiltration rate. When the rainfall intensity is low and water cannot pond on the soil surface, then infiltration rate is equal to rainfall intensity.

Exfiltration This is the opposite of infiltration (i.e., movement of water from the soil layer to the soil surface). Exfiltration could be due to pressure forcing water to move up or water moving within the soil suddenly meeting the atmosphere due to changes in topography (springs).

Percolation This process is similar to infiltration. The difference is that water movement in percolation is within the soil matrix instead of the soil–water/air interface. Percolation occurs in a vertical direction, and it only takes place in the zone above the water table.

Runoff This is the part of precipitation, snow melt, or irrigation water flowing on the earth's surface in creeks, streams, rivers, drainage ditches, sewers, and the like. Runoff could be generated as a result of excess water not infiltrating the soil or by exfiltration of the soil water.

Groundwater This is the water stored in the soil pore spaces and fractures of geologic formations that are saturated with water. The top of the water surface in unconfined aquifers is called the *water table*. It is the surface where water pressure is equal to atmospheric pressure. Water can move in all directions in the groundwater zone.

Groundwater Recharge Water percolating deep into the soil will eventually feed into the groundwater. This process of water replenishing the groundwater is called *groundwater recharge*. Conversely, if the groundwater feeds a river or a spring, then it is called *discharge*.

Snow Melt When snow melts, it either becomes surface runoff or infiltrates the soil. If the soil beneath the snow is frozen, then all the snow melt becomes runoff. Sometimes snow can evaporate directly without turning into liquid. This process of phase change directly from solid into gaseous form is called *sublimation*.

Figure 3.2 shows the global average of approximate water flow rates between the different components of the hydrologic cycle. We can

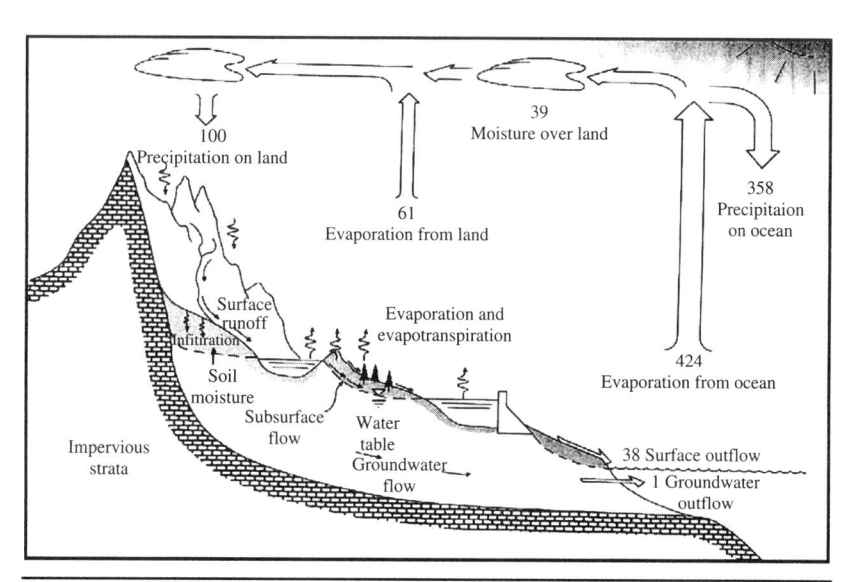

FIGURE 3.2 Hydrologic cycle and approximate water balance of the earth relative to the rate of precipitation on land (100 units = 119,000 km³/year). (*Adapted from Chow et al. 1988*)

see that more than 60 percent of the precipitation falling on the land evaporates back into the atmosphere. Evaporation from the oceans provides nearly 90 percent of atmospheric moisture. Details on some of these processes and components of the hydrologic cycle that are important to soil and water conservation are provided as follows.

3.3 Precipitation

Precipitation is the main water input to the earth's lands and oceans. The global average precipitation rate is about 1132 mm/year. This includes precipitation falling on oceans. The average global precipitation falling only on land is approximately 801 mm. However, precipitation rates show very high spatial variations on the earth, varying from less than 1 mm in the Arica Desert in Chile (during the past 59 years of instrumentation there was a period of 14 years that were totally devoid of precipitation) to about 12 m in Mount Waialeale, Hawaii. A natural question is what causes all these variations? Before checking into that let us look at the conditions for precipitation to occur. There are three general steps in the generation of precipitable water in the atmosphere.

1. Creation of saturated conditions in the atmosphere. Saturation is typically arrived by cooling of the lifting air. Temperature drops on average 1°C/100 m under dry conditions and about 0.5 to 0.7°C/100 m in moist air.
2. Condensation of water vapor into liquid water. As air cools, its capacity to hold water vapor decreases. When air becomes fully saturated, water molecules start being attracted to small particles such as clay and silt particles, smoke, sea salt, and the like.
3. Growth of small droplets by collision and coalescence until the droplets become large enough to precipitate due to gravity.

The foregoing three steps are only sufficient in generating precipitable water. To have a sustained precipitation event, a lateral supply of moist air is needed. One reason for the failure of cloud-seeding experiments to fabricate precipitation events from rainfall clouds is the lack of a lateral moist air supply.

Why do some locations on the earth receive a lot of precipitation and some locations get very little to no rain? The average precipitation on the continents of the world depends on many factors. In general those factors are

1. *Latitude:* Precipitation is higher in areas with rising air and is lower in areas with descending air. In general latitudes from 0 to 60° have rising air, and latitudes from 30 to 90° have descending air.

2. *Altitude:* Air temperature decreases with elevation. Due to this cooling effect, precipitation increases with elevation. This increase in precipitation with altitude is called the *orographic effect*.
3. *Distance from moisture sources:* Inland areas, therefore, receive less precipitation than coastal areas.
4. *Position within the continental mass.*
5. *Wind direction.*
6. *Relation to mountain ranges:* Windward sites are typically more cloudy and rainy and leeward sites are drier and sunnier.
7. *Relative temperatures of land and bordering oceans.*

The most common form of precipitation is rainfall. Rainfall has a direct impact on soil erosion. As we will see later, raindrops on soil surfaces are responsible for the initial detachment of soil particles, called rainsplash erosion. The intensity of the rain and size of the raindrop is, therefore, very important for the erosion process. Raindrops could be as large as 7 mm in diameter, but not necessarily spherical in shape. Raindrops larger than 5 mm are not very stable and usually split in the air. The velocity of raindrops depends on their size, with larger particles having higher velocity and, therefore, exerting larger forces and energy to the soil. The terminal velocity of raindrops typically varies from 5 to 9 m/s for 1- to 5-mm raindrop size, respectively.

3.3.1 Rainfall Measurement

Rainfall is traditionally measured with rain gauges. To obtain spatial variation of rainfall, measurements from many rain gauges need to be combined because measurements from these rain gauges only reflect point measurements. In areas with high spatial variation, a very dense network of rain gauges needs to be used. With the recent advancements in radar technology spatially varied rainfall distribution of an area can easily be obtained.

There are two types of rain gauges: recording and nonrecording. *Nonrecording rain gauges* are simple wedge- or funnel-shaped containers that collect rainfall over a specific period of time. Only information on rainfall depth during that time span can be obtained from such rain gauges. Nonrecording rain gauges do not record and provide rainfall intensity and the timing of the rain event. They are relatively cheap and easy to maintain. The *recording rain gauges* supply more information on the rainfall event. There are two common types: *weighing* and *tipping bucket*. The weighing type usually collects rainfall through a funnel and record the weight as a function of time. A tipping bucket consists of a container with a funnel at the top leading to a pair of small buckets. When one bucket fills, it tips, emptying the water and moving the other bucket beneath the outlet of the funnel. The time at which the bucket tips is recorded.

The U.S. National Weather Service employs a network of 158 operational Doppler radars, called Next Generation Radar (NEXRAD) to generate spatially distributed precipitation data for the entire United States (also called Weather Surveillance Radar 88 Doppler, WSR-88D). Individual radars send signals and measure reflectivity of raindrops, snow, and other forms of precipitation. These reflectivity measures are then converted to precipitation using various algorithms. Precipitation estimates from an individual radar is called Level I product. The maximum range of these radars is up to 250 miles. Therefore the service areas of individual radars often overlap. In Level II products precipitation estimates from several radars are combined to create a mosaic of precipitation estimates. These mosaics of spatially varied precipitation data are further improved with data from local rain gauges to produce Level III products. Level III products are also combined for the entire conterminous United States to produce "CONUS" NEXRAD products.

3.3.2 Mean Areal Rainfall Depth

There are two methods of estimating mean areal rainfall depth from rain gauges: The *Thiessen polygon method* and the *isohyetal method*. The former is much simpler and easier to apply. The latter is more accurate, however, it requires more data.

Thiessen Polygons Method

Figure 3.3 depicts this procedure when data from four rain gauges are available. First, the rain gauges are connected by a straight line. Then, bisection lines are drawn for the lines connecting the rain gauges. When the bisection lines are allowed to intersect, each station will be surrounded by a unique polygon. The mean areal rainfall depth is then given by the following relationship:

$$P_{avg} = \frac{1}{A}\sum_{i=1}^{n} A_i P_i \qquad (3.1)$$

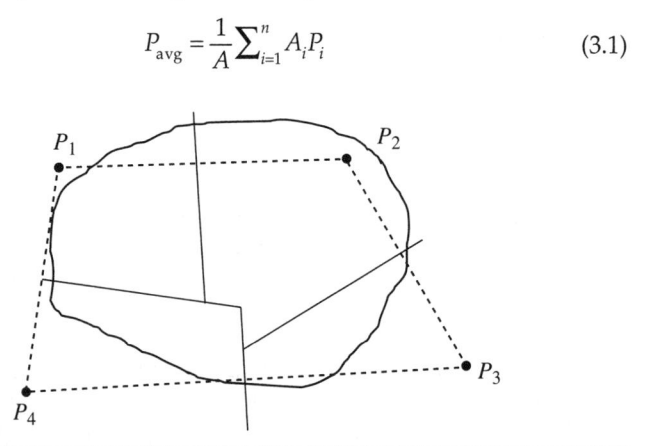

FIGURE 3.3 Schematic representation of the Thiessen polygon method.

where P_{avg} is the mean areal precipitation, A is the total area of the region (e.g., watershed), A_i is the area of the polygon encompassing station i, P_i is the rainfall depth at station i, and n is the total number of rain gauges.

Isohyetal Method

An isohyetal map is a map showing lines of equal precipitation (isohyets) similar to contour lines in a topographic map.

Figure 3.4 illustrates a hypothetical isohyetal map. The mean rainfall depth for the given watershed in the figure is computed as follows:

$$P_{avg} = \frac{1}{A}\sum_{i=1}^{n} A_i \frac{(P_{i-1} + P_i)}{2} \quad \text{(assume } P_0 = P_1 \text{)} \tag{3.2}$$

3.4 Evaporation and Evapotranspiration

Evaporation and transpiration are among the most vital processes in the hydrologic cycle. About 61 percent of all the precipitation falling on the lands evaporates back to the atmosphere (Chow et al. 1988). In semiarid lands this situation is more serious; about 96 percent of the annual precipitation evaporates. In other words at most 4 percent of the annual precipitation is available for use.

Evaporation is the processes where liquid water particles turn into vapor. In reality, water molecules at the surface are in constant movement. Although some water molecules move up from the surface to the atmosphere, some move down from the atmosphere to the surface. Net evaporation occurs when more molecules are lost from the surface to the atmosphere. Evaporation may occur from water bodies, saturated soils, or from unsaturated soils.

Transpiration is the loss of water from plants to the atmosphere through openings on their leaves (stomata). Plants have stomata for

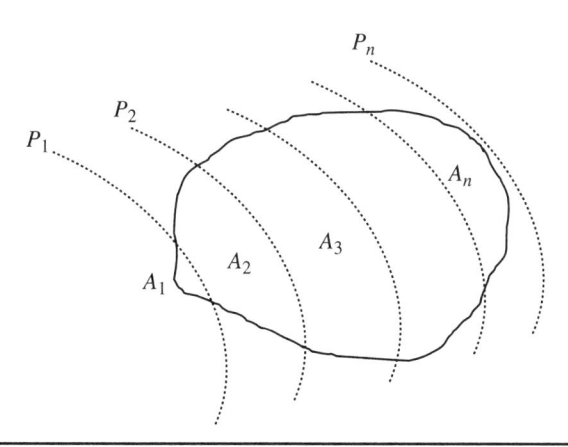

FIGURE 3.4 Schematic representation of the isohyetal method.

carbon uptake from the atmosphere. When stomata are open, plants transpire water. Almost 95 percent of the water taken by the roots is lost back to the atmosphere by transpiration. Water loss from the leaves causes the osmotic potential of cells to become negative. This increases the driving force from stem to leaves and stimulates closure of stomata.

Water yield could be increased if water loss through evaporation could be minimized without jeopardizing other natural resources. One such approach is the application of some solvents to lake surfaces to minimize water losses through evaporation. Forest cutting has shown to increase water yield significantly. Motivated by this fact, some scientists propose removal of massive forest areas to increase stream flow in rivers. However, we should also look at the other side of the coin. In spite of an increase in water yield, forest removal could have significant adverse impacts on the environment, the most obvious one being increased soil erosion. Damage to wildlife habitat and ecosystem is another aspect that also needs to be considered. The irrationality of this idea is obvious.

3.4.1 Some Relevant Concepts and Terminology

Sensible Heat Portion of internal energy proportional to temperature. This is the heat that can be sensed by contact.

Specific Heat Capacity (C_p) Measure of how internal energy changes with temperature. It is often called *specific heat*. It represents the increase in internal energy per unit mass and per unit increase in temperature. Mathematically, it can be defined as

$$C_p = \frac{\Delta E_u}{m \Delta T} \tag{3.3}$$

where E_u is the internal energy [ML^2/T^2], m is the mass [M], and T is the temperature. The specific heat of water at 20°C is approximately 4.2×10^3 J/(kg·K) or, equivalently, 1 cal/(g·K).

Latent Heat (λ) Portion of internal energy that cannot be sensed. This energy is responsible for phase changes. In other words, latent heat is the amount of internal energy either absorbed or released during phase change at a constant temperature. Because evaporation is the phase change from liquid form to gaseous form, it requires energy added to the water. Typical values of latent heat for water are

> Latent heat of vaporization (λ_v) = 2.45 × 10^6 J/kg (at 20°C)
> Latent heat of melting (λ_m) = 3.34 × 10^5 J/kg (at 0°C)
> Latent heat of sublimation (λ_s) = $\lambda_v + \lambda_m$

Solar Radiation This is the energy from the sun reaching the earth. Solar radiation is the main driving source for evaporation. When radiation hits a surface some part is reflected back and some part is absorbed. The fraction of reflected radiation is called *albedo* (α). For fresh snow it is about 0.90, and for deep waters it is about 0.06. If the incoming solar radiation is denoted as R_i, then the net radiation, which is the net input at the surface, is given by

$$R_n = R_i(1-\alpha) - R_e \qquad (3.4)$$

where R_e is radiation emitted by the body. All bodies emit radiation, which depends on their temperature. According to the Stefan–Boltzmann law

$$R_e = e \cdot \sigma \cdot T^4 \qquad (3.5)$$

where e is the emissivity (1 for a black box, 0.97 for water surfaces), σ is the Stefan-Boltzmann constant (5.67×10^{-8} W/m² K⁴), and T is the absolute temperature (K). The incoming radiation R_i is function of scattering in the atmosphere, absorption by clouds, latitude, day of the year, and time of the day.

3.4.2 Energy Balance Method for Estimating Evaporation

Consider the control volume shown in Fig. 3.5. In the figure, R_n is the net solar radiation reaching the control volume [M/T³], H is the heat conducted from the hot surface to the air [M/T³], G is the heat conducted from the hot surface to the soil [M/T³], E_L is the latent heat flux due to evaporation [M/T³], and Q is the amount of heat energy stored within the control volume. Applying the conservation principle to the control volume, we can write

$$\frac{dQ}{dt} = R_n - G - H - E_L \qquad (3.6)$$

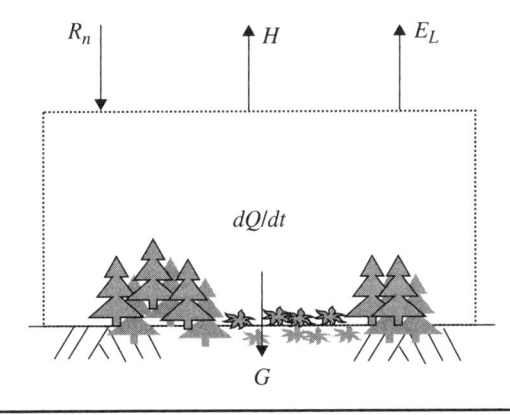

FIGURE 3.5 Control volume of a small landscape for energy balance.

Evaporation (E) is related to latent heat flux, E_L by $E = E_L/(\rho_\omega \lambda_v)$, where ρ_ω is density of water [M/L³]. Substituting this into Eq. (3.6) and rearranging results in

$$E = \left(R_n - G - H - \frac{dQ}{dt}\right) \bigg/ \rho_\omega \lambda_v \tag{3.7}$$

The ratio of sensible heat flux (H) to latent heat flux (E_L) is called the *Bowen ratio* (B) (i.e, $H = B E_L$). Substituting this ratio results in

$$E = \frac{\left(R_n - G - \dfrac{dQ}{dt}\right)}{\rho_\omega \lambda_v (1 + B)} \tag{3.8}$$

The benefit of using the Bowen ratio in evaporation computations is that B can be estimated independently as a function of temperature, vapor pressure, and air pressure with empirical relationships.

3.4.3 Dalton's Law

Dalton's law is a simple way of computing evaporation from free water surfaces:

$$E = C(e_s - e_a) \tag{3.9}$$

where E is the evaporation rate in millimeters per day, C is a constant, and e_s and e_a are saturation vapor pressures at the temperature of the water surface and actual vapor pressure of the air, respectively, both in kilopascal. For a given air temperature, there is a maximum moisture content that air can hold. The air vapor pressure at this stage is called *saturation vapor pressure*. At this pressure, evaporation and condensation rates are equal. It can be approximated as a function of temperature by the following equation:

$$e_s = 0.611 \cdot \exp\left(\frac{17.27T}{237.3 + T}\right) \tag{3.10}$$

in which e_s is in kilopascals and T is in degrees Celsius. This equation is valid for temperatures ranging from 0 to 50°C. The ratio of actual vapor pressure to saturation vapor pressure is the relative humidity (R_h):

$$R_h = \frac{e_a}{e_s} \tag{3.11}$$

The constant C in Eq. (3.9) can be estimated by $C = 112.5 + 25.1 u_{7.6}$ for shallow ponds and $C = 82.6 + 18.5 u_{7.6}$ for small lakes and reservoirs, where $u_{7.6}$ is the wind velocity at a height of 7.6 m above the water surface in meters per second (Meyer 1942). The unit for C in these equations is millimeters per month.

3.4.4 Pan Evaporation

Evaporation from free open-water surfaces is commonly measured by evaporation pans. The class A pan is the accepted standard by the U.S. Weather Bureau. These pans are 47.5 in (121 cm) in diameter and 10 in (25.4 cm) in depth. Evaporation estimates from these pans overestimate actual evaporation by 20 to 40 percent as energy enters the pan from through sides and bottom. The evaporation values obtained from these pans are typically multiplied by a correction factor of 0.7 for a more accurate estimation of actual evaporation from free open-water surfaces.

3.4.5 Evapotranspiration

Evaporation and transpiration are often combined together for convenience and termed evapotranspiration (ET). There are various methods to determine evapotranspiration (Fangmeier et al. 2005):

- Tank and lysimeter experiments
- Field plot experiments where the quantity of water applied is controlled to prevent deep percolation and runoff is measured
- Soil water studies
- Analysis of climatological data
- Integration methods where water use by plants and evaporation from water and soil surfaces are combined
- Water balance method for large areas over long time periods

Estimation of ET is important for crop management, especially irrigation. Soil moisture content is a function of water supply in forms of rainfall or irrigation, and ET. During dry periods, soil water deficit can be estimated and irrigation can be planned accordingly by estimating ET. ET depends on many factors such as amount of solar radiation reaching the surface, amount of wind, aperture of stomata, soil water content, soil type, and plant characteristics, among others (Ward and Trimble 2003). For simplification, researchers introduced the *potential evaporation* (E_p) and *potential evapotranspiration* (ET_p) concepts. Jensen et al. (1990) defines E_p as "evaporation from a surface when all surface–atmosphere interfaces are wet so there is no restriction on the rate of evaporation from the surface. The magnitude of E_p depends primarily on atmospheric conditions and surface albedo but will vary with the surface geometry characteristics, such as aerodynamic roughness." Penman (1956) was the first to define ET_p, and he defined it as "the amount of water transpired in unit time by a short green crop, completely shading the ground, of uniform height and never short of water." In general, actual ET is different than ET_p, and we need the actual value, not the potential one. In practice, ET_p is calculated first and actual ET is

Method	T^*	$R_h^\dagger$ or $e_a^\ddagger$	Latitude	Elevation	$R_s^\S$	$u^\P$
Penman (1948)	▲	▲		▲	▲	▲
Jensen-Haise (1963)	▲			▲	▲	
Blaney-Criddle (1950)	▲		▲			
Thornthwaite (1948)	▲					

*Air temperature; †relative humidity; ‡actual vapor pressure of the air; §solar radiation; ¶wind speed.

TABLE 3.1 Minimum Climatic Data Needs of Some Common ET Estimation Methods (*Adapted from Ward and Trimble 2003.*)

computed by multiplying ET_p with a crop coefficient. Crop coefficients are developed by statistical analysis of observed data.

There are various methods for estimating ET. What method to use usually depends on the available weather data for the site of interest. Table 3.1 lists some of the available methods and the minimum climatic data needs. ET is computed for a reference crop, which is usually either grass or alfalfa. This computed ET is called *reference crop* ET. Then, crop coefficients that were developed for the study area are used to compute the actual ET for the desired crop. Detailed description of ET estimation methods listed in Table 3.1 is rather lengthy. We refer interested users to Jensen et al. (1990) and Ward and Trimble (2003).

3.5 Infiltration and Runoff

Infiltration and runoff are two essential parts of the hydrologic cycle. A portion of rain reaching the soil (i.e., rainfall minus interception losses) is partitioned into two parts. Depending on rainfall intensity, and soil characteristics and conditions, some part infiltrates and the excess part that does not infiltrate runs off. Infiltrated water supplies the water needed by plants and replenishes the soil moisture. Some of the infiltrated water may further percolate and recharge the groundwater. Infiltrated water may also move horizontally in shallow soils and produce *subsurface stormflows*. Surface runoff may discharge into creeks, streams, rivers, lakes, reservoirs, or any other waterbodies. Both very high and very low runoff could have undesirable consequences, such as drought and flooding. Subsurface stormflows could also have adverse impacts on water quality. Low-flow conditions may alleviate pollutant concentrations. Similarly, high runoff rates accelerate erosion and cause soil loss and higher turbidities in waterbodies.

3.5.1 Infiltration

Infiltration is the process of water penetrating into the soil. The infiltration rate is affected by many factors such as soil moisture content (antecedent water content), soil physical characteristics (hydraulic conductivity, porosity, etc.), rainfall intensity, and condition of soil and vegetation cover.

Water movement in soil is governed by the *Darcy's law*:

$$q = -K \frac{dh}{dz} \quad (3.12)$$

where q is the water flux or discharge per unit area [L/T], K is the hydraulic conductivity [L/T], h is the total head [L], and z is the vertical distance positive upward [L]. The negative sign indicates that flow direction is from higher head to lower head. *Infiltration rate* (f) is the flow rate at which water enters the soil per unit area. It has the same unit with rainfall intensity (i) (i.e., length per unit time [L/T]). *Potential infiltration rate* or *infiltration capacity* (f_p) is the infiltration rate when there is no shortage of water supply to the soil. In other words, when rainfall intensity exceeds the infiltration rate (i.e., $i > f$), then $f = f_c$. Conversely, when $i < f$, then $f = i$. *Cumulative infiltration* (F) is the total volume of water infiltrated per unit area from the onset of the infiltration process up to time t. It has the unit of length [L]. Mathematically, cumulative infiltration is related to infiltration rate by

$$F(t) = \int_0^t f(\tau) d\tau \quad \text{or} \quad f(t) = \frac{dF(t)}{dt} \quad (3.13)$$

A theoretical infiltration curve is shown in Fig. 3.6. The infiltration rate is initially extremely high. In the figure, the infiltration rate at time $t = 0$ is shown as f_0. At the initial stages of infiltration, gravity and

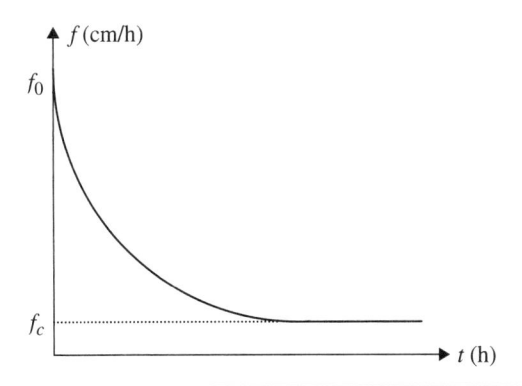

FIGURE 3.6 Theoretical infiltration curve.

soil capillary forces act together. As time goes on, the capillary effect slowly diminishes as soil pores get filled with water, and gravity forces start taking over. As time proceeds, the infiltration rate decreases rapidly, exhibiting almost an exponential–decay–function shape. The limiting infiltration rate at infinity, assuming unlimited water supply, is shown as f_c. At this point, soil becomes saturated and infiltration rate is driven only by gravity forces.

Infiltration rates measured in the laboratory have little value. Therefore, in-situ tests are developed to measure infiltration. The most common method is the use of *double-ring infiltrometers*. As its name suggests, this method requires installation of two circular rings into the soil, one inside the other. Water is filled in the inner circle and in the area in between the two rings. The outer ring serves as a control so that water in the inner circle only infiltrates vertically. The drop in water level is measured with time. Best results are obtained when the soil moisture is at field capacity.

Infiltration Models

Kostiakov Model   Mathematically, this is the simplest model. Kostiakov (1932) proposed the following equation:

$$F = at^b \tag{3.14}$$

where F is cumulative infiltration [L] at time t, and a and b are constants. Parameters a and b need to be estimated from observed infiltration data by calibration. Kostiakov's model should only be used for small times because the model does not account for the limiting infiltration rate (f_c in Fig. 3.6).

Horton Model   Horton's model (1940) represents the infiltration process better than Kostiakov's model does. However, the model parameters still have to be estimated from experimental data. The form of the equation is

$$f = f_c + (f_0 - f_c)e^{-kt} \tag{3.15}$$

The parameters are as defined in Fig. 3.6, except for k, which is the decay constant [1/T] and depends on soil and initial soil moisture content. Cumulative infiltration can be obtained by simply integrating Eq. (3.15) from 0 to t.

Philip Model   In contrast to the two previously listed empirical models of Kostiakov and Horton, Philip (1957) derived his equation from theoretical analysis based on soil physics:

$$F = S\sqrt{t} + Kt \tag{3.16}$$

where S is the sorptivity [L/T$^{1/2}$] and K [L/T] is the conductivity. Taking the derivative of F with respective to t gives

$$f = \frac{S}{2\sqrt{t}} + K \tag{3.17}$$

The first term on the right-hand side represents the suction effect, and the second term represents the gravity effect. For horizontal infiltration, suction is the only force; therefore, the last term should be dropped.

Green–Ampt Model This is a more physically based approach than the remaining models with an analytical solution (Green and Ampt 1911). Without going into details of derivation, the G–A model can be written as

$$f(t) = K\left[\frac{\Psi \Delta \theta}{F(t)} + 1\right] \tag{3.18}$$

where ψ is the wetting front suction head [L], $\Delta\theta = \theta_s - \theta_i$, θ_s is the saturated water content or porosity ($= \phi$), and θ_i is the initial or antecedent water content. In Eq. (3.18), $f(t)$ depends on $F(t)$. Therefore, to solve for $f(t)$ we first need $F(t)$. $F(t)$ can be obtained by using the relationship $f(t) = dF(t)/dt$ and integrating both sides:

$$F(t) = \Psi \Delta \theta \cdot \ln\left[1 + \frac{F(t)}{\Psi \Delta \theta}\right] + Kt \tag{3.19}$$

Note that this is a nonlinear equation in F, which can only be solved iteratively. A good starting point at the beginning of the iteration is to assume that $F = Kt$. Rawls et al. (1982) provide summaries of various soil parameter values for various soil types. Using Rawls et al. (1982) as the source, Chow et al. (1988) reported values of parameters K, ϕ, and ψ for various soil types in their widely used book *Applied Hydrology* (Chow et al. 1988).

A very important detail is that derivations of the foregoing equations are based on the assumption that there are fully ponded conditions of small water depth. Ponding occurs when the rainfall intensity is higher than the soil infiltration capacity ($i > f$). If ponding never occurs, then the infiltration rate is simply equal to the rainfall intensity. Solution for mixed conditions of ponding/nonponding is more complicated and is not covered here. Interested readers on this topic are referred to Chow et al. (1988).

3.5.2 Runoff

Runoff is the portion of precipitation flowing on overland planes after interception infiltration and evaporation losses are met. Runoff is especially important for the erosion process because sheet erosion is driven by runoff water. Runoff water sometime can infiltrate back into the soil when it flows over soil with very high hydraulic conductivity. This process is called *runon*.

There are various mechanisms of runoff generation. The most widely known one is the *infiltration excess runoff*. As discussed earlier, this happens when rainfall intensity is higher than soil infiltration capacity. The portion of rainfall not infiltrating into the soil becomes overland flow. Runoff can also be generated when rain falls on fully saturated areas, in which case all the rain falling on these wet areas becomes runoff. This type of runoff is called *saturation excess runoff*. This usually happens near streams. During a rain event, areas near the streams get saturated first. As the rain continues, the extent of these saturated area become larger. Once rainfall ceases, these saturated areas recede back and eventually diminish. The concept of the expansion and contraction of saturated areas during a rain event is termed the *variable source area* concept. Another way runoff can be generated is when shallow groundwater or water moving horizontally in the shallow soil horizon, that is, *subsurface storm flow* intersects the soil's surface.

Two quantitative measure of runoff are extremely important, both from soil and water conservation points of view: flow volume and maximum flow rate or peak flow. Next we summarize two most commonly used methods in estimating these quantities.

3.5.3 SCS Curve Number Method

The Soil Conservation Service (SCS) curve number method is probably the most widely used method in estimating runoff volume. It is simple, yet it has been shown to be very effective. Many popular watershed scale hydrologic models employ this method such as the *Soil & Water Assessment Tool* (SWAT), the *Agricultural Non-Point Source Pollution Model* (AGNPS), and the *Erosion Productivity Impact Calculator* (EPIC).

The SCS curve number (CN) method was developed by the U.S. Department of Agriculture (USDA) and is described in detail in their technical report TR-55 (U.S. Soil Conservation Service 1986). Total precipitation (P) is portioned into components of initial abstractions (I_a), excess precipitation (P_e), and continuing abstraction (F_a). The method assumes that the ratio of F_a to the potential maximum retention (S) in a watershed is equal to the ratio of P_e to the maximum potential runoff ($P - I_a$):

$$\frac{F_a}{S} = \frac{P_e}{P - I_a} \tag{3.20}$$

Rearranging this and using the equality $P = I_a + P_e + F_a$ along with the assumption $I_a = 0.2S$ leads to

$$P_e = Q = \frac{(P - 0.2S)^2}{P + 0.8S} \tag{3.21}$$

Note that the depth of the excess rainfall is equal to runoff volume, Q (depth and volume are often used interchangeably in hydrology as they can be easily converted to each other by the relationship depth = volume/area). The approximation $I_a = 0.2S$ was obtained by analyzing data from many small watersheds. SCS further defines a CN variable ranging from 0 to 100 related to S with the relationship

$$S = \frac{1000}{CN} - 10 \qquad (3.22)$$

The unit for S obtained using this relationship will be in inches. The CN is higher for less permeable soils, approaching 100 for totally impervious surfaces and lower for more pervious soils. The TR-55 handbook has comprehensive tables listing CN values for various soil- and land-cover combinations under varying hydrologic conditions. These values are for normal antecedent moisture conditions denoted as CN_2. The CN values for dry (CN_1) and wet (CN_3) conditions can be computed from CN_2 using the following relationships:

$$CN_1 = \frac{4.2\, CN_2}{10 - 0.058\, CN_2} \qquad CN_3 = \frac{23\, CN_2}{10 + 0.13\, CN_2} \qquad (3.23)$$

Dry/wet conditions are determined based on the total rainfall that fell during the previous 5-day period (P_5). If $P_5 < 0.5$ in in a dormant season or < 1.4 in in a growing season, then it is a dry condition. If $P_5 > 1.1$ in in a dormant season or $P_5 > 2.1$ in in a growing season, then it is a wet condition. Otherwise, it is a normal condition.

3.5.4 Rational Method

Peak flow rates are required for design of culverts, drainage works, soil conservation works, spillways of farm ponds, and small bridges. The rational method is the most widely used and the simplest technique in estimating peak flows from small watersheds (<10 km²):

$$Q_p = 0.278 \cdot C \cdot i \cdot A \qquad (3.24)$$

where Q_p is the peak flow rate in m³/s, i is the rainfall intensity (mm/h), A is the drainage area in km², and C is the runoff coefficient varying from 0 to 1. The runoff ratio represents the ratio of runoff to rainfall. This method assumes that rainfall continues at a uniform intensity with a duration equal to the time of concentration, which is the time required for water to travel from the remotest point of the watershed to the outlet. The runoff ratio C is the least precise variable of the rational method. In reality, C should be function of percent imperviousness, slope, soil characteristics, antecedent moisture condition, proximity of water table, and vegetation cover. For practical purposes, C is often tabulated as a function of few of these variables. At a minimum, C is listed as a function of land use type and the return period of a storm. C is higher for areas with high imperviousness and for high return periods. McCuen (1998) lists C values for type of land use, slope, hydrologic soil group, and storm return period.

3.6 Soil Erosion and Sediment Transport

Soil erosion is the process of the detachment of soil particles from the soil due to the forces exerted by raindrops, flowing water, and wind. In the United States, soil has been eroded at about 17 times the rate at which it forms: about 90 percent of U.S. cropland is currently losing soil above the sustainable rate. The situation is even worse in Asia, Africa, and South America, where soil erosion rates are estimated to be about twice as high as in the United States. In addition to the loss of fertile soils, soil erosion also has important implications for water quality and water resources. Often, eroded sediments carried by rivers to reservoirs results in a reduction in reservoir capacities. As a matter of fact, in highly erodible lands, reservoir life is dictated by the rate of sediment transport to the reservoir. Various pesticides, radioactive materials, and nutrients attach themselves to the sediments and are transported by the sediments. Therefore, management and conservation practices geared toward sediment reduction also helps to mitigate some problems in water quality.

Sediment sources could be natural or artificial. Natural sources are mainly upland areas where erosion is largely due to overland flow or erosion of ephemeral gullies. Sheet erosion results in removal of a fairly uniform layer of sediment from an area, whereas rill erosion is restricted to concentrated channel flows. Large runoff events, as in the case of floods, can lead to massive amounts of sediment removal. Artificial sources are primarily created by human activities, among which agricultural tillage has a strong influence on erosion. Highway construction, timber cutting, mining, urbanization, land development for recreational use, and animal grazing also contribute to fresh sediment sources to varying degrees. Large channels within a watershed not only transmit sediment but may also act as a sediment source because of erosion from streambeds or banks.

There are two important time scales associated with sediment movement. Depending on geomorphologic properties, nature of the sediment source, and size of the storm, sediment may move from the source-region to the watershed outlet in a single event. In such instances, the time scale is fairly short and limited to the duration of the surface runoff over the watershed. The longer time-scale problem arises when sediment travels more slowly over the watershed. Each rainfall event moves the erodible sediment closer to the watershed outlet. It may take several rainfall events before sediment that was initially eroded several events ago finally exits the watershed.

In this section we talk only about rainfall and water erosion. Although wind erosion could be significant in some areas, it will not be covered here.

3.6.1 Mechanics of Soil Detachment and Deposition

Soil erosion and sediment transport is a complex process. It involves detachment of soil particles, their transport, and sediment deposition. The complexity arises from the fact that these processes are not mutually exclusive. Journey of a soil particle may follow many paths. A particle detached from the soil may be picked up by flowing water to the stream, transported to another location, and stay suspended until it leaves the watershed outlet. Alternatively, the same particle may be deposited somewhere along the way on an overland flow plane or in a streambed. It may be later picked up again by flowing water during another storm event and deposited somewhere else or not. Therefore, the rate of soil erosion from upland areas is not necessarily a surrogate for sediment yield. *Sediment yield* is the amount of sediment leaving a watershed over a specific time period. The ratio of sediment yield to gross soil erosion is smaller than unity and is called the *sediment delivery ratio*.

Soil detachment could be triggered either by rainfall or by flowing water. Production of eroded soil by the rainfall impact is called *rainsplash erosion*. Water from raindrops, in addition to providing energy for soil detachment, also acts as a wetting source for soils. As soil becomes wetter, its shear strength decreases, and it becomes easier to detach particles from the soil. Maximum soil splash occurs when the land is covered by a very thin layer of sheet flow (Piest et al. 1975). Soil erosion caused by the shear stress exerted on the soil surface by the flowing water is called *hydraulic erosion*. Total erosion is the combination of the two. One important note here is that if the flowing water already contains a significant amount of sediment in it, that is, if it has reached its *sediment transport capacity* (more to come on this concept), then instead of further soil detachment it is possible to observe sediment deposition. Therefore, although rainsplash erosion is always positive, net hydraulic erosion could be either positive or negative depending on the sediment supply from upland areas.

Figure 3.7 is a conceptual diagram of how sediment yield can be computed in a watershed. The conceptualization assumes that the watershed is divided into smaller hydrologic units or cells in a cascading fashion. In this conceptualization, flow and sediment from each unit is routed to a downstream unit. This process continues until flow and sediment reach the watershed outlet. For each unit, theoretical rainsplash and hydraulic erosion is first computed. These two theoretical erosion values are added to erosion supplied from upland units, if there are any. Total erosion is checked against the transport capacity of the unit. If transport capacity has not yet been reached, the computed value represents the total sediment and is transported to the next (downstream) unit. Conversely, if the transport capacity is exceeded, the sediment at the amount of the transport capacity is carried off to the downstream unit. Whether there is any net erosion from the unit depends on the amount of sediment delivered from the upland cells. If that amount exceeds the transport capacity of the cell, there is no more erosion but a net deposition. If the amount is smaller than the transport capacity, there will be net erosion from the unit to supplement the sediment deficit (transport capacity minus sediment supplied by the upstream unit).

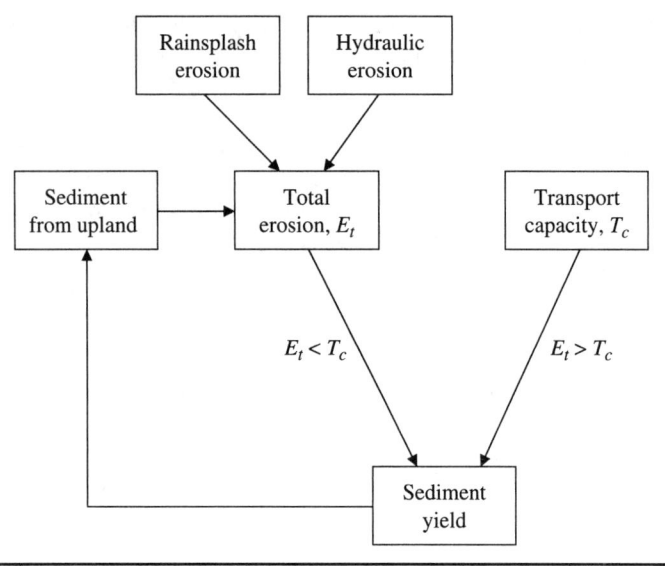

FIGURE 3.7 Conceptual framework for computing soil erosion and sediment yield in a watershed.

3.6.2 Factors Affecting Soil Erosion and Sediment Yield

Many factors affect the rate of erosion and the sediment yield observed at the watershed outlet. Climatic factors, soil texture and features, land use and land cover, topography, hydrology, and geology all affect erosion and sediment transport processes to some extent.

The effect of climate on soil erosion has been partially discussed; rainfall is the main driving force for soil erosion. Raindrop diameter, rainfall intensity, and duration of rainfall event are positively correlated to soil erosion as discussed earlier. Longer rain events increase soil moisture and make the soil more prone to erosion due to buoyancy forces. In addition to the wind itself being a factor for soil erosion, wind direction during rainfall also has a positive or negative impact on soil erosion. Rainfall hitting the soil surface with an oblique angle is known to detach soil particles more easily. Temperature effects erosion through weathering of rocks.

The physical properties of soil, soil composition, and its texture affects soil erosion in various ways. Soils rich in organic matter are known to show high resistance to erosion. Therefore, soils rich in humus have lower erodibility. Compacted and aggregated soils are less susceptible to soil erosion. It is more difficult to detach particles in smaller sizes than larger particles. Yet smaller particles are more easily transported than larger particles. That is, clay particles are more difficult to detach than sand, but clay is more easily transported.

Land use and cover are important factors affecting soil erosion and transport. Vegetation and plant cover reduces soil erosion by holding sediment particles intact and by making the detachment of soil particles either by rainsplash or by sheet flow harder. Some vegetation covers and practices, such as filter strips and riparian buffers, filter sediment from sheet flow, and therefore help mitigate the erosion problem. Soils with no land cover (i.e., barren land), have no erosion protection. By intercepting rainfall, forest canopy and ground cover decrease kinetic energy applied to the soil by raindrops. By transpiring soil water, forest canopy and ground cover increase soil storage capacity and reduce surface runoff. They also provide organic matter to the soil, which makes soils more resistive against erosion. Ground cover slows down sheet flow by increasing resistance against the flow. Land grazing, silvicultural activities, and forest operations all have negative effects on soil erosion. Similarly, construction sites are often sediment sources unless precautionary measures are taken.

Topography effects both soil erosion and sediment transport. Soils on steeper slopes are eroded easier. Furthermore, water flows in steeper slopes faster, resulting in higher shear stress. Watershed areas and drainage density (total length of channels and streams per unit area) known to affect sediment yield. As the watershed area and drainage density becomes larger, sediment delivery ratio gets smaller. This is due to the increased distance between the watershed outlet and eroded areas increasing the likelihood of sediment deposition. Also, smaller headwater watersheds are, in general, steeper compared to larger watersheds.

3.6.3 Types of Soil Erosion

The erosion process in a watershed takes place at four different spatial scales: rill, interrill, gullies, and channels. Interrill and rill erosion takes place in upland (hill slope) areas. Agricultural practices mainly affect these two types of erosion processes. Channel erosion is highly dependent on rill and interrill erosion, which are a source of sediment supply to channels. The amount of sediment supply to the channel dictates whether there will be further erosion or sediment deposition in the channel.

Rill Erosion Rills are areas in overland flow planes where water flows in a concentrated fashion. They act like microchannels. Rills are not permanent features. They may form as a result of one rain event, but another rain event may destroy them. Soil detachment, transport, and deposition in rills are due to concentrated water flow. The amount of erosion in rills depends on the distance between the rills. The greater the distance between rills, the greater the amount of rill erosion and vice versa. Net soil detachment in rills is given by the following relationship (Flanagan and Nearing 1995):

$$E_r = K_r(\tau_f - \tau_c)\left(1 - \frac{Q_s}{T_c}\right) \qquad (3.25)$$

where E_r is the rill detachment rate (kg m² s⁻¹), K_r is the rill erodibility of soil (kg s m⁻⁴) with typical values varying from 10^{-4} to 10^{-1}, τ_f is the hydraulic shear stress (Pa) given by γR_s with γ being the specific weight of water (~9.81 kN m⁻³), R being the hydraulic radius of the rill (m), and s being the slope of the rill flow. τ_c is the critical shear stress below which no rill erosion occurs (Pa), Q_s is the sediment flux in the rill (kg m⁻¹s⁻¹), and T_c is the sediment transport capacity of the rill (kg m⁻¹s⁻¹) and is given by $k_t \tau^{1.5}$. The parameter k_t here is a transport coefficient and generally varies from 0.01 to 0.1.

Interrill Erosion Interrill areas are the regions in overland flow planes between rills. Water flows in these zones as very shallow sheet flow. Soil detachments in rills are mainly due to rainsplash and, to a smaller extent, to sheet flow. Transport and deposition of eroded sediments happen via sheet flow. Soils eroded from interrill areas are carried to rills and from rills to channels. Flanagan and Nearing (1995) estimate interrill erosion as

$$E_i = K_i i q S_f C_v \tag{3.26}$$

where E_i is the interrill erosion rate (kg m² s⁻¹), K_i is the interrill erodibility of the soil (kg s m⁻⁴) with typical values between 10^5 and 10^7, i is the rainfall intensity (m/s), q is the runoff rate (m/s), C_v is the cover adjustment factor ranging from 0 to 1, and S_f is the interrill slope factor given as $S_f = 1.05 - 0.85 e^{-4\sin\theta}$ with θ being the interrill slope angle.

Gully Erosion Gullies are natural deep channels with very steep side slopes, and they carry ephemeral flows; that is, they convey water only during storm events and are dry other times. Formation of gullies is a complex process and is not yet totally understood. Gullies are often a result of accelerated erosion that is triggered by significant changes in the hydrologic regimes of upland drainage areas, such as urbanization and forest clear cutting.

Channel Erosion Channel erosion involves erosion of the streambed, the stream bank, and the flood plain. Streambed erosion produces coarse materials such as sand that moves along the streambed. Fine materials are carried out in suspended form. When the transport capacity of the stream is not exceeded, then the sediment deficit can be replenished by further erosion of the stream banks, which usually consists of fine sediments deposited in previous events and if the flow level exceeds bank-full conditions by erosion of the flood plain. When stream flows contain sediments above their transport capacity, this surplus sediment is deposited to the streambed or the flood plain.

3.6.4 Universal Soil Loss Equation

The universal soil loss equation (USLE) and its later revision, the revised universal soil loss equation (RUSLE), are widely used methods worldwide in estimating annual soil loss from small fields. The USLE was developed from more than 10,000 plot-years of basic runoff and soil loss data. It was originally described in *The Agriculture Handbook* (No. 537) published in 1965 and later revised in 1978 (Wischmeier and Smith 1978). Since then, USLE has been widely used for major conservation planning in the United States and around the world. Although the core formulation of USLE has remained the same, continued research and additional experiments have resulted in improvements of the determining factors and this new version was renamed RUSLE. In 1997, the U.S. Department of Agriculture published Handbook No. 703, which describes the RUSLE model comprehensively (Renard et al. 1997). A Windows-based version of the RUSLE model is available through the Natural Resources Conservation Service (NRCS), called RUSLE2 (Foster et al. 2000).

Both RUSLE and USLE provide estimation of average annual soil loss based on the following simple equation:

$$A = R \cdot K \cdot L \cdot S \cdot C \cdot P \tag{3.27}$$

where A is the average annual soil loss given in tons per acre. Descriptions of each parameter are provided as follows.

R This is the *rainfall–runoff erosivity factor* It is derived from long-term rain gauge data and is the kinetic energy of rainfall multiplied by 30-minute maximum rainfall intensity. Storms having total volumes less than 0.5 in (12.7 mm) often do not generate surface runoff. Therefore, such small events are, in general, not considered in computing R. The R factor shows great variations by location due to spatial variation of rainfall. Renard et al. (1997) provide detailed maps of the continental United States for the R factor. Lal (1994) also provided worldwide estimates of the R factor.

K This factor is relevant to soil and is called the *soil erodibility factor*. This factor aggregates many features of soil that could affect soil erodibility, such as soil runoff potential, susceptibility of soil to erosion, and soil transport properties. Both fine- (e.g., clay) and coarse-textured (e.g., sand) soils have relatively small K values (0.05 to 0.20). Soils high in clay content show great resistance to soil detachment, although they have high runoff generation potential. On the contrary, sandy soils have very low runoff generation potential even though these soils can be easily detached. Soils rich in silt content have the highest K values (>0.4). Soils high in organic matter content show more resistance to erosion. Renard et al. (1997) tabulated values of K values for various soils. They also provided a regression-based equation to estimate K values as a function of organic matter content, soil structure, soil permeability, particle size diameter, and so on.

LS The L and S factors—the *slope length factor* and the *slope steepness factor*, respectively—represent the effect of topography and are often considered together. When the LS value is taken as a unit, the computed erosion estimate corresponds to the USLE standard plot, which is 72.6 ft (22 m) long and has a slope of 9 percent. The LS factor is used to adjust predicted erosion for different slopes and slope lengths. Slope length here is defined as the distance from the point where overland flow starts to the point where flow is becoming concentrated or sediment deposition is started being observed. According to Renard et al. (1997), slope lengths usually do not exceed 400 ft (~122 m). Slope lengths longer than 1000 ft (~305 m) are not recommended. Soil loss increases more rapidly with slope steepness than it does with slope length.

The L factor can be calculated from

$$L\left(\frac{\lambda}{72.6}\right)^\beta \qquad (3.28)$$

where λ is the slope length in feet and β is a dimensionless number given by

$$\beta = \frac{\sin\theta}{\sin\theta + 0.269(\sin\theta)^{0.8} + 0.05} \qquad (3.29)$$

Similarly, the S factor is given as

$$S = 3(\sin\theta)^{0.8} + 0.56 \quad \text{if } L < 4\text{ m} \qquad (3.30)$$

$$S = 10.8\sin\theta + 0.03 \quad \text{if } L > 4\text{ m, } \tan\theta < 9\% \qquad (3.31)$$

$$S = 16.8\sin\theta - 0.50 \quad \text{if } L > 4\text{ m, } \tan\theta > 9\% \qquad (3.32)$$

C The *cover management factor* is used to account for the effect of cropping, crop rotation, management practices, length of growing season, and tillage practices on erosion rates (Fangmeier et al. 2005). The effect of different conservation plans on the average annual soil loss and its time distribution during construction activities, crop rotations, or other management schemes can be assessed by the cover management factor. Similar to other factors, the C factor also measures erodibility relative to a reference condition, which is an area under clean-tilled continuous-fallow conditions. The C values, again, vary by geographic regions, and most local Natural Resources Conservation Service offices have values of C factors for common crop types in that area. Forests have the lowest C values around 0.001 to 0.002, whereas an area with continuous corn or soybean with no tillage has a C value of 0.1. Conventional tillage in autumn increases C to 0.4. Cover management factors for various crop types under different management practices can be found in Renard et al. (1997).

P The *support practice factor* or the *conservation practice factor* is used to assess the impacts of support practices on the average annual soil erosion. It is the ratio of soil loss with a specific support practice to the corresponding loss with upslope and downslope tillage (Renard et al. 1997). If farming operations are up and down a slope, then P should be set to 1.0 (Fangmeier et al. 2005). If there is contouring only, the P value is approximately 0.5; if there is contour strip cropping, $P = 0.25$ (Fangmeier et al. 2005). The effects of terracing could be better represented with the slope length factor, as terracing shortens the overland flow distance and the slope length.

3.7 Best Management Practices

A best management practice, commonly referred to as BMP, is a practice or combination of practices used in preventing or reducing nonpoint source pollution or flow volume/peak in an economic and effective manner. BMPs reduce pollutant concentrations and loads in runoff by infiltration into the soil, physical infiltration by grass or other vegetation, adsorption onto soil and plants, bacterial decomposition, plant uptake, and sediment deposition (Komor 1999).

There are two types of BMPs: structural and nonstructural. *Structural* BMPs are essentially engineering solutions having physical mechanisms or structures often designed and constructed to trap or filter pollutants from runoff or reduce runoff velocities. *Nonstructural* BMPs focus on preserving open spaces, protecting natural systems, and incorporating existing landscape features such as wetlands and stream corridors into a site plan. There are no physical structures associated with these types of BMPs. Nonstructural BMPs can be achieved through education, management, and development practices. Selection of a BMP is a critical issue. BMPs should be selected by considering such factors as the pollutant to be treated, site conditions (land use, topography, slope, water table elevation, geology, and climate), and, of course, economy.

Nonpoint source pollutions could be dealt with in two ways: prevention/protection and treatment/control. Prevention is the most desirable option because it is easier and more cost effective to stop or avoid a problem before it ever happens. Examples of prevention include low-impact designs (LID), controlled land use/cover, and minimal disturbance such as preservation of wetlands. Although this is the preferred option, it is not always possible to stop or prevent pollution, and undesirable consequences are inevitable. In that case, treatment comes into the picture. Various BMPs, especially structural ones, should be considered. The goal in treatment is to reduce pollutant loads/concentrations to acceptable levels through physical, chemical, or biological processes. Trimming peak flows and storm flow volumes through various detention and retention systems, or increasing infiltration rates by reducing imperviousness, are examples of flow treatment.

Most widely used watershed models are not yet equipped with modeling BMP effects or have limited capabilities in that regard. The

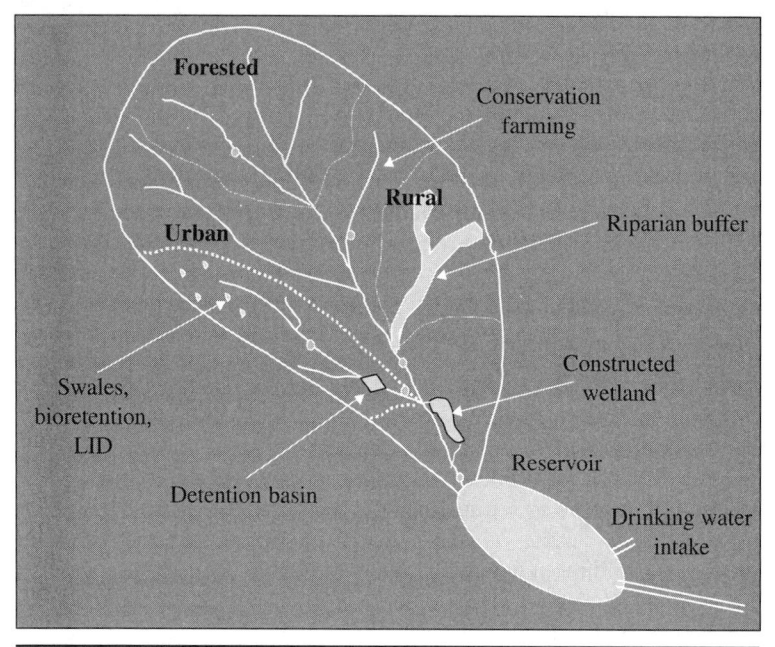

FIGURE 3.8 Simplified schematic of various BMPs on a watershed scale. (*Adapted from U.S. Department of Environmental Protection 2003.*)

Twenty Needs Report of the U.S. Environmental Protection Agency (2002) clearly stresses improved ability of models to evaluate the effectiveness of BMPs to manage many stressors, including suspended solids and sediments. Varieties of BMPs are available to trap sediments and control nutrients on a watershed scale, varying from structural such as wet and dry ponds, vegetative filter strips, riparian buffers, and wetlands to nonstructural such as conservation tillage and improved fertilizer and animal-waste management. Figure 3.8 shows a suite of BMPs located in a hypothetical watershed.

3.7.1 Commonly Used BMPs

There are hundreds of different BMPs available with each being more efficient for certain pollutants and under different conditions. It is beyond the scope of this book to list all of those BMPs and provide descriptions on their features, efficiency, how they function, and so on. Below is a compendium of commonly used BMPs compiled by the Watershed Science Institute (WSI) of the U.S. Department of Agriculture Natural Resources Conservation Service (USDA-NRCS). Detailed information on these BMPs such as their purposes, how they function, limitations, designs, maintenance, and the like can be found at http://www.wsi.nrcs.usda.gov/products/UrbanBMPs/water.html.

Construction Site Impact Reduction (Temporary Practices) Brush barrier, construction entrance/exit, construction sequence, silt fence (filter fence), storm drain inlet protection, straw bale barrier, temporary seeding, topsoiling.

Source Reduction Animal waste collection, bedding, chiseling, cisterns used to harvest water, concrete grid, modular and porous paving, curb elimination, debris removal, drain blockers, education programs, exposure reduction, green roofs, landscape management controls, minimization of pollutants, parking lots and street cleaning, protecting storm drain from hazardous waste deposition.

Erosion Control Contouring, strip cropping, conservation tillage, terracing, channel vegetation, check dam, concrete block revetment, critical area planting, erosion control blanket, gabions/gabion revetment, grade stabilization structure, grassed waterway, level spreader, lined waterway, low wall/slopeface plantings, mulching, netting and matting, native revegetation, sheet flow dispersion, sodding, stone revetment, terracing, water bar.

Water Volume Management Detention basin, dispersion trench, diversion channel, diversion structure (flow splitter), dry well, exfiltration/infiltration, infiltration trench, isolation/diversion structure, off-line infiltration system design, parking lot perimeter, infiltration trench, parking lot storage, small infiltration basin, storm water retention pond, wetland animal habitat design.

Water Quality Treatment and Constituent Entrapment (Vegetative and/or Structural) Basic biofiltration swale, bioretention system, combined infiltration/detention basin, compost filter system, constructed wetland, detention devices for dry/wet ponds, dry extended detention ponds, dry swale, filter strips, median strip infiltration trench, off-line infiltration basin, oil/grit separator, oil/water separators, organic sand filter, peat sand filter, perimeter sand filter, pocket sand filter, reversed elevation systems for parking lots and planting areas, riparian forest buffer, roadway landscape treatment system, sediment basin (water quality enhancement), side-by-side infiltration basin, surface sand filter, underground sand filter, underground trench with oil/grit chamber, under-the-swale infiltration trench, water quality volume storage tank, wet biofiltration swale, wet pond design, wet swale.

More Commonly Used BMPs

Riparian Buffers Limit sedimentation and erosion, provide shade, contribute woody debris for stream health, incorporate floodways, and allow natural stream channel movement. Buffers provide connecting corridors, allowing wildlife to move safely. By slowing down water, buffers enhance sediment deposition and reduce soil erosion. They also maintain bank stability as the roots of vegetation hold the soil. Furthermore, riparian buffers are known to reduce nutrients and bacteria from runoff and septic system effluents in rural and agricultural areas. Because vegetation provides shade, it helps to reduce stream temperatures.

Terraces Terracing is commonly used to reduce runoff and soil erosion. Because terracing requires an additional investment, it should be considered only when conventional nonstructural BMPs are found to be inadequate. Terraces reduce soil erosion by decreasing the slope length of the hill slope. They also prevent formation of gullies by attenuating overland flow (Fangmeier et al. 2005). Water retained behind terraces might serve as a supplemental water supply during dry periods.

Vegetated Waterways and Dry/Wet Swales These are shallow vegetated ditches. Runoff is sent trough a grass filter strip and may be sent through a second BMP for further treatment before being discharged. By slowing down flowing water due to increased roughness, ditches increase infiltration and reduce peak flow. Portion of suspended solids also settle due to reduced flow velocity. Vegetated waterways are used in channels where the soil is very susceptible to soil erosion due to concentrated flow (Fangmeier et al. 2005). Wet swales are preferred when the water table is high. They are most efficient when the drainage area is small, typically less than 4 to 5 hectares.

Detention Basins/Ponds These structures hold water for a certain duration and release. They are helpful in controlling floods, downstream erosion, and sedimentation, and reducing various pollutants. In the United States, areas with large impervious surfaces such as commercial areas, parking lots, airport runways, and the like, are required to have detention ponds to control flow and reduce levels of various water quality constituents, including sediment. Detention basins could be either dry or wet. Wet detention basins drain much slower and often retain a certain amount of pooled water.

Wetlands Wetlands are the kidneys of nature. They are excellent natural filters. By acting like a sponge, wetlands help to mitigate flooding events (even hurricane-generated storms). Wetlands also help in cleaning polluted waters and detaining sediment. In a sense, they are sinks for many chemicals and pollutants. Unfortunately, their benefits have not been realized until the second half of the 20th century. Today, not only are wetlands under protection by law in many countries, they are constructed to treat wastewater.

The American Society of Civil Engineers keeps a database of BMPs where they report various BMP-related statistics (input flow, concentration; output flow, concentration) from some of the studies around the United States (http://www.bmpdatabase.org). Table 3.2 provides a portion of the summary table showing how some of the selected BMPs perform in reducing pollutant.

Constituent	Sample Location	Detention Pond (n = 25)*	Wet Pond (n = 46)*	Wetland Basin (n = 19)*	Biofilter (n = 57)*	Media Filter (n = 38)*	Porous Pavement (n = 6)*
TSS (mg/L)	Influent	72.7 (41.7–103.6)	34.1 (19.2–49.1)	37.8 (18.1–53.4)	52.2 (41.4–62.9)	43.3 (27.3–59.6)	xx
	Effluent	31.0 (16.1–46.0)	13.4 (7.3–19.5)	17.8 (9.3–26.3)	23.9 (15.1–32.8)	15.9 (9.7–22.0)	16.0 (5.90–48.7)
Total P (mg/L)	Influent	0.19 (0.17–0.22)	0.21 (0.13–0.29)	0.27 (0.11–0.43)	0.25 (0.22–0.28)	0.20 (0.15–0.26)	xx
	Effluent	0.19 (0.12–0.27)	0.12 (0.09–0.16)	0.14 (0.04–0.24)	0.34 (0.26–0.41)	0.14 (0.11–0.16)	0.09 (0.05–0.15)
Diss. P (mg/L)	Influent	0.09 (0.06–0.13)	0.09 (0.06–0.13)	0.10 (0.04–0.22)	0.09 (0.07–0.11)	0.09 (0.03–0.14)	xx
	Effluent	0.12 (0.07–0.18)	0.08 (0.04–0.11)	0.17 (0.03–0.31)	0.44 (0.21–0.67)	0.09 (0.07–0.11)	xx
Total N (mg/L)	Influent	1.25 (0.83–1.66)	1.64 (1.39–1.94)	2.12 (1.58–2.66)	0.94 (0.94–1.69)	1.31 (1.19–1.42)	xx
	Effluent	2.72 (1.81–3.63)	1.43 (1.17–1.68)	1.15 (0.82–1.62)	0.78 (0.53–1.03)	0.76 (0.62–0.89)	xx

TABLE 3.2 Median of Average Influent and Effluent Concentrations of Some BMPs (Continued)

Constituent	Sample Location	Detention Pond (n = 25)*	Wet Pond (n = 46)*	Wetland Basin (n = 19)*	Biofilter (n = 57)*	Media Filter (n = 38)*	Porous Pavement (n = 6)*
Nitrate-N (mg/L)	Influent	0.70 (0.35–1.05)	0.36 (0.21–0.51)	0.22 (0.01–0.47)	0.59 (0.44–0.73)	0.41 (0.30–0.51)	xx
	Effluent	0.58 (0.25–0.91)	0.23 (0.13–0.37)	0.13 (0.07–0.26)	0.60 (0.41–0.79)	0.82 (0.60–1.05)	xx
TKN (mg/L)	Influent	1.45 (0.97–1.94)	1.26 (1.03–1.49)	1.15 (0.81–1.48)	1.80 (1.62–1.99)	1.52 (1.07–1.96)	xx
	Effluent	1.89 (1.58–2.19)	1.09 (0.87–1.31)	1.05 (0.82–1.34)	1.51 (1.24–1.78)	1.55 (1.22–1.83)	1.23 (0.44–3.44)

*Actual number of BMPs reporting a particular constituent may be greater or less than the number reported in this table, which was based on number of studies reported in database based on BMP category.

Notes: xx- Lack of sufficient data to report median and confidence interval. Values in parenthesis are the 95% confidence intervals about the median. Differences in median influent and effluent concentrations does not necessarily indicate that there was a statistically significant difference between influent and effluent. See "Analysis of Treatment System Performance," International Stormwater BMP Database (1997–2007) (Geosyntec and WWE 2007) for more detailed information.

Source: International Stormwater BMP Database June 2008 (www.bmpdatabase.org).

TABLE 3.2 Median of Average Influent and Effluent Concentrations of Some BMPs (*Continued*)

References

Blaney, H. F., and Cridle, W. D. 1950. Determining water requirements in irrigated areas from climatological and irrigation data. USDA, SCS-TP 96.

Chow, V. T., Maidment, D. R., and Mays, L. W. 1988. *Applied Hydrology*. New York, N.Y.: McGraw-Hill.

Fangmeier, D. D., Elliot, W. J., Workman, S. R., Huffman, R. L., and Schwab G. O. 2005. *Soil and Water Conservation Engineering*. 5th ed. Florence, Ky. Delmar Cengage Learning.

Flanagan, D. C. and Nearing, M. A. 1995. USDA-water erosion prediction project: Hillslope profile and watershed model documentation. *NSERL Report No 10*. West Lafayette, IN: USDA National Soil Erosion Research Laboratory.

Foster, G. R., Yoder, D. C., McCool, D. K., Weesies, G. A., Toy, T. J., and Wagner, L. E. 2000. Improvements in science in RUSLE2. *ASABE Paper 00-2147*. St. Joseph, Mich. American Society of Agricultural and Biological Engineers.

Gilley, J. E. and Flanagan, D. C. 2007. Early investments in soil conservation research continue to provide dividends. *Transactions of the ASABE* 50(5):1595–1601.

Green, W. H. and Ampt, G. A. 1911. Studies in soil physics I, the flow of air and water through soils. *Journal of Agricultural Science* 4:1–24.

Jensen, M. E. and H.R. Haise. 1963. Estimating evapotranspiration from solar radiation. Journal of Irrigation and Drainage Division, ASCE. 89:15-41.

Jensen, M. E., Burman, R. D. and Allen, R. G. 1990. *Evapotranspiration and Irrigation Water Requirements*. New York, N.Y.: American Society of Civil Engineers. p. 332.

Horton, R. E. 1940. An approach toward a physical interpretation of infiltration capacity. *Soil Science Society of American Proceedings* 5:399–417.

Komor, S. C. 1999. Water chemistry in a nutrient and sediment control system near Owasco, New York. *Water Resources Impact* 1(6):19–21.

Kostiakov, A. N. 1932. On the dynamics of the coefficient of water percolation in soils and the necessity of studying it from a dynamic view for the purposes of amelioration. *Transactions of the 6th Congress of International Society of Soil Science*, part A:17–21 (in Russian).

Lal, R. 1994. *Soil Erosion Research Methods*. Delray Beach, Fla.: St. Lucie Press. p. 352.

McCuen, R. H. 1998. *Hydrologic Anlaysis and Design*. 3rd ed. Upper Saddle River, NJ: Prentice Hall. 888 pp.

Meyer, A. F. 1942. *Evaporation from Lakes and Reservoirs*. St. Paul, Minn.: Minnesota Resources Division.

Penman H. L. 1948. Natural evaporation from open water, bare soil and grass. Proceedings of Royal Society of London, 193:120–146.

Penman H. L. 1956. Evaporation: an introductory survey. *Netherlands Journal of Agricultural Science* 1:9–29, 87–97, 151–153.

Philip, J. R. 1957. The theory of infiltration. Sorptivity and algebraic infiltration equations. *Soil Science* 84:257–264.

Piest, R. F., Kramer, L. A. and Heineman, H. G. 1975. Sediment movement from loessial watersheds. In *Present and Perspective Technology For Predicting Yields and Sources*. Washington, DC: USDA, ARS-9-40:130–141.

Rawls. W. J., Brakensiek, D. L. and Saxton, K. E. 1982. Estimation of Soil Water Properties. *Transactions of American Society of Agricultural engineers* 81–2510: 1316–1328.

Renard, K.G., Foster, G.R. Weesies, G.A., McCool, D.C., and Yoder, D.K. 1997. Predicting soil erosion by water: a guide to conservation planning with the revised universal soil loss equation (RUSLE). *Agriculture Handbook No. 703*. Washington, DC: U.S. Department of Agriculture, p. 404.

Thornthwaite, C. W. 1948. An approach toward a rational classification of climate. *Geographical Review* 38, 55–94.

U.S. Soil Conservation Service. 1986. Urban hydrology for small watersheds. *Technical Release 55*.

U.S. Department of Environmental Protection. 2002. The Twenty Needs Report: How research can improve the TMDL program. EPA841-B-02-002, Washington, DC: Office of Water. p. 43.

Ward, A. D. and Trimble, S. W. 2003. *Environmental Hydrology*. 2nd ed. Boca Raton, Fla.: CRC Press. p. 504.

Wischmeier, W. H. and Smith, D. D. 1978. Predicting rainfall erosion losses: a guide to a conservation planning. *Agriculture Handbook No. 537*. Washington, DC, U.S. Department of Agriculture.

CHAPTER 4
Models for Heat Transfer in Heated Substrates

**Manuel Ramiro Rodriguez and
Maria Dolores Fernandez**
*Universidad de Santiago de Compostela
Galicia, Spain*

4.1 Heat Transfer in Soils

4.1.1 The Soil Thermal Regime

Soil temperature is one of the factors that most influence physical, chemical, and biological processes in soil and plants. Generally, the importance of the effects of soil temperature on plants lies in the role of soil temperature as a controlling factor of the intensity of a series of processes (Buchan 1991; Jury et al. 1991) that usually reach a maximum within a given temperature range (Porta et al. 1999).

Temperature affects plant growth and development to such an extent that plant growth has often been described by a linear approximation dependent on temperature and time (thermal time). This approach can be used to predict seedling emergence time when soil temperature is taken into consideration (Finch-Savage et al. 2001). The dependence between the accuracy of soil temperature estimation and the prediction of the rate of germination has been studied for a number of species, such as *Geranium carolinium* (Washitani 1985), *Pinus densiflora* (Washitani and Saeki 1986), chickpeas (Covell et al. 1986; Ellis et al. 1986), lentils, soybeans, and cowpeas (Covell et al. 1986), millet (García-Huidobro et al. 1982), or soybeans (Andales et al. 2000). In fact, from among the wide range of factors that affect preemergence growth, three are ubiquitous: soil temperature, moisture, and mechanical impedance (Finch-Savage et al. 2001).

Besides emergence, plant development is affected by temperature in later growth phases. Thus, although most crop yield simulations are based on air temperature for thermal time calculations, it has been demonstrated that using temperature at a soil depth of 3 to 5 cm instead of air temperature to simulate maize development increases simulation accuracy (Vinocur and Ritchie 2001). Similarly, Awal and Ikeda (2002) obtained positive correlations between soil temperature and rates of development of phenophases different from germination (leaf appearance, branching, flowering, pegging, and podding).

Respiration and other plant metabolic processes, such as symbiotic nitrogen fixation, photosynthesis, or transpiration, are also affected by temperature. Results reported by some authors who analyzed variations in air and soil temperature and their effects on plants have evidenced that synchronous variation of air and soil temperature benefits nitrate metabolism (Gent and Ma 2000).

Similarly, soil temperature affects water availability to the plant. Water availability is lower in soils at very low temperatures, which affects nutrient uptake and, therefore, the spatial distribution of roots (Villalobos et al. 2002). Moreover, temperature enhances or inhibits soil microbial activity; for instance, nitrification is inhibited at low temperatures, whereas organic matter decomposition increases. Other important processes, such as the biodegradation of pesticides or other chemical and organic compounds, are dependent on soil temperature (Porta et al. 1999).

4.1.2 Factors Affecting Soil Temperature

Factors affecting soil temperature can be classified into environmental factors and factors inherent to the soil. Environmental factors comprise solar radiation, condensation, evaporation, rainfall, and vegetation, while factors inherent to the soil comprise specific heat and thermal conductivity, biological activity, soil texture, soil structure, radiation from the soil into the atmosphere, and soil moisture (Jackson and Taylor 1965). Another classification divides the factors affecting soil temperature into factors that contribute to heat availability and factors that contribute to dissipation of heat (Hanks 1992).

Solar radiation transmits a certain amount of heat to the soil. The amount of heat transmitted is dependent on atmospheric conditions, season, time of day, and latitude. When solar radiation strikes the soil, the soil is heated, reflecting a fraction of solar radiation and transmitting another fraction toward deeper layers. The proportion of these three fractions depends on the following factors:

1. *Albedo*: The fraction of incoming radiation that is reflected at the crop or soil surface. The albedo depends on the nature of the surface, the angle of the sun, and the latitude.

2. *Slope*: The amount of heat that strikes the soil reaches its maximum value when the surface is perpendicular to radiation.

3. *Exposure*: It varies according to topography, latitude, direction, and surface microtopography roughness.

4. *Soil surface color*: Color plays a key role in the absorption and reflection of solar radiation in that dark colors absorb more radiation than light colors.

5. *Variations in specific heat and thermal conductivity*: Particularly associated to water content and composition. Bulk density and moisture content are the factors that most strongly influence heat transfer in soils. Thus, increases in bulk density and moisture content raise temperature because of the increase in soil thermal conductivity and specific heat. The effect of peat incorporation in decreasing soil thermal conductivity and thermal diffusivity is diminished at low soil moisture and bulk density (Ekwue et al. 2005). For that reason, peat soils get warm slowly due to their low thermal conductivity, which results from the air filling the pores of such soils (de la Plaza et al. 1999).

6. *Water content*: The temperatures observed at different depths during a day are strongly dependent on surface phenomena, and on other factors such as soil water content below the top layers (Guaraglia et al. 2001).

7. *Surface cover*: The presence of vegetation, stubble, snow, or other elements that cover the soil surface inhibits solar radiation from reaching the soil. Even for severely compacted soils, the soil thermal regime is very sensitive to surface tilling (Sikora et al. 1990). Tillage in farmlands plays an important role not only as a soil management tool to improve the plant root environment but also as a factor that affects the water cycle and the energy balance in a region. This is because tillage greatly affects soil moisture and soil temperature in the unsaturated soil zone near the surface, which acts as an interface between the subsurface soil layer and the boundary of the atmosphere that is in contact with the surface (Moroizumi and Horino 2002).

Other physical and chemical processes cause variations in heat flow in soils. Such processes include, among others, organic matter mineralization and the wetting and drying of soil colloids. In addition, we must not forget that biological activity alters the amount of heat in the soil during the development of its vital processes.

In greenhouse substrates, where environmental conditions differ from outdoor conditions, other factors must be taken into consideration. First, the fraction of solar radiation that reaches the greenhouse substrate decreases due to the greenhouse covering material. In addition, there can be additional heat supply from warm air or substrate-heating systems. When greenhouse substrates are heated by a heating system, temperature variations are greater than in the soil outdoors and temperature distribution is fully conditioned by the geometry of the heating system. In heated greenhouse substrates, high temperature gradients occur that affect not only temperature and heat flow, but also the moisture profile, with lower moisture values in high-temperature zones (Rodriguez et al. 2006).

4.2 Soil Thermal Properties

Seed germination, crop emergence, and subsequent establishment are affected by the microclimate, which is largely influenced by soil thermal properties (Ghuman and Lal 1985). For this reason, during the last few years, considerable efforts have been devoted to developing techniques that are useful in the determination of soil thermal properties.

4.2.1 Specific Heat or Heat Capacity

The specific heat per volume or volumetric heat capacity of a soil, C_V, is the ability of the soil to store heat per unit volume and unit temperature, and can be expressed as the amount of heat required to raise a unit volume of a soil by 1°C of temperature under isobaric conditions, measured in J m^{-3} °C^{-1}:

$$C_V = \frac{1}{V}\left(\frac{dQ}{dT}\right) \qquad (4.1)$$

where V is volume (m^{-3}), Q is heat (J), and T is temperature (°C).

Similarly, the heat capacity per unit mass, measured in J kg^{-1} °C^{-1}, is defined as

$$C_M = \frac{1}{M}\left(\frac{dQ}{dT}\right) \qquad (4.2)$$

where M is mass (kg).

4.2.2 Effective Thermal Conductivity

The thermal conductivity k is defined as the ability of the soil to transfer heat by molecular conduction and is measured in W m^{-2} s^{-1} °C^{-1} (Porta et al. 1999). However, in porous media, we must consider the apparent or effective thermal conductivity k_e, which includes convection of latent heat and conduction. Because the soil is a granular medium consisting of solid, liquid, and gaseous phases, thermal conductivity is dependent on the evolution of the volumetric proportions of each component, on the size and arrangement of the solid particles, and on the contact between the solid and liquid phases (Jury et al. 1991).

There are large differences among the conductivities of the solid, liquid, and gaseous phases. Thus, the ratio of the thermal conductivities for quartz, water, and air is 333:23:1 (Jury et al. 1991; Porta et al. 1999). Because of such a large difference, the thermal conductivity of granular soils depends largely on the degree of contact of solid particles, which in turn depends on the extent to which air is displaced by water in the pore spaces between the particles (Jury et al. 1991).

Thermal conductivity depends upon many factors, which can be classified into two groups: those that are inherent to the soil and those that can be managed or controlled by humans (Abu-Hamdeh and Reeder 2000). Factors that are inherent to the soil include the texture and mineralogical composition of the soil (Wierenga and de Wit 1970), whereas factors that can be managed externally include water content and soil management (Yadav and Saxena 1973). Some of the factors that affect the thermal conductivity are listed as follows:

1. *Size of the solid particles*: The thermal conductivity of the mineral components is lowered with the decrease in particle size (Patten 1909).

2. *Degree of packing and porosity of the soil*: An increase in the soil bulk density (i.e., a decrease in porosity), improves contact between the particles, and reduces the volume of soil filled with air (van Rooyen and Winterkorn 1959), thus increasing thermal conductivity.

3. *Moisture content*: The presence of water films on the solid particles increases the contacting surface between the particles and displaces the air from the soil pores, which has the lowest conductivity (Jury et al. 1991). At very low moisture content, thermal conductivity first varies negligibly and then begins to increase from a critical moisture content whose value tends to depend on clay mass fraction (Tarnawski and Leong 2000).

4. *Temperature*: Thermal conductivity increases with temperature in wet soils, reaching values that are 3 to 5 times higher at 90°C than at room temperature (Campbell et al. 1994). This assumption is not valid for soils subject to freezing water temperatures, where an increase in thermal conductivity is observed for frozen peat (Kujala et al. 2008).

5. *Salt concentration*: The increase in salt concentration causes a decrease in the thermal conductivity of water (Abu-Hamdeh et al. 2001) and, consequently, of soil.

4.2.3 Effective Thermal Diffusivity

The effective thermal diffusivity describes the soil thermal inertia measured as temperature variation due to heat flow (Porta et al. 1999), in m² s⁻¹:

$$D_T = \frac{k_e}{C_V} \tag{4.3}$$

Because the effective thermal diffusivity is defined as the ratio of effective thermal conductivity to volumetric heat capacity, the value of effective thermal diffusivity increases with temperature when the effective thermal conductivity increases and the heat capacity remains almost constant.

4.2.4 Determination of Soil Thermal Properties

For many years, the determination of soil thermal properties has generated considerable interest among researchers, which evidences the complexity of the process. Two main approaches have been tried: (1) using indirect measurements to experimentally determine soil thermal properties, or (2) estimating soil thermal properties from other soil properties.

Experimental Methods to Determine Thermal Properties

Apart from the calorimetric methods used by de Vries (1963), many indirect methods have been proposed for estimating effective conductivity, specific heat, and thermal diffusivity of the soil. Thermal properties can be determined indirectly by measuring the rise or fall of temperature in response to heat input to a line source at the point of interest (Jackson and Taylor 1965). On this basis, many methods have been developed to determine the soil thermal properties. A classification of indirect methods is shown in Fig. 4.1.

Various authors have used the solution of the heat-conduction equation to indirectly determine the thermal properties of soil, which allows for the determination of thermal properties from quantities that can be easily measured in situ, for example, temperature, with negligible disturbances of the natural state of the soil. The type of equation used (implicit or explicit) and the solution of the expressions used (analytical or numerical) have generated various methods with different requirements, both in terms of data and of data acquisition and processing resources (Horton et al. 1983).

Wierenga et al. (1969) obtained values of soil thermal properties based on laboratory and field measurements. Later, Singh and Sinha (1977) evaluated thermal diffusivity by using a numerical solution to the heat-conduction equation based on Dirichlet-type boundary conditions and fitting the heat-conduction equation for soil by cubic splines. From that moment, exact solutions alternated with approximate solutions to the heat-conduction equation, and the accuracy and possibilities of use of both types of methods were analyzed.

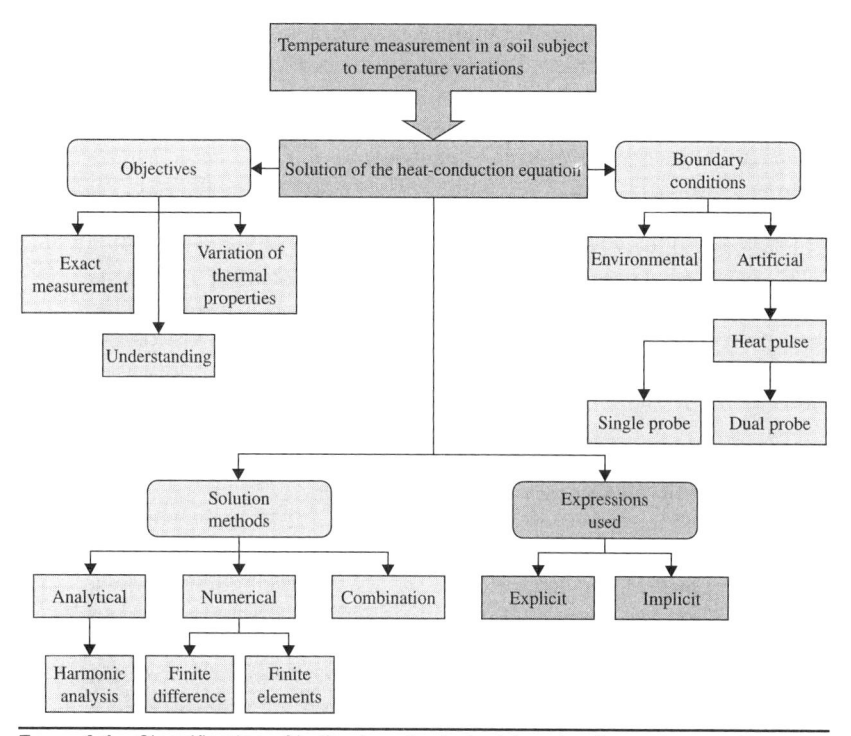

Models for Heat Transfer in Heated Substrates 131

FIGURE 4.1 Classification of indirect methods to determine thermal properties.

After having analyzed six methods for the determination of apparent thermal diffusivity in the upper 10 cm of soil based on temperature, Horton et al. (1983) found that the methods that provided explicit equations for thermal diffusivity (amplitude, phase, arctangent, and logarithm) required fewer data than the methods that provided implicit equations (harmonic and numerical). However, the results of explicit methods were less accurate than the results of implicit methods. In line with the research by Horton et al. (1983), Persaud and Chang (1985) found difficulties in determining the accuracy and the degree of fit of four indirect computation methods for the determination of mean apparent soil thermal diffusivity at two depths. Chen and Kling (1996) estimated apparent thermal diffusivity from temperature measurements taken at three depths during 5-day periods at 3-h intervals. The authors estimated apparent thermal diffusivity values from implicit expressions obtained from numerical solutions to the heat-conduction equation for soil.

Among the authors who worked on the estimation of thermal properties in nonuniform soils, Novak (1986) presented soil thermal admittance values as a power function of depth. He calculated soil thermal admittances using harmonic solutions to the one-dimensional heat-conduction equation for soil, considering values of surface roughness, wind speed, and stability. Thunholm (1988) tested four methods for estimating soil thermal conductivities at six locations in Sweden by using analytical and numerical solutions to the Fourier equation based on annual temperature variations. The analytical methods produced better results under ice and snow conditions, insofar as it was difficult to introduce variations in conductivity with depth in numerical solutions.

Nassar and Horton (1989) applied analytical methods for the indirect determination of thermal diffusivity in uniform soils to nonuniform soils, and demonstrated that these methods were not valid for nonuniform soils. After that, Nassar and Horton (1990) proposed a new method that could be used in nonuniform soils based on a harmonic analysis obtained by Fourier series. The result was a single overall diffusivity value for the whole soil layer studied. Conversely, Hurley and Wiltshire (1993) proposed changes in the expressions for bare and uniform soils that allowed for the extension of the use of such expressions to soils characterized by the variation of thermal properties with depth.

Soil thermal properties have been widely studied in terms of moisture content, texture and structure. Auvermann et al. (1992) presented a method for determining the thermal characteristics (specific heat and effective thermal diffusivity) of soils with heterogeneous moisture contents. This method was based on an explicit finite-difference model of the one-dimensional heat equation, and the Lettau regression procedure was used to obtain thermal diffusivity. Results were faithful to reality and were valid for 24-h temperature measurements and simulations.

The influence of soil structure on soil thermal properties was analyzed by Kaune et al. (1993). Soil structure was altered due to mechanical action and irrigation and drying periods. To determine soil thermal properties, the authors used indirect methods based on harmonic analysis. They found that effective thermal conductivity and effective thermal diffusivity were lower in soils with altered structures as compared to undisturbed soils, whereas the values of specific heat were similar. Noborio and McInnes (1993) studied the influence of soil texture on effective thermal conductivity. They concluded that clay–solution interactions significantly affected both properties because of the microstructural changes in clay soils. Lipiec et al. (2007) assessed the effects of tilled- and grass-covered soil on the spatial distribution of the thermal properties in the vineyard interrow with consideration of areas corresponding to machinery traffic. The mean values of thermal conductivity obtained were generally greater under tilled- than grass-covered moist soil and the inverse was true in drier soil.

The above authors used indirect methods to determine soil thermal properties from ambient temperature variations. Other authors used temperature variations caused by artificial heat input. Kasubuchi (1977) developed a twin transient-state cylindrical-probe method for determining soil thermal conductivity. He applied heat to the soil using a heating cable and then measured the temperature variations near the heat source. He proposed to simultaneously perform the test for two materials: soil and reference material (1.0 percent agar). The representation of the temperatures obtained for both materials allowed Kasubuchi (1977) to determine effective thermal conductivity without solving the heat-conduction equation.

Campbell et al. (1991) proposed the single-probe heat-pulse method, a method that was later used by other authors. Specific heat was determined by placing a line heat source and measuring temperature variations at a short distance from the source. Kaminsky (1994) proposed a very similar method to determine specific heat, effective thermal conductivity, and thermal diffusivity, which was applicable to heat conduction in materials with nonhomogeneous structure. Both methods used analytical solutions of the heat equation for cylindrical geometries with an axial heat source in a homogeneous and isotropic medium at a uniform initial temperature (Abu-Hamdeh and Reeder 2000).

Based on the method proposed by Campbell et al. (1991), Bristow et al. (1994a) developed the dual-probe heat-pulse technique. By measuring the temperature response at a short distance from the line source, and applying short-duration heat-pulse theory, Bristow et al. (1994a) could extract all the soil thermal properties, thermal diffusivity, heat capacity, and thermal conductivity, from a single heat-pulse measurement. In addition, Bristow et al. (1995) presented a computer implementation of the dual-probe heat-pulse technique for laboratory experiments. The results obtained for specific heat, effective thermal conductivity, and thermal diffusivity were identical to the results obtained using the Marquardt algorithm, and the computer implementation approach allowed for rapid and automated monitoring of these parameters.

Indirect heat-pulse methods for determining soil thermal properties, both single-probe and dual-probe methods, are affected by time- and temperature-measurement errors. The influence of noise on the temperature curves required to determine thermal properties using heat-pulse methods has been analyzed by Bristow et al. (1994a, 1994b). Noise problems that affect temperature measurement are particularly relevant in situations in which temperature rises faster (Kluitenberg et al. 1995). Moreover, the dual-probe heat-pulse technique has shown to be a good tool for determining the thermal properties of porous materials insofar as the use of nonlinear least squares curve fitting for temperature data offers highly accurate estimates of such properties (Bilskie et al. 1998).

Heat flux plates are commonly used to measure soil heat flux, a component of the surface energy balance. This method is based on the establishment of a vertical heat flow between a warm and a cold plate where the temperature gradients and the response of a soil heat flux plate are measured. The plate method is simple, but several studies have demonstrated the potential for relatively large errors. The standard-plate method has underestimated the magnitude of the heat flux by 18 to 66 percent depending on the site and type of plate due to a combination of low plate thermal conductivity, thermal contact resistance, and latent heat-transfer effects (Ochsner et al. 2006). The heat flux plate–derived heat fluxes has shown errors of up to 26 percent in coarse porous substrates due to poor contact between plate and substrate (Weber et al. 2007).

Developing reliable low-cost research material for laboratory and field tests is still a priority for many authors who develop research on soil, or in the area of environmental sciences. For this reason, many research tasks concern the development of temperature measurement probes that are suited to the procedures for indirect determination of soil thermal properties. Manohar et al. (2000) designed an experiment to measure effective thermal conductivity in soils. They constructed three thermal probes in accordance with the American Society for Testing and Materials method ASTM D 5334. The temperature–time response was logged at 1-s intervals for 1000 s, which allowed for the estimation of soil thermal conductivities. The relationship between temperature variations and time was used also in the single-probe heat-pulse method developed by Kaminsky (1994). Bristow et al. (2001) developed a small multineedle probe for measuring soil thermal properties that improved cost, robustness, and easy automation of soil instrumentation. Later, Saito et al. (2007) evaluated the effects of sensor locations and thermal properties of the heat-pulse probe-sensor body material, heater diameter and heat-pulse intensity, and vapor flow on system of measurement performance. Their results showed that significantly different temperature responses are obtained depending on the axial location of the thermistors.

Abu-Hamdeh and Reeder (2000) used the single-probe heat-pulse technique to determine the thermal conductivity of soils repacked in the laboratory under different density, moisture, salt concentration, and organic matter conditions. According to their results, effective thermal conductivity increased with increasing bulk density at a given moisture content and with increasing moisture at a given bulk density. Clay soils showed lower effective thermal conductivities than sandy soils, which verified the results obtained by Noborio and McInnes (1993). Abu-Hamdeh (2001) analyzed the influence of mineral salt concentration (NaCl, $MgCl_2$, $CaCl_2$) on the conductivity of soils altered under laboratory conditions and concluded that effective thermal conductivity decreased with the increase in salt contents.

Finally, it has been suggested that indirect methods for determining soil thermal properties from temperature may allow for simultaneous estimation of soil water retention (when combined with matric potential measurements) and unsaturated hydraulic conductivity. Simultaneous estimation of both properties is possible if numerical solutions to the heat-flow equation are used, and heat and mass flows are taken into consideration (Hopmans et al. 2002). In fact, the heat-pulse probe has received more attention as it allows in situ, simultaneous, and automated measurements of soil hydraulic and thermal properties, as well as soil water fluxes. For this reason, a number of authors have analyzed design factors that influence the accuracy of the probe (Ham and Benson 2004; Saito et al. 2007). Combining this technique for estimating soil heat properties, water flux, and water content with a Wenner array measurement of bulk soil electrical conductivity allows for simultaneous measurement of coupled water, heat, and solute transport in unsaturated porous media and provides a better understanding of flow and transport processes (Mortensen et al. 2006).

Determining Thermal Properties from Other Soil Properties

Obtaining exact measurements of the soil thermal properties and analyzing the variation of such properties is very complex. As a result, modeling soil thermal properties from soil data that can be readily measured becomes difficult. Kersten (1949) proposed a purely empirical expression for estimating effective thermal conductivity based on measurements from five soils. This expression determined the value of thermal conductivity from the physical properties of other soils (volume fraction and bulk density) and from three dimensionless empirical factors that needed to be determined for each specific soil.

De Vries (1963) proposed a method for determining the heat capacity and effective thermal conductivity of the soil. The method by de Vries (1963), with successive modifications, has been used in many models of heat transfer in soils because it allows for determining thermal properties from the physical, textural, and structural characteristics of the soil and from values of the physical properties of the soil phases. To estimate soil thermal properties, it has been assumed that the soil consists of three phases—solid, liquid, and gas—all of them contributing to the overall value of the thermal properties of the soil.

Specific heat has been computed as the weighted sum of the thermal capacities of each separate constituent, using the volume fraction of each element as a weighting factor. The specific heat of the solid phase is constant with moisture and is computed from the heat capacity per volume of the minerals that compose the solid phase and of soil organic matter. An increase in moisture reduces the volume fraction of air and subsequently increases specific heat. Considering that the specific heat of air is small as compared to the specific heat of water and solids, this term can be neglected, such that soil specific heat is a function of the volume fractions of water and solid constituents. Consequently, soil specific heat can be determined from the porosity and water content of a soil whose texture is known. The values of specific heat thus calculated corresponded well with the values measured directly by de Vries (1963) with the help of a calorimetry.

To determine soil effective thermal conductivity, de Vries (1963) modeled the soil as a continuous medium in which soil particles are dispersed and considered as regular-shaped granules. In dry soils, air is considered to be a continuous medium, and the solid and liquid phases represent the regular-shaped granules randomly arranged in space. When the soil is wet, water acts as a continuous medium in which air granules and solid particles are arranged. The soil solid phase is made up of an organic fraction and a mineral fraction. Each fraction makes its contribution according to its characteristic value of conductivity. Under these assumptions, the thermal conductivity of each constituent is weighted by a factor that is dependent on the volume fraction of each constituent, on the ratio of the thermal conductivities of the particles, and on the continuous phase and shape of the particles. The weighting factors are calculated taking into consideration the influence of three shape factors, one for each type of particle according to the phase to which it belongs.

For the solid phase, a shape factor of 0.125 has been determined for randomly arranged ellipsoidal granules despite the changes in scale caused by different particle sizes. This shape factor has proved valid for sandy and clay soils. To determine the shape factors for the discontinuous fluid phase, water or air, the model has been subdivided into four moisture regions (Hopmans and Dane 1986):

1. *Dry soil*: Air is a continuous medium. The ratio between the conductivities of the solid phase and the continuous phase is very high and the values reported by de Vries (1963) must be increased by 25 percent.

2. *Soil moisture content from permanent wilting point (PWP) to saturation*: Water is a continuous medium. Because the air shape factor ranges from 0.035 to 0.333 for PWP and saturation moisture contents, respectively, the shape factor suggested for a given moisture content is a linear approximation between both values.

3. *Soil moisture content below PWP*: Progressive drying below the permanent wilting point causes a decrease in the air shape factor from a value of 0.013 to zero moisture content.

4. *Soil moisture content below the critical water content*: The air shape factor is very low and is dependent on soil texture.

The model predicted values of soil effective thermal conductivity with errors usually less than 5 percent, except for interpolation intervals, with errors of approximately 10 percent (de Vries 1975). Johansen (1975) developed another method for determining soil effective thermal conductivity and proposed three submodels according to soil moisture content:

1. *Dry soil*: The conductivity of a dry soil depends mainly on its bulk density. To estimate the thermal conductivity of a dry soil, we can use the expression developed by Johansen (1975), which requires three dimensionless constants similar to the expression developed by Kersten (1949). The dimensionless constants required have been empirically tabulated for some soils.

2. *Saturated soil*: The solid and liquid phases increase the weight of their volume fractions. The effective thermal conductivity is a function of (a) the value of water conductivity raised to a power equivalent to the value of porosity and (b) the value of the conductivity of solid particles raised to a power equivalent to its volume fraction.
3. *Unsaturated soil*: Thermal conductivities can be obtained by linear interpolation between the value of thermal conductivity for dry soils and the value of thermal conductivity for saturated soils.

Campbell et al. (1994) modified the equation proposed by de Vries (1963) to model the variation in soil thermal conductivity for temperatures of up to 600°C. They defined a new term for fluid thermal conductivity, which allowed them to use a single expression for estimating conductivity in the gaseous phase without determining moisture ranges. This model required soil physical properties that were not needed in other models, such as the power for liquid recirculation and the mean geometric diameter of the solid particles. Some of the empirical factors proposed by Campbell et al. (1994) were aimed at characterizing the influence of water content and air on thermal conductivity, and the moisture content at which water or air gained relevance in the heat-transfer process.

Water conductivity is a function of temperature, whereas air conductivity is dependent on the temperature, pressure, and amount of water vapor, which in turn depends on temperature and pressure. Air conductivity is considered to be an apparent thermal conductivity, which is the sum of the actual conduction through the air and the latent heat of distillation across pores in the soil. For vapor diffusivity in air, Campbell et al. (1994) used the expression proposed by Fuller et al. (1966). The slope of saturation vapor pressure according to temperature corresponded to the slope developed by Richards (1971). The conductivity of solids was assumed constant with temperature.

Campbell and Norman (1998) presented a different version of the method by de Vries (1963) for determining specific heat. They suggested modeling the soil heat capacity per unit mass as the weighted sum of the heat capacities of the soil components. The weighting was a function of the density of solid particles, water density, and a dimensionless factor that related the bulk densities of the soil and of the particles.

Ochsner et al. (2001) developed a method to measure the thermal properties of medium-textured soils. They applied the method to four soil types and suggested relationships between the soil thermal properties and factors such as the proportion of air-filled pores, soil water content, or the relationship between the bulk density of the soil and the bulk density of the solid particles. Unlike de Vries (1963), they found a linear relationship between the thermal conductivity and the volume fraction of air in pores.

Gori (1983) developed an empirical model for unsaturated frozen soils that was extended by Tarnawski et al. (2000) and Tarnawski and Gori (2002). Tarnawski et al. (2000) described the development of the Kersten function, which is dependent on soil temperature and on the degree of saturation. The new Kersten function enabled the prediction of the soil thermal conductivity at different moisture contents and temperatures. The function produced good results for dry soils (moisture content below PWP) at temperatures ranging from 30 to 90°C, and for wet soils at temperatures ranging from 30 to 50°C. Tarnawski and Gori (2002) presented a more accurate model to predict thermal conductivity at high temperatures. The model was applicable between 5 and 90°C, with better results for dry soils.

In further studies, a model for determining soil thermal conductivity that was valid for dry soils at PWP moisture content was presented. Such a model improved the estimation of thermal conductivity, such that it could be incorporated into heat-transfer models that included latent heat as a component (Tarnawski and Leong 2000). Later, Tarnawski et al. (2001) developed a software package that allowed for determining the soil effective thermal conductivity at high temperatures (50 to 90°C). The relationship between conductivity and temperature was no longer linear because of the influence of mass transport, which showed strong dependence on the soil hydraulic properties.

Balland and Arp (2005) developed a new method, based on the method by Johansen (1975), to seamlessly calculate thermal conductivity for various soil conditions, from loose to compact, organic to mineral, fine to coarse textured, frozen to unfrozen, and dry to wet. This generalization was fine-tuned empirically with data from soil, gravel, and peat drawn, for frozen and unfrozen conditions from −30 to 30°C, and for variable moisture and bulk density conditions from dry to saturated. Usowicz et al. (2006) obtained thermal conductivities on silt loam in a sloping vineyard from a statistical–physical model. It was shown that the performance of the equations relating the thermal conductivity with penetration resistance and air-filled porosity was greater than it was with penetration resistance and volumetric water content. Zhang et al. (2007) developed a method for estimating the effective thermal conductivity of the soil from thermal properties and volume fractions of the components, considering that the constituent parts of the multiphase medium were all cubic in shape and randomly mixed. In addition, the difference of the thermal characteristics between the frozen and unfrozen soils was discussed. Lu et al. (2007) developed an improved model that described the relationship between thermal conductivity and volumetric water content of soils. Soil thermal conductivity could be estimated using soil bulk density, sand (or quartz) fraction, and water content. Except for the sand, the precision of their model improved considerably the Johansen (1975) model.

4.3 Models for Predicting Soil Temperature

Many models have been developed to predict soil temperature. Leaving aside statistical models, all the models were based on the transient or steady heat-conduction equation for soils. Figure 4.2 is a diagram of the models developed to predict soil temperatures.

4.3.1 Applications of the Models for Predicting Soil Temperature

Models for predicting soil temperature can be classified according to the application for which they were developed. The following factors must be taken into consideration when applying temperature-prediction models: surface cover, soil uniformity, soil water regime, management

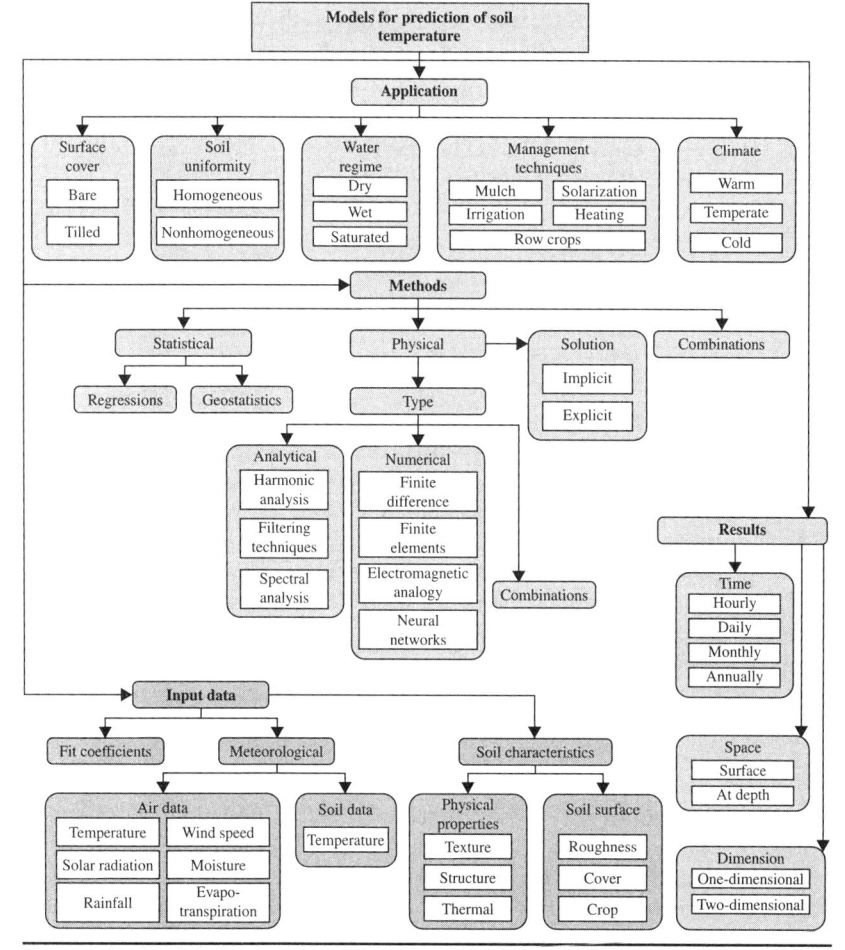

FIGURE 4.2 Models for describing soil temperature.

techniques, and climate. Moreover, some models were developed for special situations, such as fires, or soils with buried electric cables or pipes, among others.

Because the vegetation cover affects the soil thermal regime, some researchers have presented models for bare soils (Lascano and van Bavel 1983; Persaud and Chang 1983; de Vries and Philip 1986; Qin et al. 2002; Dahiya et al. 2007), whereas others have focused on the analysis of the effects of the vegetation cover on temperature, developing models that considered the effects of crops (Wiltshire 1983; Parton 1984; Novak 1986; Porter and McMahon 1987; Horton 1989; Lascano 1989; Luo et al. 1992; Renaud et al. 2001). Furthermore, other authors have performed comparative analyses to assess the effects of the vegetation cover on temperature profiles for different crops and bare soils (Kurpaska and Slipek 1996), as well as on moisture profiles under these conditions (Moroizumi and Horino 2002).

Soil heterogeneity plays an important role in the thermal properties of the soil. Therefore, the use of methods to determine soil temperature in nonuniform soils, and the complexity of using such methods, has been subject to analysis. Thus, although some authors have simplified the soil, considering it to be a homogeneous medium (Horton and Wierenga 1983), other authors have generated models that characterize the soil using different physical properties according to depth and taking soil compaction into consideration (Passerat de Silans et al. 1989; Krarti et al. 1995; Karam 2000; Fernandez et al. 2005a, 2005b).

The soil water regime greatly affects heat flow in the soil. For this reason, the temperature simulation models developed are valid only for some moisture ranges. De Vries and Philip (1986), and Kemp et al. (1992) developed models that can be applied to drying bare soils and soils in desert ecosystems, respectively, but most authors have applied their models to moisture ranges suitable for crops. Thus, Poulovassilis et al. (1998) determined the influence of irrigation on heat and water flow and proposed a model that was valid for dry and moist soils. In addition, Buonanno and Carotenuto (2000) developed a model for predicting temperature and moisture in dry or saturated soils with underground power cables.

Because anthropogenic alterations of the soil induced by crop-management techniques have a decisive influence on the soil thermal regime, many authors have been concerned with this issue. Gupta et al. (1981, 1984), Brar and Unger (1994), Novak et al. (2000), and van Donk et al. (2004) developed models for soils treated with plant residue cover. Temperatures and two-dimensional heat flows were determined in soils under row crops (Horton et al. 1984a, 1984b; Horton 1989; Hares and Novak 1992a, 1992b) by studying the effects of canopy shading on the soil. Novak (1993) analyzed the heat and temperature flows obtained from solar radiation in soils under strip tillage. In addition, a number of methods were developed to predict temperature and moisture content in greenhouse soils. Chen et al. (2006) analyzed the variation of the temperature and moisture content in soil with the increase in depth, and found that the appearance of the peak temperature in soil postponed with increase in depth. In greenhouse soils covered with rice husks, Tuntiwaranuruk et al. (2006) simulated the complexity of the temperature and moisture of a greenhouse. Nebbali et al. (2006) were concerned with estimating the ground response for some known transient ambient environment conditions (solar radiation, temperature, hygrometry, and wind speed) using a semianalytical method.

Other models have focused on situations with abnormally high soil temperatures, such as solarized soils (Cenis 1989), fires (Campbell et al. 1994), or volcanic soils (Antilen et al. 2006). These situations, in which the soil is subjected to high temperatures, occur also in thermal modeling of heated substrates (Alvarez 1996; Kurpaska and Slipek 1996; de la Plaza et al. 1999; Guaraglia and Pousa 1999; Rodriguez et al. 2004; Fernandez et al. 2005a, 2005b; Kurpaska et al. 2005; Fernandez and Rodriguez 2006; Rodriguez et al. 2006; Fernandez et al. 2007). Further research concerned with air-conditioning systems was developed by Wu et al. (2007), who analyzed earth–air–pipe systems that could be used to reduce the cooling load of buildings in summer.

The models discussed above focused on soils subjected to moderate or high temperatures, characteristic of temperate or warm climates. However, many efforts have focused on low-temperature conditions in which the water phase change has a strong impact. Hansson et al. (2004) developed a new method to account for phase changes in a fully implicit numerical model for coupled heat transport and variably saturated water flow involving conditions both above and below zero. The new function was proposed to better describe the dependency of thermal conductivity on the ice and water contents of frozen soils.

The influence of unfrozen water on the thermal properties of soils was accounted for in the one-dimensional heat-transfer model based on the finite-difference method by Ling and Zhang (2004). This method was also used by Zhang et al. (2008). The model included dynamic layering of snow, user-adjustable layering of soil, snow cover compaction and destructive metamorphisms, distinct parameterizations of the surface organic layer, and unfrozen water in frozen soil. For a freezing granite soil medium with an embedded pipeline in a closed system, Song (2006) analyzed the unsteady-state heat transfer using the commercial code ABAQUS, code that has recently transitioned to SIMULIA. These studies were focused on the development of a computational scheme by applying the effective heat capacity model to numerical procedures for predicting temperature profiles along a buried pipeline and the frozen penetration depth.

4.3.2 Methods Used to Build Models for Predicting Soil Temperature

Physical Methods

Because most models for predicting soil temperature are based on solving the transient heat-conduction equation, we need to use analytical or numerical calculation methods with implicit or explicit solutions.

Transient analytical methods have been widely implemented using the Fourier series. Horton and Wierenga (1983) estimated heat flows in homogeneous soils by using the Fourier series, and although the results were similar to the results obtained from integration methods, they obtained results that were not strictly applicable to nonuniform soils. Differences between the thermal behavior of uniform soils and nonuniform soils were also spotted by Wiltshire (1983), who calculated periodic heat flow by harmonic analysis. The harmonic analysis method was used by Cenis (1989), Alvarez et al. (1996), and Shao et al. (1998) and was combined with dynamic filtering and spectral analysis by Persaud and Chang (1983) and Matthias and Warrick (1987), who developed soil temperature models that were successfully validated. With appropriate simplifications, analytical treatment allowed for predicting temperature fields of the fluid in the pipe and the soil in the proximity of the buried pipe of earth-to-air heat exchangers installed at different depths, and used for building cooling/heating (Cucumo et al. 2008).

Despite the good results of analytical methods, many researchers have used numerical methods to analyze heat flow in soils (Hanks et al. 1971). The reason for this is that using analytical methods in the integration of the heat-transfer equation requires assuming some simplifications, whereas using numerical methods allows for introducing thermal properties that are dependent on particular soil and water conditions, and of more complex boundary conditions (Porta et al. 1999).

Horton (1989) developed a two-dimensional numerical model to predict heat and water flow in soils with row crops. Papadakis et al. (1989) estimated soil surface temperature in greenhouses using numerical methods, whereas Novak (1991) suggested the use of numerical models to improve the accuracy of temperature prediction on the soil surface after having used analytical methods to model soil surface temperature. Later, Novak (1993) proposed an analytical model with two-dimensional solutions for temperature and heat flows in soils with bare strips alternating with strips covered with plastic or other materials and used radiation surface boundary conditions.

Physically based equations solved by analytical methods combined with remote-sensing techniques permitted a more theoretically rigorous estimation of area–average soil heat flux. Muffay and Verhoef (2007) used a standard physical equation to estimate the soil heat flux in combination with a simple, but theoretically derived, equation for soil thermal inertia without the requirement for in situ instrumentation.

Different numerical methods have been used in developing models for predicting soil temperature: finite difference (Sikora et al. 1990; Hares and Novak 1992a, 1992b; Sharratt et al. 1992; de la Plaza et al. 1999; Buonanno and Carotenuto 2000; Ling and Zhang 2004; Ao et al. 2007; Keshari and Koo 2007), finite elements (Passerat de Silans et al. 1989; Renaud et al. 2001; Moroizumi and Horino 2002; dos Santos and Mendes 2005), electromagnetic analogy (Guaraglia and Pousa 1999; Karam 2000; Guaraglia et al. 2001), and neural networks (George 2001; Mihalakokou 2002).

Sikora et al. (1990) evaluated the sensitivity of soil temperature predictions to the thermal diffusivity values of a model based on a finite-different solution of the Fourier equation. The results of the model were better than experimental results under severe compaction. The authors concluded that the crop surface layer limited heat flow in the soil due to its low thermal diffusivity, which had a stronger influence than compaction values.

After having used analytical methods (Novak 1986, 1991), Hares and Novak (1992a, 1992b) applied the finite-difference solution to the heat-conduction equation to predict temperatures in soils under strip tillage with residue mulch surfaces or bare soils. They adopted an explicit solution that included a surface–energy–balance submodel. This two-dimensional model was in agreement with analytical solutions (Hares and Novak 1992a), and experimental validation of the model (Hares and Novak 1992b) produced better results for mulched soils than for bare soils.

Sharratt et al. (1992) compared finite-difference methods for the implicit solution of a Fourier equation with harmonic analysis methods. After having applied both methods to the heat-flow prediction, they performed an experimental validation of the models in fruit-tree orchards and fields covered with barley stubble and obtained better results with the numerical method. The finite-difference method has been used by various authors for heated soils (Alvarez et al. 1996; Kurpaska and Slipek 1996; de la Plaza et al. 1999). The aim of the models focused on the analysis of the performance of the heating system.

Passerat de Silans et al. (1989) developed a one-dimensional model of coupled flows solved by the finite-element method (FEM) and applied it to stratified, partially saturated bare soils. They used the Galerkin method to solve Fourier's equation and considered atmosphere boundary conditions. The model was first calibrated for validation; the results obtained after the calibration phase fitted well with values measured in the field.

The use of computer tools to solve the heat-conduction equation for soils has shown to be useful in analytical modeling and essential in numerical solutions. The continuous system modeling program (CSMP), a pioneering simulation language that was largely incompatible with PCs (Evett and Lascano 1993), was used by Horton et al. (1984a, 1984b) to model soil temperature. Later, the ENergy and WATer BALance (ENWATBAL) was included in the BASIC language to emulate CSMP functions and was used by Lascano (1989) to predict surface temperatures. Van Donk et al. (2004) modified the ENWATBAL model that simulated soil water and temperature profiles, evaporation from soil, and transpiration from crops including the effects of a mulch layer. Measured soil temperatures lagged behind those simulated, indicating that conduction was an important process of heat transfer through the mulch.

Kroes et al. (2000) used a different code, SWAP 2.0, which integrated the equations governing heat and mass transfer in soil through numerical methods and allowed for one-dimensional analysis of water movement, solute transport, and heat flows in flooded soils. Specific tools that are useful in modeling soil processes have been developed during the last few years. One example is the FEM software code developed by Thomas and Cleall (2000) to model thermal–hydraulic–mechanical (THM) behavior in unsaturated soils. Renaud et al. (2001) built a one-dimensional FEM model to predict seasonal temperature distribution in rice crops. The model was implemented in MATLAB and included the analytical computation of surface temperature by harmonic analysis.

Another computer two-dimensional *FEM* simulation model is 2DSOIL. This model simulates water flow, chemical flow, and water uptake by the plant roots, processes of physical balance, heat transfer, and gas diffusion in the soil under atmosphere environmental conditions. Timlin et al. (2002) evaluated the model to determine the error in hourly temperature data based on the maximum and minimum daily temperatures and obtained better results for hourly temperatures.

Moroizumi and Horino (2002) presented a one-dimensional model for coupled heat and water flows solved by FEM with error estimation by the Galerkin method. After the model was validated, the authors observed good agreement between measured and simulated temperature values, with worse results for matric pressure values. The authors attributed these results to the inaccuracy of the physical properties of the soil or inappropriate boundary conditions but not to the numerical method used. Qin et al. (2002) developed a model that coupled soil temperature changes simultaneously with soil moisture movement and tackled the problem by using FEM. They used the Crank–Nicolson implicit method to expand the differential equations and the Newton–Raphson method to solve the equations. The validation of the model in the South Israeli desert obtained good results. The accurate results obtained for thermal conductivity and hydraulic conductivity were particularly relevant due to their influence on the overall results.

Sun and Zhang (2004) evaluated the effect of the lower boundary position selection for the Fourier equation on heat transfer and energy balance in soil. Based on physical reasoning and the results of numerical simulation, proper depth depended on the annual heat wave damping depth, and for most soils it depended on the soil texture. In order to calculate the temperature profiles in a more accurate way, dos Santos and Mendes (2005) developed a computational code and conceived to model the coupled heat and moisture transfer in soils. The theory of Philip and de Vries was used to obtain variable thermophysical properties for two types of soil. The governing equations were discretized using the finite-volume method, and a three-dimensional model was used to describe the physical phenomena of heat and mass transfer in unsaturated moist porous soils. A finite-difference approximation of generalized solutions to the model was proposed by Thanh et al. (2007) to introduce and validate another three-dimensional linear thermal model for detecting land mines, and its convergence properties were proved.

A mathematical model based on solving the heat-transfer equation in the soil with FEM was developed by Antonopoulos (2006). The heat-transfer model was incorporated as part of the integrated model WANISIM (water and nitrogen simulation), which described soil water movement and mass transport and nitrogen transformations in the soil. Three different types of boundary conditions at soil surface were considered: soil temperature varied cosinusoidally, heat balance, and the erosion–productivity impact calculator model approach.

Saito et al. (2006) developed a numerical model in the HYDRUS-1D code that solved the coupled equations governing liquid water, water vapor, and heat transport, together with the surface water and energy balance, and provided flexibility in accommodating various types of meteorological information to solve the surface energy balance. This numerical model was used by Dahiya et al. (2007) to quantify the effect of straw mulching and rotary hoeing on the soil water and thermal regimes of a loess soil.

Karam (2000) developed a model based on electromagnetic analogy that showed good agreement with analytical solutions, with a wider application range, and with better results than the finite-difference method. Guaraglia and Pousa (1999) based their model on the solution of Fourier equation using electrical analogy. The model produced good results for radial heat conduction, which fitted the results obtained using analytical methods. This method was applied and experimentally validated for determining temperature and heat flows in sandy soils (Guaraglia et al. 2001).

The neural network model has good prospects as a prediction method but requires further development, as reported by Ferreira et al. (2002) who developed a model for predicting greenhouse air temperature. Use of this type of numerical method for the determination of soil temperature has had recent applications. Two approaches that allow for soil temperature prediction have been developed by George (2001) and Mihalakokou (2002). The latter obtained good results for prediction of daily and annual variations in soil surface temperature.

Statistical Methods and Combinations

A number of methods have been used to model soil temperatures. Statistical regression models were used by Dwyer et al. (1988, 1990), Pikul (1991), and Kluender et al. (1993). Other models (Gupta et al. 1981, 1984; Sharratt et al. 1992) have been developed by combining statistical methods for predicting surface temperature based on air temperature with the solution of the heat-conduction equation for determining temperature at depth. Geostatistics has incorporated new engineering and research tools, such as digital elevation models (DEM) for developing temperature-prediction models, such as the model developed by Kang et al. (2000).

4.4 Greenhouse Substrate Heating

Greenhouse substrate-heating systems are essentially made up of the following components:

1. *Heating elements*, which can be divided into three main categories: heating pipes with circulating warm water (Ahmed et al. 1983), heating pipes with circulating warm air (Boulard et al. 1989; Kurpaska and Slipek 1996), and electrical heating cables (Rikbost et al. 1975).

2. *Support material* (pearlstone, sand, peat or a sand-peat mixture, among others), inside which the heating elements are buried. The support material acts as a substrate and as a heat storage element.

3. *Thermal insulation* to minimize heat losses.

4. *Metal plate with high thermal conductivity* often placed immediately above the heating elements in order to achieve uniform heat distribution throughout the entire mass of the support material.

Protected cultivation has experienced many changes in the last few years, including the development of strategies that improve facilities and lower production costs without changing the crop growth cycle or the cultural techniques used. Such strategies have introduced technology in greenhouse production. This development, together with the evolution toward soilless culture techniques, has contributed to the application of substrate-heating systems.

Substrate-heating systems should be capable of guaranteeing the preset temperatures for each substrate layer according to crop requirements at each developmental stage. The systems should also be capable of providing uniform temperatures in time by mitigating in as much as possible the effects of daily and seasonal thermal fluctuation.

The high cost of the energy supply required for adequate heating demands maximum energy efficiency of the heating system. We must consider that the heat losses from the greenhouse are dependent on several factors such as wind speed, difference between indoor and outdoor temperatures, air temperature, type of cover, and greenhouse design (Sturrock 1989). In addition, heat consumption can be modeled as a linear function of the outdoor temperature (Strøm and Amsen 1981). However, heat consumption can also be estimated as a function of temperature exposure, which represents the difference between the temperatures inside and outside the greenhouse, and accumulated over a certain period. Temperature exposure can be determined from monthly means of daily maximum and minimum temperatures (Seginer and Jenkins 1987).

Any heating system used for greenhouse climate control must meet three basic conditions: safety, reliability, and flexibility. The systems used for heat transfer by conduction under the surface of greenhouse soils, or in greenhouse benches, show good prospects for use because they meet the three conditions mentioned. From a physics perspective, such systems enable the use of the air-exchange surface provided by the substrate, which is larger than the exchange surface of the traditional aerial systems (Feuilloley and Baille 1992) and make it possible to reduce greenhouse air temperature below the values considered as optimal without affecting production or quality, thus resulting in potential energy savings (Boulard and Baille 1984a; Boulard and Baille 1984b; Gosselin and Trudel 1984). However, such a temperature reduction may cause problems for the crop if the relationship between substrate temperature and greenhouse air temperature is not properly established (Jones et al. 1978).

As compared to other heating systems, use of substrate-heating systems shows additional advantages from the perspective of energy: (1) heat transfer occurs at low temperatures, which allows for correct vertical distribution of temperatures inside the greenhouse (Kurpaska and Slipek 2000); (2) substrate temperature provides sufficient heat in the root zone at low overall temperatures (Jones et al. 1978; Zeroni et al. 1984); (3) the high thermal storage capacity of the soil minimizes the effect of fluctuations in temperature supply; (4) the differential temperature between the soil surface layer and the air around it produces heat transfer into the aerial greenhouse environment; (5) use of low operating temperatures allows for the use of clean energies, such as geothermal and solar power; and (6) the system shows large thermal inertia, which means that a time lag is required before the substrate temperature increases. Similarly, the temperature decreases slowly, which is an important advantage in case of system breakdown.

Van de Braak et al. (1998) considered the slow response caused by a large thermal mass to be a significant disadvantage; a slow response time may mean that ventilation during the day will be required. Moreover, on sunny days, the temperature is more difficult to control because of the relatively fast changes that occur in the radiant thermal load compared to the rate of cooling of the soil (Hanan 1998). In warm-water distribution systems, the control of the heating system is conducted based on simple parameters such as temperature (Bailey 1985), solar radiation (Calvert and Slack 1975), or based on the simulation of the energy flow in the greenhouse (Duncan et al. 1981; Fuller et al. 1987). Different authors have developed and studied models of localized heating systems for growing substrates. Some of these models apply warm air (Boulard et al. 1989; Kurpaska and Slipek 1996), warm water (Ahmed et al. 1983; Kurpaska and Slipek 2000), or electric cables (Rikbost et al. 1975; de la Plaza et al. 1999; Rodriguez et al. 2004; Fernandez et al. 2005a, 2005b).

The aim of all these models is to supply heat to the substrate in order to achieve the temperature required for appropriate plant growth. To quantify the amount of heat that must be supplied for adequate substrate heating, the use of heat needs to be known. The amount of heat supplied to the substrate depends on many external factors (Kurpaska and Slipek 2000), among which are (1) type and physical conditions of the greenhouse substrate, (2) room temperature inside the facilities, (3) technical parameters and type of soil-heating system, and (4) properties and characteristics of use of the materials used, such as flow rate and water temperature, etc.

Many authors have been concerned with these types of heating systems, mainly warm-water heating systems, but also with electric-cable heating systems. Electric-cable heating provides great opportunities for greenhouse soil heating. Currently, other systems show cheaper operating costs. However, this drawback is largely offset by the following advantages: simplicity and low cost of initial installation, low costs of system maintenance, easy and comfortable handling, clean use, and high flexibility of the control functions. Moreover, growers can obtain an efficient use of electric energy by directly heating the growing substrate (Rodriguez et al. 2006), or by including a heat accumulator that allows for heat-energy storage during the hours when electric energy is cheaper (de la Plaza et al. 1999).

4.5 Models for Predicting Temperature in Heated Substrates

As discussed earlier in the chapter, many models have been designed to predict soil temperature under natural conditions. Yet such models do not enable the introduction of the heat supplied by the heating system, and they do not consider the effects of the heat supplied.

Therefore, a model that can be applied under conditions of artificial substrate heating is needed to improve environmental conditions for crop development. There are four ways in which such a model can be distinguished from models subject to natural conditions:

First, the geometry of the system to be modeled is more complex because of the heat supplied by the heating system and its distribution within the substrate. Second, thermal properties of heated substrates show greater variability than the thermal properties of soils under environmental conditions, which depend on other physical properties such as texture, degree of compaction, or water content (de Vries 1963; Ochsner et al. 2001). Factors that affect thermal properties can depend on two variables: space and time. In addition, studies by other authors (Campbell et al. 1994; Tarnawski et al. 2000) have reported a variation in apparent thermal conductivity depending on temperature. Considering the thermal gradients within heated substrates, the variation of apparent thermal conductivity cannot be disregarded in thermal analysis.

Third, the temporal variation of temperatures requires thermal analysis to be performed in a transient state, which allows modeling for long periods, which are determined by the crop growth cycle. It also allows modeling at short-time intervals, which can be applied as a base for control (Challa and van Straten 1993). Finally, the models should be accessible to potential users, mainly, greenhouse crop producers, technicians, and suppliers of heating installations.

A number of models that can be applied to temperature prediction in heated substrates have been developed. Alvarez et al. (1996) analyzed the thermal behavior of greenhouses with an underground heat source at a fixed depth, based on the modeling of the different energy processes involved. They developed a one-dimensional analytical model by using a Green's function solution in the Laplace domain and the fast Fourier transform (FFT) to solve the model. Model parameters were optimized by using a SIMPLEX algorithm during a 3-day simulation that showed the ability of the method to estimate greenhouse soil temperature. The method provided a reasonable description of heat flux in greenhouse substrates heated by warm-water pipes.

In this model, the soil boundary conditions were soil surface temperature and heating pipe temperature. Surface temperature was estimated from outdoor temperature data, net solar radiation on the greenhouse, and wind speed outside the greenhouse. The soil was considered as a continuous, uniform medium with constant properties at various depths. The depth of the heating system was designed as a variable that could be modified. An initial adjustment was performed in order to obtain the experimental parameters required for good performance of the model. As a result, temperature data were obtained at two soil depths (10 cm and 20 cm) at hourly or lower frequencies.

Kurpaska and Slipek (1996) developed a two-dimensional mathematical model of heat and mass exchange in garden subsoil heated by warm air released through a perforated pipe. The model studied the simultaneous development of the soil temperature and soil water profiles in the heated subsoil. The input data of the model were (1) temperature, water content, and specific heat of ambient and heated air; (2) initial temperature, water content, porosity, and thermal and hydraulic parameters of the soil; and (3) diameter and depth of the heating pipe and heat and mass transfer coefficients of the heating system.

The model was solved by the difference numerical method, and results were obtained at a given time. The following assumptions regarding the soil were made: (1) the medium was considered to be isotropic and homogeneous; (2) the gravitation potential was not considered in the description of the movement of soil water; and (3) the temperature gradient between points was not taken into account. In addition, thermophysical properties and water conductance of the soil were a function of the physical state of the soil, whereas thermal diffusivity was not dependent on temperature. The experimental validation of the model revealed reasonable correlation between estimated and measured data ($R > 0.92$) for temperature and water content, with a RMSE of 0.85°C and 0.02 kg·kg^{-1}, respectively.

Guaraglia and Pousa (1999) proposed an analog electrical model to estimate heat flow in soils heated by heaters of cylindrical geometry embedded in the soil. The analog model was solved using electrical circuit analysis software by dividing the soil into thin cylindrical layers, centered in the line that defined the heater. Values of the resistances and the capacitors were fitted to the values of the soil and the boundary conditions. The last layer was modeled by a pure resistance through which capacitors discharged, which represented heat flow into the soil. Temperatures compared well with analytical solutions and laboratory experimental data.

De la Plaza et al. (1999) developed a model of the thermal performance of a substrate heated by an electric cable and applied the model to a crop of gerbera. The model was used to predict substrate temperatures and to estimate the energy consumption of the heating system. The one-dimensional model was solved by a finite-difference numerical method, and the algorithms were implemented by using the C programming language.

The input data of the model were grouped into (1) parameters of the heating system (depth of the substrate and/or other materials used as heat accumulator, power of the heating system, heating-cable depth and spacing); (2) soil parameters (textural composition, air-entry potential, slope of the moisture curve, and saturated volumetric moisture content); (3) crop parameters (set point temperature and depth at which it must be controlled); (4) discretization parameters for the application of the finite-difference method (number of intervals into which the soil was divided for analysis); and (5) environmental parameters (initial temperatures at all depths, temperature at the substrate surface, and temperature of the heating cable during the analysis).

The model correctly simulated variations in substrate temperature and heat accumulator (sand or heatresistant brick) at different depths, both in summer, when the heating system was off, and in winter, when the localized substrate heating was switched on. The RMSE ranged from 0.44°C to 0.54°C. The model that accurately simulated the heating-cable operation predicted the energy consumption associated with the different localized heating configurations with an average error less than ±7 percent.

In the experimental installation, use of thermal insulation and heat accumulators produced energy savings in the range of 14 to 20 percent. The best configuration in terms of energy conservation was the configuration with the heating cable buried at 0.20 m, with thermal insulation and heat-resistant brick placed under the heating cable as a heat accumulator. The combination of thermal insulation and energy consumption only during the night reduced costs by 27 percent.

Kurpaska and Slipek (2000) developed a method for optimizing greenhouse substrate-heating systems with warm-water buried pipes. The method was based on steady-state heat transfer in soils and used numerical solutions to solve the two-dimensional Fourier equation in a portion of the soil in which the heating pipe was embedded.

To optimize heating system design, the authors developed two quality criteria that accounted for the thermal losses below the depth of the plant root system, and for temperature variations in the studied section. The quality criteria developed corresponded to the components of the function to be minimized, which was then used to find the optimum solutions under the experimental conditions. The optimization method considered the physical properties of soil, the characteristics of the crop, and the thermal and hydraulic characteristics of the greenhouse.

Kurpaska et al. (2005) presented a mathematical model of the greenhouse substrate-heating process, which occurs when tubes are positioned in the substrate. The two processes that take place during substrate warming, heat exchange, and mass exchange (soil water), together with cultivated plants that were heated by hot-water piping, were considered in the model. The model included the processes of transpiration, evaporation, and infiltration. Standard diffusion equations for the capillary–porous structure, and standard thermal and hydrophysical characteristics were used to describe the phenomena occurring in the substrate. Equations were solved numerically by a differential method. The parameters of the model (thermal and hydrophysical characteristics) were determined by laboratory methods, whereas heat and mass transport coefficients were calculated using correlation equations used in chemical engineering. The following parameters were determined: water content in substrate, soil water potential, soil water diffusivity coefficient in substrate, heat-diffusion coefficient, thermal conductivity of the substrate, heat-transfer coefficient from the heating pipe to the surrounding substrate, heat-transfer coefficient in free convection flow (for horizontal surface), mass-transfer coefficient in free convection flow (for horizontal surface), time, and source function.

In addition, an experiment was carried out in a greenhouse where pepper was cultivated. The substrate consisted of peat, tree bark, and pearlstone. The watering system in the soil bed was switched off for the duration of the experiment. The climatic parameters inside the greenhouse (temperature and air humidity) were monitored throughout the experiment, as well as substrate temperature, moisture, and leaf surface. Additionally, the radiation intensity (sunshine) was measured inside the greenhouse. Microclimate parameters inside the greenhouse were measured 1 m above ground level. The comparison between measured and calculated values revealed considerable convergence. The RMSE between the analyzed values did not exceed 0.73°C (temperature) and 0.003 $m^3 m^{-3}$ (moisture content). The tests demonstrated a high correlation between the predicted values and the measured values, with coefficients of correlation (R) of 0.94 (temperature) and 0.91 (moisture content). The observed differences could be due to the adoption of simplified conditions, homogeneity, and isotropy of the studied substrate.

The complexity of the physical phenomenon studied makes it difficult to offer explicit equations to determine the capacity to be installed in heated substrates. To overcome such difficulties, Rodriguez et al. (2004) used experimental techniques combined with appropriate estimation methods that led to broad-based practical knowledge of that phenomenon. The parameters on which the proposed solution was based were defined by variables such as the power of the heating cable to be used or its depth and spacing. Dimensional analysis was applied because of the complexity of the energy transfers occurring in the substrate.

The method proposed by the authors combined heat-transfer theory with data obtained experimentally through dimensional analysis, and allowed for simple estimation of the parameters that defined the heating design: depth, spacing, and power per unit length of the electric cable. A number of variables were required for estimation, among which were thermal properties of the substrate, ambient temperature, and substrate temperatures required to grow the relevant crop.

Later, Fernandez et al. (2005a) developed a two-dimensional FEM model using a general-purpose finite element code (ANSYS) that was capable of describing the thermal state of substrates heated by electric cable based on the geometry of the heating system and on the thermal properties of the substrate and insulator. This model allowed for (1) the introduction of different properties of compaction and moisture, and mixture of materials in the modeled substrate, (2) the transient analysis of the system, showing the two-dimensional distribution of temperatures at each step, and allowing heat-flow assessment, and (3) the use of a minimum number of boundary conditions: initial distribution of temperatures in the system formed by the substrate, insulator, and greenhouse air temperature, substrate surface temperature, and heating-cable temperature.

Under natural conditions, the model showed good agreement with experimental data, regardless of the values and methods used to determine the thermal properties of the substrate, and of the introduction of substrate properties according to depth (compaction, moisture, and presence of a differentiated surface layer). However, when an artificial heat source was present, simulation accuracy depended on several factors. After having tested different methods to estimate the thermal properties of the substrate, the authors found that the method by Campbell et al. (1994) generated the best response and that considering the variation of substrate properties at depth affected the accuracy of simulation results.

The final model was validated by using an experimental test consisting of nine geometric configurations for substrate electric cable heating using a sand substrate (Fernandez et al. 2005b). Temperature was measured at nine locations and moisture content was measured at three locations inside the substrate. The experimental validation was conducted after having simulated each geometric configuration during 3 days, under various environmental and operational conditions. Predictions for the root zone (0 to 150 mm depth) were generally acceptable with slight overestimations. The mean value of the RMSE at the vertical of the heating cable was 0.30°C, whereas its maximum value did not exceed 0.61°C. For locations between cables, prediction was less exact, and RMSE values ranged from 0.12 to 1.00°C, with a mean value of 0.57°C. Overestimations in the root zone occurred mainly when room temperatures were high. Drying of the surface layer could account for such overestimations. Drying would produce a sharp decrease in the effective thermal conductivity of the upper layer of the substrate, which was not considered in the model.

The coefficient of determination reached values close to 1. Errors were acceptable, with maximum values observed in zones different from the root development zone, which was less relevant for the study. Therefore, the model properly described the thermal state of heated substrates and provided a useful tool for designing and monitoring this type of substrate-heating system.

By combining dimensional analysis with the FEM-based physical model described above, Fernandez and Rodriguez (2006) expanded the range of application of the dimensioning expressions used for electric cable–heating systems and achieved cable spacings of 350 mm for a 225-mm depth. Their research was structured in two parts, according to the mathematical model used: (1) application of a FEM-based physical model to simulate different configurations of the heating system and (2) application of dimensional analysis to the results of the model in order to obtain the relevant dimensioning expressions.

Because the model was originally designed for validation, it was modified such that it conformed to simulation. Thus, the temperature at all the nodes was estimated in a new part of the model, a static analysis was used as the initial condition, and the greenhouse air temperature, the substrate surface temperature, and the heat flow emitted by the heating element were used as conditions at each step of the analysis.

Dimensioning expressions obtained from the model allowed for the estimation of the power per unit length of heating cable required to achieve a given temperature in the root zone. For that purpose, two reference points were used, both of them 75 mm deep. The first reference, which was the most favorable, was located at the vertical of the heating cable, and the second one was located at an intermediate spacing between cables. Because temperature is generally higher at the vertical of the heating cable, the power per unit length required at that point was lower, but the temperature values generated were below the temperature selected for the reference depth. Such differences in temperature were quantified to allow for a most suitable design.

The authors used the same environmental conditions and identical operating conditions to model the performance of heating installations with different heating-cable spacings (Fernandez et al. 2007). These experimental conditions allowed for an efficient comparison of energy efficiency and temperature distribution, which would not be possible under environmental conditions because of the level of detail of the model. The simulation allowed the authors to analyze the effect of heating-cable spacing on temperature distribution in the root zone. In addition, they obtained data of energy consumption and applied heat flow use which are the two key factors to consider when choosing a heating system.

4.6 An Analysis of Electric Cable Heating Systems

A heat exchange process is caused by temperature differences between the warm area of heating elements and soil particles. This results in a nonuniform region of temperature within the adjacent soil substrate. As a result of this temperature gradient, heat flow occurs, soil water potential changes, and mass transport takes place (Hanks 1992). Consequently, heat and mass exchange occurs in the soil bed, which increases temperature and decreases water content. We can observe that the influence of radiation practically disappears at a depth of 0.15 m. The increase in temperature is the result of the influence of two independent flows of heat affecting the substrate: the flow originating from heating ducts and sunshine radiation. Reduction in moisture content was caused by three forces driving the entire process: transpiration, infiltration, and evaporation (Kurpaska et al. 2005).

In unheated substrates, heat flows occur mainly in the vertical direction, and heat exchange occurs through the substrate surface.

Consequently, the soil absorbs heat when ambient temperature is high and releases heat when temperature decreases. However, because this situation is completely different in heated substrates, we focus on temperature distribution and heat flows in a greenhouse bench with thermal insulation and substrate electric cable heating. Under these conditions, switching the heating element on considerably increases temperature in the surroundings of the heating element, thus generating high-temperature circular-section areas. In the rest of the substrate, temperatures are distributed into horizontal layers, with an undulated area that comprises the warmest areas around the heating element (Fig. 4.3).

While the heating system is switched on, the highest heat flows are radially distributed around the heating element and are more intense in the upward direction (Fig. 4.4). The lowest heat flows occur under

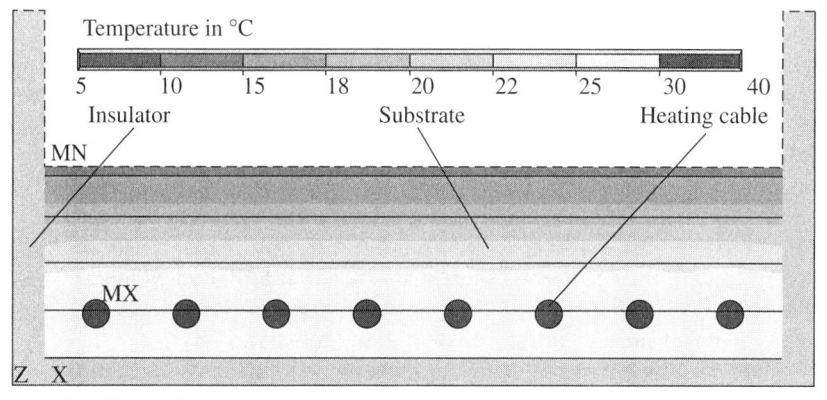

FIGURE 4.3 Temperature distribution in a cross-section of a heated substrate when the heating element is switched on.

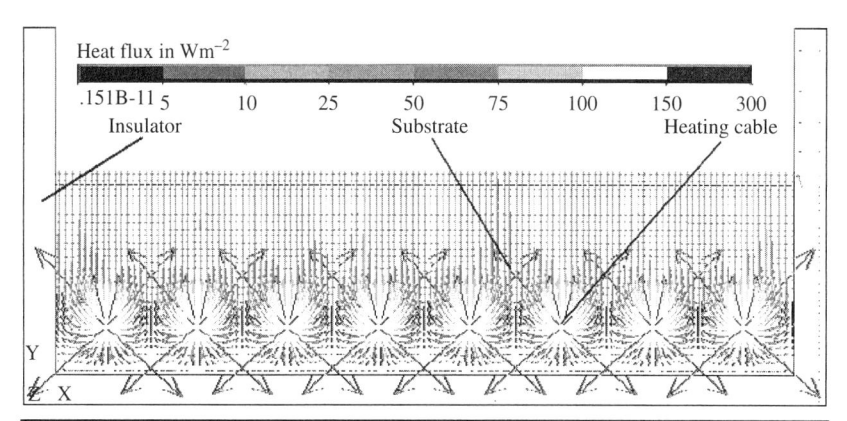

FIGURE 4.4 Heat flow distribution in a cross-section of a heated substrate while the heating element is switched on.

the heating elements and tend to increase between the heated points because of the presence of insulation on the bottom of the greenhouse bench. Above the heating element, heat flows vertically toward the substrate surface and is transferred into the air. After the heating element has been switched off, the flow of heat gradually changes direction and becomes vertical in the whole substrate. Heat will flow upward or downward depending on substrate surface temperature.

The energy used by the heating system is a function of three factors: geometry of the system, ambient temperature, and substrate temperature. Thus, an increase in ambient temperature involves a decrease in energy demand, whereas an increase in the root-zone temperature will increase energy demand.

Heating-cable spacing is one of the geometric variables that most affects the energy consumption of the heating system. Heat supply decreases with the increase in spacing. Short spacings produce high temperatures on the plane of installation of the heating cable. Consequently, heat flows that are transferred to the lower part of the substrate stay at the bottom of the substrate and are prevented from going up to the root zone. This implies an increase in the energy at short spacings because part of the heat transferred by the heating cable does not arrive to the plane of reference. For wider spacings, the heating-cable zone shows high temperatures, whereas temperatures are lower in intermediate zones of the heating cable than in the bottom layers of the substrate. Such temperature gradients cause an ascent toward the root zone of the heat flows that initially headed to the bottom of the substrate. The substrate acts as an accumulator, capable of transferring the stored heat to the root zone.

Heat supply increases with the increase in heating-cable depth because the distance from the heat source to the plane of reference of temperature increases. However, a depth of installation very near the surface may demand more heat because of the occurrence of high temperature gradients between the heating cable and the environment, which produces high heat flows outside the substrate, with subsequent energy losses. Moreover, part of the heat supplied by the heating cable produces the heating of the bottom of the substrate, where the temperatures achieved are near the reference temperatures. As a result, the contribution to heating in the root zone is very poor.

The influence of the maximum operating temperature of the heating cable is remarkable. The number of times the cable is switched on decreases considerably with an increase in the maximum operating temperature of the cable; a high-temperature fluctuation occurs in the substrate. The spacing between the heating cable and the reference plane has a direct effect on the number of times the system is switched on. A shorter spacing results in more frequent switching on, which in turn results in greater temporal uniformity of substrate temperatures mainly because the heating cable operates at lower temperatures. Moreover, an increase in the spacing between the

Models for Heat Transfer in Heated Substrates 157

heating cable and the reference plane causes greater difficulty for the heating system to reach the target temperature, even for higher temperatures of the heating cable. Similarly, as with depth, increased spacing results in an increase in the number of times the heating system is switched on.

The moisture profile of an unheated soil shows an increase in moisture with depth. However, moisture levels of heated soils decrease in the zones where the heating cable is installed because of the high temperatures observed (Fig. 4.5). With a decrease in temperature, moisture content increases. Moreover, when the heating cable is on, the temperature gradient causes an exchange of water from the depth of installation of the heating cable to the root zone. When the heating cable is off, the exchange of water occurs in the opposite direction. Referring to Fig. 4.5, we can see that during the initial period (0 to 14.5 h), the heating cable is switched on and heating-cable temperature is around 25°C (T 150). Consequently, moisture at a 150-mm depth (H 150) is lower than moisture at a 75-mm depth (H 75). During an off period, 14.5 to 18.5 h, the moisture values measured at the mentioned depths are inverted, thus generating a typical moisture profile, in which moisture increases with depth.

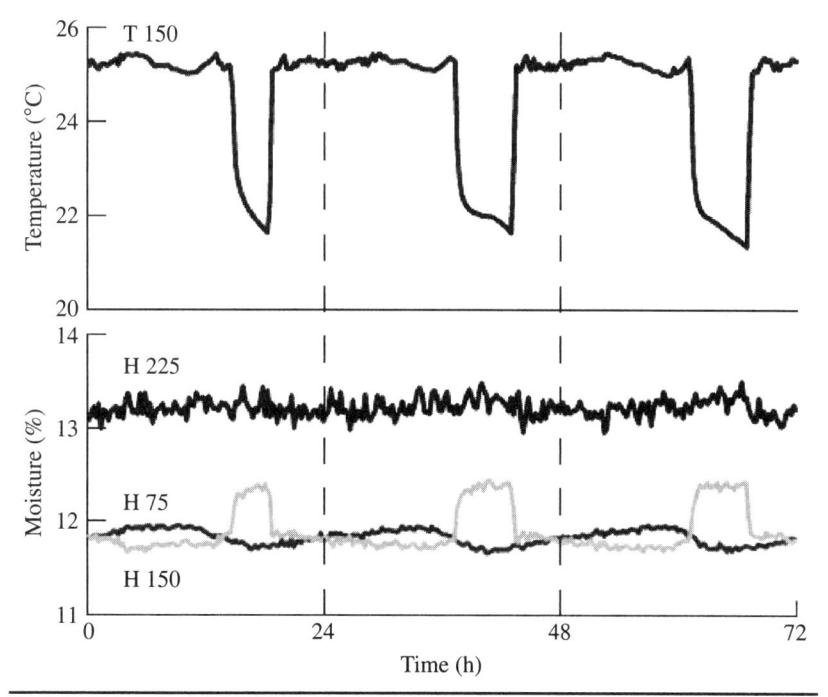

Figure 4.5 Evolution of heating-cable temperatures at 150-mm depth (T 150), and moisture values at depths of 75 mm (H 75), 150 mm (H 150), and 225 mm (H 225) in a heated substrate.

Water moves to the zones with the lowest temperatures, which helps reduce the temperature differences in the root zone. With the increase in the depth of installation of the heating cable, the amount of water flowing toward the root zone decreased considerably. Such a decrease is enhanced by the increase in heating-cable spacing.

Substrate heating in greenhouses shows important benefits for crop development. Yet the high-energy costs of these systems call for energy-saving strategies. This requires an appropriate dimensioning of the system that allows us to predict substrate temperature distribution and energy performance. Based on the expressions proposed by Fernandez and Rodriguez (2006), we obtained a number of power-per-area curves that are required for an electric cable–heating system (Fig. 4.6).

The power per area required to obtain a preset temperature at a root depth vertically above the heating cable is much more affected by spacing between cables than by air temperature. For a depth of 225 mm, the power per area decreased with the increase in spacing. Such a reduction accounted for 71 percent for a 350-mm spacing as compared to a 100-mm spacing. Similarly, the increase in air temperature caused decreases in power, such that a rise in temperature from 6 to 18°C decreased energy by 15 to 16 percent, depending on cable spacing.

An interesting fact is worth noting here; the geometry of the heating system causes vertical variations in temperature, as in unheated soils, and horizontal variations in temperature. Therefore, the location of the temperature measurement point must be considered for control of the heating system. Thus, the power required increases when measurements are performed on the axis of heating-cable spacing, where the lowest temperatures are observed. On average, the

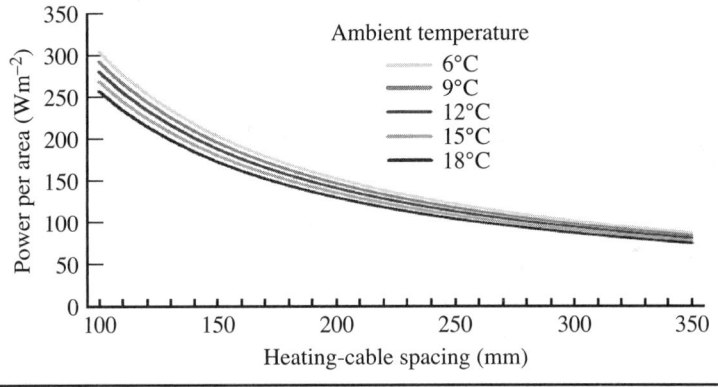

FIGURE 4.6 Demand of power per unit area to reach 20°C in the root zone, measured at the vertical of the heating cable at a depth of 75 mm and as a function of spacing, for a heating cable buried at a depth of 225 mm and considering different air temperatures.

increases in the power required amount to 4.4 percent for the analyzed temperatures and geometries.

Generally, the increase in spacing causes an overall decrease in temperatures that mainly affects the zone between cables near the substrate surface. For a heating cable buried at a depth of 225 mm, the largest differences are found for spacings between 100 and 150 mm, with a decrease of 1.02°C in the value of the average temperature that mainly affects the surface layer of the substrate (< 75 mm deep). For larger spacings between cables, between 150 and 350 mm, an increase of 50 mm in heating-cable spacing causes slight decreases in temperature, as shown in Fig. 4.7.

The time of operation of the heating cable increases with spacing. Conversely, energy consumption decreases with increased heating-cable spacing. By analyzing installations with the heating cable buried at a depth of 225 mm and at different spacings (100 to 350 mm), it has been observed that the increase in spacing causes a decrease in energy consumption, reaching a value of 33.4 percent. The largest difference in energy consumption is found between spacings of 100 and 150 mm, with a value of 1323 kJ during a 24-h period, whereas the differences for spacings larger than 150 mm are smaller, with values in the range 300 to 750 kJ during a 24-h period. Such variations are represented in Fig. 4.7, where we can clearly observe that the energy savings induced by the increase in heating-cable spacing cause slight decreases in temperature at spacings larger than 150 mm.

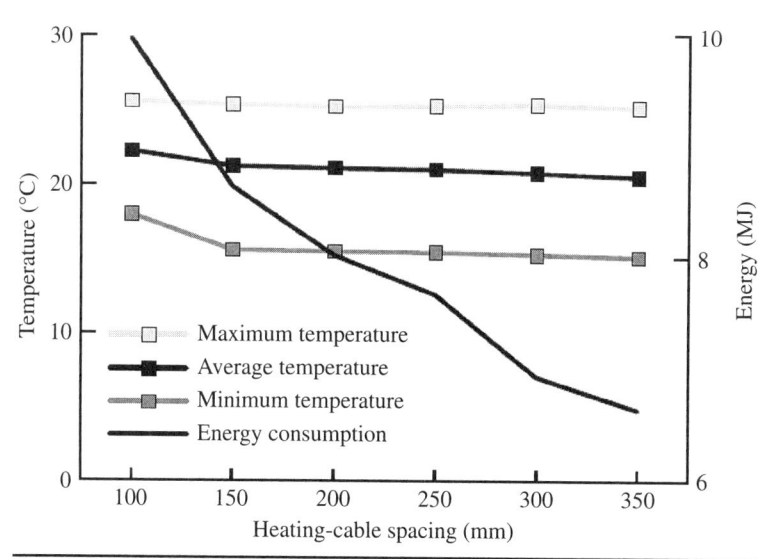

FIGURE 4.7 Variation in maximum, average, and minimum temperatures in the root zone and in energy consumed during 24 h by a heating cable buried at a depth of 225 mm as a function of spacing (Fernandez et al. 2007).

References

Abu-Hamdeh, N. H. 2001. Measurement of the thermal conductivity of sandy loam and clay loam soils using single and dual probes. *Journal of Agricultural and Engineering Research* 80(2):209–216.

Abu-Hamdeh, N. H. and Reeder, R. C. 2000. Soil thermal conductivity: Effects of density moisture salt concentration and organic matter. *Soil Science Society America Journal* 64(4):1285–1290.

Abu-Hamdeh, N. H. Reeder, R. C., and Khdair, A. I. 2001. A comparison of two methods used to evaluate thermal conductivity for some soils. *International Journal of Heat and Mass Transfer* 44(5):1073–1078.

Ahmed, A. E., Hamdy, M. Y., Roller, W. L., and Elwell, D. L. 1983. Technical feasibility of utilising reject heat from power stations in greenhouses. *Transactions of the American Society of Agricultural and Biological Engineers* 26(1):200–206.

Alvarez, J., Sobron, F., and Bolado, S. 1996. Solution for heat flow in soil with a heat source at a fixed depth. *Soil Science Society of America Journal* 60(4):1028–1035.

Andales, A. A. Batchelor, W. D., and Anderson, C. E. 2000. Modification of a soybean model to improve soil temperature and emergence date prediction. *Transactions of the American Society of Agricultural and Biological Engineers* 43(1):121–129.

Antilen, M., Fudym, O., Vidal, A. J., Foerster, E. Moraga, N. and Escudey, M. 2006. Mathematical modelling of temperature profile of volcanic soils affected by an external thermal impact. *Australian Journal of Soil Research* 44(1):57–61.

Antonopoulos, V. Z. 2006. Water movement and heat transfer simulations in a soil under ryegrass. *Biosystems Engineering* 95(1):127–138.

Ao, Y. H. Wen, J., and Lu, S. H. 2007. A study of the water and energy transfer at the soil surface in cropped field on the loess plateau of northwestern China. *Environmental Geology* 52(3):595–603.

Auvermann, B. W., McFarlan, M. J., and Hill, D. W. 1992. In-situ determination of soil thermal characteristics. *Transactions of the American Society of Agricultural and Biological Engineers* 35(3):833–839.

Awal, M. A. and Ikeda, T. 2002. Effects of changes in soil temperature on seedling emergence and phenological development in field-grown stands of peanut (*Arachis hypogaea*). *Environmental and Experimental Botany* 47(2):101–113.

Bailey, B. J. 1985. Wind dependent control of greenhouse temperature. *Acta Horticulturae* 174:381–386.

Balland, V. and Arp, P. A. 2005. Modeling soil thermal conductivities over a wide range of conditions. *Journal of Environmental Engineering and Science* 4(6):549–558.

Bilskie, J. Horton, R. R., and Bristow, K. L. 1998. Test of a dual-probe heat-pulse method for determining thermal properties of porous materials. *Soil Science* 163(5):346–355.

Boulard, T. and Baille, A. 1984a. Using warm water for forcing greenhouse crops .1. Microclimatic and thermal aspects. *Agronomie* 4(3):213–220.

Boulard, T. and Baille, A. 1984b. Using warm water for forcing greenhouse crops .2. Effects of soil warming on growth and development of bulbous flowers (*gladiolus, iris, tulip, lily*). *Agronomie* 4 (3):221–230.

Boulard, T., Razafinjohany, E. and Baille, A. 1989. Heat and water vapour transfer in a greenhouse with an underground heat storage system. Part I. Experimental results. *Agricultural and Forest Meteorology* 45(3–4):75–184.

Brar, G. and Unger, P. 1994.Soil–temperature simulation with varying residue management. *J. Agron. Crop Science—Zeitschrift Fur Acker Und Pflanzenbau* 172(1):1–8.

Bristow, K. L., Bilskie, J. R., Kluitenberg, G. J., and Horton, R. 1995. Comparison of techniques for extracting soil thermal properties from dual-probe heat-pulse data. *Soil Science* 160(1):1–7.

Bristow, K. L., Kluitenberg, G. J., Goding, C. J. and Fitzgerald, T. S. 2001. A small multi-needle probe for measuring soil thermal properties, water content and electrical conductivity. *Computers and Electronics in Agriculture* 31(3):265–280.

Bristow, K. L., Kluitenberg, G. J. and Horton, R. 1994a. Measurement of soil thermal properties with a dual-probe heat-pulse technique. *Soil Science Society of America Journal* 58(5):1288–1294.

Bristow, K. L., White, R. D., and Kluitenberg, G. J. 1994b. Comparison of single and dual probes for measuring soil thermal properties with transient heating. *Australian Journal of Soil Research* 32(3):447–464.

Buchan, G. D. 1991. Soil temperature regime. In *Soil Analysis. Physical Methods*, K. A. Smith and C. E. Mullins, eds. New York: Marcel Dekker. 551–612.

Buonanno, G. and Carotenuto, A. 2000. A comparison of thermal performances of underground power cables in dry and saturated soils. *Journal of Porous Media* 3(1):1–10.

Calvert, A. and Slack, G. 1975. Light-dependent control of day temperature for early tomato crops. *Acta Horticulturae*. 51:163–168.

Campbell, G., Calissendorff, S. C., and Williams, J. H. 1991. Probe for measuring soil–specificheat using a heat-pulse method. *Soil Science Society of America Journal* 55(1):291–293.

Campbell, G. S., Jungbauer, J. D., Bidlake, W. R., and Hungerford, R. D. 1994. Predicting the effect of temperature on soil thermalconductivity. *Soil Science* 158(5):307–313.

Campbell, G. S. and Norman, J. M. 1998. *An Introduction to Environmental Biophysics* 2nd ed. New York: Springer–Verlag. 316.

Cenis, J. L. 1989. Temperature evaluation in solarized soils by Fourier analysis. *Phytopathology* 79(5):506–510.

Challa, H. G. and van Straten, G. 1993. *Optimal diurnal climate control in greenhouses as related to greenhouse management and crop requirements* In The Computerized Greenhouse. Automatic Control Application in Plant Production, Y. Hashimoto, ed. New York: Academic Press, 119–137.

Chen, D. L. and Kling, J. 1996. Apparent thermal diffusivity in soil: estimation from thermal records and suggestions for numerical modelling. *Physical Geography* 17(5):419–430.

Chen, W., Liu, W., and Liu,B. C. 2006. Numerical and experimental analysis of heat and moisture content transfer in a lean-to greenhouse. *Energy and Buildings* 38(2):99–104.

Covell, S., Ellis, R. H., Roberts, E. H., and Summerfield, R. J. 1986. The influence of temperature on seed germination rate in grain legumes. I. A comparison of chickpea lentil soyabean and cowpea at constant temperatures. *Journal of Experimental Botany* 37(178):705–715.

Cucumo, M., Cucumo,S., Montoro, L., and Vulcano, A. 2008. A one-dimensional transient analytical model for earth-to-air heat exchangers taking into account condensation phenomena and thermal perturbation from the upper free surface as well as around the buried pipes. *International Journal of Heat and Mass Transfer* 51:506–516.

Dahiya, R., Ingwersen, J., and Streck, T. 2007. The effect of mulching and tillage on the water and temperature regimes of a loess soil: Experimental findings and modeling. *Soil and Tillage Research* 96:52–63.

de la Plaza, S., Benavente, R., García, J., Navas, L., Luna, L., Durán, J., and Retamal, N. 1999. Modelling an optimal design of an electric substrate heating system for greenhouse crops. *Journal of Agricultural and Engineering Research* 73(2):131–139.

de Vries, D. A. 1963. *Thermal Properties of Soils*. In *Physics of Plant Environment*. W. R. van Wijk, ed. Amsterdam: North–Holland. 210–235.

de Vries, D. A. 1975. *Heat Transfer in Soils*. In *Heat and Mass Transfer in the Biosphere Part 1 Transfer Processes in the Plant Environment*, D. A. de Vries and N. H. Afgan, eds. New York: Wiley. 5–28.

de Vries, D. A. and Philip, H. R. 1986. Soil heat flux thermal conductivity and the mull-alignment method. *Soil Science Society of America Journal* 50:12–18.

dos Santos, G. H. and Mendes, N. 2005. Unsteady combined heat and moisture transfer in unsaturated porous soils. *Journal of Porous Media* 8(5):493–510.

Duncan, G. A., Loewer, O. J., and Colliver, D. G. 1981. Simulation of energy flows in a greenhouse: Magnitudes and conservation potential. *Transactions of the American Society of Agricultural and Biological Engineers* 24(4):1014–1021.

Dwyer, L., Bootsma, A., and Hayhoe, H. 1988. Performance of three regression-based models for estimating monthly temperatures in the Atlantic region of Canada. *Canadian Journal of Soil Science* 68(2):223–235.

Dwyer, L., Hayhoe, H., and Culley, J. 1990. Prediction of soil temperature from air temperature for estimating corn emergence. *Canadian Journal of Plant Science* 70(3):619–638.

Ekwue, E. I., Stone, R. J. Maharaj, V. V., and Bhagwat, D. 2005. Thermal conductivity and diffusivity of four trinidadian soils as affected by peat content. *Transactions of the American Society of Agricultural and Biological Engineers* 48(5):1803–1815.

Ellis, R. H., Covell, H. S. Roberts, E. H., and Summerfield, R. J. 1986. The influence of temperature on seed germination rate in grain legumes. II. Intraespecific variation on chickpea (*Cicer arietanum* L.) at constant temperatures. *Journal of Experimental Botany* 37(183):1503–1515.

Evett, S. R. and Lascano, R. J. 1993. A mechanistic evapotranspiration model written in compiled basic. *Agronomy Journal.* 85(3):763–772.

Fernandez, M. D. and Rodriguez, M. R. 2006. Application of a heat transfer model for heated substrates in greenhouses to the dimensioning of spacing between electric heating cables. *Biosystems Engineering* 94(4):573–585.

Fernandez, M. D., Rodriguez, M. R., and Diaz, F. 2007. Modeling heat transfer in substrates heated by electric cable depending on heating cable spacing. *Transactions of the American Society of Agricultural and Biological Engineers* 50(2):607–614.

Fernandez, M. D., Rodriguez, M. R., Maseda, F., Velo, R., and Gonzalez, M. A. 2005a. Modelling the transient thermal behaviour of sand substrate heated by electric cables. *Biosystems Engineering* 90(2):203–215.

Fernandez, M. D., Rodriguez, M. R., Maseda, F., and Velo, R. 2005b. Validation of temperature simulation based on finite element analysis in substrates heated by electric cable. *Transactions of the American Society of Agricultural and Biological Engineers* 48(3):1241–1251.

Ferreira, P. M., Faria, E. A., and Ruano, A. E. 2002. Neural network models in greenhouse air temperature prediction. *Neurocomputing* 43(51–75).

Feuilloley, P. and Baille, A. 1992. Principes généraux d'ulisation des eaux tièdes pour le chauffage des serres. *Informations Techniques du CEMAGREF* 87:1–8.

Finch–Savage, W., Phelps, E. K., Steckel, J. R., Whalley, W. R., and Rowse, H. R. 2001. Seed reserve–dependent growth responses to temperature water potential in carrot (*Daucus Carota* L.). *Journal of Experimental Botany* 52(364):2187–2197.

Fuller, E. N., Schettler, P. D., and Giddings, J. C. 1966. A new method for prediction of binary gas–phase diffusion coefficients. *Industrial and Engineering Chemistry* 58:19–22.

Fuller, R. J., Meyer, C. P., and Sale, P. J. M. 1987. Validation of a dynamic model for predicting energy use in greenhouses. *Journal of Agricultural Engineering Research* 38(1):1–14.

García–Huidobro, J. L., Monteith , L., and Squire , G. R. 1982. Time temperature and germination of pearl millet (Penisetum typhoides S&H) I. Constant temperature. *Journal of Experimental Botany* 33 133 288–296 (1982).

Gent, M. P. N. and Ma, Y. Z. 2000. Mineral nutrition of tomato under diurnal temperature variation of root and shoot. *Crop Science* 40(6):1629–1636.

George, R. K. 2001. Prediction of soil temperature by using artificial neural networks algorithms. *Nonlinear Analysis: Theory, Methods and Applications* 47(3):1737–1748.

Gosselin, A. and Trudel, M. J. 1984. Effect of soil heating on plant productivity and energy conservation in northern greenhouses. *Acta Horticulturae* 148:853–858.

Ghuman, B. S. and Lal, R. 1985. Thermal conductivity thermal diffusivity and thermal capacity of some Nigerian soils. *Soil Science* 139(1):74–80.

Gori, F. A. 1983. Theoretical model for predicting the effective thermal conductivity of unsaturated frozen soils. *Proceedings of the Fourth International Conference on Permafrost* Washinton DC: National Academy of Sciences Press. 363–368.

Guaraglia, D. O. and Pousa, J. L. 1999. An electrical model of heat flow in soil. *Soil Science Society of America Journal* 63(3):457–463.

Guaraglia, D. O., Pousa, J. L., and Pilan, L. 2001. Predicting temperature and heat flow in a sandy soil by electrical modeling. *Soil Science Society of America Journal* 65(4):1074–1080.

Gupta, S., Larson, W., and Allmaras, R. 1984. Predicting soil temperature and soil heat flux under different tillage surface residue conditions. *Soil Science Society of America Journal* 48(2):223–232.

Gupta, S., Larson, W., and Radke, J. 1981. Predicting temperatures of bare and residue covered soils with and without a corn crop. *Soil Science Society of America Journal* 45(2):405–412.
Ham, J. M. and Benson,E. J. 2004. On the construction and calibration of dual–probe heat capacity sensors. *Soil Science Society of America Journal* 68(4):1185–1190.
Hanan, J. J. 1998. *Greenhouses. Advanced Technology for Protected Horticulture* Boca Ratón FL: CRC Press. 684.
Hanks, R. J. 1992. *Applied Soil Physics. Soil Water and Temperature Applications.* New York: Springer–Verlag. 176.
Hanks, R. J., Austin, D. D., and Ondreche, W. T. 1971. Soil temperature estimation by a numerical method. *Soil Science Society of America Journal* 35 (5):665–669.
Hansson, K., Šimůnek, J., Mizoguchi, M., Lundin, L. C., and van Genuchten, M. T. 2004. Water flow and heat transport in frozen soil: Numerical solution and freeze–thaw applications. *Vadose Zone Journal* 3(2):693–704.
Hares, M. A. and Novak, M. D. 1992a. Simulation of surface energy balance and soil temperature under strip tillage. I. Model description. *Soil Science Society of America Journal* 56(1):22–29.
Hares, M. A. and Novak, M. D. 1992b. Simulation of surface energy balance and soil temperature under strip tillage. II. Field test. *Soil Science Society of America Journal* 56 (1):29–36.
Hopmans, J. W. and Dane, J. H. 1986. Thermal conductivity of two porous media as a function of water content temperature and density. *Soil Science* 142(4):187–195.
Hopmans, J., Šimůnek, J., and Bristow, K. 2002. Indirect estimation of soil thermal properties and water flux using heat pulse probe measurements: geometry and dispersion effects. *Water Resources Research* 38(1):77–90.
Horton, R. 1989. Canopy shading effects on soil heat and water flow. *Soil Science Society of America Journal* 53(3):669–679.
Horton, R., Aguirre-Luna, O., and Wierenga, P. 1984a. Observed and predicted two-dimensional soil temperature distributions under a row crop. *Soil Science Society of America Journal* 48(5):1147–1152.
Horton, R., Aguirre-Luna, O., and Wierenga, P. 1984b. Soil temperature in a row crop with incomplete surface cover. *Soil Science Society of America Journal* 48(6):1225–1232.
Horton, R. and Wierenga, P. 1983. Determination of the mean soil temperature for evaluation of heat flux in soil. *Agricultural Meteorology* 28(4):309–319.
Horton, R., Wierenga, P., and Nielsen, D. R. 1983. Evaluation of methods for determining the apparent thermal diffusivity of soil near the surface. *Soil Science Society of America Journal* 47(1):25–32.
Hurley, S. and Wiltshire, R. J. 1993. Computing thermal diffusivity from soil temperature measurements. *Computers and Geosciences* 19(3):475–477.
Jackson, R. D. and Taylor, S. A. 1965. Heat transfer. In *Methods of Soil Analysis Part 1.* 5th ed. C. A. Black, ed. Madison, WI: American Society of Agronomics. 349–356.
Johansen, O. 1975. Thermal conductivity of soils Ph.D. diss. University of Trondheim, Norway.
Jones, D. A. G., Sandwell, I., and Talent, C. J. W. 1978. The effect of soil temperature when associated with low air temperatures on cropping of early tomatoes. *Acta Horticulturae* 76:167–172.
Jury, W. A., Gardner, W. R., and Gardner, W. H. 1991. *Soil Physics.* 5th ed. New York: Wiley. 328.
Kaminsky, W. 1994. *Heat conduction in materials with nonhomogeneous structure.* In *Experiments and Heat Transfer and Thermodynamics*, R. A. Granger, ed. New York: Cambridge University Press. 19–22.
Kang, S., Kim, S., Oh, S., and Lee, S. 2000. Predicting spatial and temporal patterns of soil temperature based on topography surface cover and air temperature. *Forest Ecology and Management* 136(1–3):173–184.
Karam, M. A. 2000. A thermal wave approach for heat transfer in a nonuniformal soil. *Soil Science Society of America Journal* 64 (4):1219–1225.

Kasubuchi, T. 1977. Twin transient state cylindrical probe method for the determination of the thermal conductivity of soil. *Soil Science* 124(5):255–258.
Kaune, A., Turk, T., and Horn, R. 1993. Alteration of soil thermal–properties by structure formation. *Journal of Soil Science* 44(2):231–248.
Kemp, P. R., Cornelius, J. M. and Reynolds, J. F. 1992. A simple model for predicting soil temperatures in desert ecosystems. *Soil Science* 153(4):280–287.
Kersten, M. S. 1949. *Thermal Properties of Soils*, Bulletin 28. Washington DC: University of Minnesota Engineering Experiment Station. 227.
Keshari, A. K. and Koo,M. H. 2007. A numerical model for estimating groundwater flux from subsurface temperature profiles. *Hydrological Processes* 21:3440–3448.
Kluender, R., Thompson, L. and Staygerwald, D. 1993. A conceptual–model for predicting soil temperatures. *Soil Science* 156(1):10–19.
Kluitenberg G. J., Bristow, K. L., and Das, B. S. 1995. Error analysis of the heat pulse method for measuring soil heat capacity diffusivity and conductivity. *Soil Science Society of America Journal* 59(3):719–726.
Krarti, M., Claridge, D., and Kreider, J. 1995. Analytical model to predict nonhomogeneous soil–temperature variation. *Journal of Solar Energy Engineering, Transactions of the American Society of Mechanical Engineers* 117(2):100–107.
Kroes, J. G., Wesseling, J. C., and van Dam, J. C. 2000. Integrated modelling of the soil–water–atmosphere–plant system using the model SWAP 2.0. An overview of theory and an application. *Hydrological Processes* 14(11–12):1993–2002.
Kujala, K., Seppala, M., and Holappa, T. 2008. Physical properties of peat and palsa formation. *Cold Regions Science and Technology* 52(3):408–414.
Kurpaska, S. and Slipek, Z. 1996. Mathematical model of heat and mass exchange in a garden subsoil during warm-air heating. *Journal of Agricultural and Engineering Research* 65(4):305–311.
Kurpaska, S. and Slipek, Z. 2000. Optimization of greenhouse substrate heating. *Journal of Agricultural and Engineering Research* 76(2):129–139.
Kurpaska, S., Slipek, Z., Bozek, B., and Fraczek, J. 2005. Simulation of heat and moisture transfer in the greenhouse substrate due to a heating system by buried pipes. *Biosystems Engineering* 90(1):63–74.
Lascano, R. J. 1989. Simulation of the energy and water balance of a field crop: Comparison between measured and calculated surface temperatures. *Proceedings of the 1989 Summer Computer Simulation Conference*. San Diego, CA: The Computer Simulation Society. 649–654.
Lascano, R. J. and van Bavel, C. H., 1983. Experimental verification of a model to predict soil moisture and temperature profiles. *Soil Science Society of America Journal* 47(3):441–448.
Ling, F. and T. J. Zhang 2004. A numerical model for surface energy balance and thermal regime of the active layer and permafrost containing unfrozen water. *Cold Regions Science and Technology* 38(1):1–15.
Lipiec, J., Usowicz, B., and Ferrero, A. 2007. Impact of soil compaction and wetness on thermal properties of sloping vineyard soil. *Int. J. Heat Mass Transf.* 50(19–20):3837–3847.
Lu, S., Ren, T. S., Gong, Y. S., and Horton, R. 2007. An improved model for predicting soil thermal conductivity from water content at room temperature. *Soil Science Society of America Journal* 71(1):8–14.
Luo, Y., Loomis, R., and Hsiau, T. 1992. Simulation of soil–temperature in crops. *Agricultural and Forest Meteorology* 61(1–2):23–38.
Manohar, K., Yarbrough, D. W., and Booth, J. R. 2000. Measurement of apparent thermal conductivity by the thermal probe method. *Journal of Testing and Evaluation* 28(5):345–351.
Matthias, A. D. and Warrick, A. W. 1987. Simulation of soil–temperatures with sparse data. *Soil Science* 144(6):394–402.
Mihalakokou, G. 2002. On estimating soil surface temperature profiles. *Energy Building* 4(3):251–259.
Moroizumi, T. and Horino, H. 2002. The effects of tillage on soil temperature and soil water. *Soil Science* 167(8):548–559.

Mortensen, A. P., Hopmans, J. W., Mori, Y. and Šim nek, J. 2006. Multi-functional heat pulse probe measurements of coupled vadose zone flow and transport. *Advances in Water Resources* 29(2):250–267.
Muffay, T. and Verhoef, A. 2007. Moving toward a more mechanistic approach in the determination of soil heat flux from remote measurements—I. A universal approach to calculate thermal inertia. *Agricultural and Forest Meteorology* 147:80–87.
Nassar, I. N. and Horton, R. 1989. Determination of the apparent thermal diffusivitiy of a non-uniform soil. *Soil Science* 147(4):238–244.
Nassar, I. N. and Horton, R. 1990. Determination of soil apparent thermal diffusivitiy from multiharmonic temperature analysis for non–uniform soils. *Soil Science* 149(3):125–130.
Nebbali, R., Makhlouf, S., Boulard, T., and Roy, J. C. 2006. A dynamic model for the determination of thermal boundary conditions in the ground of a greenhouse. *Acta Horticulturae* (ISHS) 719:295–302.
Noborio, K. and McInnes, K. J. 1993. Thermal conductivity of salt–affected soils. *Soil Science Society of America Journal* 57(2):329–334.
Novak, M. D. 1986. Theoretical values of daily atmospheric and soil thermal admittances. *Boundary–Layer Meteorology* 34(1–2):17–34.
Novak, M. D. 1991. Analytical solutions to predict the long-term surface energy balance components and temperatures of a bare soil. *Water Resources Research* 27(10):2565–2576.
Novak, M. D. 1993. Analytical solutions for two-dimensional soil heat flow with radiation surface boundary conditions. *Soil Science Society of America Journal* 57(1):30–39.
Novak, M. D., Chen, W., and Hares, M. A. 2000. Simulating the radiation distribution within a barley–straw mulch. *Agricultural and Forest Meteorology* 102 (2/3): 173–186.
Ochsner, T. E., Horton, R., and Ren, T. 2001. A new perspective on soil thermal properties. *Soil Science Society of America Journal* 65(6):1641–1647.
Ochsner, T. E., Sauer, T. J., and Horton, R. 2006. Field tests of the soil heat flux plate method and some alternatives. *Agronomy Journal* 98(4):1005–1014.
Papadakis, G., Frangoudakis, A., and Kyritsis, S. 1989. Soil energy balance analysis of a solar greenhouse. *Journal of Agricultural and Engineering Research* 43(4):231–243.
Parton, W. 1984. Predicting soil temperatures in a shortgrass stepped. *Soil Science* 138(2):93–101.
Passerat de Silans, A., Bruckler, L., Thony, J. L., and Vauclin, M. 1989. Numerical modelling of coupled heat and water flows during drying in a stratified bare soil comparison with fields observations. *Journal of Hydrology* 105(1–2):109–138.
Patten, H. E. 1909. Heat transference in soils. Washington DC: *U. S. Dept. of Agriculture Bureau of Soils Bulletin 59*, U. S. Department of Agriculture:54.
Persaud, N. and Chang, A. C. 1983. Estimating soil temperature by linear filtering of measured air temperature. *Soil Science Society of America Journal.* 47(5):841–847.
Persaud, N. and Chang, A. C. 1985. Computing mean apparent soil thermal diffusivity from daily observations of soil temperature at two depths. *Soil Science* 139(4):297–304.
Pikul, J. L. 1991. Estimating soil surface temperature from meteorological data. *Soil Science* 151(3):187–195.
Porta, J. , López–Acevedo, M., and Roquero, C. 1999. *Edafología para la agricultura y el medio ambiente.* 2nd ed. Madrid: Ediciones Mundi–Prensa. 849.
Porter, N. and McMahon, T. 1987. A computer simulation model for soil temperatures in Australian cereal cropping. *Soil and Tillage Research* 10(2):131–145 (1987).
Poulovassilis, A., Kerkides, P., Alexandris, S., and Rizos, S. 1998. A contribution to the study of the water and energy balances of an irrigated soil profile. A. Heat flux estimates. *Soil and Tillage Research* 45(1–2):189–198.
Qin, Z., Berliner, H. P., and Karnieli, A. 2002. Numerical solution of a complete surface energy balance model for simulation of heat fluxes and surface temperature under bare soil environment. *Applied Mathematics and Computation* 130(1):171–200.
Renaud, F., Scott, H., and Brewer, D. 2001. Soil temperature dynamics and heat transfer in a soil cropped to rice. *Soil Science* 166(12):910–920.

Richards, J. M. 1971 Simple expression for the saturation vapor pressure of water in the range −50–140. *British Journal of Applied Physics* 4(4):L15–L18.

Rikbost, K., Boersma, A. L., Mack, H. J., and Schmisseur, W. E. 1975. Yield response to soil warming. *Agronomy Journal* 67(6):733–745.

Rodriguez, M. R., Fernandez, M. D., Maseda,F., and Velo, R. 2006. Influence of depth and spacing of an electric cable heating system in a sand substrate. *Applied Engineering in Agriculture* 22(3):443–450.

Rodriguez, M. R., Fernandez, M. D., Maseda, F., Velo, R., and Gonzalez , M. A. 2004. Electrical wire heating in sand substrates: Influence of depth and spacing on power per unit length. *Biosystems Engineering* 89(2):187–195.

Saito, H., J., Šimůnek, J., Hopmans, W., and Tuli, A. 2007. Numerical evaluation of alternative heat pulse probe designs and analyses. *Water Resources Research* 43(7):W07408.

Saito, H., Šimůnek, J., and Mohanty, B. P. 2006. Numerical analysis of coupled water vapor and heat transport in the vadose zone. *Vadose Zone Journal* 5(2):784–800.

Seginer, I. and Jenkins, B. M. 1987. Temperature exposure of greenhouses from monthly means of daily maximum and minimum temperatures. *Journal of Agricultural Engineering Research* 37(3):191–208.

Shao, M., Horton, R., and Jaynes, D. B. 1998. Analytical solution for one-dimensional heat conduction-convection equation. *Soil Science Society of America Journal* 62(1):123–128.

Sharratt, B. S., Campbell, J. S., and Glenn, D. M. 1992. Soil heat-flux estimation based on the finite-difference form of the transient heat-flow equation. *Agricultural and Forest Meteorology* 61(1–2):95–111.

Sikora, E., Gupta, S. C., and Kossowski, J. 1990. Soil temperature predictions from a numerical heat-flow model using variable and constant thermal diffusivities. *Soil and Tillage Research* 18(1):27–36.

Singh, S. R. and Sinha, B. K. 1977. Soil thermal diffusivity determination from overspecification of boundary data. *Soil Science Society of America Journal* 41(5): 831–834.

Song, W. K. 2006. Thermal transfer analysis of a freezing soil medium with an embedded pipeline. *Journal of Cold Regions Engineering* 20(1):20–36.

Strøm, J. S. and Amsen, M. G. 1981. Heat consumption model for greenhouse nurseries. *Acta Horticulturae* 115:503–510.

Sturrock, J. W. 1989. Wind protection of greenhouses. *Acta Horticulturae* 245:94–101.

Sun, S. F. and Zhang, X. 2004. Effect of the lower boundary position of the Fourier equation on the soil energy balance. *Advances in Atmospheric Science* 21(6):868–878.

Tarnawski, V. R. and Gori, F. 2002. Enhancement of the cubic cell soil thermal conductivity model. *International Journal of Energy Research* 26(2):143–157.

Tarnawski, V. R. and Leong, W. H. 2000. Thermal conductivity of soils at very low moisture content and moderate temperatures. *Transport in Porous Media*. 41(2):137–147.

Tarnawski, V. R. Leong, W. H., and Bristow, S. K. L. 2000. Developing a temperature–dependent Kersten function for soil thermal conductivity. *International Journal of Energy Research*. 24(15):1335–1350.

Tarnawski, V. R., Leong, W. H., Wagner, B., and Gori, F. 2001. An expert system for estimating soil thermal and transport properties. *Strojniški vestnik–Journal of Mechanical Engineering*. 47(8):390–395.

Thanh, N. T., Sahli, H., and Hao, D. N. 2007. Finite-difference methods and validity of a thermal model for landmine detection with soil property estimation. *IEEE Transactions in Geoscience Remote Sensing*. 45(3):656–674.

Thomas, H. R. and Cleall, P. J. 2000. A validation exercise for THM modelling in unsaturated soil. *Proceedings of the European Congress on Computational Methods in Applied Sciences and Engineering (ECCOMAS)* Barcelona. p. 12.

Thunholm, B. 1988. Methods for estimating thermal diffusivity of the soil based on annual temperature variations. *Swedish Journal of Agricultural Research*. 18(4):179–184.

Timlin, D. J., Pachepsky, Y., Acock, B. A., Šimůnek, J., Flerchinger, G., and Whisler, F. 2002. Error analysis of soil temperature simulations using measured and estimated hourly weather data with 2DSOIL. *Agricultural Systems*. 72(3):215–239.

Tuntiwaranuruk, U., Thepa, S., Tia, S., and Bhumiratana, S. 2006. Modeling of soil temperature and moisture with and without rice husks in an agriculture greenhouse. *Renewable Energy*. 31(12):1934–1949.

Usowicz, B., Lipiec, J., and Ferrero, A. 2006. Prediction of soil thermal conductivity based on penetration resistance and water content or air-filled porosity. *International Journal of Heat and Mass Transfer*. 49(25–26):5010–5017.

van de Braak, N. J., Kempkes, F. L. K., Bakker, J. C., and Breuer, J. J. G. 1998. Application of simulation models to optimise the control of thermal screens. *Acta Horticulturae*. 456:391–398.

van Donk, S. J., Tollner, E. W., Steiner, J. L., and Evett, S. R. 2004. Soil temperature under a dormant bermudagrass mulch: Simulation and measurement. *Transactions of the American Society of Agricultural Engineers*. 47(1):91–98.

van Rooyen, M. and Winterkorn, H. F. 1959. Structural and textural influences on thermal conductivity of soils. *Proceedings of the 38th Annual Meeting of the Highway Research* Board. Washington DC: National Highway Research Board:576–621.

Villalobos, F. J., Mateos, L., and Ferreres, E. 2002. *Fitotecnia. Bases y tec–logías de la producción agrícola*. Madrid: Ed. Mundi–Prensa. 496.

Vinocur, M. G. and Ritchie, J. T. 2001. Maize leaf development biases caused by air–apex temperature differences. *Agronomy Journal*. 93(4):767–772.

Washitani, I. 1985. Germination rate depending on temperature of *Geranium carolinium* seeds. *Journal of Experimental Botany*. 36(163):330–337.

Washitani, I. and Saeki, T. 1986. Germination responses of *Pinus densiflora* seeds to temperature light and interrupted imbibition. *Journal of Experimental Botany*. 37(182):1376–1397.

Weber, S., Graf, A., and Heusinkveld, B. G. 2007. Accuracy of soil heat flux plate measurements in coarse substrates—Field measurements versus a laboratory test. *Theories of Applied Climatology*. 89(1–2):109–114.

Wierenga, P. J., Nielsen, D. R., and Hagan, R. M. 1969. Thermal properties of a soil based upon field and laboratory measurements. *Soil Science Society of America Journal*. 33(3):354–561.

Wierenga, P. G. and de Wit, C. T. 1970. Simulation of heat transfer in soils. *Soil Science Society of America Journal*. 34:845–848.

Wiltshire, R. J. 1983. Periodic heat conduction in a non-uniform soil. *Earth Surface Processes and Landforms*. 8(6):547–555.

Wu, H. J., Wang, S. W., and Zhu, D. S. 2007. Modelling and evaluation of cooling capacity of earth-air-pipe systems. *Energy Conversion and Management*. 48(5):1462–1471.

Yadav, M. R. and Saxena, G. S. 1973. Effect of compaction and moisture content on specific heat and thermal capacity of soils. *Journal of the Indian Society of Soil Science*. 21:129–132.

Zeroni, M., Nir, A., and Kopel, R. 1984. Root zone heating as an element in energy conservation in a seasonal heat storage greenhouse *Acta Horticulturae*. 148:859–864.

Zhang, H. F., Ge, X. S., Ye, H., and Jiao, D. S. 2007. Heat conduction and heat storage characteristics of soils. *Applied Thermal Engineering*. 27:369–373.

Zhang, Y., Wang, S., Baff, A. G., and Black, T. A. 2008. Impact of snow cover on soil temperature and its simulation in a boreal aspen forest. *Cold Regions Science and Technology*. 52(3):355–370.

CHAPTER 5
Geographic Information System-Based Watershed Modeling Systems

Puneet Srivastava
Biosystems Engineering
Auburn University, Auburn, Alabama

Latif Kalin
School of Forestry and Wildlife Science
Auburn University, Auburn, Alabama

5.1 Introduction

Agricultural, urban, forest, and mining nonpoint source (NPS) pollutants continues to impact and degrade surface and groundwater quality. Among the laws that brought NPS pollution to international attention was the U.S. Clean Water Act (CWA) of 1972 and its amendments of 1987. More specifically, Section 319 of the 1987 CWA amendments authorized planning and limited implementation funds to U.S. states for the assessment of NPS problems and development and implementation of programs for their control (Logan 1990). Furthermore, the U.S. Environmental Protection Agency (USEPA), the federal agency that administers the CWA, required states to develop a list of high-priority or critical bodies of water [called the 303(d) list] on a watershed-by-watershed basis to the maximum practicable extent and to develop total maximum daily loads (TMDLs) for pollutants in these waters. A TMDL is defined as the maximum amount of pollutant that a body of water can receive and still meet the water quality standards. TMDLs include both point source and NPS discharges that arise from a watershed or the environs of a watercourse (Ward and Benaman 1999). The CWA requires development of TMDLs for all waters on the 303(d) list by developing restoration scenarios. The creation of this list is authorized by Section 303(d) of the CWA. The ultimate goal of a TMDL development and implementation can be stated as removal of bodies of water from the 303(d) list by attaining water quality standards. Eventually, the list of impaired bodies of water and established TMDLs compiled by states, territories, and authorized tribes must be approved by USEPA.

These requirements made by USEPA lead to the assessment of NPS water quality problems and identification of critical contributing areas in several states. For example, in the U.S. state of Pennsylvania, Hamlett et al. (1995) identified and ranked critical NPS contributing areas on a watershed-by-watershed basis for the entire state. In Virginia, Tim et al. (1992) used an integrated approach, coupling a water quality computer simulation model and the Virginia Geographic Information System (GIS) to delineate critical areas of NPS pollution in the Nomini Creek watershed located in Westmoreland County, Virginia. Evans and Myers (1990) developed a methodology using GIS and DRASTIC to evaluate groundwater pollution potential in regional areas larger than 13,000 ha. DRASTIC is a widely used ground-water vulnerability mapping method. DRASTIC is named for the seven factors considered in the method: *D*epth to water, net *R*echarge, *A*quifer media, *S*oil media, *T*opography, *I*mpact of vadose zone media, and hydraulic *C*onductivity of the aquifer (Aller et al. 1987).

5.2 Watershed Models

5.2.1 Need for Watershed Models

The CWA also required states to develop management plans to remediate NPS pollution problems. With limited resources available (in terms of time, labor, and money), it is imperative that control and implementation programs focus on critical contributing areas and adequately consider the impacts of alternative management, land use, and conservation approaches (e.g., conservation tillage, contour cropping, strip cropping, and fertilizer management) on NPS pollution. Evaluating NPS pollutant reduction effectiveness of alternative management, land use, and conservation practices at a watershed scale through experiments and monitoring systems is not feasible because of the enormous cost, time, and labor involved. Modeling studies based on experimental data are often the only viable means of providing timely inputs to management decisions with the least cost. Therefore, modeling NPS pollutant fate and transport processes at a watershed scale is fundamental to addressing contamination of surface and ground waters. Watershed-scale NPS models are currently used for a variety of purposes, including design of soil conservation practices, water table management, prevention of chemical pollution of surface bodies of water and groundwater, protection of aquatic biota, and development of TMDLs. In the years ahead, worldwide, watershed models will play an increasing important role in managing NPS pollutants.

5.2.2 Origin of Watershed Models

Because NPS pollutant transport is mainly driven by meteorological events, the early to mid-20th century saw the development of mathematical descriptions of individual hydrologic components [e.g., infiltration, runoff, evapotranspiration (ET), and interception]. The digital revolution of 1960s witnessed the integration of individual hydrologic components (Singh and Woolhiser 2002) into functional models that can be applied at various spatial and temporal scales. Initially, the models were developed and applied at point and field scales. However, it was quickly realized that, to truly address NPS pollution, watershed-scale models are needed. The development of watershed-scale hydrologic and NPS models in the United States began in response to the CWA (Arnold and Fohrer 2005). Examples of these models include the Agricultural Non-Point Source Pollution Model (AGNPS) (Young et al. 1987), Annualized AGNPS (AnnAGNPS) (Bingner and Theurer 2003), Hydrologic Simulation Program—Fortran (HSPF) (Bicknell et al. 2001), the Kinematic Erosion Model (KINEROS) (Woolhiser et al. 1990), and the Soil and Water Assessment Tool (SWAT) (Neitsch et al. 2002).

5.2.3 Characterization of Watershed Models

Models in general and watershed models in particular can be characterized as mechanistic or empirical (based on the cognitive value of a model), stochastic or deterministic (based on the character of results obtained), linear or nonlinear (based on the mathematical properties of the operator function), event or continuous simulation, and lumped or distributed parameter model (Haan et al. 1982).

Truly mechanistic models are those in which governing physical, chemical, and biological laws and the model structure are well known and can be described by mathematical equations. Empirical models are used when model structure and governing laws are unknown or the mechanistic model is so complicated that simplification of model behavior is needed. In reality, most current watershed-scale models have mechanistic and empirical components. Further, most make an attempt to model physical, chemical, and biological processes that occur on land or in bodies of water (e.g., streams, ponds, lakes, or reservoirs). These models are best described as process-based models. If any of the variables in a process-based model is regarded as a random variable having a probability distribution function, the model is called a stochastic model. However, if all of the variables are free from random variations, then the model is a deterministic model.

Although most present-day models are deterministic models, as many of the input variables, model parameters, and modeled processes are stochastic in nature, stochasticity is introduced through model uncertainty analysis.

Because watershed-scale models have mostly nonlinear components, based on the mathematical properties of the operator function, they can best be regarded as nonlinear models. Linear models, such as the unit hydrograph theory, are based on two simple principles: the principle of proportionality and the principle of superposition. The former can be stated as follows: if $f(x)$ is a solution of a system, then $c \cdot f(x)$ is also a solution of the same system, with c being a constant. The latter principle implies that if $f_1(x)$ and $f_2(x)$ are both solutions of the same system, then $f_1(x) + f_2(x)$ is also a solution of the same system.

Based on the length of simulation, models can be characterized as event or continuous. An event model represents runoff event occurring over a period of time ranging from hours to several days, whereas a continuous simulation model can operate over an extended period of time, simulating flows and conditions during both runoff periods and nonrunoff periods. Continuous models keep a continuous account of watershed characteristics. Watershed models can also be characterized as lumped or distributed parameter models. A distributed parameter model directly uses the areal variations in watershed characteristics (e.g., soils, land use, slope, and rainfall); a lumped parameter model cannot do this. Most watershed-scale models, however, are lumped to some extent. Based on the criteria listed above, currently used popular watershed-scale hydrologic and NPS models can be characterized as nonlinear, process based, deterministic, distributed parameters, and events or continuous simulations. In this chapter, we limit our discussion to these types of models.

5.2.4 Important Components of Watershed Models

Although descriptions of processes vary in different watershed-scale models, most models allow various physical, chemical, and biological processes to be simulated in a watershed. Furthermore, because water balance is the driving force behind accurate prediction of movement of sediment, nutrients, and pesticides, accurate simulation of various components of hydrologic cycle is important for watershed models. The following water balance equation, especially in continuous simulation models, is often used (Neitsch et al. 2005):

$$SW_t = SW_0 + \sum_{i=1}^{t}(R_{day} - Q_{surf} - E_a - w_{seep} - Q_{gw})$$

where SW_t = the final soil water content (mm H_2O)
SW_0 = the initial soil water content on day i (mm H_2O)
t = the time (days)

R_{day} = the amount of precipitation on day i (mm H_2O)
Q_{surf} = the amount of surface runoff on day i (mm H_2O)
E_a = the amount of evapotranspiration on day i (mm H_2O)
w_{seep} = the amount of water entering the vadose zone from the soil profile on day i (mm H_2O)
Q_{gw} = the amount of return flow on day i (mm H_2O)

To simulate watershed hydrology, a hydrologic cycle is usually divided into the land phase (large dashed box in Fig. 5.1) and the routing phase (small dashed box, same figure). The land phase of the hydrologic cycle transports water, sediment, nutrient, and pesticide loads from a land surface to a stream, whereas the routing phase transports them through stream channels of the watershed to the outlet. Figure 5.1 is a schematic of hydrologic processes/water transport pathways simulated in one of the most widely used watershed models, the SWAT model. These individual processes have been described in more detail in Chap. 3.

To account for areal variations in watershed characteristics (e.g., soils, land use, slope, and rainfall), a distributed parameter model subdivides the watershed into subwatersheds, grid cells, or hydrologic response units (HRUs). Runoff is predicted separately in each smaller unit and routed to obtain total runoff for a watershed. The subdivision of a watershed increases accuracy and gives a more accurate physical description of the hydrologic processes (Neitsch et al. 2005).

As discussed above, water balance is the driving force behind accurate prediction of movement of sediment, nutrients, and pesticides. For example, in many of the watershed-scale models, runoff volume and peak runoff rate are used to simulate erosion and sediment yield. Watershed models also track transport and transformation of various forms of nutrients [nitrogen (N) and phosphorus (P)] in a watershed. In the soil, transformation of N is governed by the N cycle, whereas transformation of P is governed by the P cycle. To simulate point sources, nutrients can be added to the main channel and transported downstream through the stream flow. Inorganic and organic forms of nutrients applied can be taken by plants, adsorbed by soils, move to streams and lakes/reservoirs through surface runoff or lateral subsurface flows, or percolate to deeper groundwater. Pesticide movement is controlled by its solubility, degradation half-life, and soil organic carbon adsorption coefficient.

The loadings of water, sediment, nutrients, and pesticides from the landscape are routed through a watershed's stream network. While keeping track of the mass flow of pollutants in the stream channel, most watershed models also account for transformation of pollutants in the stream. Many sophisticated watershed models also simulate movement and transformation of pollutants in lakes and reservoirs, although not as comprehensively as the reservoir models.

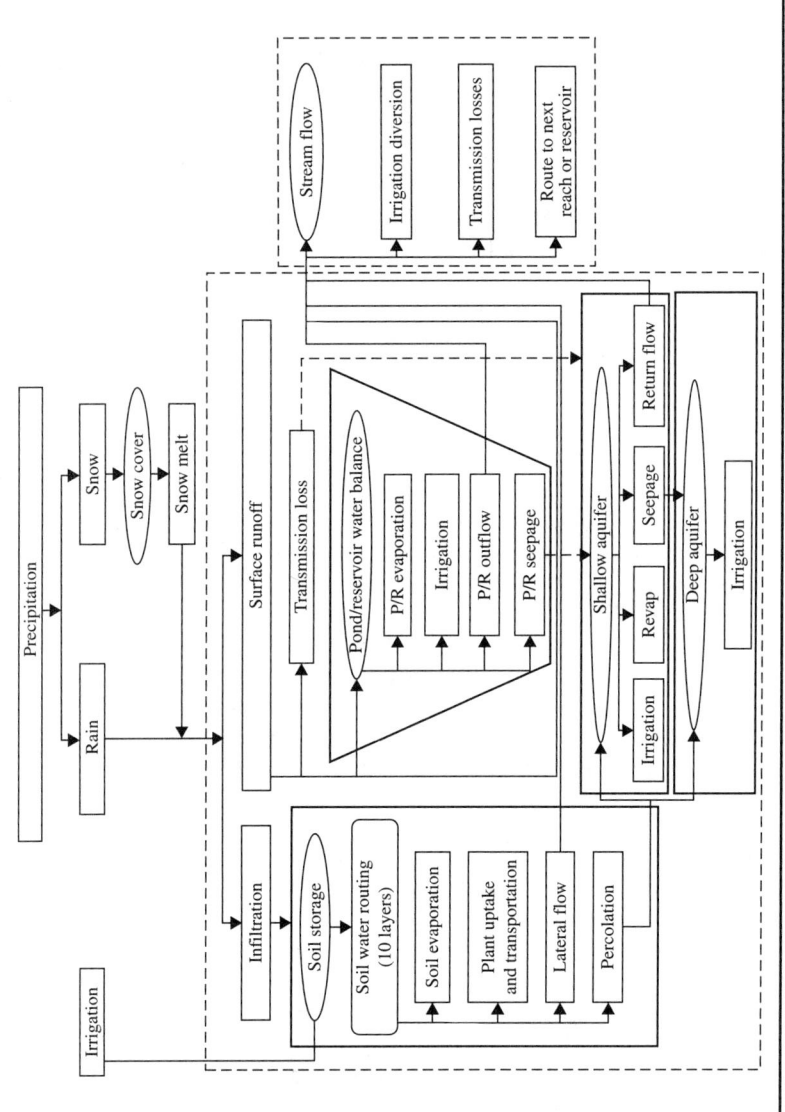

FIGURE 5.1 Schematic showing common hydrologic processes/water movement pathways simulated in watershed-scale NPS models. *(Redrawn after Neitsch et al. 2005.)*

5.2.5 Examples of Commonly Used Watershed Models

Currently, a number of NPS models exist that are designed to address specific problem domains. Some of the more widely used watershed-scale NPS models are described in this section.

SWAT: Soil and Water Assessment Tool

The SWAT model (Neitsch et al. 2002) is a modified version of the simulator for water resources in rural basins (SWRRB) and routing outputs to outlets (ROTO) models for application to large, complex rural basins and uses new routing structures. Simulator for water resources in rural basins—water quality (SWRRBWQ) was a continuous simulation, daily time–step computer model developed to simulate hydrologic and nutrient-transport processes in rural basins. It was designed to predict the effect of management decisions on water, sediment, nutrients, and pesticide yields at the subbasin or basin outlet. In SWRRBWQ a basin could be divided into a maximum of 10 subbasins to account for differences in soils, land use, crops, topography, vegetation, or weather. SWRRBWQ also had a water quality component that tracked the fate of pesticides and P from their initial application on the land to their final deposition in a lake. SWRRBWQ could be used to model the effect of farm-level management systems such as crop rotations, tillage, planting date, irrigation scheduling, and fertilizer and pesticide application rates and timing. The modified universal soil loss equation (MUSLE) was used to determine sediment yield. Nutrient, pesticide, and sediment yields at the basin outlet were determined after accounting for channel transmission losses and deposition in the subbasins.

SWAT is an extended and improved version of SWRRB, running simultaneously in several hundred subbasins to predict the effects of management practices on sediment and chemical yields from large river basins. SWAT has the ability to simulate surface flow, subsurface flow, sediment, nutrients, pesticides, and bacteria in addition to various best management practices (BMPs): agricultural practices, ponds, and tile drains, for example. Management practices are handled within the MUSLE. Soil Conservation Service (SCS) curve numbers can also be varied throughout the year to take variations in management conditions into account. SWAT divides the watershed into HRUs that have uniform properties. Edge-of-filter strips may be defined in an HRU. The filter strip trapping efficiency for sediment is calculated empirically as a function of the width of the filter strip. When calculating sediment movement through a body of water, SWAT assumes that the system is completely mixed. Settling occurs only when the sediment concentration in the water body exceeds the equilibrium sediment concentration specified by the user. The sediment concentration at the end of a day is determined based on an exponential decay function. SWAT also simulates the buildup and washoff mechanisms similar to the Storm Water Management

Model (SWMM). SWAT has its own GIS interface that uses the Environmental Systems Research Institute's (ESRI) ArcView 3.X and ArcGIS GIS softwares. SWAT is also linked to the water quality model QUAL2E for in-stream nutrient processes.

AGNPS and AnnAGNPS Pollutant Loading Model

AnnAGNPS, developed by the USDA-ARS at the National Sedimentation Laboratory, Oxford, Mississippi, is a batch-process continuous-simulation, distributed parameter, watershed-scale pollutant–loading computer model developed in ANSI Fortran 90. This model (Young et al. 1987; Bingner and Theurer 2003) is mainly an expansion of the capabilities in the AGNPS single-event model.

AGNPS, a watershed-scale, distributed parameter, event-based model, was designed to simulate runoff, sediment, and nutrient transport. Square cell representation was used in AGNPS to represent field boundaries. AGNPS used the USDA SCS runoff curve number and unit hydrograph method for calculating flow volume and peak flow, respectively. The modified USLE was used for erosion, and a simple correlation of extraction coefficients was used for nutrient and sediment. Capabilities were also provided in AGNPS to simulate gully, wastewater treatment plants, and feedlot point source pollution as well as a number of agricultural practices. Flow, sediment, and nutrients could be calculated in every cell within the watershed.

AnnAGNPS divides a watershed into homogeneous land areas with respect to soil type, land use, and land management. The areas can be of any size and any shape, including hydrologically-based and square grids. AnnAGNPS simulates movement of surface water, sediment, N, P, and pesticides leaving the land areas as well as their travel through the watershed. Portions of water, sediment, N, P, and pesticides reach the watershed outlet while the remainder is deposited within the watershed. Calculations are done on a daily time step and pollutant loads generated from land areas are routed through the stream systems on a daily basis.

AnnAGNPS is suitable for evaluating long-term NPS pollution from agricultural watersheds. AnnAGNPS is also suitable for comparing the effects of alternative cropping and tillage systems, fertilizer, pesticides, and irrigation application rates and feedlot management systems on levels of NPS pollutants. AnnAGNPS is also capable of simulating point source loads. AnnAGNPS generates both event- and source-accounting outputs. Source accounting indicates the fraction of a pollutant loading at the watershed outlet or at a particular reach coming from a user-identified source location. Outputs generated by AnnAGNPS are water, sediment, nutrients, and pesticides at the user-defined watershed source location (specific cell, reach, feedlots, point sources, and gullies).

AnnAGNPS uses the SCS curve number method for determining runoff volume and peak discharge is calculated using a procedure similar to the TR-55 method. The sheet and rill erosion model is based on RUSLE. RUSLE is an erosion model that predicts longtime average annual loss of sediment resulting from raindrop splash and runoff from specific field slopes in specified cropping and management systems. HUSLE is used to determine the delivery ratio for the sheet and rill erosion for each cell to its receiving reach. The delivery ratio for the individual particle-size classes is proportioned according to their respective fall velocities. The resulting sediment is called the sediment yield to the stream system. Sediment reach routing by class of particle size is based on a modified Einstein deposition equation using the Bagnold suspended sediment formula for transport capacity. If the upstream sediment load is greater than the transport capacity for the respective particle-size class, then degradation is assumed.

Mass balance of N, P, and organic carbon is computed for each cell on a daily basis. Uptake of N and P by plants, application of fertilizers, residue decomposition, and downward movement of N and P are the major components considered in AnnAGNPS. The output from each field includes sediment N, dissolved N, sediment P, dissolved P, sediment organic carbon, and dissolved organic carbon. AnnAGNPS partitions N and P into organic and mineral parts, and a separate mass balance is computed for each. The N and organic carbon cycles represented in AnnAGNPS are simplifications that track only major N transformations of mineralization from humified soil organic matter and plant residues, crop residue decay, and fertilizer and plant uptake. A simple crop-growth stage index is used to model plant uptake of N and P.

Runoff, ET, and percolation are considered to affect soil moisture. Runoff is calculated using the SCS curve number method, and potential ET is calculated using the Penman equation. Actual ET is estimated as a function of potential ET and available soil moisture content. Percolation is assumed to occur at the rate of the hydraulic conductivity corresponding to the soil moisture content using the Brooks–Corey equation.

AnnAGNPS also includes winter routing for snow, snowmelt, and frozen soil by maintaining a daily thermal energy balance to track the soil and snowpack temperatures. Rainfall is supplemented by snowmelt. The depth of the uppermost frozen soil layer is used to adjust the curve number and soil erodibility for potential sheet and rill erosion.

Several GIS interfaces using ArcView 3.X and ArcGIS have been developed for the AGNPS and AnnAGNPS models.

Hydrologic Simulation Program—Fortran

Hydrologic Simulation Program—Fortran (HSPF) (Bicknell et al. 2001), supported by both USEPA and the U.S. Geological Survey (USGS), performs long-term simulations of the hydrologic and associated water-quality processes on pervious and impervious land surfaces and in streams and well-mixed impoundments. It is incorporated into the Better Assessment Science Integrating Point and Nonpoint Sources (BASINS) and Watershed Modeling System (WMS) modeling systems (described later) and is extensively used for TMDL studies. The model contains hundreds of process algorithms developed from theory, laboratory experiments, and empirical relations from instrumented watersheds. There are three basic modules: Pervious Land Segment Module (PERLND), a watershed loading model for pervious surfaces; Impervious Land Segment Module (IMPLND), a watershed loading model for impervious surfaces; and, the Reach/Reservoir Routing Module (RCHRES), a one-dimensional receiving water model for simulating in-stream transport and transformation processes. HSPF is based on the Stanford Watershed Model, Agricultural Runoff Management (ARM), and Non-Point Source (NPS) models. HSPF uses simple storage-based equations for flow routing. Flows in streams are one dimensional. HSPF is one of the few comprehensive models of watershed hydrology and water quality that allows for integrated simulation of land and soil contaminant runoff processes with in-stream hydraulic and sediment–chemical interactions. HSPF uses continuous rainfall and other meteorologic input parameters to compute streamflow hydrographs and pollutographs. HSPF simulates interception, soil moisture, surface runoff, interflow, baseflow, snowpack depth and water content, snowmelt, ET, groundwater recharge, dissolved oxygen, biochemical oxygen demand, temperature, pesticides, conservatives, fecal coliforms, sediment detachment and transport, sediment routing by particle size, channel routing, reservoir routing, constituent routing, pH, ammonia, nitrite-nitrate, organic N, orthophosphate, organic P, phytoplankton, and zooplankton. HSPF can simulate one or many pervious or impervious unit areas discharging to one or many river reaches or reservoirs. Frequency-duration analysis can be done for any time series. Time steps from 1 minute to 1 day, divided equally within 1 day can be used. Time period of simulation can be from few minutes to hundreds of years. HSPF is generally used to assess the effects of land-use change, reservoir operations, point or NPS treatment alternatives, flow diversions, etc. A number of programs are available for pre- and postprocessing of data for statistical and graphical analysis of data saved to the Watershed Data Management (WDM) file. The most famous application of the HSPF model is the Chesapeake Bay Project.

KINEROS-2

KINEROS-2, an improved version of KINEROS (Woolhiser et al. 1990) model, is an event-based model because it lacks a true soil moisture redistribution formulation for long rainfall interval. Also, it does not consider ET losses. The model is primarily useful for predicting surface runoff and erosion for a rainfall event over small agricultural and urban watersheds. Smith et al. (1995) suggest using a watershed size smaller than 1000 ha for best results. Runoff is calculated based on the Hortonian approach using a modified version of the Smith–Parlange (Smith and Parlange 1978) infiltration model. KINEROS-2 requires the watershed to be divided into homogeneous overland flow planes and channel segments and routs water movement over these elements in a cascading fashion. An implicit finite difference numerical scheme in a one-dimensional framework is used for mass balance and kinematic wave approximations to the Saint–Venant equations. KINEROS-2 accounts for erosion resulting from raindrop energy and by flowing water separately. A mass balance equation is solved to describe sediment dynamics at any point along a surface flow path. Erosion is based on maximum transport capacity determined by Engelund–Hansen equation (1967). A first-order uptake rate is to determine rate of sediment transfer between soil and water. KINEROS-2 can be used under the Automated Geospatial Watershed Assessment system (AGWA) (described later), which provides a GIS interface for data preparation and visualization of results.

MIKE SHE

MIKE SHE, from the Danish Hydraulic Institute (DHI) is an integrated hydrological modeling system for simulating surface and groundwater flows. The user-friendly tools provided by MIKE SHE can simulate the entire land phase of the hydrologic cycle and allows components to be used independently and customized to local needs. MIKE SHE emerged from the System Hydrologique European (SHE) as developed and extensively applied since 1977 onward by a consortium of three European organizations: the Institute of Hydrology (the United Kingdom), Société Grenbloise d'Etudes et d'Application Hydrauliques (SOGREAH) (France), and DHI Water.Environment. Health (Denmark). Since then, DHI has continuously put forth efforts in development and research on MIKE SHE. MIKE SHE can be used for the analysis, planning, and management of a wide range of water resources and environmental problems related to surface water and groundwater. Examples include surface water impact from groundwater withdrawal, conjunctive use of groundwater and surface water, wetland management and restoration, river basin management and planning, environmental impact assessments, aquifer vulnerability mapping with dynamic recharge and surface water boundaries, groundwater management, floodplain studies, impact studies for changes in land use, climate, and agricultural practices.

MIKE SHE is a watershed model and is distributed parameter and physically based. It is capable of handling both single events and continuous simulations. The watershed is divided into square grid cells. Overland flow routing is based on two-dimensional diffusive wave equations, where options vary for channel flow from simple Muskingum routing to the Higher Order Dynamic Wave formulation of the Saint–Venant equations. Groundwater flow is solved with the three-dimensional full Richards' equation. Stream–groundwater interactions are considered. In general, depending on the size of the watershed, simulations can be computationally very intensive. MIKE SHE can be used in combination with MIKE-11 for river hydraulics. This modeling package, however, is proprietary. The GIS interface for MIKE SHE uses ArcView 3.X GIS.

SWMM

The USEPA's Storm Water Management Model (SWMM) model (Huber and Dickinson 1988) is a comprehensive computer model for analyzing water quantity and quality issues associated with urban runoff. Both, single events and continuous simulation can be performed on catchments having storm sewers, or combined sewers and natural drainage, to predict flows, stages, and pollutant concentrations. The structure of SWMM contains several computational blocks including Runoff, Transport, Storage/Treatment, and Extended Transport or Extran. The Runoff Block is used for generating runoff and quality constituents from rainfall (plus simple routing of flow and quality), the Transport Block for kinematic wave routing and for additional dry-weather flow and quality routing, the Storage/Treatment Block for reservoir routing and simulation of treatment and storage quality processes, and the Extended Transport or Extran Block for hydraulic routing of flow (no quality routing) using the Saint–Venant equations. All aspects of the urban hydrologic and quality cycles, including rainfall, snowmelt, surface and subsurface runoff, flow routing through the drainage network, storage, and treatment can be simulated using the SWMM model. The Rain Block is used to process hourly and 15-min precipitation time series for input to continuous simulation. Although historically designed to address urban runoff quality issues, the model is often used for hydrologic and hydraulic analyses. SWMM Version 4 is microcomputer based (DOS compatible). For hydrologic simulation in the Runoff Block, data requirements include area, imperviousness, slope, roughness, width (a shape factor), depression storage, and infiltration parameters for either the Horton or Green–Ampt equations for up to 100 subcatchments (number of subcatchments, pipes, etc. are variable depending on the compilation). Flow routing can be performed in the Runoff, Transport, and Extran Blocks in increasing order of sophistication. Extran can also simulate dynamic boundary conditions (e.g., tides). Quality processes, initiated in the Runoff Block, include options for constant concentration, regression of load versus flow, and buildup/washoff. Options such as street cleaning, erosion, and quality contributions from precipitation, catchbasins, adsorption, and base flow can also be simulated. USEPA Nationwide Urban Runoff Program data are often used as starting values for quality computations. Basic SWMM output consists of hydrographs and pollutographs (concentration vs. time) at any desired location in the drainage system. Depths and velocities are also available as are summary statistics on surcharging, volumes, continuity, and other quantity parameters. Additional quality output includes loads, source identification, continuity, residuals (e.g., sludge), and other parameters. Some of the main limitations of the SWMM model include lack of subsurface quality routing (a constant concentration is used), lack of interaction of quality processes (apart from adsorption), difficulty in simulating wetlands quality processes, and a weak scour deposition routine in the Transport Block. The current edition, version 5.0 (Rossman 2008), is a complete rewrite of the previous release and uses the Windows operating system to provide an integrated environment for editing study area input data; running hydrologic, hydraulic, and water quality simulations; and viewing the results in a variety of formats. These formats include color-coded drainage area and conveyance system maps, time series graphs and tables, profile plots, and statistical frequency analyses. Although a Windows-based graphical interface is present for this model, it has not been integrated with GIS. SWMM has been extensively used in United States, Canada, Europe, and Australia.

5.3 Geographic Information Systems (GIS)

A GIS is also known as a number of alternative names, including Land Information System, Spatial Information System, Geoscience Information System, Geomatics, and Geographically Referenced Information System. It is a powerful collection of tools for capturing and converting, storing and managing, manipulating and analyzing, editing and updating, and displaying and outputting spatial and nonspatial data. Several important components, including hardware, software, data, methods, and people, are integral components of a GIS. A number of disciplines have contributed to the development and advancement of a GIS, including geography, cartography, remote sensing, photogrammetry, surveying, geodesy, statistics, operations research, computer science, and mathematics. The power of GIS comes from linking real-world locations with the attribute data associated with those locations. These locations could be a point, a line, or a polygon feature, or a unit in space. This powerful linking allows us to (1) tie data analyses to real-world locations, (2) visually represent data, and (3) makes, at will, sophisticated spatial analysis feasible, which are not feasible using only maps or only data. Some of the basic questions that can be asked of a GIS include the following. Where is it (i.e., location)? What has changed since (i.e., trend)? Which is the best way (i.e., routing)? What is the pattern (i.e., pattern)? and What if (i.e., modeling)? Currently, a number of vendors provide GIS software including Environmental Systems Research Institute (ESRI); Strategic Locations Planning, Inc.; ERDAS, Inc.; U.S. Army Corps of Engineers; Mapinfo Corporation; and Intergraph Corporation.

Some of the most widely used software provided by these companies include ArcGIS, ArcView, ERDAS Imagine, GRASS, Mapinfo, and TIGRIS.

These softwares have ability of capture, manage, manipulate, analyze, and output spatially referenced data such as points, lines, and polygons (using vector data structure) or a unit space (using tessellation or raster data structure). A vector data structure represents geographic features and objects as points, lines, and polygons; the information (attributes) associated with these features are stored in relational database management systems. Examples of features relevant to watershed modeling that use vector data structures include streams (line features); watershed outlets, point sources, water quality and streamflow measurement locations, etc. (point features); and soil types and land use/land cover (polygon feature). Most of the vector data structures also store topological relationships (either explicitly or implicitly), describing a feature's spatial relation with neighboring features, specifically its connectivity and adjacency. The topological relationship provides a GIS a powerful facility for efficient spatial analyses.

A raster (or tessellation) data structure divides space into two-dimensional units of space. Square grids (square pixels) and triangulated irregular networks (TIN) are most common raster data structures used for storing raster data relevant for watershed modeling. In a square grid, each square cell contains a representative value of the attribute being mapped. In a raster data structure, a point is represented as a cell, a line by a string of connected cells, and areas by groups of adjacent cells. Examples of raster datasets used for watershed modeling include digital elevation models (DEMs) and satellite-derived land use/land cover maps.

Using the vector and raster data sets, a number of watershed characteristics can be derived for distributed parameter watershed modeling. Examples include watershed and subwatershed boundaries, stream networks, slope, aspect, and soil and land use for each individual watershed elements (i.e., subwatersheds or hydrologic response units). These derived characteristics are important state variables (i.e., the state of the watershed for a period for which the watershed modeling is being performed) to a distributed parameter model. Once a watershed model is run, a GIS can help manage and spatially display the vast amount of data generated by a model.

5.4 GIS and Watershed Models

Regardless of the level of sophistication, application of a dynamic watershed model to preserve water quality on a site-specific basis requires efficient acquisition, storage, organization, reduction, and analysis of model input data accompanied by manipulation, interpretation, reporting, and display of model output data (DePinto et al. 1994). Input data required by the models are currently available in varying spatial and temporal scales, resolutions, uncertain quality (because of errors associated with measuring, digitizing, editing, etc.), and a multiplicity of formats. The large quantity of complex data can be overwhelming, and model users may spend disproportionate amounts of time on data manipulation and systems integration to perform each application.

The only solution to this problem is a GIS-interfaced tool that supports flexible interactive analysis and visualization. The use of GIS can make the task of compiling necessary spatial data, nonspatial data, and required hydrologic parameters for modeling watershed runoff and water quality more manageable (Bhasker et al. 1992). Interfacing GIS with a NPS pollution model results in a powerful tool that is capable of storage, acquisition, organization, reduction, and analysis of input and output data as well as visualization of input and model simulated data, thereby reducing the time required for interpreting, reporting, and displaying model results. The interfacing of GIS and hydrologic and NPS models also enables users to identify critical areas of pollution and to perform various "what if?" scenarios to better understand the effect of various alternative management strategies on NPS pollution.

In the past, a number of researchers have attempted to interface spatially distributed NPS, and other environmental models, with GIS. For example, Roaza et al. (1993) interfaced a finite element model of an aquifer with GIS to develop a system for optimal management of resources for a sand and gravel aquifer in Escambia County, Florida. Ross and Tara (1993) interfaced a commercial GIS [Tydac Technologies, Inc., Spatial Analysis System (SPANS)], a public domain surface-water model [USEPA-supported HSPF], a groundwater model [MODFLOW (McDonald and Harbaugh 1988)], and an evapotranspiration code to aid P-mining–reclamation design. Hay et al. (1995) interfaced a hydroclimatic model, GIS, Scientific Visualization System (SVS), and an advanced statistical tool (STAT–StatSci S-PLUS) as part of the U.S. Geological Survey's Gunnison River Basin Climate Study. Bromberg et al. (1995) developed the Scientific Geographic Information System (SGIS) integrating Genamap (GIS), S-Plus (Statistical Tool), Geo-EAS (Geostatistical Tool), and the CENTURY model for ecosystem modeling. Geter et al. (1995) described the development of the Hydrologic Unit Water Quality Model (HU/WQ) GIS interface. This interface was developed for four Agricultural Research Service (ARS) pollutant-loading models: AGNPS (Young et al. 1987), SWRRBWQ (Arnold et al. 1990), EPIC (Sharpley and Williams 1990), and GLEAMS (Knisel et al. 1992).

The interface was designed for use by the U.S.D.A. Natural Resource Conservation Service (NRCS) State Office personnel as a screening tool in formulating water quality project plans for impaired watersheds and for evaluating the impacts of conservation and management practice implementation on NPS pollution. Invariably, all these researchers concluded that interfacing GIS with NPS models provided a tool for efficient management of temporally and spatially referenced data and allowed flexible interactive model runs, analyses, and visualization.

5.4.1 Approaches for Interfacing GIS with Watershed Models

Interfacing GIS with distributed parameter hydrologic models is playing an increasingly important role in the management of water quality and water resources as well as in designing, calibrating, modifying, and comparing these models. There are several strategies and approaches to interface spatially distributed environmental models and GIS. They range from simple pre- and postprocessor linkages through shared data files to building models as complex analytical functions into fully functional GIS, or embedding the required GIS functionality into spatially distributed models (Fedra 1995). The interfacing of models, especially watershed models, is not a novel concept anymore and has been tried numerous times since the 1990s. However, as pointed out by Vckovski et al. (1999) and Martin et al. (2005), the lack of consistent data protocols and the lack of interest by commercial software providers in creating universal data transfer standards has stifled the development of standardized frameworks for terminology, data exchange formats and interfacing procedures. The following terminology describe interfacing efforts: "couple," "link," "combine," "interface," "model within," and "integrate" (Martin et al. 2005).

As described by Martin et al. (2005), linking typically means that data generated by GIS is used as input by the models, and output is transferred back to GIS for display and spatial analysis. This is typically done through ASCII and binary data file transfers. This approach limits users to take full advantage of the spatial analysis capabilities of a GIS because it is mainly used for pre- and postprocessing for input and output data. In combining, information is passed between the model and GIS via memory-resident data models rather than external files (Liao and Tim 1997). Martin et al. (2005) stated that "This approach improves computational performance and interactivity between the two software systems, translating into a more sophisticated modeling environment." Although data transfer between models is automated and hidden from the user, the interactivity remains somewhat limited.

The most sophisticated approach for interfacing GIS with a predictive model is termed "integration" or "embedding." In this approach, functional components of one system are incorporated within the other system, thus eliminating the need for intermediate transfer software (Liao and Tim 1997). In this approach, a seamless integration is developed through sharing of processes and data, thus reducing redundancy (Martin et al. 2005) and increasing computational performance. A true integration, however, is difficult, as a lot of communication is required between GIS programmers and model developers. To the best of our knowledge, although limited attempts have been made to integrate simpler water resources models, no attempt to truly integrate a watershed model with GIS has been made due to complexities involved in developing these systems.

As pointed out by a number of researchers (e.g., Tim et al. 1996; Burrough 1997; Liao and Tim 1997) and cited by Martin et al. (2005), interfacing strategies are limited by the lack of compatible data structures, software requirements, and model–GIS functionality requirements. Linking approach underutilizes the functional capabilities of GIS by using is just as a display medium. Combining and integration approaches are more sophisticated but are hindered by huge development costs. Often, the interfacing strategy is limited by research objectives, expertise of developers, and availability of resources. Most current, state-of-the-art GIS-based WMSs can, at best, be described as combined systems.

5.4.2 Challenges with Interfacing

Perhaps the biggest challenge to interfacing a GIS with a watershed model is the lack of a time dimension within the GIS. The absence of time dimension limits a user's ability to readily model, within GIS, spatial variability over time (Martin et al. 2005). The approach to overcoming this is to visualize a time series of historic surveys, remote-sensing data, or future time variations predicted by models using a series of overlays that may be analyzed using statistical approaches (Croft and Kessler 1996).

The relational database structure of GIS also limits the collusion of GIS and some predictive models. The database relation, through a common key item between two sets of databases, is a weak connection between the two entities. Martin et al. (2005) stated that "When compared to the mathematical rigor of a hydrologic model, spatial relationships do not effectively capture the governing hydrologic algorithms. Differential equations utilized in a typical hydrologic model thus have limited operability within a GIS data structure. Accordingly, hydraulic models utilizing advanced algorithms or complex mathematical structures are currently incapable of being fully integrated into a GIS relational database."

5.4.3 Recent State-of-the Art GIS–Integrated Watershed Modeling Systems

Recently, efforts have been made to bring several models under the same framework using GIS interfaces that pre- and postprocess data for several models through the use of a single graphical user interface (GUI). Examples of such attempts are the USEPA Better Assessment Science Integrating Point and Nonpoint Sources (BASINS) (USEPA, 2008) modeling system, WMS, developed by Brigham Young University (Environmental Modeling Systems, Inc. 2007), and AGWA, developed by the USDA Agricultural Research Services (ARS) Southwest Watershed Research Center in cooperation with the USEPA Office of Research and Development.

BASINS

According to the USEPA, "BASINS is a multipurpose environmental analysis system for use by regional, state, and local agencies in performing watershed- and water-quality–based studies" (USEPA 2008). The system was designed to address three objectives:

1. To facilitate examination of environmental information,
2. To support analysis of environmental systems, and
3. To provide a framework for examining management alternatives.

The primary goal of the BASINS system was to support development of TMDLs required by Section 303(d) of the CWA. Because traditional approaches to watershed-based assessments involve many separate steps for data preparation, developing maps and tables, and applying and interpreting models, and result in a lack of integration, coordination, and time-intensive execution, BASINS was designed to bring key data and analytical components "under one roof" (USEPA 2008).

BASINS bring together a suite of interrelated components for performing various aspects of environmental analysis (Fig. 5.2). Components include (1) nationally derived databases with Data Extraction tools and Project Builders, (2) assessment tools (TARGET, ASSESS, and Data Mining) that address large- and small-scale characterization needs, (3) utilities to facilitate organizing and evaluating data, (4) tools for Watershed Delineation, (5) utilities for classifying DEMs, land use, soils, and water quality observations, (6) Watershed Characterization Reports that facilitate compilation and output of information on selected watersheds, (7) an in-stream water quality model (QUAL2E), (8) two watershed loading and transport models (HSPF and SWAT), and (9) PLOAD, a simplified GIS-based model that estimates NPS loads of pollution on an annual average basis.

GIS-Based Watershed Modeling Systems 187

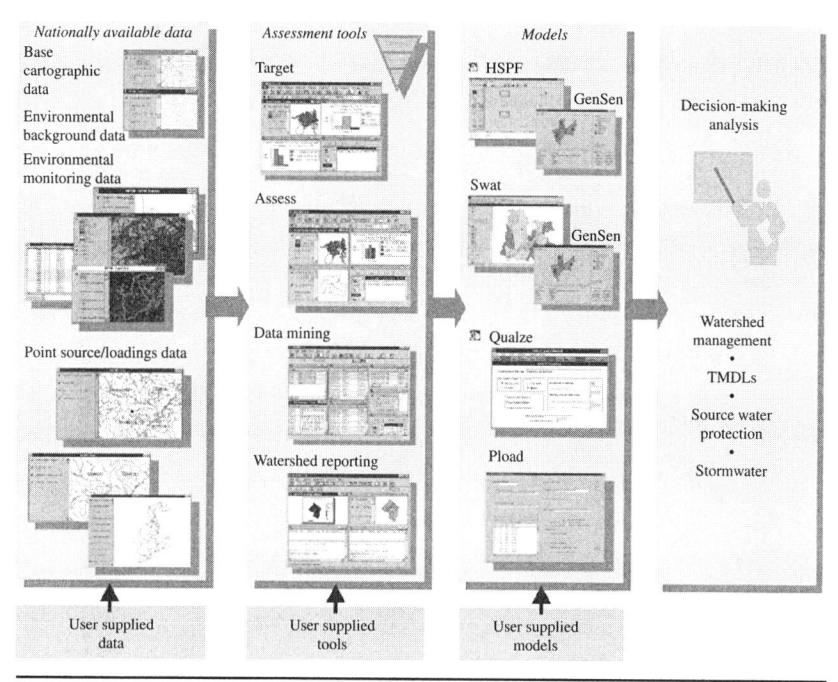

FIGURE 5.2 System overview of BASINS Version 3.0 (U.S. Environmental Protection Agency 2001).

BASINS was first introduced in 1996 and since then several versions of BASINS have been released. BASINS Version 1.0, 2.0, 3.0, and 3.1 were developed using ArcView 2.X and 3.X environments. The simulation models run in a Windows environment using data input files generated in ArcView. In 2007, USEPA released a newer version, BASINS 4.0. The most significant change in this version is the use of open-source GIS software architecture using Mapwindow GIS (www.mapwindow.com). With this significant change, BASINS became available to everyone and became independent of any proprietary GIS platform and open for useful "plug-ins." BASINS 4.0 includes WinHSPF and PLOAD models. The new Parameter Estimation (PEST) tool in WinHSPF automates the model calibration process and allows users to quantify the uncertainty associated with specific model predictions. WinHSPF links to the AQUATOX model for integrating watershed analysis with effects on aquatic biota in receiving waters. It also allows a new BASINS feature, the Windows-based Climate Assessment Tool, for assessing potential impacts of changing climate on stream flows and pollutant loads (USEPA 2008).

AGWA

AGWA is a multipurpose GIS-based hydrologic analysis system use by watershed, water resource, land use, and biological resources managers and scientists in performing watershed- and basin-scale studies. AGWA was developed by the USDA-ARS Southwest Watershed Research Center, in cooperation with the USEPA Office of Research and Development Landscape Ecology Branch, to address four objectives:

1. To provide a simple, direct, and repeatable method for hydrologic model parameterization.
2. To use only basic, attainable GIS data.
3. To be compatible with other geospatial watershed–based environmental analysis software.
4. To be useful for scenario development and alternative futures simulation work at multiple scales.

AGWA provides functionalities to conduct all phases of a watershed assessment for two widely used watershed hydrologic models: SWAT for large watersheds and long-term simulations, and KINEROS-2 for small watersheds (<100 km^2) for event-based studies (Fig. 5.3). AGWA is designed as a tool for performing relative assessment (change analysis) resulting from land cover/use change. Areas identified through large-scale assessment with SWAT as being most susceptible to change can be evaluated in detail at smaller scales with KINEROS-2. Data used in AGWA include DEMs, land cover grids, soils data, and precipitation data.

Currently, AGWA is available as AGWA 2.0 for ArcGIS 9.x and AGWA 1.5 for ArcView 3.x. The interface is similar to USEPA's BASINS. There are five major tasks: (1) watershed delineation, (2) land cover and soils parameterization, (3) writing a precipitation file for model input, (4) writing parameter files and running the chosen model, and (5) viewing results.

WMS

WMS is another comprehensive graphical modeling environment for all phases of watershed hydrology and hydraulics. WMS includes powerful tools to automate modeling processes. The current version, WMS 8.1, supports hydrologic modeling with the Hydrologic Engineering Center model HEC-1 (HEC-HMS; HEC-Hydrologic Modeling System to simulate surface runoff from a single precipitation event), TR-20 (to compute surface runoff from natural or synthetic rainstorm events), TR-55 (simplified method to compute storm runoff in a small, urbanized watershed), the Rational Method (for peak discharge), NFF (National Flood Frequency), MODRAT (Modified Rational Method), and HSPF. Hydraulic models supported include HEC-RAS (HEC-River Analysis System; one-dimensional hydraulic model), SMPDBK (Simplified Dam-Break, for predicting downstream flooding produced by a dam failure), and CE-QUAL-W2

GIS-Based Watershed Modeling Systems 189

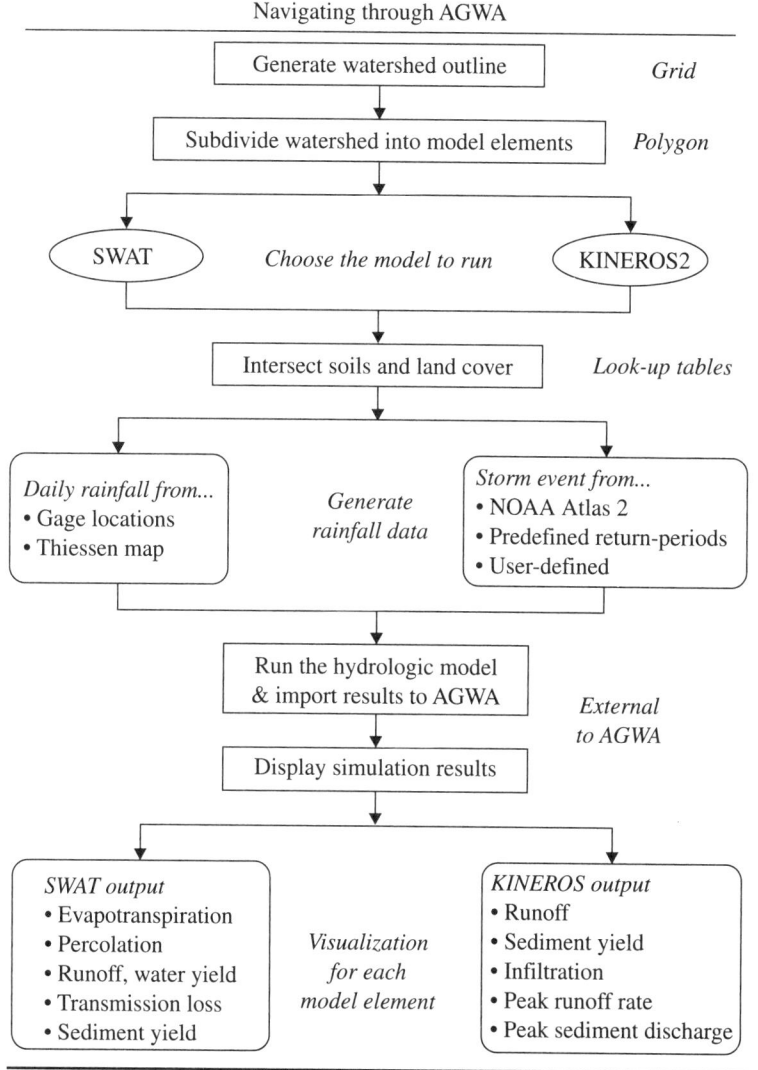

```
Navigating through AGWA
```

	Generate watershed outline	*Grid*
	↓	
	Subdivide watershed into model elements	*Polygon*

(SWAT) *Choose the model to run* (KINEROS2)

Intersect soils and land cover — *Look-up tables*

Daily rainfall from...
- Gage locations
- Thiessen map

Generate rainfall data

Storm event from...
- NOAA Atlas 2
- Predefined return-periods
- User-defined

Run the hydrologic model & import results to AGWA — *External to AGWA*

Display simulation results

SWAT output
- Evapotranspiration
- Percolation
- Runoff, water yield
- Transmission loss
- Sediment yield

Visualization for each model element

KINEROS output
- Runoff
- Sediment yield
- Infiltration
- Peak runoff rate
- Peak sediment discharge

FIGURE 5.3 Overview of AGWA tools and processes (Environmental Modeling Systems, Inc. 2007).

(a two-dimensional hydraulic model used for water quality analysis in rivers and reservoirs). In addition, two-dimensional integrated hydrology (including channel hydraulics and groundwater interactions) can be modeled with Gridded Surface Subsurface Hydrologic Analysis (GSSHA).

Developed in cooperation with the Waterways Experiment Station (WES), WMS provides graphical tools for use in the delineation of watersheds and flood plains. Hydrologic models may be set up and viewed in a user-friendly graphical environment. The WMS software package is divided logically into well-integrated task-oriented modules including map module, GIS module, terrain data module, two-dimensional grid module, drainage module, hydrologic-modeling module, river-modeling module, and scatter-point module. The WMS modules can perform automated watershed delineation, floodplain modeling and mapping, stochastic modeling, and two-dimensional (distributed) hydrologic modeling. Typical applications include flood forecasting (depth and velocity over the entire two-dimensional domain), thunderstorms (localized rainfall) flood analysis, surface ponding and infiltration analysis, and groundwater/surface water interaction modeling. WMS utilizes ArcGIS GIS software.

5.5 Future of GIS-based Watershed Modeling Systems

Widely used current GIS-based watershed modeling systems provide a common GUI for a variety of models. One of the major limitations of this approach is that the user is still able to use only one complete model, with all of its components, at a time. This approach does not allow combining submodels from different models to construct a new model that is most appropriate in representing the hydrologic conditions of a specific watershed. In future, development of modular structures under a common GIS-based GUI for sets of popular models will allow more accurate prediction of watershed responses. Such a modular structure should also allow various parts of the watersheds to be modeled using a unique combination of submodels. For example, the parts of watershed where infiltration excess runoff mechanism dominate should be allowed to use the rainfall-runoff models that are based on infiltration excess theory (e.g., curve number approach, Green–Ampt infiltration equation, and Horton infiltration equation), whereas the parts dominated by saturation excess mechanisms should be allowed to use models that are based on saturation excess theory.

Development and expansion of customized GIS modules that are capable of representing and incorporating time series input and output datasets will greatly assist modelers, researchers, and decision makers. For example, land-use in a watershed model is currently modeled using a dataset that is fixed in time. To quantify the effect of changes in land-use, multiple model setups are constructed for fixed times to represent land-use change over a period of time. However, the fact is that land-use changes occur continuously. In the GIS community, effort is underway to capture land-use change in real time. Once this approach is sufficiently developed, GIS-based modeling systems should be configured in such a way to allow incorporation of time series land-use input. Similarly, incorporation of time series output would be crucial for modelers and decision makers.

Certain modeling steps that are becoming necessary for comprehensive application of watershed models include sensitivity analysis, calibration and validation, uncertainty analysis, and BMP optimization. Currently available GIS-based modeling systems include some of these steps but are currently not very sophisticated and user friendly. In the future, incorporating these modeling steps in a GIS-based interface will be critical.

With the invention of Internet GIS, more and more GIS functionalities are currently being deployed through a Web browser. In the watershed modeling area as well, researchers have started to use Web browsers for screening-level watershed modeling studies. In future, it would be possible to deploy a comprehensive watershed model through a Web browser.

Using GIS has greatly simplified the task of watershed modeling, which has resulted in efficient use of models, even by nonmodelers. In future, GIS will make watershed modeling even simpler. However, this has resulted in misuse of models. With new developments in GIS-based modeling systems, it would be even more critical to train future modelers in such a way that they have a thorough grounding in underlying physical, chemical, and biological processes. This would result in not only efficient but also accurate and effective modeling efforts.

References

Aller, L., Bennett, T., Lehr, J.H., Petty, R.J. and Hackett, G. 1987. DRASTIC: a standardized system for evaluating ground water pollution potential using hydrogeologic settings. U.S. Environmental Protection Agency, Robert S. Kerr Environmental Research Laboratory, Ada, OK. Office of Research and Development, EPA/600/2–87/035. p. 641.

Arnold, J. G. and Fohrer, N. 2005. SWAT2000: Current capabilities and research opportunities in applied watershed modeling. *Hydrological Processes*. 19:563–572.

Arnold, J. G., Williams, J. R., Nicks, A. D., and Sammons, N. B. 1990. *SWRRB, a Basin Scale Simulation Model for Soil and Water Resources Management*. College Station, TX: Texas A & M University Press.

Bhasker, N. R., James, W. P., and Devulapalli, R. S. 1992. Hydrologic parameter estimation using geographic information system. *Journal of Water Resources Planning and Management*. 118(5):492–512.

Bicknell, B. R., Imhoff, J. C., Kittle, Jr., J. L., Jobes, T. H., and Donigian, Jr., A. S. 2001. *Hydrological Simulation Program—FORTRAN Version 12. User's Manual*. Aqua Terra Consultants, Mountain View, CA.

Bingner, R. L. and Theurer, F. D. 2003. *AnnAGNPS Technical Processes Documentation. Version 3.2*.

Bromberg, J. G., McKeown, R., Knapp, L., Kittel, T. G. F., Ojima, D. S., and Schimel, D. S. 1995. Integrating GIS and CENTURY model to manage and analyze data. In *GIS and Environmental Modeling: Progress and Research Issues*, eds. Goodchild, M. F. et al. eds. Fort Collins, CO: GIS World Books:429–431.

Burrough, P. A. 1997. Environmental modeling with geographical information systems. In *Innovations in GIS*, ed. Kemp, Z. London:Taylor and Francis. 43–153.

Croft, F. and Kessler, B. 1996. Remote sensing, image processing, and GIS: Trends and forecasts. *Journal of Forestry*. 4(6):31–35.

Chapter Five

DePinto, J. V., Calkins, H. W., Densham, P. J., Atkinson, J. F., Guan, W., H., Lin, and Rodgers, P. W. 1994. An approach for integrating GIS and watershed analysis models. *Microcomputers in Civil Engineering.* 9:251–262.

Engelund, F. and Hansen, E. 1967. *A Monograph on Sediment Transport in Alluvial Streams.* Copenhagen: Teknisk Vorlag.

Environmental Modeling Systems, Inc. 2007. Watershed Modeling System (WMS). Available at http://www.ems-i.com/. Accessed February 2007.

Evans, B. M. and Myers, W. L. 1990. A GIS–based approach to evaluating regional groundwater pollution potential with DRASTIC. *Journal of Soil and Water Conservation* 45(2):242–245.

Fedra, K. 1995. Distributed models and embedded GIS: Integration strategies and case studies. In *GIS and Environmental Modeling: Progress and Research Issues,* eds., Goodchild, M.F., et al. Fort Collins, CO: GIS World Books: 413–417.

Geter, W. F., Smith, P., Drungil, C., Shepherd, R., and Kuenstler, B. 1995. Hydrologic unit water quality model GIS interface to four ARS water quality models for use by Soil Conservation Service. *International Symposium on Water Quality Modeling,* Kissimmee, FL, April 2–5:341–347.

Haan, C.T., Johnson, H. P., and Brakensiek, D. L. 1982. Hydrologic modeling of small watersheds. *ASABE Monograph Number 5.* American Society of Agricultural and Biological Engineers, St. Joseph, MI.

Hamlett, J. M., Petersen, G. W., Evans, B. M., Deichert, L. A., Messier, S. R., Anderson, M. C., and Baumer, G. M. 1995. *Statewide Screening of Groundwater Nitrate Pollution Potential from Agricultural Lands in Pennsylvania.* Final Report ME 1235, Pennsylvania Department of Environmental Resources, Bureau of Land and Water Conservation, The Pennsylvania State University, University Park, PA.

Hay, L., Knapp, L., and Bromberg, J. 1995. Integrating GIS, scientific visualization systems, statistics, and an orographic precipitation model for a hydroclimatic study of the Gunnison river basin. In *GIS and Environmental Modeling: Progress and Research Issues,* eds. M. F. Goodchild et al. Fort Collins, CO: GIS World Books: 235–246.

Huber, W. C. and Dickinson, R. E. 1988. *Storm Water Management Model, Version 4, User's Manual.* EPA/600/3–88/001a (NTIS PB88–236641/AS), U.S. Department of Environmental Protection, Athens, GA.

Knisel, W. G., Leonard, R. A., and Davis, F. M. 1992. Draft. *GLEAMS Version 2.0, Part 1: Model Documentation.* U.S. Department of Agriculture, Agriculture Research Service, Southeast Watershed Research Laboratory, Tifton, GA. 87 pp.

Liao, H. and Tim, U. S. 1997. An interactive modeling environment for non-point source pollution control. *Journal of the American Water Resources Association* (JAWRA) 33(3):591–603.

Logan, T. J. 1990. Agricultural best management practices and groundwater protection. *Journal of Soil and Water Conservation* 45(2):201–206.

Martin, P. H., LeBoeuf, E. J., Bobbins, J. P., Daniel, E. B., and Abkowitz, M. D. 2005. Interfacing GIS with water resource models: A state–of–the–art review. *Journal of the American Water Resources Association* (JAWRA) 41(6):1471–1487.

McDonald, M. and Harbaugh, A. W. 1988. *A Modular Three–Dimensional Finite Difference Groundwater Flow Model.* Reston, VA: U.S. Geological Survey. Techniques of Water Resources Investigation. TWI6–AI.

Neitsch, S. L., Arnold, J. G., Kiniry, J. R., Williams, J. R., and King, K. W. 2002. *Soil and Water Assessment Tool Theoretical Documentation—Version 2000.* Available at http://www.brc.tamus.edu/swat/doc.html. Accessed December 2008.

Neitsch, S. L., Arnold, J. G., Kiniry, J. R., and Williams, J. R. 2005. *Soil and Water Assessment Tool Theoretical Documentation—Version 2005.* Available at http://www.brc.tamus.edu/swat/doc.html. Accessed December 2008.

Roaza, H. R., Roaza, M., and Wagner, J. R. 1993. Integrating geographic information systems in groundwater applications using numerical modeling techniques. *Water Resources Bulletin* 29(6):981–988.

Ross, M. A. and Tara, P. D. 1993. Integrated hydrologic modeling with geographic information systems. *Journal of Water Resources Planning and Management* 119(2):29–140.

Rossman, L. A. 2008. *Storm Water Management Model. User's Manual*. Version 5.0. EPA/600/R–05/040. National Risk Management Research Laboratory, Office of Research and Development, USEPA, Cincinnati, OH.

Sharpley, A. N. and Williams. J. R. 1990. *EPIC—Erosion Productivity Impact Calculator: 1. Model Documentation*. Technical Bulletin No. 1768, U.S. Department of Agriculture. 235 pp.

Singh, V. P. and Woolhiser, D. A. 2002. Mathematical modeling of watershed hydrology. *Journal of Hydrologic Engineering* (4):270–292.

Smith, R. E., Goodrich, D. C., and Quinton, J. N. 1995. Dynamic, distributed simulation of watershed erosion: The KINEROS2 and EUROSEM models. *Journal of Soil and Water Conservation* 50(5):517–520.

Smith, R. E. and Parlange, J. Y. 1978. A parameter-efficient hydrologic infiltration model. *Water Resources Research* 14(3):533–538.

Tim, U. S., Mostaghimi, S. and Shanholtz, V. O. 1992. Identification of critical non-point pollution source areas using geographic information systems and water quality modeling. *Water Resource Bulletin* 28(5):877–887.

Tim, U. S., Jain, D., and Liao, H., 1996. Interactive modeling of ground-water vulnerability within a geographic information system environment. *Ground Water* 34(4):618–627.

U.S. Environmental Protection Agency. 2001. *Better Assessment Science Integrating Point and Non-Point Sources (BASINS). Version 3.0. User's Manual*. EPA-823-B-01-001. Office of Water.

U.S. Environmental Protection Agency. 2008. *BASINS: Better Assessment Science Integrating Point and Non-Point Sources*. Online at www.epa.gov/waterscience/basins. Accessed November 2008.

Vckovski, A., Brassel, K. E., and Schek, H. J., 1999. *Interoperating Geographic Information Systems*, Second International Conference on Interoperating Geographic Information Systems, INTEROP 1999, Zurich, Switzerland. New York: Springer–Verlag.

Ward, G. H., and Benaman, J. 1999. Models for TMDL *Application in Texas Watercourses: Screening and Model Review*. Report submitted to the Texas Natural Resource Conservation Commission, Austin, TX. Online report CRWR–99–7.

Woolhiser, D. A., Smith, R. E., and Goodrich, D. C. 1990. *KINEROS, A Kinematic Runoff and Erosion Model: Documentation and User Manual*. U.S. Department of Agriculture, Agricultural Research Service, ARS-77. 130 pp.

Young, R. A., Onstad, C. A., Bosch, D. D., and Anderson, W. P. 1987. *AGNPS: Agricultural Non-Point Source Pollution Model. A Watershed Analysis Tool*. U.S. Department of Agriculture, Conservation Research Report 35. 80p.

CHAPTER 6
Design of Sustainable Water Management Systems

R. Sri Ranjan

Department of Biosystems Engineering
University of Manitoba
Winnipeg, Manitoba, Canada

6.1 Introduction

Soil and water engineering, one of the core branches within biosystems engineering, deals with the interaction between soil, water, and plants. Plants require water and nutrients for sustained growth, and biosystems engineers design irrigation and drainage systems that can provide an ideal soil environment for sustainable crop production. Thus, designing sustainable water management systems is a major area within biosystems engineering. Biosystems engineers are knowledgeable about soil physical and chemical properties and plant physiology to understand the interaction of the plant roots with the soil. Irrigation systems are designed to deliver water and nutrients at the right quantities to enable optimum plant growth. Drainage systems remove excess water and salts from the soil and maintain optimum aeration of the plant root zone.

If the irrigation water quality is poor, there is possibility of salt accumulation within the root zone that will render the soil unproductive. This process of salt accumulation within the root zone is known as salinization. There is ample evidence in history where civilizations have disappeared because of crop failure due to salinization of the soil. Proper water management can reduce the buildup of salts within the root zone. A well-designed water management system can remove excess salts by applying excess water to wash the salts below the root zone. The deep percolating water below the root zone is removed by drainage systems and treated off-site to remove the salts. Biosystems engineers also design engineered wetlands that contain plants capable of removing salts from the drainage return flows. An overview of the design of a sustainable water management system is presented in this chapter.

6.2 Physical Properties of Soil

In order to understand the soil and water interaction, it is important to learn about basic soil water retention and movement within the soil. Soils are made up of different-sized soil particles consisting of sand, silt, and clay, along with humus, that is formed by the degradation of organic residues. Plant roots take up water that occupies the pore space between the soil particles. The water is held within the pore space of the soil particles by capillary forces. Capillary forces exist at the interface formed by the pore water, soil particle, and pore air. Capillary pressure is defined as the difference in pressure between pore air and pore water across the meniscus (Hillel 1998). Figure 6.1 shows the relationship between the pore size and the relative force required to remove the water. The smaller the radii of curvature of the meniscus, the greater is the capillary pressure across it. This requires greater negative pressure to remove the water from within these

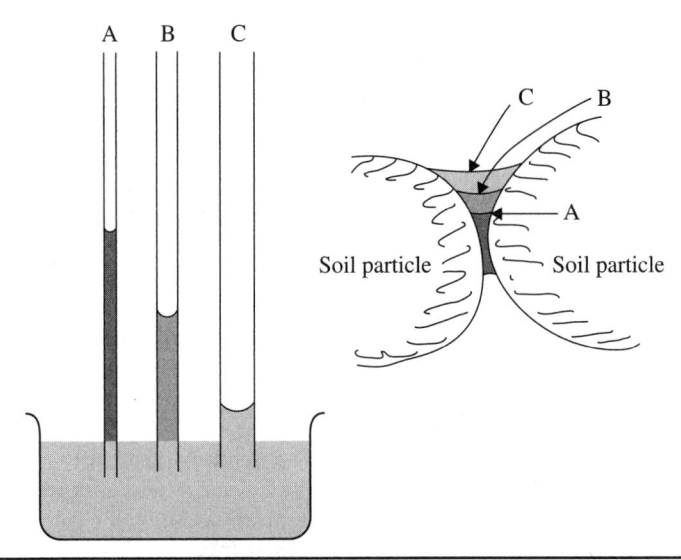

FIGURE 6.1 Relationship between capillary pressure and water content of soil.

pores. When a soil is saturated with water, the larger pores drain first due to gravitational forces. As the larger pores drain, the menisci begin to retreat toward smaller pores. As the pores get smaller, the capillary forces become larger and it becomes increasingly difficult to drain freely. The volume of water within the pore space as a ratio of the total volume of the soil is defined as volumetric water content.

When soil is thoroughly wetted and allowed to drain under the force of gravity, all of the large pores will drain. The volumetric water content after two days of free drainage of saturated soil is defined as the *field capacity* (FC) (Soil Science Society of America 1984). All of the remaining water is held within the pore space by capillary action. As plants begin to uptake this water, the menisci continue to recede to release the water. However, as the menisci become smaller, the capillary forces increase making it more difficult for plants to uptake the remaining water. If this process continues, there will reach a stage when the lowest pair of leaves of a particular kind of plant wilts and fails to recover when placed back in a water-saturated environment. The volumetric water content at which this happens is called the *permanent wilting point* (PWP) (Soil Science Society of America 1984). The volume of water between FC and PWP is available for plant growth.

6.3 Water Management

6.3.1 Plant Water Relations

Plants take up water through their roots, which is transported through the stems to the leaves and eventually lost to the atmosphere via the stomata. This process is known as *transpiration*, which is affected, by soil water content and atmospheric conditions. Water will also evaporate from the soil and the plant surface. These two processes, transpiration and evaporation, are combined and termed *evapotranspiration* (ET) and is often difficult to distinguish (FAO 1998). Dry and windy conditions promote higher ET. During hot weather conditions, the evapotranspiration process helps to cool the plant. However, if the evapotranspiration rate is larger than the plant water uptake, the plant will be subjected to water stress, signaling the plant to conserve water by closing its stomata. The plant will also show signs of wilting whereby the leaves droop, reduce exposure to solar radiation, and minimize ET.

Consequently, the gaseous exchange between the atmosphere and the leaf cells are reduced. The wilted leaves do not capture the solar radiation effectively. The reduced gaseous exchange and wilting of the leaves will lead to a reduction in yield. Therefore, it is important to maintain adequate root–zone moisture levels to help keep the stomata open.

6.3.2 Assessing Plant Water Needs

When the soil water content is maintained between the FC and the permanent wilting point (PWP), the water is accessible to plant roots. The water content within this range is known as available moisture (AM) (ASAE 1990). However, as the water content of the soil decreases from FC to PWP, the plant roots progressively access water from smaller and smaller pores. As a result, the capillary pressure increases and the plants have to work increasingly harder to take up water from the soil. Ideally, if the soil water content is maintained near FC, the plants will uptake water easily. This will require more-frequent replenishment of the soil water by irrigation. Depending on the type of water application method used to supply water to the plants, it may not be practical to apply water more frequently. Generally, the soil water content is allowed to deplete about half the available water (ASAE 1990). The water content from FC to the midway point between FC and PWP is known as readily available moisture (RAM). Biosystems engineers design irrigation systems based on supplying the RAM in a single irrigation sequentially to all the fields and plan to come back to the same field just before the RAM is depleted.

Many methods are used to assess plant water needs. Biosystems engineers have helped develop many of these methods that have helped meet plant water needs accurately. The different methods can be broadly classified as soil based, weather based, and plant based, which are described as follows.

Soil-Based Methods

Soil moisture sensors are strategically embedded within the root zone and the data are manually read by the farmer or transmitted telemetrically to a central station. The farmer will use the current soil moisture status to determine the level of soil moisture depletion of a particular field and plan the next depth and timing of irrigation. Some of the methods measure the soil water content by measuring the electrical resistance, which is proportional to the water content. The dielectric constant of water, dry soil, and air are 80, 2-7, and 1, respectively (Topp et al. 1980). This difference in dielectric constant can be used in a dielectric mixing model based on volumetric ratios to determine the volume of water present in a soil sample. Topp et al (1980) presented an empirical equation to determine the volumetric water content of soil based on apparent dielectric constant measured by time-domain reflectometry (TDR). This is a nondestructive method whereby a probe is inserted into the soil and a reading can be obtained using a TDR. Sri Ranjan and Domytrak (1997) have presented the design of TDR miniprobes that nondestructively measure the soil water content in a soil volume the size of a finger. These miniprobes are ideal for determining soil water depletion from different parts of the plant root zone.

The soil water content is inversely proportional to the capillary pressure of the soil, and this is depicted by soil moisture characteristic curves. Tensiometers have been used to measure the capillary pressure of the soil and the soil moisture characteristic curve of that particular soil is used to determine the water content. Generally, tensiometers have been used to trigger irrigation when the soil reaches a predetermined capillary pressure.

Weather-Based Methods

As described previously, weather plays a crucial role in controlling the rate of ET. Weather stations that collect maximum and minimum temperature, wind speed, solar radiation, rainfall, and relative humidity can also transmit this data telemetrically to a central station. Empirical relationships have been developed relating the ET of well-watered grass to the weather variables (FAO 1998). The weather data are entered into these empirical equations to predict the daily reference ET rates for grass as the reference crop. For other crops, crop coefficients have been developed to adjust the reference ET to actual ET. The sum of this daily depletion (actual ET) since the last irrigation is used as a measure of the soil water depletion level, and the next depth of irrigation is planned based on this information. One major drawback of this method is the lack of disconnect between the soil water content and the ET rate. As described previously, plant ET will not proceed at the maximum rate if the soil water content is below FC; this phase is known as the *soil-limiting phase* (FAO 1998). The weather-based methods do not account for this lowered ET. The empirical equations were developed using grass growing in lysimeters, which were well-watered. A lysimeter is a large container in which a crop is grown in the middle of a field; load cells monitor the entire weight of the soil, plant, and water. The stage of growth of the plant also determines the ET rate, and this is accounted for by incorporating another coefficient.

Plant-Based Methods

Leaves show different signs of water stress. The leaf angel may change in response to water stress. Leaves will also lose turgor and appear to be wilted. The color of the leaf will also change, which may not be easily discernible by the naked eye. More details are presented as follows.

Leaf Water Potential Method As the soil begins to dry, the plant leaves have to generate a higher negative pressure to help uptake the water. By measuring the leaf water potential, one can assess the water stress experienced by the plant. Turner (1988) reported a method for determining the potential of the leaf water. In this method, a leaf is cut and placed in to a chamber with the cut petiole exposed to the atmosphere. When a leaf under water stress is cut, sap will retreat into the leaf tissue, and the cut portion of the petiole will appear dry. However, when that leaf is placed within the chamber and the chamber is pressurized slowly, the sap is squeezed back out through the leaf petiole that is exposed to the atmosphere. The pressure at which the sap begins to reappear on the surface of the cut petiole is considered to be equal to the leaf water potential. The higher the leaf water potential, the more the plant is water stressed. Due to the destructive nature of this method, it is restricted to research only.

Leaf Color Analysis Method   Methods have been developed to analyze the color spectrum of leaf images obtained by digital cameras (Zakaluk and Sri Ranjan 2007; Zakaluk and Sri Ranjan 2008). It is well known that when grass turns bluish-green it is experiencing severe water stress, and it is time for irrigation. However, digital image analysis can detect subtle differences in reflectance in response to water stress much earlier than is visible to the naked eye. Thus, irrigation could be triggered at a much earlier stage of soil water depletion before the plant suffers severe water stress. With advances in digital image acquisition technology, the inexpensive hardware necessary to implement this method is within reach of irrigation equipment makers. Biosystems engineers have a major role to play in developing this further for wide adoption by farmers. It will lead to conservation of water and savings in nutrients.

6.4 Need for Irrigation

The primary purpose of irrigation is to supply the crop water demand beyond what is supplied by precipitation. Assuming an average of 5 mm/day ET and a 100-day ET demand period, areas receiving more than 500 mm of precipitation during the growing season do not require irrigation. Areas receiving less than 250 mm of precipitation definitely need irrigation. Areas that receive precipitation between 250 mm and 500 mm will require irrigation to meet their supplemental needs. In addition, there are secondary benefits to irrigation as well.

6.4.1 Meeting the Crop Water Demand

The primary purpose of irrigation is to supply water to plants. Areas receiving precipitation in the form of snow will have soil moisture accumulating within the root zone due to spring snowmelt infiltration. Soil moisture storage at the beginning of the season is used by plants during the initial germination stage. This is also the most critical stage because of the shallow depth of roots that have a smaller soil volume to store the water. The smaller volume of soil will require more frequent irrigation than that is required during later stages of growth with deeper roots. However, the water stored within the soil profile will migrate toward the root zone to replenish the water removed by seedlings. Thus, a certain level of soil moisture depletion is allowed as part of the irrigation planning to help lengthen the irrigation interval. A longer irrigation interval will help maximize the use of irrigation equipment and help cover a much larger area of land for the given size of the equipment. As plants develop and mature, the root zone volume also increases, allowing a larger water storage volume. Thus, the irrigation system is usually designed to meet the water demand during this stage.

6.4.2 Other Benefits of Irrigation

There are additional benefits to irrigation such as frost protection, fertigation, minimizing soil salinization, crop cooling, and effluent disposal, described here.

Frost Protection

During early spring, an unusual warming up may trigger bud formation in perennial trees. When cold weather returns, the newly formed buds will be damaged. By spraying water to the tree canopy, the microclimate can be cooled to delay bud formation. Another mechanism protects plants against frost damage in early summer. When temperature dips below freezing point in early summer, water is misted onto the crop canopy to form a thin layer of ice that acts as an insulator of the buds and young leaves. Water is sprayed onto the plants until the atmospheric temperature rises above the freezing point. Even though irrigation has been used for frost prevention, atmospheric air inversion methods have been used to conserve water in water-scarce areas. Giant fans are operated to create this air inversion and keep the crop canopy warm.

Minimizing Salinization

All irrigation water has some salts in dissolved form, and this salt is accumulated within the root zone during ET. Continued accumulation of salts will change the soil's physical properties and render them unproductive. This salt accumulation can be prevented by applying excess water, which washes the salts below the root zone. A drainage system to remove the excess salt-laden water will help maintain a favourable root zone environment. This excess water, beyond plant water requirements, is designated as *leaching requirement* (LR). Further description of leaching is presented at the end of the chapter.

Crop Cooling

During hot summer days, plants may show signs of wilting to minimize ET. However, this will also affect the photosynthetic activity of the plant, leading to a reduction in yield. By spraying water onto the crop canopy, the temperature can be reduced by evaporative cooling. The latent heat energy required to evaporate the water droplets will be taken from the atmosphere causing the temperature to decrease leading to the cooling effect. The plants will continue to photosynthesize under these cooler conditions.

Effluent Disposal
Irrigation systems have also been used as a means of disposing wastewater from industrial processes and animal production operations. The wastewater is initially treated in holding ponds, and the supernatant water is pumped to different cells to achieve further degradation. The water from the last treatment cell is used for irrigating pastures as a means of disposal.

6.5 Irrigation Systems

6.5.1 Introduction
There are several different types of irrigation systems, which have evolved over the years, depending on the complexity of the terrain and the availability of water. The following sections describe the different systems and their applications. History is replete with evidence of well-designed irrigation systems in use. The finding of major ancient civilizations has also shown evidence of well-constructed water storage reservoirs and canal delivery systems.

6.5.2 Types of Irrigation Systems
Irrigation systems can be broadly classified into surface irrigation, sprinkler irrigation, drip irrigation, and subsurface irrigation, depending on how water is applied. The principles governing the design of these systems also differ because of the differences in how water is delivered to the root zone. It is important to understand the water-saving features of these systems before selecting a particular system for a given irrigation application.

6.5.3 Irrigation Efficiency
Irrigation efficiency is defined as the amount of water beneficially stored within the plant root zone as a fraction of the applied quantity of water. Generally, surface irrigation systems with 50 to 60 percent efficiency have been in use in many parts of the world due to its simplicity in design. Biosystems engineers have made advances in the design of sprinkler irrigation systems, which have efficiencies as high as 80 percent. Further advances in irrigation have been made with the development of drip irrigation systems, which attained 90 to 95 percent efficiency (ASAE 1990).

6.5.4 Design of Irrigation Systems
Depending on soil type, land slope, and water availability, different types of irrigation systems are used. The choice is also determined by

the type of crop and potential return on investment. The following sections briefly describe these methods.

Surface Irrigation Systems

In its most primitive form, wild flooding has been practiced with only 50 percent of the applied water beneficially stored within the plant root zone. In sloping terrain, basin irrigation has been used to allow more time for infiltration into the soil. The graded border system is used in hilly terrain, where the water is allowed to advance over the sloping land while it continues to infiltrate vertically into the root zone. Such systems require a very good understanding of the soil's infiltration characteristics. Biosystems engineers have modeled such systems to develop a design that will maintain uniform infiltration depth along the length of the field. Furrow irrigation systems allow water to flow along the length of the field, whereas plants growing on the ridges between the furrows will take up the water uniformly. The uniformity of depth of infiltration along the length of the furrow is of major concern. Biosystems engineers have improved on the design of these systems and developed an automated method called *cablegation system* (Kemper et al. 1981). Figure 6.2 shows a cablegation system designed and set up by the author for irrigating corn in Coimbatore, India. The cablegation system applies water sequentially to a few furrows at a time. The system applies water at

FIGURE 6.2 A cablegation system irrigating corn in Coimbatore, southern India.

the maximum rate at the beginning of the irrigation and cuts back the flow when the advancing waterfront reaches the end of the field. All dry soils initially absorb water rapidly, and once the soil near the surface begins to swell, infiltration declines. Therefore, a well-designed system should have the capability to cut back the water flow to the furrow after the initial surge in application. The cablegation system achieves this automatically based on gravity flow.

Sprinkler Irrigation Systems

Sprinkler irrigation systems have been very popular due to their low cost and the ability to irrigate rolling terrain. Figure 6.3 shows a part of a sprinkler irrigation system. Figure 6.4 shows an aerial view of center-pivot irrigated fields. At the center of the field, the water is pumped through a pipe supported above the crop canopy on towers on independently driven wheels, hence the circular shape of the irrigated field. Drop tubes attached to the pipe bring the water down to the canopy level and sprays it though rotator-type nozzles. The drop tubes are telescopic and can be extended to reach the crop canopy and retracted up as the plant grows taller. In the design process, the spray nozzle is selected first based on the basic infiltration rate of the soil. Sprinklers are selected so that the water application rate never exceeds the basic infiltration rate of the soil to avoid ponding and runoff. The pressure required to break up the water stream into droplets is supplied by the irrigation pump. The sprinkler nozzles are connected to the main line via lateral pipes. As water flows through

FIGURE 6.3 Center-pivot irrigation system showing independently driven wheels and drop tubes with rotators spreading water.

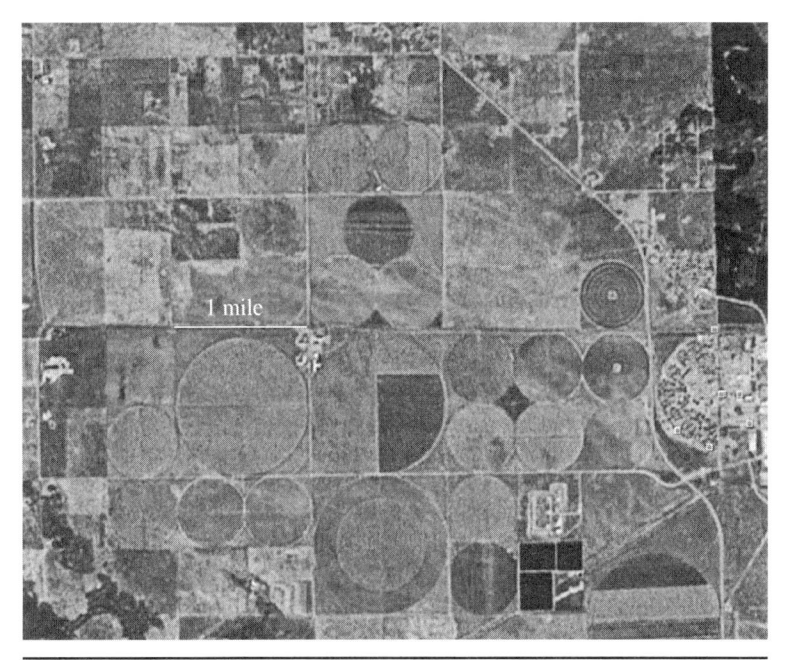

Design of Sustainable Water Management Systems

FIGURE 6.4 An aerial view of a field irrigated by center-pivot systems in Southern Manitoba, Canada.

the main line and laterals, friction losses will lead to a reduction in water pressure generated by the irrigation pump at the source. Although larger-diameter pipes have lower friction losses, the higher capital cost of larger pipes has to be taken into consideration in relation to the higher operating cost of pressurizing the water. Generally, the allowed pressure loss in the lateral is taken as 20 percent of the operating pressure of the sprinkler head. The main line and laterals also have fittings to divert and control the flow of water, which lead to friction losses. Biosystems engineers consider the fluid mechanics of flow through pipes and fittings to optimally design the piping system and select the required size of the pump.

Drip Irrigation Systems

The drip irrigation system was the next major advance in irrigation water conservation achieving water application efficiencies as high as 90 percent (ASAE 1990). In this method, the water is applied at a rate that will permit water to spread within the root zone by capillary action. If the water is applied at a rate higher than the basic infiltration rate of the soil, the water will spread on the soil surface and infiltrate the soil over a larger diameter. Alternatively, if water is applied at a lower rate than the basic infiltration rate, the water will tend to percolate more vertically. Therefore, selection of the type of drip emitter and its placement is very important. Drip systems operate under low pressure and use capillary-sized channels within the emitter to control the flow rate. However, this leads to clogging of the emitters. Therefore, drip systems are designed with filtration systems to remove suspended sediments. Water quality is also of major concern due to the formation of salt deposits in the emitters, which alter the hydraulic flow characteristics of the emitter. The water needs to be chemically treated to prevent salt precipitation. The flow though the emitters is controlled by the pressure of water within the lateral at the location of the emitter. Biosystems engineers have designed emitters that have pressure-compensating mechanisms, which maintain a similar flow rate even under a large variation in water pressure.

Subsurface Irrigation Systems

This is the most advanced irrigation system with water application efficiencies as high as 95 percent (ASAE 1990). The water is applied directly into the root zone via porous pipes. Because the plant root is directly wetted in the root zone, the plant canopy does not get wet. This reduces the incidence of diseases, which are common in high-humidity conditions associated with wet canopy. Just as in drip irrigation systems, subsurface irrigation systems also require water filtration and chemical treatment to prevent clogging and salt precipitation. Figure 6.5 shows a cotton field subsurface irrigated with porous pipes buried within the root zone. Note the absence of weeds in the furrows due to limited availability of water.

FIGURE 6.5 Cotton field irrigated by subsurface irrigation system in central Arizona.

6.6 Drainage Systems

Although the previous sections described the importance of supplying the plant water needs by irrigation, it is also important to maintain proper aeration status in the root zone. Drainage systems allow the rapid removal of excess water from within the zone. Biosystems engineers are knowledgeable about the physical properties of soil and are able to model the flow of water through soils. They are also trained to identify the source of water problems and to design appropriate drainage systems.

6.6.1 Importance of Drainage

Waterlogged soils will adversely affect the plant roots by reducing the aeration status of the root zone. The primary purpose of drainage system is to remove the excess water from within the root zone or prevent the water from entering. There are additional benefits to drainage as well. Water has 5 times more heat capacity than dry soil particles. As a result, it takes less heat to warm dry soils than wet soils. In a temperate climate, solar energy is used to warm the soil during the spring. Therefore, the soil that is drained warms up much earlier than undrained soil. Early warming up will advance the planting season and allow the planting of longer season crops that have higher yields. By lengthening the growing season, the number of days available for machinery use is increased. This will allow more land to be cultivated with the available machinery and labor.

Wet soils have a low weight-bearing capacity compared to dry soils. As a result, farm machinery operating under very wet conditions will result in puddling of the soil, leading to loss of soil structure. When the water content of the soil is below that of the lower plastic limit, the soil will be able to support machinery (Smedema et al. 2004). Therefore, well-drained soils will allow soil tillage operations to begin much earlier than undrained soils. In addition to early warming up, the ability to allow early tilling and late harvesting operations lead to extended time for machinery use. This increases the number of days machinery can be used and reduces the cost of running the machinery. These additional benefits have to be taken into account when weighing the need for draining a particular farmland.

6.6.2 Methods for Dealing with Excess Water

The first task of the biosystems engineer assigned to solve a drainage problem is to identify the source of the excess water. Aerial photographs taken under flooded conditions or a contour map of the land surface can identify the extent of the problem. If the water is coming from outside the boundaries of the field as surface water flow, it can be prevented from entering the field by the construction of dikes to keep the water out. Alternately, interceptor drains can be installed to divert the water before it enters the field. If the water is present on most of the land surface because of snowmelt or rainfall, reshaping the land may be necessary to improve surface water runoff. The following sections describe different drainage methods in detail.

Land Forming

When puddles of water exist in many parts of a field, it might be more appropriate to reshape the land surface by land forming. Land forming is the process where the undulating land surface is reshaped by moving soils from higher areas into depressions. First, a contour survey is carried out on a grid, based on which contour maps are drawn for the area. Based on the general slope of the land, two design slopes in orthogonal directions are selected to minimize the cut and fill. Once the cut and fill at each location on the grid is determined, an assessment of the depth of the cut and fill is made. It is important to preserve the topsoil during land forming because it takes about 25 to 50 years to form a 25-mm layer of topsoil in a temperate climate. Generally, topsoil is scraped up and piled before the land forming begins. The land is reshaped according to the design cut/fill requirements. Land forming is done in two passes running perpendicular to each other. Because cutting the soil releases the overburden pressure, the soil fluffs up in areas where the soil has been cut. Similarly, the soil will settle over time in areas where it has been filled. Therefore, the two passes are done over two seasons to allow time for the soil to settle. Once the reshaping is completed, the topsoil is spread over the land in preparation for planting.

Although land forming is more expensive, in the end it will improve machine efficiency by making it easier to navigate the field without having to cross drainage ditches. Unlike drainage ditches, well-shaped land requires hardly any maintenance, thus saving a considerable amount of money in drainage maintenance costs.

Surface Drainage

If only a few areas are waterlogged in the field, it might be easier and cost effective to construct drainage ditches just to drain the low spots. Water has to be drained toward an outlet, which could be a project drainage ditch, a lake, or a river. The selection of ditch route should also take into account field operations and machinery efficiency. A few ditch routes are chosen and a profile survey is conducted to determine the land elevation along the alternative ditch routes. The preliminary ditch design is done within depth and slope constraints. Generally, the maximum depth of a ditch is 1 m to allow for machinery crossing. The minimum depth of cut is 0.15 m to allow for a sufficient cross-sectional area for water to flow. Over time, the ditches get silted up and the cross-sectional area decreases, leading to a reduction in drainage capacity.

The steeper the slope, the greater the velocity of flow and the lesser the cross-sectional area required to transmit a given flow rate.

However, steeper slopes and increased velocities of water will cause soil erosion. Therefore, the maximum slope should not exceed 0.6 percent in agricultural soils to prevent soil erosion. When the land surface profile is generally flat, a minimum slope needs to be maintained to allow water flow to take place. As the ditch slope decreases, the velocity of water decreases, allowing suspended sediments to settle. This will cause frequent silting up of ditches and reduction in water conveyance capacity, which imposes additional ditch cleanup work. To avoid this problem, a minimum slope of 0.1 percent is recommended. Depending on the land surface profile, the slope might change along the length of the ditch. The slope change should be gradual to avoid the scouring action of the change in velocity of water flowing in the drainage ditch.

Sometimes, it would appear that a ditch design would not be possible without violating the above criteria. If the land profile is too steep, the ditches could be constructed in sections, at the required maximum slope, with the different elevations connected by *drop structures* consisting of rocks or wood to break the energy of the falling water. The ease of machinery operation should be taken into account when such designs are proposed.

Subsurface Drainage

Unlike surface drainage ditches, subsurface drainage tiles are buried below the ground surface and do not interfere with machinery operation. The tiles are relatively maintenance free. The word *tiles* originates from the use of 30-cm-long clay tiles that were tapered to facilitate insertion in series during installation. At present, corrugated plastic pipes are more popular because of their ease of installation, low cost, and durability. However, the phrase "tile drainage" is still used to refer to such systems as well. Corrugated plastic pipes are installed at depths ranging from 1.0 to 1.3 m below the ground surface. Specially designed drainage installation machinery is used for field installation (Fouss and Fausey 2007). The depth of installation is controlled by a laser plane set at the desired slope. A laser beam detector attached to the installation arm will move up and down to compensate for the undulating land surface while maintaining a uniform slope along the length of the subsurface drain. All subsurface drains are connected to a large collector drain, which diverts the water to the drainage outlet.

Biosystems engineers have developed drainage design methods and computer models to help determine optimum spacing of depth of installation for a given soil (Skaggs 2007).

6.7 Salinity Control

The consequences of salinization of agricultural land have been recorded in history (Hillel 2000). Wherever irrigation and drainage is practiced, careful precautions should be taken to monitor salt accumulation within the root zone. As described previously, excess irrigation water is used to wash the salts below the root zone, and this excess water is removed from root zone by drainage systems. Environmental regulations set the water quality standards of drainage return flows and require desalination of the water before it is emptied into a river or a lake. Alternatively, return flows could be impounded in a constructed wetland where the salts are taken up by specific plant species. Biosystems engineers design such engineered wetlands that help clean up the water and prevent environmental degradation of waters in lakes and rivers, which are described as follows.

6.7.1 Introduction to Salinization

Whenever irrigation water is applied to plants, the plants take up the water for ET and leave behind salts within the root zone. Over time, the accumulation of salts will begin to alter the physical properties of the soil. Generally, clay soil particles tend to disperse in the presence of sodium ions, leading to the destruction of the soil structure (U.S. Salinity Laboratory Staff 1954).

Biosystems engineers have to determine the origin of the salts when proposing a solution. The salts may already exist within the parent material. As soil salinity increases, the electrical conductivity of the soil also increases. Electromagnetic methods are available to map the electrical conductivity from the surface without destroying the soil profile (Wittler et al. 2006). Equipment is available to map soil salinity to a depth of 6 m from the ground surface. This equipment can be linked to global positioning systems (GPS) to georeference the data. These data can then be mapped using geographic information systems (GIS). The subsurface salinity can be mapped using GIS, and hot spots of salt concentration can be identified.

6.7.2 Methods to Control Salinization

Dealing with Saline Parent Material

The water table should be kept below the root zone to avoid the upward movement of water from the deeper saline layers of soil. Drainage systems can help keep the water table much below the root zone. Water is applied in excess of irrigation requirement to wash the salts accumulating within the root zone below it. Drainage systems can help remove this excess water.

Dealing with Saline Irrigation Water

The salinity of the water is assessed by measuring the electrical conductivity (EC). Water with an EC greater than 4 dS/m is unsuitable for irrigation (U.S. Salinity Laboratory Staff 1954). Water with an EC less than 0.25 dS/m is considered to be of excellent quality for irrigation. When the EC of irrigation water is greater than 0.25 dS/m, excess water should be applied to wash the salts below the root zone.

If the soil drainage is poor, drainage systems have to be installed to remove the water leaching below the root zone.

6.7.3 Dealing with Water Quality Issues

When the drainage water outflow is diverted to an outlet, it has to meet the water quality standards. Depending on the quality of the drainage water, excess water may need to be applied to dilute the salt concentration below the permissible limit. Sometimes, the water needs to be treated in a desalination facility before it can be disposed into an outlet. However, this may be too costly to be economically viable.

Engineered wetlands can be used to hold the drainage return flows. Halophytic plants that can grow in saline environments can be planted in these wetlands to help remove the salts. Halophytic plants have the capability to uptake salts and accumulate the salts within the plant biomass. Periodic harvesting of the biomass can remove the salt load from engineered wetlands. Biosystems engineers learn about plant physiology to understand this process and design better wetlands.

6.8 Summary

Biosystems engineers are knowledgeable about the physical and chemical properties of soil as well as plant–soil–water relations, and plant physiology. They understand how to optimize the conditions to maximize plant production while sustaining the environment. Their training in fluid mechanics and material sciences will help them design efficient irrigation and drainage systems. In addition, new and innovative instrumentation is used to assess the need for irrigation and drainage and protect the quality of the drainage water. Biosystems engineers play a major role in expanding food and fiber production through irrigated agriculture, reclaiming salinity-affected land, and protecting water quality in rivers and lakes.

References

ASAE (American Society of Agricultural Engineers). 1990. *Management of Farm Irrigation Systems*, eds. G. J. Hoffman, T. A. Howel, and K .H. Solomon. St. Joseph, MI: American Society of Agricultural Engineers. 1038 pp.

FAO (Food and Agriculture Organization). 1998. *Crop Evapotranspiration*, eds. R. G. Allen, L. S. Perara, D. Raes, and M. Smith. Irrigation and Drainage Paper No. 56. , Rome, Italy. 300 pp.

Fouss, J. L. and Fausey, N. R. 2007. Research and development of laser-beam automatic grade-control system on high-speed subsurface drainage equipment. *Transactions of the ASABE* 50(5):1663–1667.

Hillel, D. 1998. *Environmental Soil Physics*, San Diego, CA:Academic Press.

Hillel, D. 2000. *Salinity Management for Sustainable Irrigation: Integrating Science, Environment, and Economics*, Washington, DC: The World Bank Group. 104 pp.

Kemper, W. D., Heinemann, W. H., and Kincaid, D. C. 1981. Cablegation 1. Cable controlled plugs in perforated supply pipes for automatic furrow irrigation. *Transactions of the ASABE* 24(6):1526–1532.

Skaggs, R. W. 2007. Criteria for calculating drain spacing and depth. *Transactions of the ASABE* 50(5):1657–1662.

Smedema, L. K., Vlotman, W. F., and Rycroft, D. W. 2004. *Modern Land Drainage— Planning, Design and Management of Agricultural Drainage Systems*, New York: A. A. Balkema Publishers. 446 pp.

Soil Science Society of America. 1984. *Glossary of Soil Science Terms*, Madison, WI: Soil Science Society of America.

Sri Ranjan, R. and Domytrak, C. J. 1997. Effective volume measured by TDR miniprobes. *Transactions of the ASABE* 40(4):1059–1066.

Topp, G. C., Davis, J. L., and Annan, A. P. 1980. Electromagnetic determination of soil water content: Measurement in coaxial transmission lines. *Water Resources Research* 16:574–582.

Turner, N. C. 1988. Measurement of plant water status by the pressure chamber technique, *Irrigation Science* 9(4):289–308.

U.S. Salinity Laboratory Staff. 1954. *Agriculture Handbook No. 60: Diagnosis and Improvement of Saline and Alkali Soils*, ed. L.A. Richards. Washington DC: U.S. Department of Agriculture.

Wittler, J. M., Cardon, G. E., Gates, T. K., Cooper, C. A., and Sutherland, P. L. 2006. Calibration of electromagnetic induction for regional assessment of soil water salinity in an irrigated valley. *Journal of Irrigation and Drainage Engineering* 132(5):436–444.

Zakaluk, R. and Sri Ranjan, R. 2007. Artificial neural network modelling of leaf water potential for potatoes using RGB digital images: A greenhouse study. *Potato Research* 49(4):255–272.

Zakaluk, R. and Sri Ranjan, R. 2008. Predicting leaf water potential of potato using spectral reflectance indices. *Canadian Biosystems Engineering* 50:7.1–7.12.

CHAPTER 7
Biomass Pyrolysis and Bio-Oil Refineries

Manuel Garcia-Pérez
Department of Biological Systems Engineering
Washington State University
Pullman, Washington

7.1 Introduction

Replacing petroleum-based industrial chemicals and fuels with those from renewable sources is a very worthy and alluring objective. The reduction of greenhouse emissions, the lessening of dependence on nonrenewable resources, the increase in markets for farm products, and the potential for new industries in widely distributed geographical locations are outcomes that are attractive to any country. Biomass is virtually the only renewable, "carbon neutral" widely distributed natural resource capable of supplying carbon-based transportation fuels as well as some chemicals presently obtained from nonrenewable sources such as coal and petroleum. Today, nearly 2 million tons of fuelwood are consumed daily by almost half the world population for cooking and heating homes. Although this energy represents less than 10 percent of the total energy consumed worldwide daily, it accounts for as much as 95 percent of the energy supply in some of the poorest nations (Wood and Baldwin 1985).

The pyrolysis of biomass (cellulose, hemicelluloses, lignin, extractives, triglycerides, etc.) results in the formation of crude bio-oil, char, and gases (Moses 1994; Czernik and Bridgwater 2004; Bridgwater et al. 2001; Roy et al. 2001; Meier and Faix 1999). It is certainly not a new technology in that pyrolysis has been used for centuries to produce charcoal. Charcoal was the first synthetic material produced by humans.

It was used more than 38,000 years ago by the Cro-Magnon people to create the magnificent drawings in the Grotte Chauvet (Antal and Gronli 2003). Egyptian mummy makers used the tars and the pyrolygneous acids resulting from biomass pyrolysis to practice their craft. In the nineteenth and early twentieth centuries, the biomass pyrolysis (wood distillation) was a profitable business producing soluble tars, creosote oil, methanol, and acetone (Klar and Rule 1925). By the 1930s, the wood-distillation industry entered a period of steady decline due to the fierce competition from cheaper petroleum-derived products.

Fast pyrolysis technologies were developed in the 1970s to maximize the production of crude bio-oils. One of the main advantages of producing crude bio-oil is that this energy-densified liquid can be transported from the processing unit to distant power stations or to bio-oil refineries (Peláez-Samaniego et al. 2008). Thus, it is possible to visualize the creation of a new model of biomass economy (see Fig. 7.1) formed by distributed pyrolysis units located close to biomass resources and centralized refineries where second-generation transportation fuels and high-value chemicals can be produced, taking advantage of the economies of scale. Fast pyrolysis can convert up to 75 mass% of the biomass into crude bio-oils. Forty percent of these oils could be further transformed into green gasoline and green diesel via bio-oil hydrotreatment (Holmgren et al. 2008; Elliott 2007). An advantage of this new model of biomass economy is that it can make good use of the existing infrastructure created by the petroleum industry.

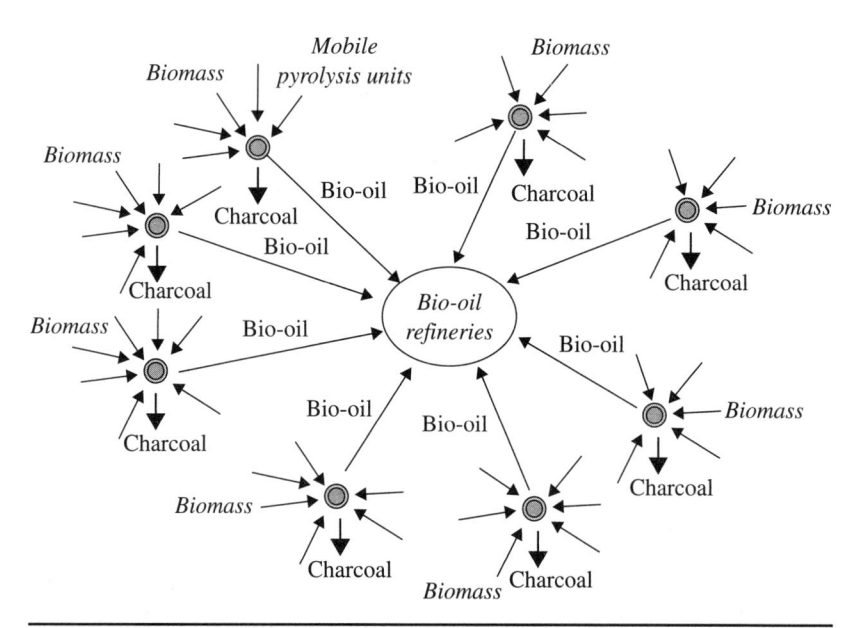

Figure 7.1 New model of biomass economy formed by distributed pyrolysis units and bio-oil refineries.

In the United States alone it is possible to sustainably produce 1.3 billion dry tons of biomass annually (Perlack et al. 2005). Processing 75 percent of this biomass via fast pyrolysis and bio-oil hydrotreatment could result in the production of 273 million tons of green gasoline and green diesel annually with the potential to replace up to 66 percent of current U.S. gasoline consumption (410 million tons/year). The impact of this new model of biomass economy in rural areas could be significant. For example, 54,000 new mobile pyrolysis units (at 50 tons/day each) will be needed in the United States to process 975 million tons of biomass annually. Thousands of additional jobs could be created to operate and maintain the bio-oil refineries. The design, manufacture, and assembly of all components needed to build these systems could have a multiplicative effect, generating thousands of additional jobs in the broader economy.

The current status, opportunities, and challenges to convert biomass into second-generation transportation fuels via pyrolysis and bio-oil refineries have been discussed in this chapter. Excellent reviews on state of the art of biomass pyrolysis technologies and on the application of bio-oils as sources of fuels and chemicals have been published (Bridgwater and Peacocke 1994; Diebold and Bridgwater 1999; Meier and Faix 1999; Oasmaa and Czernik 1999; Piskorz et al. 1999; Radlein 1999; Bridgwater et al. 2001; Czernik and Bridgwater 2004; Mohan et al. 2006).

7.2 Biomass Composition

The composition of a biomass will influence the yield and properties of the resulting crude bio-oils, charcoals, and gases. Trees can be classified as hardwoods (Angiospermae) or softwoods (Gymnospermae). Biomasses are composed of *cellulose, hemicelluloses, lignin, extractives,* and *mineral materials* (also known as ash) (Campbell 1983; Klass 1998). The cellulose, hemicelluloses, and lignin are bio-polymers that form the structure of biomass cell walls.

7.2.1 Cellulose

Cellulose is a linear polysaccharide with a flat, linear configuration composed of between 10,000 and 15,000 glucose units linked by glycosidic bonds (see Fig. 7.2) (Campbell 1983; Klass 1998). Cellulose structure is reinforced by inter- and intrachain hydrogen bonds. Cellulose can be found in amorphous and crystalline forms. Six convertible polymorphs of cellulose (I, II, III$_I$, III$_{II}$, IV$_I$, and IV$_{II}$) have been identified. The polymorph I, or native, cellulose is the only form found in nature. Although it was initially thought that cellulose I was single phase, it is now known that it is always a mixture of two allomorphs (Iα and Iβ). These allomorphs differ in their hydrogen bonding patterns (O'Sullivan 1997). Cellulose Iβ is more abundant in higher plants.

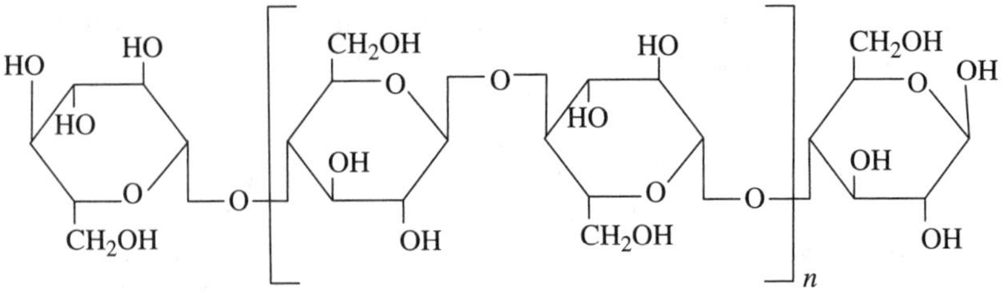

FIGURE 7.2 Cellulose structure.

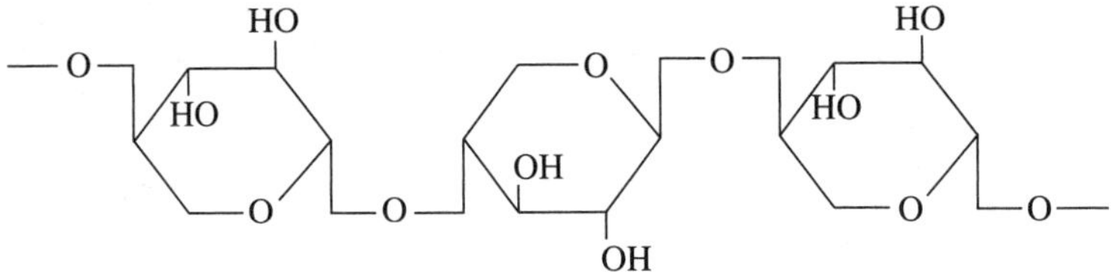

FIGURE 7.3 Hemicellulose structure.

7.2.2 Hemicellulose

In contrast with cellulose, which is a homopolymer, hemicellulose is an amorphous branched heteropolysaccharide (Sjöström 1993). It is a mixture of mainly pentoses (chiefly xylose), with lesser amounts of arabinose, glucose, mannose, and galactose. Thus, hemicelluloses also have a hexose component (Campbell 1983). This diversity of sugars and their amorphous structure are the main causes for hemicellulose's reduced thermal resistance. Figure 7.3 is a simplified scheme for hemicelluloses consisting only of D-xylan units linked in the 1- and 4-positions.

7.2.3 Lignin

Lignin is a heterogeneous polymer consisting of phenylpropane units (see Fig. 7.4) linked through various ether and C–C bonds. Lignin is usually found in chemical associations with cellulose and hemicellulose. It acts as a binder of three-dimensional structures that protects cellulose and hemicellulose against chemical and microbial attacks (Sjöström 1993).

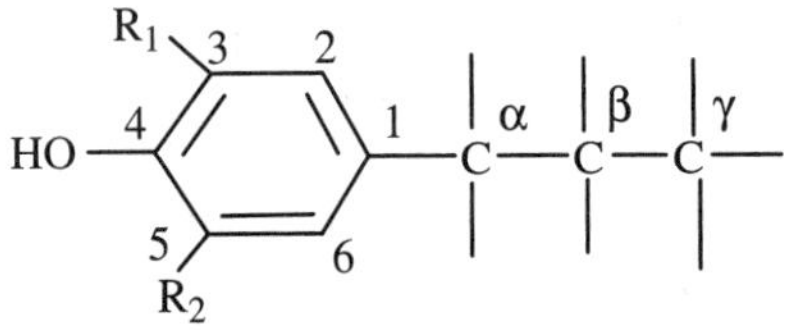

Hydroxyphenyl:	$R_1 = H$,	$R_2 = H$
Guaiacyl:	$R_1 = H$,	$R_2 = OCH_3$
Syringyl:	$R_1 = OCH_3$,	$R_2 = OCH_3$

FIGURE 7.4 Structure of the carbon skeleton of lignin monomeric units.

7.2.4 Extractives

Extractives are nonstructural compounds formed by secondary metabolism. They are both lipophilic and hydrophilic organic compounds that can be extracted by hot water or with the aid of organic solvents. Extractives confer odor, color, and durability to the wood and protect the plants against predators. Bark contains 4 to 5 times more extractives than bark-free wood.

7.2.5 Ash

The mineral materials (ash) are composed mainly of carbonates, sulfates and phosphates of silica and metals from the first and second group of the Periodic Table (Ca, Mg, Na, and K). These compounds act as catalysts for polycondensation reactions leading to char formation.

7.3 Overview of Biomass Pyrolysis Reactions

As the temperature of the biomass particle rises, thermal decomposition reactions start to take place on a cylindrical reaction front, which advances toward the core of the particle at a speed determined by the particle size, biomass thermal conductivity, and the temperature of the environment (see Fig. 7.5). The cleavage of cellulose, hemicellulose, lignin, and extractives chemical bonds occurs in well-defined temperature ranges (Boroson et al. 1989a, 1989b; Evans and Milne 1987a, 1987b). In the literature, these reactions are called *primary thermal decomposition (pyrolysis) reactions*. These reactions are commonly studied by thermogravimetry. Two or three peaks assigned to each of the biomass macrocomponents (extractives, cellulose, hemicellulose,

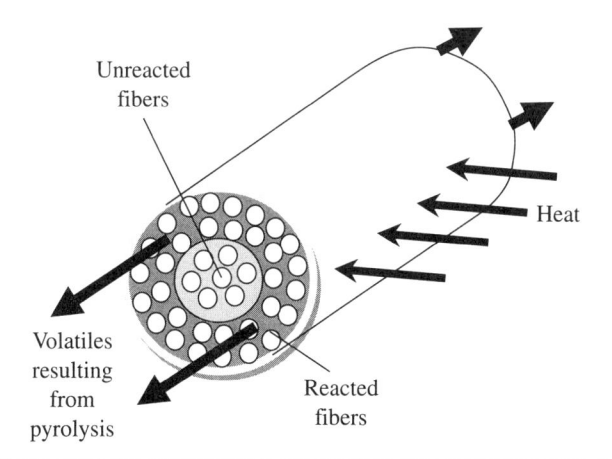

FIGURE 7.5 Representation of a biomass particle made up of hollow cylindrical cells.

and lignin) appear (Koufopanos et al. 1991; Caballero et al. 1997; Varhegyi et al. 1997; Orfao et al. 1999). The products of these primary reactions could undergo additional depolymerization, fragmentation, or polycondensation in the solid phase or in a melted state to form molecules of lower molecular weight and charcoal (Graham et al. 1984; Agblevor et al. 1994). The resulting volatile species will continue reacting as they diffuse inside the pores of the particles (*intraparticle secondary reactions*) either homogeneously in the gas phase or heterogeneously with the partially converted solid biomass or with the char (Hastaoglu and Berruti 1989). The intensity of these secondary reactions depends on the time-temperature history to which the products of primary reactions will be subjected before collection in the condensers (Bridgwater et al. 1999).

7.3.1 Mechanism of Primary Reactions

Cellulose

The mechanism of cellulose thermal degradation has been the source of passionate discussions for many years (Piskorz et al. 2000). Several models have been proposed to explain the complex cellulose thermal decomposition pathways. The pioneering model of Broido and Shafizadeh, shown in Fig. 7.6, is an important reference because it introduced for the first time the concept of active cellulose to explain the behavior of cellulose pyrolysis. This concept was developed as a consequence of experimental evidences showing the formation of cellulose at a lower degree of polymerization at temperatures as low as 220°C (Golova 1975). Broido et al. (1973) found that crystalline cellulose undergoes a large change in degree of polymerization before weight loss occurs. Bradbury et al. (1979) noted that at low heating temperatures (259 to 312°C), an initiation period occurs. Thus, the model of Broido and Shafizadeh (Fig. 7.6; Broido and Nelson 1975; Wooten et al. 2004) postulates that the cellulose is converted into a more active form and that this is a rate-limiting step followed by the formation of either char, gases, or tar.

An excellent overview of several of the semiglobal mechanisms for primary pyrolysis of lignocellulosic materials can be found elsewhere (Di Blasi 1998). Figure 7.7 is an overview of a more detailed model describing more cellulose thermal-degradation reactions. Much experimental evidence suggest that cellulose thermal degradation proceeds by the following seven major reactions: (1) hydrolysis reactions to produce active cellulose (Broido et al. 1973; Golova 1975;

Cellulose ⟶ Active cellulose ⟶ Volatile tar
⟶ Char + gases

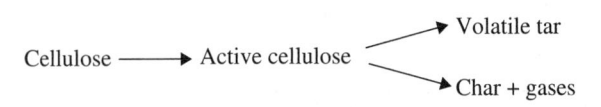

FIGURE 7.6 The Broido–Shafizadeh model.

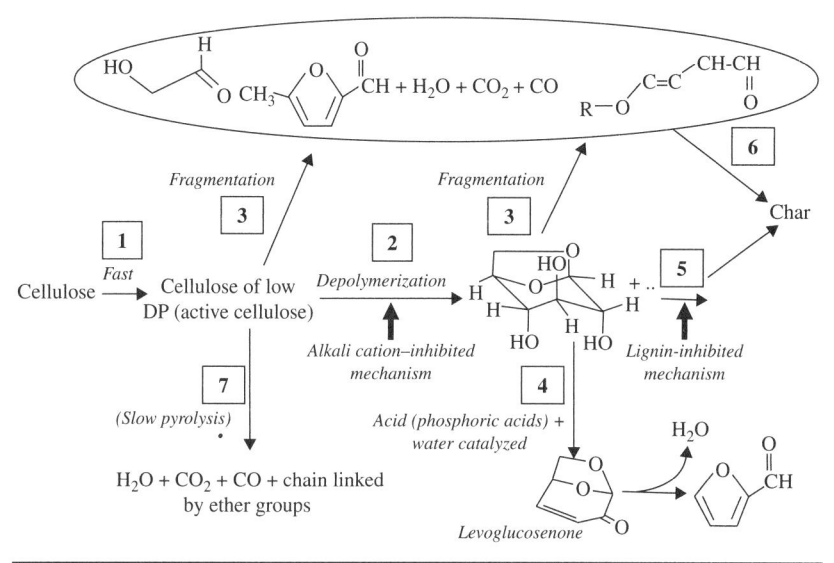

FIGURE 7.7 Cellulose thermochemical-degradation reactions (Garcia-Pérez and Metcalf 2008).

Piskorz et al. 1986; Kim et al. 2001; Wooten et al. 2004; Mamleev et al. 2007; Zickler et al. 2007); (2) depolymerization of active cellulose to produce mono- and oligosugars (Radlein 1987; Radlein et al. 1991a, 1991b; Piskorz et al. 2000); (3) fragmentation or open-ring reactions controlled by the presence of alkalines (Broido and Kilzer 1963; Radlein et al. 1991a, 1991b; Piskorz et al. 1986, 1989a, 1989b, 2000; Arisz et al. 1990; Lomax et al. 1991; Julien et al. 1993; Evans and Milne 1987a, 1987b; Richards 1987); (4) acid-catalyzed dehydratation reactions (Radlein et al. 1991a, 1991b; Dobele et al. 2001, 2003; Kawamoto et al. 2003a, 2003b, 2007a; Kawamoto and Saka 2006); (5) polymerization of anhydrosugars (Wooten et al. 2004; Hosoya et al. 2006, 2007a, 1997b); (6) cross-linking reactions of fragmentation products to produce char; and (7) cross-linking reactions with evolution of water, which is typical of slow heating rate regimes (Kilzer and Broido 1965).

Low heating rates tend to favor the cross-linking reactions leading to the formation of larger yields of char and water [see reaction (7), Fig. 7.7]. Conversely, high heating rates will favor depolymerization reactions resulting in higher bio-oil yields.

Hemicellulose

Xylan thermal-degradation reactions follow a similar pathway to the one described for cellulose. Xylan possesses ion-exchange sites, which make it particularly susceptible to the uptake of impurity cations. The main products of the thermal decomposition of hemicelluloses found in the bio-oils are acetic acid, furans, and mono- and oligopentoses.

Because of their amorphous structure, hemicelluloses have lower thermal stability than cellulose. Their thermal degradation starts at temperatures as low as 200°C. The yields of furfural increase when the substrate is impregnated with acid catalysts (Radlein et al. 1991a, 1991b).

Lignin
Thermal degradation of lignin occurs over a wide range of temperatures (200 to 600°C), resulting in the formation of monomeric phenols, guaiacols and syringols, formic acid, formaldehyde, methanol, carbon dioxide, and water. Demethoxylation of lignin is the source of methanol. In general, the thermal degradation of lignin can be described by a competitive mechanism involving depolymerization and condensation/carbonization reactions as suggested by Kawamoto et al. (2007b). During fast pyrolysis, most of the lignin is converted into monomers, dimers, and trimers. The dimers and trimers are known as pyrolytic lignin. Many of these dimers and trimers are released in the form of aerosols (Garcia-Pérez et al. 2008).

7.4 Single-Particle Models
Understanding how the thermal degradation of individual biomass particles proceeds is critical to design and to operate more efficient pyrolysis reactors. The conversion of a single biomass particle is controlled by the physicochemical properties and dimensions of biomass particles as well as by the reaction conditions (reactor temperature, reacting environment, external heat transfer coefficient). The residence time of biomass particles inside a pyrolysis reactor needed to achieve complete conversion will depend on the devolatilization and heating rates. Total conversion time can be calculated by single-particle models. The values of total devolatilization time for small biomass particles (diameter less than 2 mm) in fluidized-bed reactors reported in the literature are between 10 and 70 s (Kumar et al. 2006).

Many models have been developed over the last 60 years to describe the thermal degradation of single biomass particles. In general, it is accepted that for particles with length-to-diameter ratio smaller than 3 should be described by two-dimensional models. One-dimensional models should be used when describing biomass particles with length-to-diameter ratios larger than 3. Excellent reviews on the single-particle models used to calculate biomass devolatilization time can be found elsewhere (Kersten et al. 2005, Di Blasi 2008).

When designing a pyrolysis reactor it is very important to distinguish between the reactor temperature and the temperature at which primary thermochemical reactions occur. The reactor temperature is usually higher than the particle temperature, which controls the reaction rate of primary thermochemical reactions (Bridgwater et al. 1999).

Three controlling mechanisms or extreme cases have been used to describe the devolatilization of biomass particles: (1) external heat transfer controlled, (2) kinetically controlled, and (3) internal heat transfer controlled. Each mechanism is briefly explained here.

External Heat Transfer–Controlled Regime   This occurs when the rate of external heat transfer is very slow compared with internal heat transfer. Thus, the temperature inside the biomass particle will be essentially uniform but will gradually climb over time. Accordingly, the reaction rate will be uniform across the particle (Pyle and Zaror 1984).

Kinetically Controlled Regime   The reaction rate is very slow compared with the heat transfer to and within the biomass particle. The particle temperature will be very similar to the reactor temperature. The thermal degradation of biomass particles with diameters less than 2 mm in fast pyrolysis reactors happens in this regime (Pyle and Zaror 1984).

Internal Heat Transfer–Controlled Regime   If the internal heat transfer is very slow compared with the reaction rate, there will be large gradients of temperature inside the particles. The reaction zone will be very narrow; consequently, the process can be regarded as if a charcoal front is advancing into the virgin biomass at "a rate controlled by the velocity at which the thermal wave propagate" (Pyle and Zaror 1984). Most of the models tested (Pyle and Zaror 1984) suppose that the reaction happens at temperatures around 430°C. This regime occurs when large biomass particles are converted in slow pyrolysis reactors.

Pyle and Zaror (1984) proved that the controlling mechanism during the thermal degradation of a single biomass particle can be identified if the following three dimensionless numbers are known:

$$\text{Biot number: Bi} = hR/K \tag{7.1}$$

$$\text{Pyrolysis number: Py} = K/(k \rho c_p R^2) \tag{7.2}$$

$$\text{Pyrolysis number: Py}' = h/(k \rho c_p R) \tag{7.3}$$

where h = external heat transfer coefficient (J/s m² K)
R = particle radius (m)
K = thermal conductivity of biomass (J/s m K)
k = rate of biomass thermal-degradation reactions (1/s)
ρ = biomass bulk density (kg/m³)
c_p = biomass specific heat (J/g K)

The Biot number compares the rate of heat transfer to the surface by convection with the internal conduction resistance (Graham et al. 1984). Large woody materials processed in pyrolysis reactors will have large

	Approximate Range of Validity		
Model	Bi	Py	Py'
Non-controlling conditions		All values	
External heat transfer	<1	>1	>1
Kinetics	<1	>10	>10
Internal heat transfer	>50	<10^{-3}	<<1

TABLE 7.1 Criteria for Identifying the Controlling Mechanism during Biomass Devolatilization

Biot number values. The criteria to identify the controlling mechanism for a single biomass particle devolatilization are listed in Table 7.1 (Pyle and Zaror 1984).

The poor thermal conductivity of biomass particles (0.1 W/m K along the fibers and 0.05 W/m K cross-fiber) limits the maximum heating rates that can be achieved. Large particles (diameter more than 2 mm) can only achieve high heating rates if the char is continuously removed mechanically (Bridgwater et al. 1999).

7.5 Pyrolysis Technologies

The reactors used for biomass pyrolysis can be classified into slow and fast pyrolysis. Although slow pyrolysis reactors are mainly used to produce charcoal, fast pyrolysis is the technology of choice to maximize bio-oil yields. Yields of oil as high as 80 mass% can be obtained with this technology. Slow pyrolysis, on the other hand, will result in much lower yields (30 to 50 mass%) of a liquid composed of two phases (decanted oil or tars and pyroligneous acid) (Wood and Baldwin 1985).

7.5.1 Slow Pyrolysis

Most of the charcoal producers in developing nations employ temporary earthen kilns in which the wood pile is buried under a layer of soil. Charcoal makers are in charge of opening and closing vent holes during the whole production period to control the amount of oxygen supplied. Slow pyrolysis happens when the heating rate is less than 10°C/s, the pyrolysis temperatures are below 500°C, and gases and solids stay inside the reactor for a long time (Graham et al. 1984). A typical yield of products for these technologies is liquids, 30 mass%; charcoal, 35 mass%; and gases, 35 mass%. Slow pyrolysis typically achieves energy efficiencies between 20 and 30 percent.

Perhaps the most advanced slow pyrolysis reactors developed so far were those built and operated by the wood-distillation industry (Klar and Rule 1925). Advanced batch and semicontinuous reactors such as the American, Schwartz, Ljungberg, Ottelinska, Reichenbach, Leschhorn, Swedich, Bosnian Meiler, and Meyer kilns were used by this industry.

Brazil is currently, the world largest charcoal producer with an annual production output of around 10 million tons of charcoal most of which is used in the iron and steel industry (Peláez-Samaniego et al. 2008). In spite of the size and importance of this industry, carbonization reactors in operation are basically traditional low-efficiency processes with very low recovery of volatile factions. The yields of charcoal obtained are around 25 mass%. The main pyrolysis reactors in operation in Brazil are (1) internally heated by controlled combustion of the raw material (autothermal systems) and (2) externally heated by combustion of firewood, fuel oil, or natural gas (Peláez-Samaniego et al. 2008).

Internally heated kilns sacrifice up to 20 percent of the biomass in an initial heating step in which air is supplied to heat up the kiln. The yields of gases and vapors are as high as 60 percent. Some of the most common configurations of this type of kilns are alvenaria kilns and metallic kilns (Rosillo-Calle et al. 1996; Pinheiro 2001; Kammen and Lew 2005; de Meira et al. 2005).

Externally heated kilns are commonly used to produce large volumes of charcoal. Heat is provided by hot combustion gases produced by an external combustion chamber where an auxiliary fuel is burned. Although the control of the process is better and the resulting charcoal has better quality, these systems are more expensive to build and operate (Peláez-Samaniego et al. 2008).

7.5.2 Fast Pyrolysis

Although fast pyrolysis is related to the traditional carbonization industry, polycondensation reactions leading to the formation of charcoal are greatly mitigated, and the selectivity toward fragmentation and depolymerization reactions favored by heating the biomass particles at rates between 10 and 10,000°C/s. Thus, these reactors are able to convert up to 80 mass% of the biomass into a single oily phase (Bridgwater et al. 1999). Fast pyrolysis was developed in the 1970s and 1980s as a response to some of the challenges created by the first and second petroleum crises.

Low yields of gases (around 15 mass%) are obtained when the residence time of pyrolysis vapors is below 2 s (Graham et al. 1984). An exhaustive analysis of the present status of biomass pyrolysis technologies is not possible within the scope of this chapter. Excellent reviews on biomass fast pyrolysis technologies can be found elsewhere

Reactor Type	Advantages	Disadvantages
Ablative	Can process feedstocks with large particles sizes. Has compact design	Mechanical char abrasion is very important. Heat supply is difficult
Circulating fluid bed	Has high heat-transfer rates. The solid is recycled	Char abrasion is very important. Needs very high volumes of carrier gases
Fluid bed	Has high heat-transfer rates	Char abrasion is not very important. Uses large volumes of carrier gases. Particle size limit <2 mm
Vacuum moving bed pyrolysis	Can process feedstocks with large particles. Does not have carrier gas	System is operated under vacuum. Has a relatively low external heat transfer coefficient

TABLE 7.2 Characteristics of Some Fast Pyrolysis Reactors

(Meier and Faix 1999; Bridgwater and Peacocke 2000; Czernik and Bridgwater 2004; Kersten et al. 2005; Mohan et al. 2006).

The most important fast pyrolysis reactors developed so far are (1) fluidized beds, (2) circulating fluid beds, (3) ablative pyrolysis, and (4) vacuum pyrolysis (Scott et al. 1999). Table 7.2 shows some of the most important features, advantages, and disadvantages of these reactors (Bridgwater et al. 1999, 2001).

Bubbling fluidized beds are the most popular fast pyrolysis reactors in operation today. These reactors are easy to operate and can be readily scaled up (see Fig. 7.8) (Vanasse et al. 1988; Bridgwater et al. 1999; Huber and Dumesic 2006). Almost 90 percent of the heat is transferred by direct contact between the biomass and the fluidized-bed material (usually sand). The convective heat transfer from the gas only represents around 10 percent of the total heat transferred (Bridgwater et al. 1999). Basically, these systems are formed by a pyrolysis reactor; one or two cyclones to separate the char particles and condensers where the oils are collected (see Fig. 7.8). The separation efficiencies of cyclones are not high enough to retain all the charcoal Thus, unless hot filtration is used, char particles will find their way into the bio-oil. The alkalis contained in the charcoal will be leached out, additionally decreasing bio-oil thermal stability. The charcoal or the gases can be used to supply the energy needed in the process. Better temperature control, efficient heat transfer, and short residence times for vapors are the main advantages of fluidized beds. The main drawback of this technology is the use of large volumes of carrier gas and very small particles (diameter less than 2 to 3 mm). This type of reactor is not well adapted to operate as part of mobile units. The high gas-to-biomass

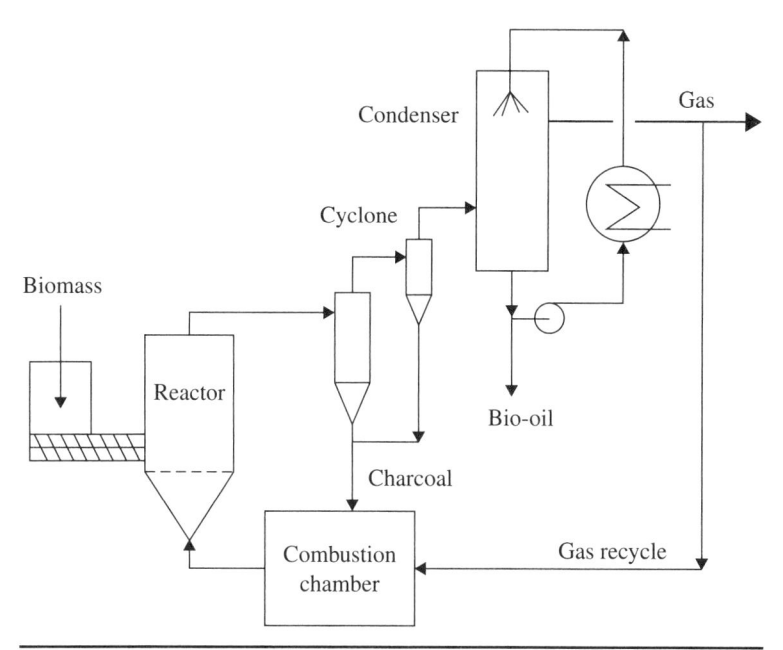

FIGURE 7.8 Conceptual fluid bed fast pyrolysis process.

feed ratio results in lower thermal efficiency, which is typically on the order of 60 to 70 percent (Huber and Dumesic 2006).

Circulating fluid beds pyrolysis reactors operate in a regime called "fast fluidization," which lies between turbulent fluidization and pneumatic transport (Basu 2006). At relatively high velocities the particles are elutriated as a gas–solid suspension. The partially converted particles are recovered by a cyclone and returned to the base of the reactor. These reactors are commonly operated at extremely high heating rates (1000 to 10,000°C/s), high temperatures (greater than 600°C) and very short vapor-residence time (less than 0.5 s) (Graham et al. 1984).

Ablative pyrolysis reactors are compact and intensive systems in which large biomass particle are pressed against a hot surface. The heating rates achieved are very high because the hot surface continuously abrades the product char off the particle. Thus, fresh biomass is always exposed to the hot surface. These systems do not use carrier gases (Bridgwater et al. 1999).

Although the heating rates achieved in *vacuum pyrolysis* reactors are not as high as those attained by fluidized-bed reactors, they do generate high yields of oils because the vacuum removes the pyrolysis products from the hot zones fast enough to substantially reduce the secondary cracking reactions in the vapor phase (Bridgwater et al. 1999). The reactor patented by the group of Professor C. Roy at Laval University, Québec, Canada, used molten salt heated in an external

7.6 Crude Bio-Oils

Fast pyrolysis can convert up to 70 mass% of the biomass into crude bio-oils. This section describes the physicochemical characteristics of these oils as well as their potential uses as fuels and chemicals.

7.6.1 Chemical Composition of Crude Bio-Oils

The chemical composition of bio-oil depends on feedstock composition, pyrolysis conditions, and upgrading technology used in its production. The elemental composition of bio-oil on an anhydrous basis is not very different than that of the feedstock: carbon content, 44 to 47 mass%; hydrogen content, 6 to 7 mass%; oxygen content, 46 to 48 mass%; and nitrogen content, 0 to 0.2 mass%. The content of water is usually between 10 and 30 mass%.

More than 300 organic compounds have been identified in these oils (Evans and Milne 1987a, 1987b; Diebold et al. 1999; Boucher et al. 2000 Oasmaa et al. 2003a, 2003b; Ba et al. 2004a, 2004b; Czernik et al. 2004, Garcia-Pérez et al. 2002, 2007b). A detailed chemical characterization based on the quantification of individual species is almost impossible to achieve and has limited practical application. A characterization strategy based on the quantification of macrofractions or chemical families instead of individual species is more useful to get an overview of bio-oil composition (Garcia-Pérez et al. 2007b). The fitting of thermogravimetry (DTG) curves, like the one seen in Fig. 7.9, suggests that these materials could be broadly described as a mixture of seven macrogroups (Garcia-Pérez et al. 2007a, 2007b, Garcia-Pérez et al. 2008). The overall differences in the chemical composition of

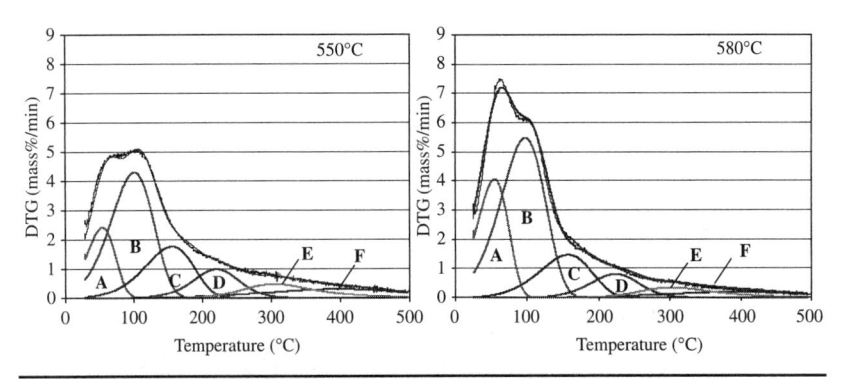

FIGURE 7.9 Deconvolution of DTG curves of bio-oils produced at two pyrolysis temperatures (Garcia-Pérez et al. 2008).

these two oils can be easily appreciated. A detailed description of the methods used to fit the different chemical families can be found elsewhere (Garcia-Pérez et al. 2008).

Family A is formed by very volatile organic compounds (mainly hydroxylacetaldehyde, formic acid, and methanol), which result from biomass fragmentation reactions.

Family B consists of water and other volatile compounds with similar boiling points (acetic acid, acetol, and propionic acid). The water usually represents between 5 to 30 mass% of the whole bio-oil and is the single most abundant specie found in these liquids.

Family C consists of compounds with moderate volatility with boiling points between 100 and 250°C. This family can represent up to 45 mass% of crude bio-oils and is dominated by furans and monophenols resulting from depolymerization reactions. These compounds have between 5 and 13 atoms of carbon and are commonly divided into two subfractions. The hydroxyl, hydroxymethyl, and oxo-substituted furans, furanones, pyranones, and cyclopentents obtained from the thermal degradation of cellulose and hemicellulose are commonly called neutral fractions and can account for between 10 and 30 mass% of bio-oil. Some of the compounds found in this subfraction are cyclopentanone, 2-furancarboxaldehyde, ethanone-1-(2-furanyl), methyltetrahydrofuran, 2-furanone, furfuryl alcohol, and methyltetrahydrofuran. The second subfraction is composed of monophenols having between 6 and 12 atoms of carbon. This fraction is weakly acidic and includes compounds such as phenol, catechol, 2,3-dimethylphenol, hydroquinone, 1,2-benzenediol, syringaldehyde, and 3-ethyl-phenol (Garcia-Pérez et al. 2006a, 2007a, 2008). These compounds can account for up to 15 mass% of the whole oil.

Species with moderate volatility having a boiling point between 200 and 300°C very similar to the kerosene constitute *family D*. This fraction usually accounts for between 5 and 35 mass% of crude bio-oils. Five major groups of compounds can be found in this family: (1) polyaromatics, (2) aliphatic hydrocarbons, (3) fatty acids and its methyl esters, (4) sterols, and (5) sugars. The polyaromatics found in bio-oils have between 10 and 28 atoms of carbon. Some of the most common polyaromatics found in bio-oils are naphthalene, 1,2-dihydroxy3-methylnaphtalene, phenanthrene, 3,6-dimethyl-phenanthrene, and 1-methylanthracene. The aliphatic hydrocarbons found in bio-oils have between 13 and 32 atoms of carbon. Examples of these compounds are 1-octadecene, nonadecane, eicosane, heneicosane, and 1-docosene. Fatty acids and their methyl esters generally have between 14 and 27 atoms of carbon. Examples of compounds forming this group are tetradecanoic acid, hexadecanoic acid, hexadecanoic and acid methyl ester. Sterols are also obtained from biomass extractives; some examples are β-sitosterol, 3-ol, and stigmata-4-en-3-one. Sugars are another important group in family D and usually account for around 10 mass% of the whole oil. The main sugar found in bio-oils is 1,6-anhydro-β-D-glucopyranose (levoglucosan) accounting generally for around 2 mass% of bio-oils. Other sugars found are cellobiosan, 1,6-anhydroglucofuranosem, and fructose. Anhydrosugars with molecular weights of around 500 g/mol account for up to 10 mass% of the whole oil (Garcia-Pérez et al. 2007a, 2008).

Families E and F account for up to 35 mass% of crude bio-oils and mainly consist of oligomers with molecular masses between 600 and 10,000 g/mol. These families are the result of pyrolysis reactions leading to the formation of dimers and trimers and of aging reactions (Garcia-Pérez et al. 2007, 2008). Family E is formed by lignin-derived oligomers commonly found in water-insoluble CH_2Cl_2 soluble fractions. Although the exact composition of the compounds forming *peak F* is not well known, some authors have attributed this peak to cellulose- and hemicellulose-derived oligomers (Garcia-Pérez et al. 2007, 2008).

Table 7.3 compares the chemical composition of crude bio-oils and crude petroleum oils in terms of the percentage of compounds with equivalent boiling points. Bio-oil composition in terms of volatility compares very well with the composition of crude petroleum oils.

Although crude bio-oils contain oxygenated hydrocarbons from the thermal degradation of cellulose, hemicellulose, and lignin abundant in earthy biomasses, petroleum is composed of hydrocarbons with a very low-oxygen content. These hydrocarbons are generated from the preserved remains of prehistoric zooplankton and algae that have been settled to the sea bottom in large quantities under anoxic conditions. Zooplankton and algae are very rich in triglycerides and proteins.

Physicochemical Properties of Bio-Oil

Some of the most common physicochemical properties of bio-oils are presented in Table 7.4. The water content of bio-oils varies between 5 and 30 mass%. This water results from pyrolysis reactions (pyrolytic water) and from biomass moisture. Water contents over 30 percent results in the formation of two separate phases (an aqueous phase and an oily phase). Bio-oil density is approximately 1.2 times higher than that of distilled water. It decreases almost linearly with the tem-

Boiling Point (°C)	Petroleum Fraction	Mass%	Crude Bio-Oil Fraction	Mass%
30–250	Naphtha	15–30	Family A, B, C	20–65
200–300	Kerosene	4–12	Family D	5–35
300+	Gas oil and residual oil	35–65	Family E and F	10–35

TABLE 7.3 Comparison between the Composition of Crude Bio-Oils and Petroleum

Property	Values	Standard Method
Moisture content (mass%)	15–30	ASTM D1744, ASTM E203, ASTM D95
pH	2.0–3.8	No standard
Density (kg/m^3)	1208–1238 (20°C) 1199–1229 (30°C) 1190–1211 (40°C) 1184–1207 (50°C)	ASTM D4052
Kinematic viscosity (cSt)	50–672 (20°C) 35–300 (40°C) 5–200 (50°C)	ASTM D445-88
Elemental analysis (mass%, anhydrous basis) C H N S Ash O (by difference)	48.0–63.5 5.2–7.2 0.07–0.39 0.00–0.05 0.03–0.3 32–46	ASTM D5291-92
High heating value (HHV) (anhydrous basis) (MJ/kg)	20–24.3	ASTM D 2382/DIN 51900
High heating value (HHV) (as produced) (MJ/kg)	15–18	
Flash point (°C)	64–120	ASTM D93-90
Solid content (as methanol insoluble)	0.17–1.14	No standard
Conradson carbon residue (mass%)	18–23	ASTM D189

TABLE 7.4 Typical Bio-Oil Properties and Standard Methods

perature (thermal expansion coefficient of around $0.8.10^{-3}$ g/mL °C). The low pH of bio-oils is mainly due to the presence of low-molecular-weight carboxylic acids. The calorific value of bio-oils is about 40 percent that of petroleum, so approximately 1.8 kg or 1.5 L of bio-oils are needed to provide the same energy as 1 kg of petroleum fuel.

Bio-Oil Multiphase Structure

The multiphase structure of crude bio-oils can be attributed to the presence of (1) char particles, (2) waxy materials, (3) aqueous droplets, (4) droplets of extractives derived compounds, (5) nanoparticles of oligomers in a matrix of holocellose-derived compounds, and water

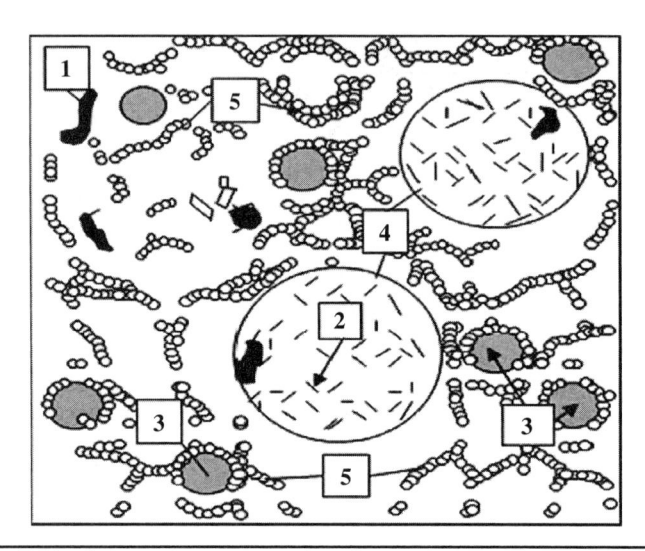

FIGURE 7.10 Schematic representation of a physical model illustrating the multiphase structure bio-oils.

(see Fig. 7.10) (Garcia-Pérez et al. 2006a, 2006b). The existence of nanoparticles formed mainly by the association of units of oligomers has been recently probed by small-angle neutron scattering (SANS) (Fratini et al. 2006) and by dynamic rheological studies (Garcia-Pérez et al. 2006a, 2006b).

The complex multiphase properties of bio-oils (Oasmaa and Czernik 1999; Oasmaa et al. 2003a, 2003b; Oasmaa and Kuoppala 2004; Ba et al. 2004a, 2004b) could create serious problems during storage and handling (Boucher et al. 2000a, 2000b). The layering or separation of bio-oil phases could be a problem in installations that do not have facilities to homogenize these liquids (Oasmaa and Czernik 1999). The presence of an aqueous upper layer phase in bio-oils is attributed to the inability of water to dissolve in the oily matrix (Oasmaa and Czernik 1999; Boucher et al. 2000a, 2000b; Garcia-Pérez et al. 2006a, 2006b). A separate phase also can be formed due to the presence of wood extractives (Boucher et al. 2000a, 2000b; Oasmaa et al. 2003a, 2003b; Oasmaa and Kuoppala 2004; Garcia-Pérez et al. 2004). Specifications must ensure that bio-oils are supplied as homogeneous liquids.

The main problem encountered in storing bio-oils is the gradual increase in viscosity and molecular weight with the formation of separate phases in relatively short periods of time (Diebold and Czernik 1997; Diebold et al. 1999; Ba et al. 2004a, 1994b; Garcia-Pérez et al. 2006a, 2006b). Aging also happens in petroleum but at a much lower rate. For example, although the shelf life of petroleum exceeds 6 months, some bio-oils can only be stored for a few months or for a few weeks in hot environments.

According to Diebold (1999) the most important reactions happening during bio-oil aging are (1) organic acids with alcohols to form esters and water, (2) organic acids with olefins to form esters, (3) aldehydes and water to form hydrates, (4) aldehydes and alcohols to form hemiacetals or acetals and water, (5) aldehydes to form oligomers and resins, (6) aldehydes and phenolics to form resins and water, (7) aldehydes and proteins to form oligomers, (8) organic sulphur to form oligomersunsaturated compounds to form polyolefins, and (9) air oxidation to form more acids and reactive peroxides that catalyze the polymerization of unsaturated compounds. Reactions (1) through (5) are reversible processes, which mean that the change in temperature of relative amounts of water and other reactive compounds will upset the equilibrium and initiate compositional changes (Diebold et al. 1999). Methanol and ethanol are added to stabilize the oils because these alcohols interrupt molecular–building reactions responsible for bio-oil aging (Diebold and Czernik 1997, Diebold 1999). From a practical point of view, a minimum of 14 days of storage should be possible; preferably the bio-oil must have at least 6 months of storage life (at normal pressure and temperature) (Diebold et al. 1999; van de Kamp 2000).

7.6.2 Fuel Applications of Crude Bio-Oils

Contrary to the practices of the petroleum industry, where most of the combustion tests are carried out using fractions, all the tests reported so far (Banks et al. 1992; Lee et al. 1993; Barbueci et al. 1995; van de Kamp 2000, Baglioni et al. 2001; Oasmaa et al. 2001; Venderbosch et al. 2001, Chiaramonti et al. 2003; Stamatov et al. 2006) for bio-oil, have been performed using crude bio-oils or with added solvents (mainly alcohols). The lack of bio-oil refining technologies is the main reason for this situation.

Fuel Properties of Crude Bio-Oils

The characteristics of crude bio-oil that make this fuel difficult to use are low heating value, high water content, immiscibility with petroleum-derived fuels, high ignition temperatures, phase separation, high content of solids, relatively high viscosity, low volatility, high corrosiveness, high Conradson carbon numbers, chemical instability, high alkaline content (high content of K and Na) (Oasmaa and Czernik 1999; Czernik and Bridgwater 2004). As Table 7.4 shows the viscosity of crude bio-oil varies between 5 and 200 cSt at 50°C (Mohan et al. 2006). These values of viscosity are higher than those found for typical fuel No. 2 (between 2 and 5 cSt at 50°C) but lower than that found for fuel No. 6 (up to 500 cSt at 50°C). The high viscosity of crude bio-oil limits its uses in applications designed for fuel No. 2 (Diebold et al. 1999).

Many of the undesirable fuel properties in crude bio-oils are generally associated with low-molecular-weight organic molecules (acids, aldehydes) and water (families A and B) and oligomers (families E and F). Furans and phenols, sugars, diphenols, and some extractive-derived compounds (families C and D) contribute more positively to the fuel properties of bio-oils (Garcia-Pérez et al. 2007a, 2007b, 2008).

A handicap facing both the crude bio-oil producers and manufacturers of boilers, gas turbines, and diesel engines is the lack of classification and specification for crude bio-oils. To label an organic liquid as a fuel, it must fulfill a set of requirements in relation to combustion, storage, handling, and safety (to the engine and its environment) (Moses 1994; Diebold et al. 1999; Oasmaa and Czernik 1999). The IEA Pyrolysis Activity Group proposed a bio-oil classification strategy similar to the one used for petroleum fuels (Oasmaa and Czernik 1999). Consequently, bio-oils were classified as light bio-oils (viscosity similar to ASTM No. 2), light medium bio-oils (viscosity similar to ASTM No. 4), medium bio-oils (viscosity similar to PORL 100), and heavy bio-oil (viscosity similar to fuel No. 6). This classification has been poorly applied but remains as a reference of how the bio-oils could be classified.

Combustion of Crude Bio-Oils

Bio-oil combustion tests has been ongoing ever since the development of biomass pyrolysis technologies. Many combustion tests at atmospheric pressure in flame tunnels and boilers have been performed in several laboratories: Massachusetts Institute of Technology (Shihadeh et al. 1994), Canada Centre for Mineral and Energy Technology (Banks et al. 1992; Lee 1993), Ente Nazionale per l'Energia Elettrica (Rossi et al. 1993; Barbueci et al. 1995), Coordination Geoinformation Services (Salvi and Salvi Jr. 1991), Red Arrow Products Co. (Freel et al. 1990; Freel and Huffman 1994), Neste Oil (Gust 1994, 1997), VTT Energy (Oasmaa et al. 2001), Monash University (Stamatov et al. 2006), and International Flame Research Foundation (IFRF) (van de Kamp and Smart 1991, 1993).

An advantage of converting biomass into liquid fuels is the possibility of using these fuels in highly efficient engines to produce electricity. Gas turbines and diesel engines offer higher thermal efficiencies than the Rankine cycles (efficiency of around 26 percent) (Roy and Morin 1998). These higher efficiencies have been the main driving force for testing crude bio-oil in engines since the 1990s (Kasper et al. 1983; Solantausta et al. 1993, 1994; Shihadeh et al. 1994, Gros 1995; Jay et al. 1995; Andrews et al. 1996, 1997; Suppes et al. 1996; Leech 1997; Frigo et al. 1998; Bertoli et al. 2000; Lopez-Juste and Salvá Morfot 2000; Ormrod and Webster 2000; Shihadeh and Hochgreb 2000; Strenziok et al. 2001; Chiararamonti et al. 2003). An excellent review on bio-oil combustion tests in gas turbines and diesel engines can be found elsewhere (Czernik and Bridgwater 2004; Chiaramonti et al. 2007).

Nevertheless, some problems in the use of crude bio-oil in diesel engines and gas turbines still remain unsolved. Although boilers can operate with a great variety of fuels (Moses 1994), little variation is allowed in the properties of fuels for gas turbines and diesel engines. The residence times in the combustion chambers of these engines are much smaller than for boilers and the ash content, and alkalinity must be strictly controlled. Formation of carbon residues in turbine blades and in diesel engine exit valves is an important problem that needs to be addressed (Shihadeh and Hochgreb 2000). Most of these carbonaceous materials are formed from the oligomeric fraction of bio-oils (families E and F).

The combustion in a diesel engine is caused by self-ignition of the fuel injected at a high pressure into a chamber filled with a constant mass of compressed and preheated air. The cetane number is a measure of the ease of self-ignition in a diesel engine. Low cetane numbers between 0 and 27 have been reported for crude bio-oils (Suppes et al. 1996; Chiaramonti et al. 2007). Pilot-ignited medium- or low-speed diesel engines can be fueled with a low-cetane fuel like bio-oils (Gros 1995; Diebold et al. 1999; Ormrod and Webster 2000; Chiaramonti et al. 2003; Czernik and Bridgwater 2004). It has also been reported that crude bio-oils can autoignite at cylinder charge temperatures above 600°C (Leech 1997; Shihadeh and Hochgreb 2000). Increasing the compression ratio up to 22:1 has been suggested as a way to deal with the poor self-ignitability of crude bio-oils (Bertoli et al. 2000). Despite a longer ignition delay, diesel engines can run smoothly using bio-oils. Thermal efficiencies seem to be more or less similar to those obtained using diesel fuels (Gros 1995; Jay et al. 1995; Shihadeh and Hochgreb 2000).

It is important to point out that no one has ever built and commercialized a gas turbine or diesel engine designed to handle bio-oils. The only place where bio-oils are used regularly as fuel is the Red Arrow Products Co. pyrolysis plant in Wisconsin, but, in this case, bio-oils are burned in a conventional boiler to generate steam (Freel et al. 1996; Czernik and Bridgwater 2004). The commercial use of bio-oils for power generation in relatively large boilers, gas turbines, and diesel engines seems to be an achievable goal in the short term.

Upgrading Crude Bio-Oils

The term *upgrading* is commonly used in describing chemical and physical methods used to improve the properties of crude bio-oils (Bridgwater et al. 2001). Some of the technologies being studied are hot gas filtration (Solantausta et al. 2000), formation of diesel/bio-oil microemulsions (Ikura et al. 1998; Baglioni et al. 2001; Chiaramonti et al. 2003), blending with polar solvents (Diebold and Czernik 1997; Boucher et al. 2000a, 2000b), and hydrogenation or catalytic vapor cracking (Elliott and Baker 1987). Most of these upgrading methods are still at the laboratory stage (Maggi and Elliott 1997). For most operating pyrolysis units, crude bio-oil upgrading is limited to the removal of char and, in some cases, to the addition of polar solvents (e.g., methanol and ethanol). Solvent addition can impact bio-oil properties by two mechanisms: (1) physical dilution and (2) esterification and acetilazation, preventing chain growth responsible for the increase in molecular weight and viscosity.

Blending pure pyrolysis oil with high cetane–oxygenated compounds is another effective method for upgrading bio-oils for diesel engine applications. Tests on blends of pyrolysis oils with up to 56.8 mass% of dyglyme (diethylene glycol dimethyl ether), an oxygenated compound with a very high cetane number (112 to 130), have been reported (Chiaramonti et al. 2007). Minor differences were found in the overall combustion process with diesel fuels. It has been reported that mixtures of 72 percent of pyrolysis oils, 24 percent methanol, and 4 percent cetane enhancer (tetraethyl glycol dinitrate) also show similar performances to that of diesel oil (Suppes et al. 1996). Mixtures with ethanol and nitrate alcohols have also been studied (Czernik and Bridgwater 2004, Solantausta et al. 1993). Ethers have been identified as good fuels for diesel engines. Diglyme (diethylene glycol dimethyl ether) is suitable for blending with bio-oils (Bertoli et al. 2000). The maximum amount of bio-oil allowed for stable operation was 44.1 percent (by weight) in these studies. The economic feasibility of these approaches is limited by the high cost of cetane improvers, solvents, and emulsifiers.

Biodiesel is another renewable fuel derived from triglycerides that can be used as cetane improvers for bio-oils. Heating value, density, and viscosity of biodiesels are comparable to those of No. 2 diesel from petroleum. Biodiesel is also a very good solvent. Producing bio-oil/biodiesel blends can be a simple system for using crude bio-oils as diesel fuel extenders (Garcia-Pérez et al. 2007b).

7.6.3 Producing Chemicals from Bio-Oils

Crude bio-oils are a potential source of chemicals. Although most of the crude bio-oil are expected to be processed as transportation fuels or as a source of energy, chemicals could also make an important contribution to the profitability of this technology, mainly because chemicals are much more expensive than fuels. In the case of crude petroleum oils, about 88 percent is processed as fuel; the other 12 percent is converted into other materials such as plastic. Detailed reviews on existing and potential short-term applications of crude bio-oils as a source for higher-value products can be found elsewhere (Radlein 1999; Czernik and Bridgwater 2004).

Three approaches have been studied to obtain chemicals from crude bio-oils: using crude bio-oils, using fractions, and isolating individual compounds (see Fig. 7.11).

Products from Whole Bio-Oils
Useful products can be obtained from crude bio-oil by taking advantage of its most abundant functional groups: carbonyl, carboxyl, and

Biomass Pyrolysis and Bio-Oil Refineries

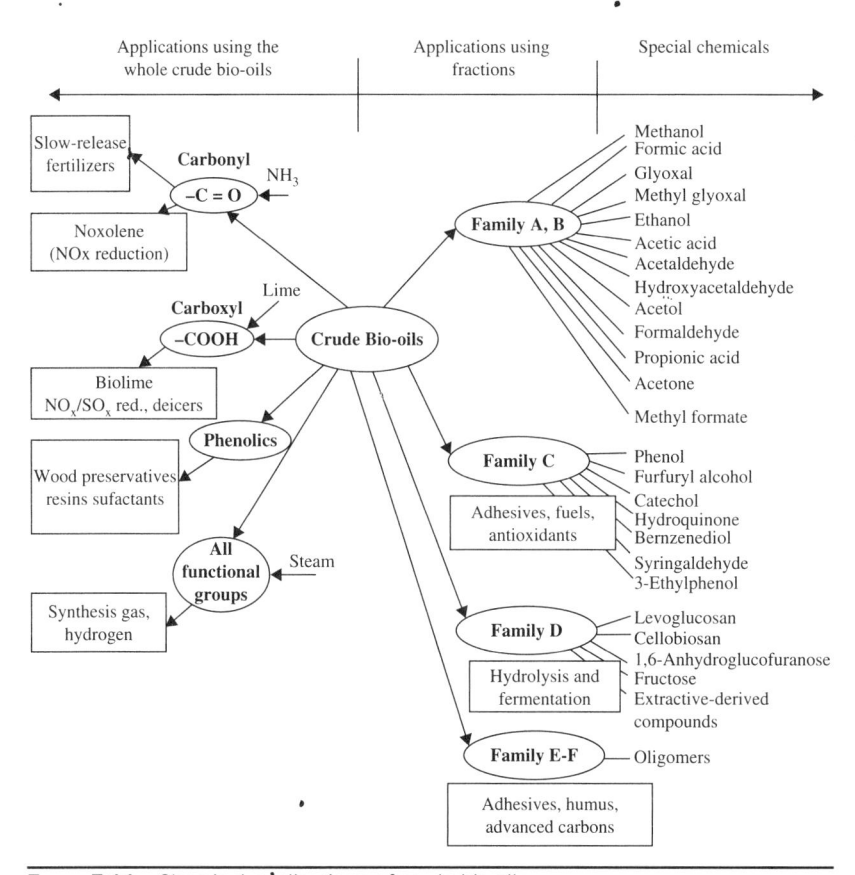

FIGURE 7.11 Chemical applications of crude bio-oils.

phenolic. These functional groups react in such a way that the nonreacting part of the bio-oil does not have to be separated from the final product (Radlein 1999; Czernik and Bridgwater 2004). The characterization approach proposed by Radlein (1999) and Nicolaides (1984) is very useful in estimating the content of these functional groups in bio-oils. They proposed to consider bio-oil from the viewpoint of the distribution of its principal functional groups. The *carboxyl* (1.2 to 2.1 mol/kg organics) (–COOH), *carbonyl* (1.8 to 6.2 mol/kg organics) (–C=O), *hydroxyl* (0.7 to 1.4 mol/kg organics) (–OH), *phenolic* (1.8 to 3.0 mol/kg organics), and *methoxyl* (0.7 to 2.1 mol/kg organics) (–OCH$_3$) were chosen due to their strong impact on bio-oil chemistry. These results suggest that the carbonyl groups are the most abundant groups in bio-oils.

Carboxylic acids and *phenols* react with lime to form calcium salts and phenolates (Oehr 1995; Czernik and Bridgwater 2004). A product called BioLime was developed by the DynaMotive Energy Systems Corp., Vancouver, BC, Canada, and has been successfully tested in capturing SO_x in coal combustors. The lack of availability of low-cost lime is the limiting step for the commercialization of this technology.

The *carbonyl groups* present in bio-oils can react with ammonia, urea, or other $-NH_2$-containing materials to form imide and amide bonds. Slow-release fertilizers containing up to 10 mass% of nitrogen can be obtained. These fertilizers have lower leachability and contribute to good soil conditioning due to the formation of humic type materials (Robson 2000; Czernik and Bridgwater 2004).

The presence of *phenolic groups* and *terpenoids* makes crude bio-oils very attractive in the production of wood preservatives replacing creosote (Freel and Graham 2002). Some terpenoid and phenols are known to act as insecticides and fungicides. Biocarbo (http://www.biocarbo.com), in association with Vallourec & Mannesmann Tubes (V&M) in Brazil, produces glues and a sealing material from decanted tars.

Chemicals Obtained from Crude Bio-Oil Fractions

Products of better quality, with better performance could be obtained if the bio-oils were refined and separated into fractions in which the functional groups of interest are concentrated. Adding water to bio-oils results in the formation of a viscous oily phase rich in lignin-derived compounds and an aqueous phase rich in carbohydrate-derived compounds. The aqueous phase is used commercially in effective meat browning agents. Low-molecular-weight aldehydes (especially hydroxyacetaldehyde) and phenolic compounds are effective meat-browning agents and responsible for the smoky flavor. Many food-flavoring compositions have been patented and commercialized by the Red Arrow Products Co. (Underwood and Graham 1991; Underwood 1990; Czernik and Bridgwater 2004).

Environmentally friendly road deicers (Oehr et al. 1993; Oehr 1995; Czernik and Bridgwater 2004) can be obtained from the reaction of volatile organic acids (mostly acids such as formic, acetic, and propionic) abundant in water-soluble fractions with calcium salts. The technology used to obtain these products is very similar to the one used by the old wood-distillation industry to remove the acids before distilling the resulting water phase to produce methanol spirits (Klar and Rule 1925; Czernik and Bridgwarter 2004). The resulting liquid residue after the removal of methanol is an oily phase rich in sugars.

A water-insoluble fraction is rich in oligomers sometimes called pyrolytic lignin. Pyrolytic lignin has been used as a phenol replacement for the production of phenol–formaldehydes resins (Himmelblau 1991; Chum and Kreibich 1993; Roy et al. 2000; Giroux et al. 2001). Thirty to 50 mass% of phenol can be replaced by pyrolytic lignin in novalak and resole formulations. Pyrolytic lignin is less reactive than phenols but shows lower toxicity and is cheaper (Czernik and Bridgwater 2004).

The U.S. National Renewable Energy Laboratory (NREL) has patented a separation method to produce resins from the phenols and furans present in bio-oils (family C). The method developed by NREL

(Chum and Kreibich 1993) involves dissolving bio-oils in ethyl acetate followed by vacuum filtration to remove the char. Upon standing, the ethyl acetate/pyrolysis oil separate into two phases: organic-rich, ethyl acetate soluble (top) and ethyl acetate insoluble (bottom). The ethyl acetate soluble fraction was extracted with aqueous $NaHCO_3$. The phenols and neutrals remained (mostly family B) remained in the ethyl acetate solution. The resulting oil after the ethyl acetate is removed is used successfully to replace part of the phenol in the production of phenol-formaldehyde resins (Chum and Kreibich 1993). Excellent reviews of other strategies for bio-oil separation can be found elsewhere (Fagernäs 1995; Oasmaa et al. 1997; Mohan et al. 2006).

Specialty Chemicals

The production of specialty chemicals using pyrolysis seems more attractive if especially selected feedstocks are used; for example, cellulose, hemicellulose, or lignin-rich material, as opposed to wood or waste residues. In that case, the liquid contains a higher concentration of the desired product, and isolation is easier and cheaper (Czernik and Bridgwater 2004).

Bridgwater (2005) presented a list of compounds that are economic to recover and listed the maximum concentration reported in the literature after carefully controlling the biomass composition and the pyrolysis conditions used. The compounds identified were: levoglucosan (30 mass%), hydroxyacateldehyde (15 mass%), acetic acid (10 mass%), formic acid (9 mass%), acetaldehyde (8 mass%), furfuryl alcohol (5 mass%), catechol (5 mass%), methyl glyoxal (4 mass%), ethanol (3 mass%), cellobiosan (3 mass%), 1,6 anhydroglucofuranose (3 mass%), fructuose (2 mass%), glyoxal (3 mass%), formaldehyde (2 mass%), phenol (2 mass%), propionic acid (2 mass%), acetone (2 mass%), methyl-cyclopentene-oil-one (2 mass%), methyl formeate (2 mass%), hydroquinone (2 mass%), acetol (2 mass%), angelica lactone (2 mass%), syringaldehyde (1 mass%), methanol (1 mass%), and 1-hydroxy-2-butanone (1 mass%).

Hydroxyacetaldehyde is the only of these chemicals that is currently produced from bio-oils. It is the most active meat-browning agent in liquid smoke. Red Arrow Products (Stradal and Underwood 1995) and Resource Transforms International (RTI) (Majerski et al. 2001) patented methods for isolating this compound based on crystallization. Levoglucosan and levoglucosenone are not typical components of bio-oil produced for fuel applications but can be generated in high yields by similar pyrolysis processes from demineralized cellulose or biomass (Czernik and Bridgwater 2004). Levoglucosenone can be produced at high yields (24 mass%) using fluidized-bed pyrolysis of phosphoric acid–impregnated cellulose (Radlein 1999; Czernik and Bridgwater 2004).

The recovery of methanol, acetic acid, and acetone from the water-soluble fractions of bio-oils and the production of turpentine and tars from the oily insoluble fraction was an economically viable activity until the 1940s (Klar and Rule 1925). The technologies to recover these compounds are well known but not currently used because most of these compounds can be produced at a lower cost from other feedstocks derived from natural gas, crude oil, or coal.

7.6.4 Production of Transportation Fuels

There are five major concepts under study today to produce transportation fuels from bio-oils: (1) gasification of the whole bio-oils at a centralized unit followed by the synthesis of Fischer–Tropsch fuels (van Rossum et al. 2007), (2) steam reforming of whole bio-oils or the water-soluble fractions to produce hydrogen (Czernik et al. 2002; Takanabe et al. 2006; Basagiannis and Verykios 2007; Bleeker et al. 2007; Davidian et al. 2007; Iojoiu et al. 2007; Ramos et al. 2007), (3) hydrotreatment of whole bio-oils or the water-insoluble fractions to produce green gasoline and green diesel (Baker and Elliott 1988; Yu and Zhang 2003; Zhang et al. 2006; Elliott 2007; Mahfud et al. 2007a, 2007b), (4) hydrolysis, neutralization, detoxification, and fermentation of pyrolytic sugars (Prosen et al. 1993; Yu and Zhang 2003, 2004; Bhatia 2005; Muyafuji et al. 2005; Helle et al. 2007), (5) aqueous phase catalytic processes to produce hydrogen and/or alkanes (Huber and Dumesic 2006).

Bio-Oil Gasification

Synthesis gas can be produced by reacting bio-oil with oxidizing agents (steam, oxygen) at temperatures over 600°C. Several nickel-based catalysts have been tested to enhance the selectivity toward the formation of synthesis gas (van Rossum et al. 2007). During bio-oil gasification, an important fraction of bio-oils is converted to coke. These coke-formation problems can be handled by using fluidized-bed reactors. Van Rossum et al. (2007) proposed the use of a twin-bed setup for the coupling of endothermic reactions (cracking, reforming) and exothermic reactions (coke burning). The proposed system uncouples the atomization/cracking of the oil and the catalytic conditioning of the produced gases (van Rossum et al. 2007). This synthesis gas can be further converted to produce green diesel via Fischer–Tropsch reactions over ion, cobalt, nickel, and ruthenium catalysts.

Steam Reforming

Studies on steam reforming of bio-oil and several of its model compounds to produce hydrogen have shown promising results (Garcia et al. 2000; Takanabe et al. 2006; Basagiannis and Verykios 2007; Czernik et al. 2007; Davidian et al. 2007; Iojoiu et al. 2007). The main problem observed is the rapid formation of carbon deposits, which seriously limits reforming times to less than 4 h, after which regeneration is needed (Davidian et al. 2007). The use of two reactors in parallel operation in reforming-regeneration successive cycles has been proposed. Although fluidized-bed reactors have shown good results reducing coke formation and catalyst deactivation, more attrition-resistant catalysts should be developed (Davidian et al. 2007).

Hydrolysis, Neutralization, Detoxification, and Fermentation of Pyrolytic Sugars

The most abundant anhydrosugars resulting from cellulose depolymerization reactions are levoglucosan (1,6-anhydro-β–D-glucopyranose) and cellobiosan (β-D-glucopyranosyl-(1-4)-1, 6-anhydro-D-glucopyranose). These sugars can easily be separated from most lignin-derived compounds by adding water to the oils. The sugars recovered in the resulting aqueous phase can be hydrolyzed to form glucose at temperatures between 90 and 110°C and concentrations of sulfuric acid around 500 mM. The carboxylic acids and phenols that also solubilize in the aqueous phase are known to be toxic to the yeasts. Muyafuji et al. (2005) proved that increasing the pH to values over 5.5 and removing the phenols over activated carbon can result in a detoxified solution. The sugars can then be readily fermented to ethanol using yeasts (*Saccharomyces cerevisiae*).

Aqueous Phase Catalytic Processes

The University of Wisconsin has patented a technology for the reforming or hydrogenation of aqueous solutions containing biomass-derived oxygenated organic compounds (Cortright and Dumesic 2005). This process could be potentially employed to convert bio-oil water soluble fractions into hydrogen and alkanes ranging from C_1 to C_{15}. Alkanes with molecular weight ranging from C_1 to C_6 can be produced by aqueous phase dehydration/hydrogenation (APD/H). This APD/H process involves a bifunctional pathway in which the sugars are repeatedly dehydrated by a solid acid (SiO_2–Al_2O_3) or a mineral acid (HCl) catalyst and then hydrogenated on a metal catalyst (Pt or Pd) (Huber and Dumesic 2006). Liquid alkanes ranging from C_7 to C_{15} (naphtha) can be produced from carbohydrates by combining the dehydration/hydrogenation process with an upstream aldol condensation step to form C–C bonds (Huber and Dumesic 2006).

Hydroprocessing

Extensive studies for the hydroprocessing of fast pyrolysis oils have been conducted for more than 25 years (Churin et al. 1987; Elliott and Baker 1987; Gagnon and Kaliaguine 1988; Piskorz et al. 1989a, 1989b; Baldauf et al. 1994; Centeno et al. 1995; Elliot 2007).

Piskorz et al. (1989a, 1989b) studied the hydrotreatment of the so-called pyrolytic lignin (decanted oil) with positive results. After hydrotreatment, the decanted oils are converted to light hydrocarbons (60 to 65 mass%), water (20 mass%), and gases (8 to 10 mass%). The heavy residue that formed was only 1 to 2 mass%. The light compounds have an H/C ratio of 1.5 with 0.46 mass% of oxygen. As much as 50 percent of the pyrolytic lignin was converted to compounds in the gasoline range. The results obtained by Piskorz et al. (1989) are very interesting because it proved the feasibility of hydrotreating decanted oils. In fact, the oils obtained from the slow pyrolysis reactors and by high-pressure liquefaction are also decanted oils with relatively low contents of polar compounds and tend to behave satisfactorily during hydrotreatment. New hydrotreatment studies with pyrolytic lignin (decanted oil) were carried out in 2005 by UOP LLC, a Honeywell company, and Pacific Northwest National Laboratory (PNNL) (Marinangeli et al. 2005). Two catalysts were tested. Although PNNL used Pd on a carbon catalyst in a continuous pilot plant, UOP used a Ni–Mo catalyst in a batch autoclave. The results obtained were very similar to those obtained by Piskorz et al. (1989). The yield of upgraded oils varies between 40 and 55 percent with an oxygen removal of 69 to 93 percent. Between 30 and 60 percent of the bio-oil was converted to naphtha.

UOP, PNNL, and NREL have recently shown (Holmgren et al. 2008) that although it is more difficult, the hydrotreatment of whole bio-oils is also viable. There was 21 mass% of the oil converted to distillable products with boiling points similar to naphtha and another 21 percent was converted to compounds with boiling points similar to diesel. The rest of the bio-oil was converted to CO_2, water, and light hydrocarbons. Thus, around 28 mass% of the biomass can be converted into fungible transportation fuel. These values are very high if we take into account that current efforts at enzymatic hydrolysis only aim at producing 63.2 gallons of ethanol per ton of biomass processed. The fast pyrolysis/hydrotreatment pathway could yield 148 gallons of ethanol equivalent per ton of biomass. Recent economic analyses performed by UOP (Marinangeli et al. 2005) suggest that the hydrotreatment of a bio-oil water-insoluble fraction is economically viable at petroleum prices over $50 per barrel. Although hydrotreatment of fast pyrolysis oils still requires more development to enable large commercial operations, it is certainly one of the most promising alternatives to convert biomass into transportation fuels.

7.7 Bio-Oil Refineries

The use of concepts like *bio-based economy and biorefineries* is not new, but their importance has only been recognized recently. It is now generally accepted that biomass-derived products will have better opportunities to compete with our dominant petroleum economy.

The equilibrium between the cost of transportation and savings associated with the economies of scale will determine the feasibility of building centralized refineries near consumer centers, or distributed rural refineries closer to the biomass resources. The main goal of a biorefinery is to produce high-value low-volume (HVLV) and low-value high-volume (LVHV) marketable products at competitive cost using a series of unit operations (Fernando et al. 2006). Bridgwater (2005) cited the separation of phenolics for the production of resins with the use of the remaining aqueous phase to produce hydrogen at NREL, and the production of liquid smoke and other specialty chemicals with the burning of the remaining organics in boilers—only two examples of the bio-oil–based refineries implemented. However, many other concepts could be developed following some of the separation schemes described by Fagernas (1995), Oasmaa et al. (1997), and Mohan (2006). The development of several new bio-oil–based refinery concepts is possible today thanks to the important progress made in the last 20 years in the production, separation, and uses of bio-oils (Czernik and Bridgwater 2004; Bridgwater 2005).

The potential exists to integrate several of the products described in the previous section into a new bio-oil biorefinery concept allowing the full utilization of crude bio-oils. For example, the organic compounds present in family A could be separated in a scheme similar to the one used in the old wood-distillation industry (Klar and Rule 1925) to produce salts from the acids (Oehr et al. 1993) and distillated spirits rich in methanol. These compounds could be also reformed to produce part of the hydrogen needed to hydrotreat bio-oils. The hydroxyacetaldehyde could be recovered using a concept very similar to the one presently used by Red Arrow Products (Himmelblau et al. 1991). The water-soluble fraction rich in sugars, furans, and polar oligomers could be used as a feedstock for the production of more hydrogen either using vapor or liquid reforming (Czernik et al. 2002; Rioche et al. 2005; Huber and Dumesic 2006) or as a feedstock for the production of hydrocarbons through aqueous phase catalytic processing or via bio-oil hydrotreatment (Huber and Dumesic 2006). Phenol formaldehyde resins could be produced using the pyrolytic lignin precipitated from the bio-diesel insoluble fraction (Himmelblau et al. 1991; Chum and Kreibich 1993; Roy et al. 2000; Giroux et al. 2001).

New separation strategies will have to be developed to cope with the unique properties of bio-oils. For example, the difficulties of distilling a reactive material with a tendency to polymerize, as well as the lack of viable products from biopitches, has been commonly cited as hurdles against distillation at the core of bio-oil–based biorefineries. Several new products, such as polyurethane elastomers and biocarbon electrodes, have been recently developed from biopitches (Carvalho et al. 1999; Coutinho et al. 2000; Araujo and Pasa 2002, 2003, 2004; Rocha et al. 2002), making distillation-based refineries more attractive. The use of advanced distillation techniques such as molecular distillation remains practically unexplored.

Although there is considerable room for innovation and creativity to develop new concepts for bio-oil refineries, using the infrastructure created by the petroleum industry seems to be the fastest way to deploy this technology. In this regard, bio-oil hydrotreatment to produce fungible fuels in centralized or rural refineries is likely to receive more attention in the next few years.

7.8 Summary and Conclusions

The scientific and technical progresses made in the last 30 years led to the birth of companies commercializing pyrolysis units. It is still premature to predict how many more years it will take to develop commercial upgrading technologies, international fuel standards, advanced combustion systems, bio-oil refineries, and more products, allowing the full commercialization of crude bio-oils. The amount of scientific and technological progress that needs to be done is still substantial, but all the results so far obtained clearly indicate that the thermochemical pathway (fast pyrolysis-bio-oil refineries) is very promising to convert biomass into fuel and chemicals. The potential use of crude bio-oils as source of fuels and chemicals depends on the cooperation between research centers and universities, bio-oil producers, gas turbine/diesel engine manufacturers, the energy and chemical sector representatives, and investors. Liquid bio-oil has the considerable advantage of being a storable and transportable material as well as a potential source of a number of valuable chemicals that offer the attraction of much higher added value than fuels. These characteristics offer clear advantages to bio-oils as potential feedstock for new biorefineries. The development of new concepts for bio-oil–based refineries is an important step in the implementation of a global biomass economy.

References

Agblevor, F. A., Blesler, S., and Evans, R. J. 1994. Inorganic compounds in biomass feedstock: Their role in char formation and effect on the quality of fast pyrolysis oils. *Proceedings of the Biomass Pyrolysis Oil, Properties and Combustion Meeting*. September 26–28, Estes Park, CO, 77–89.

Andrews, R. G., Fuleki, D., Zukowski, S., and Patnaik, P. C. 1996. Results of industrial gas turbine tests using a biomass derived fuel. *Developments in Thermochemical Biomass Conversion Proceedings*. 1: 495–506.

Andrews, R. G., Zukowski, S., and Patnaik, P. C. 1997. Feasibility of firing an industrial gas turbine using a bio-mass derived fuel. In *Developments in Thermochemical Biomass Conversion*, eds. A. V. Bridgwater and D. G. B. Boocock. London, UK: Blackie Academic & Professional. 495–506.

Antal, J. M. and Gronli, M. 2003. The art, science and technology of charcoal production. *Industrial and Engineering Chemical Research* 42:1619–1640.

Araujo, R. C. S. and Pasa, V. M. D. 2002. Thermal study of polyurethane elastomers based on biopitch–PEG–MDI system. *Journal of Thermal Analysis and Calorimetry* 67(2):313–319.

Araujo, R. C. S. and Pasa, V. M. D. 2003. Mechanical and thermal properties of polyurethane elastomers based on hydroxyl-terminated polybutadienes and biopitch. *Journal of Applied Polymer Science* 88(3):759–766.

Araujo, R. C. S. and Pasa, V. M. D. 2004. New Eucalyptus tar-derived polyurethane coatings. *Progress in Organic Coatings* 51(1):6–14.

Arisz, P. W., Lomax, J. A., and Boon, J. J. 1990. High–performance liquid chromatography/chemical ionization mass spectrometric analysis of pyrolysates of amylose and cellulose. *Analytical Chemistry* 62:1519.

Ba, T., Chaala, A., Garcìa–Pérez, M., Rodrigue, D., and Roy, C. 2004a. Colloidal properties of bio-oils obtained by vacuum pyrolysis of softwood bark. Characterization of water soluble and water insoluble fractions. *Energy & Fuels* 18:704–712.

Ba, T., Chaala A., Garcia–Pérez, M., and Roy, C. 2004b. Colloidal properties of bio-oils obtained by vacuum pyrolysis of softwood bark. Storage stability. *Energy & Fuels*, 18:188–201.

Baker, E. G. and Elliott, D. C. 1988. Catalytic hydrotreating of biomass-derived oils. *ACS Symposium Series* 376:228–240.

Baglioni, P., Chiaramonti, D., Bonini, M., Soldani, I., and Tondi, G. 2001. Bio-crude-oil/diesel oil emulsification: Main achievements of the emulsification process and preliminary tests on diesel engine. In *Progresses in Thermochemical Biomass Conversion*, ed. A. V. Bridgwater. Oxford, UK: Blackwell Science. 1525–1539.

Baldauf, W., Balfanz, U., and Rupp, M. 1994. Upgrading of flash pyrolysis oil and utilization in refineries. *Biomass & Bioenergy* 7(1–6):237–244.

Banks, G. N., Wong, J. K. L., and Whaley, H. 1992. Combustion evaluation and heat transfer characterization of fast pyrolysis product, Division Report ERL 92-35 (CF) CANMET, Energy Mines and Resources Canada, Ottawa, Canada.

Barbueci, P., Costanzi, F., Lagasacchi., S., Mosti, A., and Rossi, C. 1995. Bio-fuel oil combustion in a 0.5 MW furnace. *Proceedings of the Second Biomass Conference of the Americas* NREL/CP 200-8098, Golden, CO. 1110–1120.

Basagiannis, A. C. and Verykios, X. E. 2007. Steam reforming of the aqueous fraction of *bio-oil* over structured Ru/MgO/Al$_2$O$_3$ catalyst. *Catalysis Today* 127(1–4): 256–264.

Basu, P. 2006. *Combustion and Gasification in Fluidized Beds*. New York, NY: Taylor & Francis. 35.

Bertoli, C., D'Alessio, J., Del Giacomo, N., Lazzaro, M., Massoli, P, and Moccia, V. 2000. Running light-duty DI diesel engines with wood pyrolysis oil. SAE Technical Paper Series, International Fall Fuels and Lubricants Meeting and Exposition, Baltimore, MD, October 16–19. Document number: 2000-01-2975.

Bhatia, K. K. 2005. Recovery and purification of anhydro sugar alcohols from a vapour stream. U.S. Patent 6,867,296 B2 (WIPO 2003/089420).

Biocarbo Industria e Comercio Ltd., Itabirito, Brazil. Available at http://www.biocarbo.com, last accessed June, 2009.

Bleeker, M. F., Kersten, S. R. A., and Veringa, H. J. 2007. Pure hydrogen from pyrolysis oils using the steam–iron process. *Catalysis Today* 127:278–290.

Boroson, M. L., Howard, J. B., Longwell, J. P., and Peters, W. A. 1989a. Product yield and kinetics from the vapour phase cracking of wood pyrolysis tars. *AICHE Journal* 35(1):120–128.

Boroson, M. L., Howard, J. B., Longwell, J. P., and Peters, W. A. 1989b. Product yield and kinetics from the vapour phase cracking of wood pyrolysis: Heterogeneous cracking of wood pyrolysis tars over fresh wood char surfaces. *Energy and Fuels* 3:735–740.

Boucher, M. E., Chaala, A., Pakdel, H., and Roy, C. 2000a. Bio-oils obtained by vacuum pyrolysis of softwood bark as a liquid for gas turbines. Part I. Properties of bio-oils and its blends with methanol and pyrolytic aqueous phase. *Biomass & Bioenergy* 19(5):337–350.

Boucher, M. E., Chaala, A., Pakdel, H., and Roy, C. 2000b. Bio-oils obtained by vacuum pyrolysis of softwood bark as a liquid for gas turbines. Part II. Stability and aging and its blends with methanol and pyrolytic aqueous phase. *Biomass & Bioenergy* 19(5):351–361.

Bradbury, A. G. W., Sakai, Y., and Shafizadeh, F. 1979. A kinetic model for pyrolysis of cellulose. *Journal of Applied Polymer Science* 23:3271–3280.

Bridgwater, A. V. 2005. Fast pyrolysis based biorefineries. Presentation made to the American Chemistry Society, Washington, DC, 31 August.

Bridgwater, A. V., Czernik S., and Piskorz, J. 2001. An overview of fast pyrolysis. In *Progress in Thermochemical Biomass Conversion*, ed. A. V. Bridgwater. West Sussex, UK: Oxford, UK: Blackwell Science. 977–997.

Bridgwater, A. V. and Peacocke, G. V. C. 1994. Engineering developments in fast pyrolysis for bio-oils. *Proceedings of Biomass Pyrolysis Oil Properties and Combustion Meeting* Sep. 26–28 Estes Park, CO, 110–127.
Bridgwater, A. V. and Peacocke, G. V. C. 2000. Fast pyrolysis processes for biomass. *Renewable and Sustainable Energy Reviews* 4(1):1–73.
Bridgwater, A. V., Meier, D., and Radlein, D. 1999. An overview of fast pyrolysis of biomass. *Organic Geochemistry* 30:1479–1493.
Broido, A., Javier–Son, A. C., Ouano, A. C., and Barrall, E. M. 1973. Molecular weight decrease in the early pyrolysis of crystalline and amorphous cellulose. *Journal of Applied Polymer Science* 17:3627.
Broido, A. and Kilzer, F. J. 1963. A critique of the present state of knowledge of the mechanisms of cellulose pyrolysis. *Fire Research Abstract Reviews* 5:157.
Broido, A. and Nelson, M. A. 1975. Char yield on pyrolysis of cellulose. *Combustion and Flame* 24:263–268.
Caballero, J. A., Conesa, J. A., Font. R., and Marcilla, A. 1997. Pyrolysis kinetics of almond shells and olive stones considering their organic fractions. *Journal of Analytical and Applied Pyrolysis* 42(2):159–175.
Campbell I., *Biomass, Catalysts and Liquid Fuels*, Lancaster, UK: Technomic Publishing Co. Inc., 1983, 38.
Carvalho, R. M. de, Kubota, L. T., and Rohwedder, J. J. 1999. Carbon fibers: Electroanalytical applications as electrodic material. *Química Nova* 22(4):591–599.
Centeno, A., Laurent, E., and Delmon, B. 1995. Influence of the support of CoMo sulfide catalysts and of the addition of potassium and platinum on the catalytic performances for the hydrodeoxygenation of carbonyl, carboxyl, and guaiacol-type molecules. *Journal of Catalysis* 154(2):288.
Chiaramonti, D., Bonini, M., Fratini, E., Tondi, G., Garter, K., Bridgwater, A.V., Grimm, H .P., Soldaini, I., Webster, A., and Baglioni, P. 2003. Developments of emulsions from biomass pyrolysis liquid and diesel and their use in engines. Part 1. Emulsion production. *Biomass & Bioenergy* 25:85–99; Part 2: Test in diesel engines. *Biomass & Bioenergy* 25:101–111.
Chiaramonti, D., Oasmaa, A., and Solantausta, Y. 2007. Power generation using fast pyrolysis liquids from biomass. *Renewable Sustainable Energy Reviews* 11(6):1056–1086.
Chum, H. L. and Kreibich, R. E. 1993. Process for preparing phenol formaldehyde resin products derived from fractionated fast-pyrolysis oils. U.S. Patent 5, 091, 499, 1993.
Churin, E., Grange, P., and Delmon, B. 1987. Upgrading of bio-oils by hydrotreatments. *Preprints Division of Petroleum Chemistry, American Chemical Society* 32(1–2):328.
Cortright, R. D. and Dumesic, J. A. 2005. Low temperature hydrocarbon production from oxygenated hydrocarbons. U.S. Patent 6,953,873, filed May 9, 2003, and issued Oct. 11, 2005.
Coutinho, A. R., Rocha, J. D., and Luengo, C. A. 2000. Preparing and characterizing biocarbon electrodes. *Fuel Processing Technology* 67(2):93–102.
Czernik, S. and Bridgwater, A. V. 2004. Overview of Applications of biomass fast pyrolysis oil. *Energy & Fuels* 18(2):590–598.
Czernik, S., French, R., Feik, C., and Chornet, E. 2002. Hydrogen by catalytic steam reforming of liquid by products from biomass thermochemical processes. *Industrial and Engineering Chemical Research* 41:4209–4215.
Czernik, S., Evans, R., and French, R. 2007. Hydrogen from biomass-production by steam reforming of biomass pyrolysis oil. *Catalysis Today: Recent Advances in Catalytic Production of Hydrogen from Renewable Sources* 129(3–4):265–268.
Davidian, T., Gulhaume, N., Iojoiu, E., Provendier, H., and Mirodatos, C. 2007. Hydrogen production from crude bio-oil by a sequential catalytic process. *Applied Catalysis B, Environmental* 73:116–127.
Di Blasi, C. 2008. Modeling chemical and physical processes of wood and biomass pyrolysis. *Progress in Energy and Combustion Science* 34:47–90.
Di Blasi, C. 1998. Comparison of semi-global mechanisms for primary pyrolysis of lignocellulosic fuels. *Journal of Analytical and Applied Pyrolysis* 47(1):43–64.

Diebold, J. 1999. A review of the chemical and physical mechanisms of the storage stability of fast pyrolysis bio-oils. Available at http://www.p2pays.org/ref/19/18946.pdf. Accessed May 5, 2009.
Diebold, J. P. and Bridgwater, A. V. 1999. Overview of fast pyrolysis of biomass for the production of liquid fuels. In *Fast Pyrolysis of Biomass: A Handbook*. vol. 2, ed. A. Bridgwater, Newbury, UK: CPL Press. 14–32.
Diebold, J. P., Milne, T. A., Czernik, S., Oasmaa, A., Bridgwater, A. V., Cuevas, A., Gust, S., Huffman, D., and Piskorz, J. 1999. Proposed specifications for various grades of pyrolysis oils. In *Fast Pyrolysis of Biomass: A Handbook*. vol. 2. ed. A. V. Bridgwater. Newbury, UK: CPL Press. 102–113.
Diebold, J. P. and Czernik, S. 1997. Additives to lower and stabilize the viscosity of pyrolysis oils during storage. *Energy & Fuels* 11:1081–1091.
Dobele, G., Dizhbite, T., Rossinskaja, G., Telysheva, G., Meier, D., Radtke, S., and Faix, O. 2003. Pre-treatment of biomass with phosphoric acid prior to fast pyrolysis. A promising method for obtaining 1,6-anhydrosaccharides in high yields. *Journal of Analytical and Applied Pyrolysis* 68–69:197–211.
Dobele, G., Meier, D., Faix, O., Radtke, S., Rossinskaja G., and Telysheva, G. 2001. Volatile products of catalytic flash pyrolysis of celluloses. *Journal of Analytical and Applied Pyrolysis* 58–59:453–463.
Elliott, D. C. and Baker, E. 1987. Hydrotreating biomass liquids to produce hydrocarbon fuels. In *Energy from Biomass and Wastes X*, ed. D. Klass. Chicago, IL: Institute of Gas Technology. 765–784.
Elliott, D.C. 2007. Historical developments in hydro-processing bio-oils. *Energy & Fuels* 2007(21):1792–1815.
Evans, R. J. and Milne, T. A. 1987a. Molecular characterization of the pyrolysis of biomass I. Fundamentals. *Energy & Fuels* 1(2):123–137.
Evans, R. J. and Milne, T. A. 1987b. Molecular characterization of the pyrolysis of biomass II. Applications, *Energy & Fuels* 1(4):311–319.
Fagernäs, L. 1995. Chemical and physical characterization of biomass-based pyrolysis oils. Literature review. VTT Technical Research Centre Notes, Oulu, Finland.
Fernando, S., Adhikari S., Chandrapal C., and Murali N. 2006. Biorefineries: Current status, challenges, and future direction. *Energy & Fuels* 20: 1727–1737.
Fratini, E., Bonnini, M., Oasmaa, A., Solantausta, Y., Teixeira, J., and Baglioni, P. 2006. SANS analysis of the microstructure evolution during the aging of pyrolysis oils from biomass. *Langmuir* 22(1):306–312.
Freel, B. A. and Graham, R. G. 2002. Bio-oil preservatives. U.S. Patent 6,485,841, filed Oct. 30, 1998 and issued Nov. 26, 2002.
Freel, B. A., Graham, R. G., and Huffman, D. R. 1990. The scale-up and development of rapid thermal processing (RTP) to produce liquid fuels from wood, Ontario Ministry of Energy Report (CF), Toronto, Canada.
Freel, B. A. and Huffman, D. R. 1994. Applied bio-oil combustion. *Proceedings Workshop on Biomass Pyrolysis Oil Properties and Combustion*, Estes Park, CO, 309–315.
Freel, B. A., Graham, R. G., and Huffman, D. R. 1996. Commercial aspects of rapid thermal processing (RTM). In *Bio-oil Production and Utilization*, eds. A. V. Bridgwater and E. Hogan Newbury, UK: CPL Press. 86–95.
Frigo, S., Gentilli, R., Tognotti, L., Zanforlin, S., and Benelli, G. 1998. Feasibility of using wood flash-bio-oil in diesel engines, SAE Technical Paper Series no. 982529.
Gagnon, J. and Kaliaguine, S. 1988. Catalytic hydrotreatment of vacuum pyrolysis oils from wood. *Industrial and Engineering Chemistry Research* 27(10): 1783–1788.
Garcia–Pérez, M., Chaala, A., and Roy, C. 2002. Vacuum pyrolysis of sugarcane bagasse. *Journal of Analytical and Applied Pyrolysis* 65:111–136.
Garcia–Pérez, M., Chaala, A., Pakdel, H., Kretschmer, D., Hughes P., and Roy C. 2004. The complex multiphase structure of bio-oils. Paper presented at the *Second World Conference on Biomass for Energy, Industry and Climate Protection*, Rome, Italy, May 10–14.

Garcia–Pérez, M., Chaala, A., Pakdel, H., Kretschmer, D., Rodrigue, D., and Roy, C. 2006a. Multiphase structure of Bio–oils. *Energy & Fuels* 20:364–375.
Garcia–Pérez, M., Chaala, A., Pakdel, H., Kretschmer, D., Rodrigue, D., and Roy, C. 2006b. Aging of bio-oils obtained from the vacuum pyrolysis of wood industry residues. *Energy & Fuels* 20:786–795.
Garcia–Pérez, M., Adams, T. T., Goodrum, J. W., Geller. D. P., and Das, K. C. 2007a. Production and fuel properties of pine chip bio-oil. Biodiesel blends. *Energy & Fuels* 21(4):2363–2372.
Garcia–Pérez, M., Chaala, A., Pakdel, H., Kretschmer, D., Hughes P., and Roy C. 2007b. Characterization of bio-oils in chemical families. *Biomass & Bioenergy* 31(4):222–242.
Garcia, L., French, R., Czernik, S., and Chornet, E. 2000. Catalytic steam reforming of bio-oils for the production of hydrogen: effects of catalyst composition. Applied Catalysis A: General 201, 225–239.
Garcia–Pérez, M., Wang, S., Shen, J., Rhodes, M. J., Lee,W.–J., and Li. C.–Z. 2008. Effects of temperature on the formation of lignin derived oligomers during the fast pyrolysis of mallee woody biomass. *Energy & Fuels* 22:2022–2032.
Garcia–Pérez, M. and Metcalf, J. 2008. Formation of polyaromatic hydrocarbons and dioxins during pyrolysis: Review of biomass. A review of the literature with descriptions of biomass composition, fast pyrolysis technologies and thermochemical reactions, Final Report. Washington State Department of Ecology, contract no. C0800247, Olympia, WA.
Giroux, R., Freel, B., and Graham, R. 2001. Natural resin formulation. U.S. Patent 6,326,461, filed July 29, 1999, issued Dec. 4, 2001.
Golova, O.P. 1975. Chemical effects of heat on cellulose. *Russian Chemical Reviews* 44:687–697.
Graham, R. G., Bergougnou, M. A., and Overend, R. P. 1984. Fast pyrolysis of biomass. *Journal of Analytical and Applied Pyrolysis* 6:95–135.
Gros, S. 1995. Pyrolysis liquid as diesel fuel. Wärtsilä Diesel International. In Seminar on Power Production from Biomass II, March 27–28, Espoo, Finland.
Gust, S. 1994. Combustion experiences of flash pyrolysis fuel in intermediate size boilers, Project Progress Reports (CF), Porvoo, Finland.
Gust, S. 1997. Combustion experiences of flash pyrolysis fuel in intermediate size boiler. In *Developments in Thermochemical Biomass Conversion*, eds. A. V.Bridgwater and D. G. B. Boocock, London, UK: Blackie Academic & Professional 481–488.
Hastaoglu, M. A. and Berruti F. 1989. A gas–solid reaction model for flash wood pyrolysis. *Fuel* 68(11):1408–1415.
Helle, S., Bennett, N. M., Lau, K., Matsuio, J.H., and Duff, S. J. B. 2007. A kinetic model for the production of glucose by hydrolysis of levoglucosan and cellobiosan from pyrolysis oils. *Carbohydrate Research* 342:2365–2370.
Himmelblau, A. 1991. Method and apparatus for producing water–soluble resin and resin product made by the method. U.S. Patent 5,034,498, filed July 19, 1989 and issued July 23, 1991.
Holmgren, J., Marinangeli, R., Nair, P., Elliott. D., and Bain, R. 2008. Converting pyrolysis oils to renewable transportation fuels: Processing challenges and opportunities. UOP LLC, a Honeywell company, paper no. AM-08-81.
Hosoya, T., Kawamoto, H., and Saka, S. 2006. Thermal stabilization of levoglucosan in aromatic substances. *Carbohydrate Research* 341:2293–2297.
Hosoya, T., Kawamoto, H., and Saka, S. 2007a. Cellulose–hemicellulose and cellulose–lignin interactions in wood pyrolysis at gasification temperature. *Journal of Analytical and Applied Pyrolysis* 80:118–125.
Hosoya, T., Kawamoto, H., and Saka, S. 2007b. Pyrolysis behaviours of wood and its constituent polymers at gasification temperature. *Journal of Analytical and Applied Pyrolysis* 78:328–336.
Huber, G. W. and Dumesic, J. A. 2006. An overview of aqueous phase catalytic process for production of hydrogen and alkanes in a biorefinery. *Catalysis Today* 111:119–132.
Ikura, M., Slamak, M., and Sawatzky, H. 1998. Pyrolysis liquid in diesel oil microemulsions. U.S. Patent 5,820,640, filed July 9, 1997 and issued Oct. 13, 1998.

Iojoiu, E., Domine, M. E., Davidian, T., Guihaume, N., and Mirodatos, C. 2007. Hydrogen production by sequential cracking of biomass-derived pyrolysis oil over noble metal catalysts supported on ceria-zirconia. *Applied Catalysis A: General* 323:147–161.

Jay, D. C., Sipilä, K. H., Rantanen, O. A., and Nylund, N. O. 1995. Wood pyrolysis for diesel engines. *Proceedings of the 17th Annual Fall Technical Conference of the ASME Internal Combustion Engine Division, Volume 3*, September 24–27, Milwaukee, WI, 51–59.

Julien, S., Chornet, E., and Overend, R. P. 1993. Influence of acid pre-treatment (H_2SO_4, HCl, HNO_3) on reaction selectivity in the vacuum pyrolysis of cellulose. *Journal of Analytical and Applied Pyrolysis* 27:25–43.

Kammen, D. M. and Lew, D. J. 2005. Review of technologies for the production and uses of charcoal. Renewable and appropiate energy laboratory report. National Renewable Energy Laboratory, Golden, CO.

Kasper, J. M., Jasas, G. B., and Trauth, R. L. 1983. Use of pyrolysis-driven fuel in a gas turbine. ASME Paper No. 83–GT–96.

Kawamoto, H., Saito, S., Hatanaka, W., and Saka, S. 2007a. Catalytic pyrolysis of cellulose in sulfolane with some acid catalysts. *Journal of Wood Science* 53:127–133.

Kawamoto, H., Horigoshi, S., and Saka, S. 2007b. Pyrolysis reactions of various lignin model dimers. *Journal of Wood Science* 53:168–174.

Kawamoto, H., Murayama, M., and Saka, S. 2003a. Pyrolysis behaviour of levoglucosan as an intermediate in cellulose pyrolysis: polymerization into polysaccharide as a key reaction to carbonized product formation. *Journal of Wood Science* 49:469–473.

Kawamoto, H., Hatanaka, W., and Saka, S. 2003b. Thermochemical conversion of cellulose in polar solvent (sulfolane) into levoglucosan and other low molecular–weight substances. *Journal of Analytical and Applied Pyrolysis* 70:303.

Kawamoto, H. and Saka, S. 2006. Heterogeneity in cellulose pyrolysis indicated from the pyrolysis in sulfolane. *Journal of Analytical and Applied Pyrolysis* 76:280–284.

Kersten, S. R. A., Wang, X., Prins, W., and Swaaij, M. P. M. 2005. Biomass pyrolysis in a fluidized bed reactor. Part 1. Literature review and model simulations. *Industrial & Engineering Chemistry Research* 44:8773–8785.

Kilzer, F. J. and Broido, A. 1965. Speculations on the nature of cellulose pyrolysis. *Pyrodynamics* 2:151.

Kim, D.–Y., Nishiyama, Y., Wada, M., Kuga, S., and Okano, T. 2001. Thermal decomposition of cellulose crystalites in wood. *Holzforschung* 55:521.

Klar, M., and Rule, A. 1925. *The Technology of Wood Distillation*. London, UK: Chapman & Hall Ltd.

Klass, D. L. 1998. *Biomass for Renewable Energy, Fuels and Chemicals*. San Diego, CA: Academic Press.

Koufopanos, C. A., Papayannakos, N., Maschio, G., and Lucchesi, A. 1991. Modelling of the pyrolysis of biomass particles. Studies on kinetics, thermal and heat transfer effects. *The Canadian Journal of Chemical Engineering* 69(4):907–915.

Kumar, R. R., Kolar, A. K., and Leckner, B. 2006. Effect of fuel particle shape and size on devolatilization time of Casuarina wood. In *Sciences in Thermal and Chemical Biomass Conversion*, eds. A. V. Bridgwater and D. G. B. Boocock. Chippenham, UK: CPL Press. 1251–1264.

Lee, S. W. 1993. Preliminary combustion evaluation of wood derived fast pyrolysis liquids using a residential burner. Division Report ERL 93–29 (CF) CANMET, Energy Mines and Resources Canada, Ottawa, Canada.

Leech, J. 1997. Running a dual fuel engine on pyrolysis oil. In *Biomass Gasification and Pyrolysis, State of the Art and Future Prospects*, eds. M. Kaltschmitt and A. V. Bridgwater A.V. Newbury, UK: CPL Press. 495–497.

Lomax, J. A., Commadeur, J. M., Arisz, P. W., Boon, J. J. 1991. Characterization of oligomers and sugar rings–cleavage products in the pyrolyzate of cellulose. *Journal of Analytical and Applied Pyrolysis* 19:65–79.

López–Juste, G. and Salvá Monfort, J. J. 2000. Preliminary test on combustion of wood derived fast pyrolysis oils in a gas turbine combustor. *Biomass & Bioenergy* 19(2):119–128.

Maggi, R. and Elliott, D. 1997. Upgrading overview. In *Developments in Thermochemical Biomass Conversion*, eds. A. V. Bridgwater and D. G. B. Boocock. London, UK: Blackie Academic & Professional. 575–588.

Mahfud, F. H., Bussemaker, S., Kooi, B. J., Ten Brink, G. H., and Heeres, H. J. 2007a. The application of water-soluble ruthenium catalysts for the hydrogenation of the dichloromethane soluble fraction of fast pyrolysis oil and related model compounds in a two-phase aqueous–organic system. *Journal of Molecular Catalysis A: Chemical* 277(1–2):127–136.

Mahfud, F. H., Ghijsen, F., and Heeres, H. J. 2007b. Hydrogenation of fast pyrolyis oil and model compounds in a two–phase aqueous organic system using homogeneous ruthenium catalysts *Journal of Molecular Catalysis A: Chemical* 264(1–2):227–236.

Majerski, P., Piskorz, J., and Radlein, D. 2001. Production of glycolaldehyde by hydrous thermolysis of sugars. U.S. Patent 7094932, filed Nov. 8, 2001, issued Aug. 22, 2006, PCT CA01/01562.

Mamleev, V., Bourbigot, S., and Yvon, J. 2007. Kinetic analysis of the thermal decomposition of cellulose: The main step of mass loss. *Journal of Analytical and Applied Pyrolysis* 80(1):151–165.

Marinangeli, R., Marker, T., Petri, J., Kalnes, T., McCall, M., Mackowiak, D., Jerosky, B., et al. 2005. Opportunities for biorenewables in oil refineries. Final Technical Report. Department of Energy award number: DE–F636–056015085.

Meier, D. and Faix, O. 1999. State of the art of applied fast pyrolysis of lignicellulosic materials—A review. *Bioresource Technology* 68:71–77.

de Meira, A. M., Brito, J. O., and Rodriguez, L. C. E. 2005. Estudo de aspectos técnicos, econômicos e sociais da produção de carvão vegetal no Município de Pedra Bela, São Paulo, *Brasil Revista Árvore* 29(5):809–817.

Mohan, D., Pittman, C. U., and Steel, P. H. 2006. Pyrolysis of wood/biomass for bio–oil: A critical review. *Energy & Fuels* 20:848–889.

Moses, C. 1994. Fuel-specification consideration for biomass liquids. *Proceedings on Biomass Pyrolysis Oils. Properties and Combustion Meeting*, Sept. 26–28, Estes Park, CO, 362–382.

Muyafuji, H., Nakata, T., Ehara, K., and Saka, S. 2005. Fermentability of water–soluble portion to ethanol obtained by supercritical water treatment of lignocellulosics. *Applied Biochemistry and Biotechnology* 124(1–3):963–971.

Nicolaides, G. M. 1984. The chemical characterization of pyrolysis oils. M.A.Sc. thesis University of Waterloo, Waterloo, Ontario, Canada.

Oasmaa, A. and Czernik, S. 1999. Fuel oil quality of biomass pyrolysis oils—State of the art for end users. *Energy & Fuels* 13:914–921.

Oasmaa, A. and Kuoppala, E. 2004. Fast pyrolysis of forest residue. 3. Storage stability of liquid fuel. *Energy & Fuels* 17:1075–1084.

Oasmaa A., Kuoppala, E., Gust, S., and Solantausta, Y. 2003a. Fast pyrolysis of forest residue. 1. Effect of extractives on phase separation of pyrolysis liquids *Energy & Fuels* 17(1):1–12.

Oasmaa A., Kuoppala E., and Solantausta Y. 2003b. Fast pyrolysis of forest residue. 2. Physicochemical composition of product liquid. *Energy & Fuels* 17:433–443.

Oasmaa A., Kytö M., and Silipä K. 2001. Pyrolysis oil combustion Tests in an Industrial Boiler. In: *Proceeding in Thermochemical Biomass Conversion*, ed. A. V. Bridgwater, Oxford, UK: Blackwell Science. 1468–1481.

Oasmaa, A., Leppämäki, E., Koponen, P., Levander, J., and Tapola, E. 1997. Physical characterization of biomasa–based pyrolysis liquids. Application of standard fuel oil analyses. VTT Publications 306. VTT Technical Research Centre of Finland.

Oehr, K. 1995. Acid emission reduction. U.S. Patent, 5,458,803, filed Sept. 30, 1993, issued Oct. 17, 1995.

Oehr, K. H., Scott, D. S., and Czernik, S. 1993. Method of producing calcium salts from biomass. U.S. Patent 5, 264, 623, filed Jan. 4, 1993, issued Nov. 23, 1993.

Orfao, J. J. M., Antunes, F. J. A., and Figueiredo, J. L. 1999. Pyrolysis kinetics of lignocellulosic materials. Three independent reactions model. *Fuel* 78:349–358.

Ormrod, D. and Webster, A. 2000. Progresses in utilization of bio-oils in diesel engines. *PyNe Newsletter* 10:15.
O'Sullivan, A. C. 1997. Cellulose: the structure slowly unravels. *Cellulose* 4: 173–207.
Peláez-Samaniego, M. R., Garcia–Pérez, M., Cortez, L. B., Rosillo-Calle, F., and Mesa, J. 2008. Improvements of Brazilian carbonization industry as part of the creation of a global biomass economy. *Renewable and Sustainable energy Reviews* 12:1063–1086.
Perlack, R. D., Wright, L. L., Turhollow, A. F., Graham, R. L., Stokes, B. J., Erback, D. C. 2005. Biomass as feedstock for a bioenergy and bioproducts industry: The technical feasibility of a billion-ton annual supply. DOE/GO-102005-2135. Oak Ridge, TN: U.S. Department of Energy.
Pinheiro, P. C. 2001. Fundamentos e practica da carbonizacao da biomassa. 1st International Congress on the use of Biomass for the production of Metals and Electricity, Belo Horizonte, Brazil (In Portuguese).
Piskorz, J., Majerski, P., Radlein, D., Vladars-Usas, A., and Scott, D. S. 2000. Flash pyrolysis of cellulose for production of anhydro-oligomers. *Journal of Analytical and Applied Pyrolysis* 56:145–166.
Piskorz, J., Peacocke, G. V. C., and Bridgwater, A. V. 1999. IEA pyrolysis fundamentals review. In *Fast Pyrolysis of Biomass: A Handbook*, eds. A. V. Bridgwater, et al. Newbury, UK: CPL Press. 33–50.
Piskorz, J., Radlein, D., and Scott, D. S. 1986. On the mechanism of the rapid pyrolysis of cellulose. *Journal of Analytical and Applied Pyrolysis* 9:121–137.
Piskorz, J., Majerski, P., and Scott, D. S. 1989a. Conversion of lignins to hydrocarbon fuels. *Energy & Fuels* 3(6):723–726.
Piskorz, J., Radlein, D., Scott, D. S., and Czernik, S. 1989b. Pretreatment of wood and cellulose for production of sugars by fast pyrolysis. *Journal of Analytical and Applied Pyrolysis* 16:127–142.
Prosen, E. M., Radlein, D., Piskorz, J., Scott, D. S., and Legge, R. L. 1993. Microbial utilization of levoglucosan in wood pyrolysate as a carbon and energy source. *Biotechnology and Bioengineering* 42:538–541.
Pyle, D. L. and Zaror, C. A. 1984. Heat transfer and kinetics in the low temperature pyrolysis of solids. *Chemical Engineering Science* 39(1):147–158.
Radlein, D. 1987. On the presence of anhydro-oligosaccharides in the syrups from the fast pyrolysis of cellulose. *Journal of Analytical and Applied Pyrolysis* 12: 39–49.
Radlein, D. 1999. The production of chemicals from fast pyrolysis bio-oils. *Fast Pyrolysis of Biomass: A Handbook*, ed. A. V. Bridgwater. Newbury, UK: CPL Press. 164–188.
Radlein, D., Mason, S. L., Piskorz, J., and Scott, D. S. 1991a. Hydrocarbons from the catalytic pyrolysis of biomass. *Energy & Fuels* 5:760.
Radlein, D., Piskorz, J., and Scott, D. S. 1991b. Fast pyrolysis of natural polysaccharides as a potential industrial process. *Journal of Analytical and Applied Pyrolysis* 19:41–63.
Ramos, M. C., Navascues, A. I., Garcia, L., and Bilbao, R. 2007: Hydrogen production by catalytic steam reforming of acetol, a model compound of bio-oil. *Industrial and Engineering Chemistry Research* 46(8):2399–2406.
Richards, G. N. 1987. Glycolaldehyde from pyrolysis of cellulose. *Journal of Analytical and Applied Pyrolysis* 10:251–255.
Rioche, C., Kulkarni, S., Meunier, F. C., Breen, J. P., and Burch, R. 2005: Steam reforming of model compounds and fast pyrolysis bio-oil on supported noble metal catalysts. *Applied Catalysis B, Environmental* 61:130–139.
Robson, A. 2000. Dynamotive 2000 progress report. *PyNe Newsletter* 10:6.
Rocha, J. D., Coutinho, A. R., and Luengo, C. A. 2002. Biopitch produced from eucalyptus wood pyrolysis liquids as a renewable binder for carbon electrode manufacture. *Brazilian Journal of Chemical Engineering* 19(2):127–132.
Rosillo-Calle, F., de Rezende, M. A. A., Furtado, P., and Halla, D. O. 1996: *The Charcoal Dilemma: Finding a Sustainable Solution for Brazilian Industry*. London, UK: Intermediate Technology Publications.

Rossi, C., Frandi, R., Bonfitto, E., Jacoboni, S, Pistone, L, and Mattiello, M. 1993. Combustion tests of bio–oils from biomass slow pyrolysis. In *Advances in Thermochemical Biomass Conversion*, ed. A.V. Bridgwater. London, UK: Blackie Academic & Professional. 1205–1213.

Roy, C., Blanchette, D., de Caumia, B., Dubé, F., Pinault, J., Bélanger, É, and Laprise, P. 2001. Industrial scale demonstration of the Pyrocycling™ process for the conversion of biomass to biofuels and chemicals. *Proceedings of the First World Conference and Exhibition on Biomass for Energy and Industry*, Sevilla, Spain, June 5–9, eds. S. Kyritsis et al. London, UK: James & James Ltd. 1032–1035.

Roy, C., Lu, X., and Pakdel, H. 2000. Process for the production of phenolic rich pyrolysis oils for use in making phenol-formaldehyde reseol resin. U.S. Patent 6,143,856, filed Feb. 5, 1999, issued Nov. 7, 2000.

Roy, C. and Morin, D. 1998. Efficient electricity production from biomass through IPCC power plants. *Proceedings of the Fourth International Conference on Greenhouse Gas Control Technologies*. Interlaken, Switzerland, Aug. 30–Sep. 2, eds. B. Eliasson, P. Riemer, and A. Wokaun. Oxford, UK: Elsevier Science Ltd. 729–734.

Salvi, G. and Salvi, G. Jr. 1991. Pyrolytic products assessment study. Report from the Commission of European Communities, Contract No. EN3B–0191–1 (CH).

Scott, D. S., Majerski, P., Piskorz, J., and Radlein, D. 1999. Second look at fast pyrolysis of biomass—The RTI process. *Journal of Analytical and Applied Pyrolysis* 51(1–2):23–37.

Shihadeh, A. and Hochgreb S. 2000. Diesel engines combustion of biomass pyrolysis oils. *Energy & Fuels* 14:260–274.

Shihadeh, A., Lewis, P., Manurung, R., and Beér, J. 1994. Combustion characterisation of wood–derived flash pyrolysis oils in industrial scale turbulent diffusion flames. *Proceedings of Biomass Pyrolysis Oil Properties and Combustion Meeting*. Estes Park, CO, Sept. 26–28, NREL–CP–430–7215, 281–295.

Sjöström, E. 1993. *Wood Chemistry: Fundamentals and Applications*. 2nd ed. New York, NY: Elsevier.

Solantausta, Y., Gust, S., Hogan, E., Massoli, P., and Sipilä, K. 2000. Bio fuel oil upgrading by hot filtration and novel physical methods. Contract JOR3–CT98–0253. Final report 1998–2000. Research funded in part by the European Commission in the framework of the Non Nuclear Energy Programme Joule III.

Solantausta, Y., Nylund, N.–O., and Gust, S. 1994. Use of pyrolysis oil in a test diesel engine to study the feasibility of a diesel power plant concept. *Biomass & Bioenergy* 7:297–306.

Solantausta, Y., Nylund, N.–O., Westerholm, M., Koljonen, T., and Oasmaa, A. 1993. Word pyrolysis oil as fuel in a diesel power plant. *Bioresources Technology* 46:177–188.

Stamatov, V., Honnery, D., and Soria, J. 2006. Combustion properties of slow pyrolysis bio-oil produced from indigenous Australian species. *Renewable Energy* 31(13):2108–2121.

Stradal, J. A. and Underwood, G. 1995. Process for producing hydroxyacetaldehyde. U.S. Patent 5,393,542, filed July 16, 1993, issued Feb. 28, 1995.

Strenziok, R., Hanse, U., and Künster, H. 2001. Combustion of bio-oil in a gas turbine. In *Progress in Thermochemical Biomass Conversion*, ed. A.V. Bridgwater. Oxford, UK: Blackwell Science. 1452–1458.

Suppes, G. J., Natarajan, V. P., and Chen, A. 1996. Autoignition of select oxygenate fuels in simulated diesel engine environment, Paper (74e) presented at the AIChE National Meeting, New Orleans, LA, Feb. 26, 1996.

Takanabe, K., Aika, K., Inazu, K., Baba, T., Seshan, K., and Lefferts, L. 2006. Steam reforming of acetic acid as a biomass derived oxygenate: Bifunctional pathway for hydrogen formation over Pt/ZrO$_2$ catalysts. *Journal of Catalysis* 243 (2):263–269.

Underwood, G. L. 1990. Commercialization of fast pyrolysis products. In *Biomass Thermal Processing*, eds. E. Hogan, Robert, J., Grassi G., and A. V. Bridgwater. Newbury, UK: CPL Press. 226–228.

Underwood, G. L. and Graham, R. G. 1991. Methods of using fast pyrolysis liquids of making a high browning liquid smoke composition. U.S. Patent 5,039,537, filed June 6, 1991 and issued August 4, 1992.

Vanasse, C., Chornet, E., and Overend, R. P. 1988. Liquefaction of lignocellulosics in model solvents: Creosote oil and ethylene glycol. *Canadian Journal of Chemical Engineering* 66(1):112–120.

van de Kamp, W. L. 2000. The combustion of pyrolytic biomass oils with coal for electricity generation in the MEMWEG station. Technical Report. IFRF Research Station b.v.

van de Kamp, W. L. and Smart J. P. 1991. Evaluation of the combustion characteristics of pyrolytic oils derived from biomass. *Proceedings of the Sixth European Conference on Biomass for Energy, Industry and Environment*, Athens, Greece, April 22–26. eds. G. Grassi, A. Collina, and H. Zibetta. London, UK: Elsevier Applied Science. 1131–1137.

van de Kamp W. L and Smart J. P. 1993. Atomization and combustion of slow pyrolysis biomass oil. In *Advances in Thermochemical Biomass Conversion*, ed. A. V. Bridgwater. London, UK: Blackie Academic & Professional. 1265–1274.

van Rossum, G., Kersten, S. R. A., and Van Swaaij, W. P. M. 2007. Catalytic and noncatalytic gasification of pyrolysis oil. *Industrial and Engineering Chemistry Research* 46 (12):3959–3967.

Varhegyi, G., Antal, M., Jakab, E., and Szabo, P. 1997. Kinetic modeling of biomass pyrolysis. *Journal of Analytical and Applied Pyrolysis* 42(1):73–87.

Wood, T. S. and Baldwin, S. 1985. Fuelwood and charcoal use in developing countries. *Annual Review of Energy* 10:407–429.

Wooten, J. B., Seeman, J. I., and Hajaligol, M. R. 2004. Observation and characterization of cellulose pyrolysis intermediates by ^{13}C CPMAS NMR. A new mechanistic model. *Energy & Fuels* 18:1.

Yu, Z. and Zhang, H. 2003. Pre-treatment of cellulose pyrolysate for ethanol production by *Saccharomyces cerevisiae*, *Pichia* sp. YZ–1 and *Zymamonas mobilis*. *Biomass & Bioenergy* 24:257–262.

Yu, Z. and Zhang, H. 2004. Ethanol fermentation of acid–hydrolyzed cellulosic pyrolysate with *Saccharomyces cerevisiae*. *Bioresource Technology* 93:199–204.

Zhang, Q., Chang, J., Wang, T. J., and Xu, Y. 2006. Upgrading bio-oil over different solid catalysts. *Energy & Fuels* 20(6):2717–2720.

Zhang, S.–P., Yan, Y.–J., Ren, Z., and Li, T. 2003. Study of hydrodeoxygenation of bio-oil from the fast pyrolysis of biomass. *Energy Sources* 25(1):57–65.

Zickler, G. A., Wagermaier, W., Funari, S.S., Burghammer. M., and Paris, O. 2007. In situ X-ray diffraction investigation of thermal decomposition of wood cellulose. *Journal of Analysis and Applied Pyrolysis* 80:134–140.

CHAPTER 8
Performance and Emissions of Biodiesel and Ethanol in Engines

Ahindra Nag
Natural Product Laboratory
Department of Chemistry
Indian Institute of Technology
Karagpur, India

8.1 Introduction

The resources of petroleum as fuel are dwindling day by day and increasing demand of fuels pose a challenge to science and technology. Petroleum diesel comes in the category of nonrenewable fuel and will last for a limited period of time. The U.S. Department of Energy has predicted that the role of bioenergy will eventually represent 5.75 percent of transportation fuels by 2010.[1] Recently, President George W. Bush's "20 in 10 program," launched in January 2007, proposed that 15 percent of the transportation fuel demand would be met with biofuels in 2018. Other countries have seriously considered the massive use of bioenergy in the future. China,[2] Germany,[3] Austria,[4] and Sweden[5] have set goals of using 10 to 15 percent of their internal primary energy supply through bioenergy up to the year 2020, where Vietnam has set a target of using bioenergy (i.e., 38.8 percent as future energy consumption).[6]

A renewable fuel such as biodiesel is an alternative fuel of diesel. Biodiesel is a renewable energy source that can readily replace a small part of petroleum diesel and may help to reduce greenhouse gas emissions, promote sustainable rural development, and improve income distribution. Furthermore, production of biodiesel reduces dependence on foreign supplies, saves a lot of funds on imports of diesel, and is helpful in counteracting inflation.[7] It is widely agreed that biodiesel decreases emissions of hydrocarbons (HC), carbon monoxide (CO), particulate matter (PM), and sulphur dioxide (SO_2). Only nitrogen oxides (NO_x) are reported to increase, which is due to the oxygen content in the biodiesel.[8–10] NO_x is also said to be carbon neutral because it contributes no net carbon dioxide to the atmosphere.[11–13]

Processing vegetable oils to biodiesel[1,8,10] and its applications has several advantages such as the following:

1. Technologies for extraction and processing are very easy and simple, as conventional equipment with low energy inputs are needed.
2. Fuel properties are close to diesel.
3. Vegetable oils are renewable in nature.
4. Being liquid, biodiesel offers ease of portability and possesses stability and low handling hazards.
5. The by-product leftovers after extraction of oil is rich in protein and can be used as animal feed or solid fuel.
6. Cultivation of oil-producing plants is flexible to a wide range of geographic locations and climatic conditions.
7. Biodiesel can be used directly in compression ignition engines without substantial modifications of the engine.
8. Biodiesel does not emit sulphur oxides.
9. Biodiesel reduces the exhaust emissions in diesel engines without significantly affecting engine performance.

Different oils have been used in different countries as raw materials for biodiesel production (e.g., soybean oil is commonly used in the United States, rapeseed and sunflower oils are used in many European countries, and coconut and palm oils are used in Malaysia).[14–17] Most of these oils are edible and are costlier than nonedible oils as needed for human consumption. Thus, nonedible oils will be economical to use in internal combustion engines. Furthermore, the main economic criteria are considered to be manufacturing cost and price of raw feedstock. Manufacturing costs include direct costs for oil extraction, reagents, and operating supplies, as well as indirect costs related to insurance and storage. Fixed capital costs involved in the construction of processing plants and auxiliary facilities, such as distribution and retailer costs, must also be taken into consideration.

To promote biodiesel consumption, several countries exempted biodiesel from the fuel excise tax. Among them, the European Union approved the biodiesel tax exemption program in May 2002; the financial EU law funds biofuels through excise exemption over a period of 3 years (Article 21 Finance Law 2001). The U.S. Senate Finance Committee approved an excise tax exemption for biodiesel in 2003. Moreover, the legislation provides a 1-cent reduction in the diesel fuel excise tax for each percentage of biodiesel blended with petroleum diesel up to 20 percent. However, the tax exemption will one day come to an end, and other steps will be needed to continue to promote the social inclusion and the economic attraction of biodiesel. This could be facilitated by the selection of low-cost raw materials, such as nonedible oils, used in frying oil or animal fat, and use of a lower-cost transesterification process.

The literature[1,6,18] already shows that many researchers have worked on nonedible vegetable oils such as Jatropha, mahua, karanja, neem, and so on and edible vegetable oils such as rapeseed, sunflower, soybean palm, and others separately to study the performance and emission characteristics of these oils in a diesel engine.

Among several indigenous plant species—Jatropha, karanja, and putranjiva—are plants that bear seeds of nonedible oils and are going to be of interest all over the world. These plants can be cultivated easily in barren soils of stony, sandy, and clayish soil textures with less maintenance and low moisture content. These plants are medium-sized and are found abundantly in the Northern Hemisphere. Million tons of seeds of Jatropha, karanja, and putranjiva go to waste annually, which can be useful to reduce the fuel crisis. Also, growing these plants can provide employment, improve the environment, and enhance the quality of rural life.

Biodiesels consist of methyl esters of fatty acids that are produced by catalytic transesterification of triglycerides of vegetable oils with alcohol. In this process, triglyceride molecules are successively converted into di- and monoglycerides and finally into fatty acid alkyl esters and glycerol.[18] These monoesters are known as *biodiesels*.

$$\begin{array}{c}\text{OCOR}\\\text{OCOR}\\\text{OCOR}\end{array} + \text{MeOH} \underset{K1}{\overset{K1}{\rightleftharpoons}} \begin{array}{c}\text{OH}\\\text{OCOR}\\\text{OCOR}\end{array} + \text{RCOOMe}$$

Triglyceride (TG) → Diglyceride (DG) + Fatty acid methyl ester (Fame)

$$\begin{array}{c}\text{OH}\\\text{OCOR}\\\text{OCOR}\end{array} + \text{MeOH} \underset{K2}{\overset{K2}{\rightleftharpoons}} \begin{array}{c}\text{OH}\\\text{OH}\\\text{OCOR}\end{array} + \text{RCOOMe}$$

DG → Monoglyceride (MG) + (Fame)

$$\begin{array}{c}\text{OH}\\\text{OH}\\\text{OCOR}\end{array} + \text{MeOH} \underset{K3}{\overset{K3}{\rightleftharpoons}} \begin{array}{c}\text{OH}\\\text{OH}\\\text{OH}\end{array} + \text{RCOOMe}$$

MG → Glycerol + (Fame)

The alcohols commonly used are methanol and ethanol. Several parameters, such as type of catalyst (alkaline, acid, or enzyme), oil/alcohol ratio, temperature, reaction time, and purity of the reactants (the main contaminants are water and free fatty acids) have influenced transesterification reaction.

As per a survey of the literature,[10,18] one drawback of biodiesel is that it is more prone to oxidation than petroleum-based diesel fuel. In its advanced stages, this oxidation can cause the fuel to become acidic and to form insoluble gums and sediments that can plug the fuel filter. It is observed that long-term use of pure vegetable oil and its blend with diesel in diesel engine may result in carbon deposits in the combustion chamber, coking and trumpet formation on the injectors to such an extent that fuel atomization becomes difficult, oil ring sticking, thickening, and gelling formation. Lubricating oil contamination showed high piston, liner, and bearing wear, indicating that pure vegetable oils are acceptable only for short-term use. However, engine durability is an issue during extensive use of pure vegetable oil blends because of carbon deposits and fueling system problems of critical engine components resulting in premature engine failure. Some investigators[18] have suggested that a fuel additive or a fuel blend with less vegetable oil is needed for engine durability. Pure vegetable oil is needed for filtering and chemical treatment to reduce viscosity and to improve combustion and flow properties. In recent years, researchers and automobile industries have been trying to design various parts of engine components so that pure vegetable oils can be used 100 percent without any unwanted difficulties. Elsbett AG from Germany has designed a piston/combustion chamber in a compression ignition engine to give excellent running using pure vegetable oils. Pure vegetable oils are used mainly in large stationary engines. Large stationary engines run on lower grades of fuel oil, including thick crude oil full of impurities. The injector is designated to spray thick oils efficiently. The slow speed and size of these engines give more time for the fuel to burn completely.

Goering et al.[19] suggested that vegetable oils are too viscous for prolonged use in direct-injected diesel engines that resulted in poor fuel atomization and inefficient mixing with air, contributing to incomplete combustion. Due to different chemical and physical properties of vegetable oils, vegetable oils accumulate and remain as charred deposits when they contact engine cylinder walls. The problem of charring and deposits of particles on the injector and cylinder wall can be partially overcome by transesterification of the oils. The process of transesterification removes glycerol from the triglycerides and replaces it with radicals from the alcohol used for the conversion process.[20]

8.1.1 Transesterification of Vegetable Oils

Gummy materials that are present in vegetable oils inhibit the rate of reaction.

Degumming is an economical chemical process of acid treatment by which the gum of the vegetable oil is removed to improve the viscosity and cetane number (CN) of vegetable oil up to a certain limit.[18] The oils are degummed by phosphoric acid. The quantity of the acid required during this process is very important in optimizing results such as reduction of viscosity.[21] Three percent phosphoric acid is mixed with oil at 40°C with vigorous stirring. After being stirred for half an hour, the mixture is kept for 1 week so that the reacted gum materials may settle. Solid gum materials that can be used as soil fertilizer are separated by decantation, and the degummed oil is washed 3 times to remove the remaining acid followed by filtering through a packed bed that is filled by activated carbon. The filtered oil is then ready for transesterification.[21]

Generally, transesterification of vegetable oils is done by acid, base, and enzyme catalysts. Preparation of biodiesel is optimized through a series of reactions. In the transesterification process the oil (4.0 g) is poured into a three-necked round-bottom flask and is heated, for 1 hour at a temperature of 100°C, with continuous stirring, to remove excess moisture. In the acid-catalyzed process, sulphuric acid (1 percent of oil and methanol) is mixed well with methanol. Oil is poured into a round-bottom flask that is fitted with a thermometer and a condenser; the flask is then placed on a magnetic stirrer. The acid–methanol mixture is added slowly in the flask. The methanol–catalyst solution is kept for 2 hours at 67°C with vigorous stirring. The mixture is cooled by ice water. Two layers are formed in which the upper layer is the product layer and the lower layer is the glycerol layer. The upper layer is placed into a separating flask and is washed 3 times with distilled water to remove the acid and the methanol. The washed oil is stirred with anhydrous sodium sulfate and is kept for 3 hours at room temperature for dehydration. The dehydrated oil is used for its properties, performance, and emission measurement.

In a base-catalyzed process, sodium methoxide (1.5 percent, that is, w/w of the total reactant) is put in the methanol and then is dissolved by stirring. The solution is added slowly in the oil present in the round-bottom flask. The reaction procedure and purification of the product are done similar to the acid process.

In the enzymatic process, specified ratios of oil and methanol are poured into a conical flask fitted with a standard joint-stoppered cock. The enzyme, *Candida antarctica*, is taken as 1 percent (w/w) of the oil as the catalyst. The conical flask is placed in an incubator shaker for stirring at 600 rpm. The reaction mixture is drawn out after a specified duration of time and is filtered to remove the catalyst. Then, the product is washed with water to remove the methanol and the glycerol immediately. The washed product is treated with anhydrous sodium sulfate for 2 hours to remove excess moisture.

8.1.2 Analyzing the Products

The fatty acid compositions of nonedible oils and their transesterified products can be analyzed by a CN10543004 gas chromatograph (Restek International, Bellefonte, Pennsylvania) with a flame-ionization

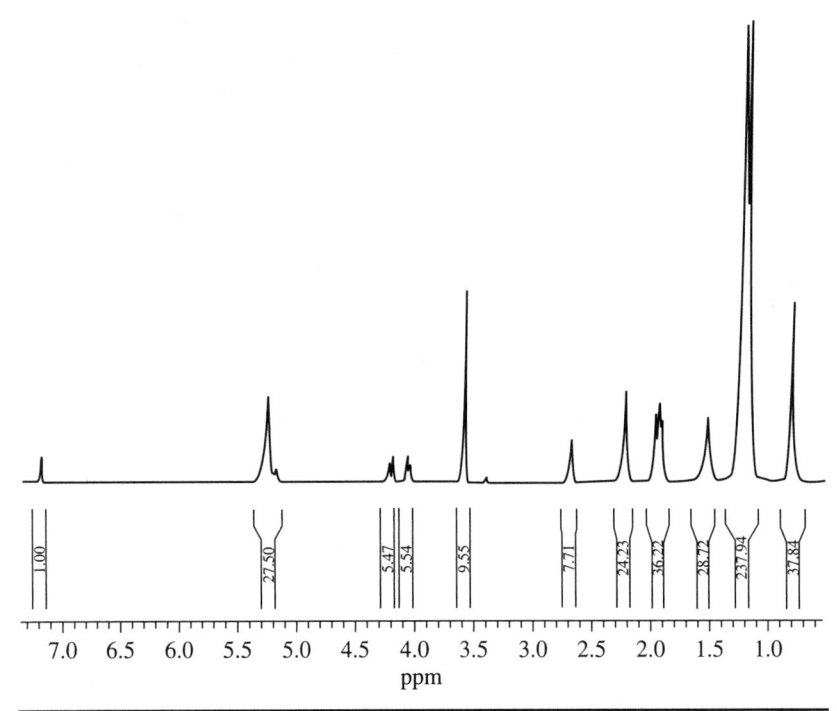

FIGURE 8.1 NMR spectra of transesterified Jatropha oil with methanol.

detector. The used capillary is 30 m long with an internal diameter of 0.25 mm. The carrier gas is nitrogen at a flow rate of 1 mL/min. The injection port temperature is 150°C and the ionization detector temperature is 170°C.

The transesterified product can also be analyzed by nuclear magnetic resonance (NMR). A reaction aliquot is taken with $CDCl_3$ in a NMR tube and is placed in a NMR spectroscope (400 MHz, Bruker BioSpin, Fällanden, Switzerland) for analysis. The percentage conversion of reactant to product is analyzed by ^{1}H NMR (Fig. 8.1), considering the peak of methoxy groups in the methyl esters at $\delta = 3.7$ ppm (singlet) and of the α-carbonyl methylene groups of fatty ester derivatives at $\delta = 2.3$ ppm. The percentage of conversion is equal to $(2 \times A_{\delta = 3.7}/3 \times A_{\delta = 2.3}) \times 100$ where A is the integrated peak area at different chemical shifts.

8.1.3 Property Measurement of Transesterified Oils

The suitability of any material as fuel, including biodiesel, is influenced by the nature of its major components as well as that of minor components arising from production or other sources. The nature of these components ultimately determines the fuel and its physical properties. Cold-temperature operation is very critical due to high viscosity that causes fuel system problems such as starting failure, unacceptable emission level, and injection pump failure. The engine shuts down for long duration due to accelerated gum formation where the fuel contacts the bare metal. This might further impair the engine or injection system.[19]

Calorific value (ASTM D4809) and viscosity can be measured by a bomb calorimeter and a Redwood viscometer, respectively. The flash point and the fire point can be determined using the Pensky Martens apparatus closed-cup method according to ASTM D93. The pour point is measured according to ASTM D98. The carbon residue is measured by the Conradson method (ASTM D189), and the CN is determined according to ASTM D4738. The CN of a diesel fuel (DF) is related to the ignition delay (ID) time, that is, the time that passes between injection of the fuel into the cylinder and onset of ignition. The shorter the ID time, the higher the CN and vice versa.

Standards have been established worldwide for CN determination, for example, ASTM D613 in the United States and the International Organization for Standardization (ISO) standard ISO 5165 internationally.

Recently, cetane studies on fatty compounds were conducted using the Ignition Quality Tester (IQT) (Advanced Engine Technology Ltd., Ottawa, Ontario, Canada).[21] The IQT is a further, automated development of a constant-volume combustion apparatus (CVCA)[20]. The CVCA was originally developed for determining CNs more rapidly, with greater experimental ease, better reproducibility, reduced use of fuel, and therefore cheaper than the ASTM D613 method using a cetane engine. The IQT method, which is the basis of ASTM D6890, was shown to be reproducible and the results competitive with those derived from ASTM D613. The IQT, ID, and CN are related by the following equation:[20]

$$CN_{IQT} = 83.99 \times (ID - 1.512)^{-0.658} + 3.547 \qquad (8.1)$$

However, the results for fatty compounds with the IQT are comparable to those obtained by other methods.[20]

The difference in properties for different vegetable oils are due to variations of fatty acid composition and other associated compounds such as coloring matters and odorant compounds, etc. The fuel properties of three transesterified vegetable oils (karanja, Jatropha, and putranjiva) and diesel are shown in Table 8.1. The kinematic viscosities of 100 percent of fatty acid methyl esters (FAME) at 40°C are almost same as of diesel.[21] The flash point and the fire point are better than diesel for the engine application. It has been observed that the calorific values and CNs of the three transesterified oils (karanja, putranjiva, and Jatropha) are less than that of diesel but are comparable with other available biodiesels as reported by Rakopoulos et al.[22]

	Biodiesel (100% FAME)			
Properties	Putranjiva	Karanja	Jatropha	Diesel
Viscosity in cSt (at 40°C)	5.81	5.81	5.42	5.03
Cetane number	40.2	35.6	41.0	46.3
Calorific value (kJ/kg)	3,9254	38,119	39,065	42,707
Pour point (°C)	2	5	2	–12
Specific gravity (at 25°C)	0.883	0.899	0.878	0.834
Flash point (°C)	152	190	171	78
Fire point (°C)	155	192	175	85
Carbon residue (%)	0.10	1.20	0.15	0.10

TABLE 8.1 Fuel Properties of Three Biodiesels and Diesel Fuel

When we compare the three oils, Jatropha shows better results in specific gravity, viscosity, CN, and pour point, whereas putranjiva is found to be better in calorific value, flash point, and carbon residue.

8.1.4 Measurement of Performance and Emission

To measure emissions, an automotive exhaust analyzer and a smoke meter are generally used. Automotive exhaust analyzer functions on the principle of the nondispersive infrared (NDIR) method. The operational principle of the smoke meter analyzer is attenuation of a LED pure green light beam. The Ricardo variable compression diesel engine is used to run with blends of (10, 30, 50, 70, and 100 percent) transesterified oils, with diesel at different loads (0 to 2.7 kW), constant injection times, constant speeds, and constant compression ratios.

Ziejewski and Kaufman[23] reported the results of an endurance test using 25 to 75 percent alkali-refined sunflower oil with diesel and 25 to 75 percent of safflower oil with diesel on a volume basis. The major problems experienced were premature injection, determination of nozzle performance, and heavier carbon deposits in the grooves of the piston ring. There was no significant problem with engine operation when a blend of safflower oil was used. That investigation revealed that chemical differences between vegetable oil and diesel had a very important influence on long-term engine performance.

8.1.5 Comparing the Effect of Loads on Biodiesel Performance

Raheman and Phadatare[24] carried out the experiments with different oils and suggested that there were variations of the engine torque with

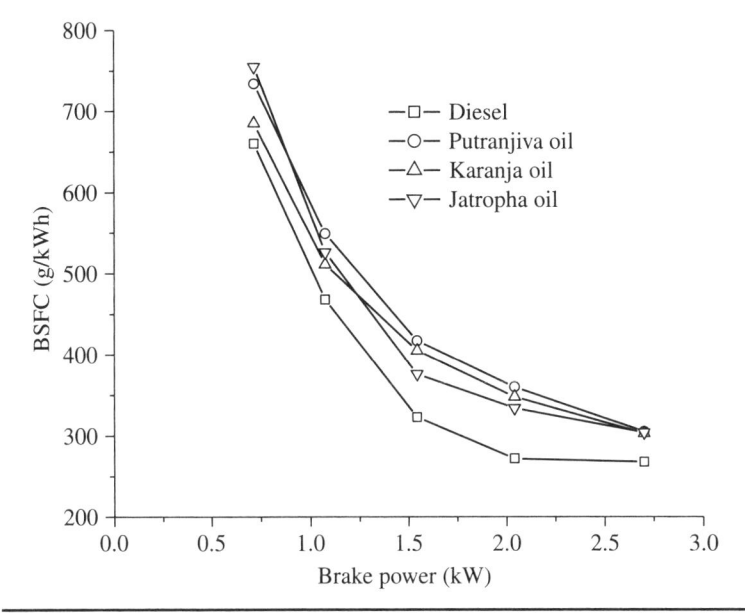

FIGURE 8.2 Brake-specific fuel consumption versus brake power of diesel fuel, 100 percent biodiesel of Jatropha, karanja, and putranjiva at 1200 rpm, 45° angle before top dead centre (bTDC), and 20 compression ratio.

loads for different fuels. The torque increased with increase in load. This was due to an increase in fuel consumption with an increase in load. Mathur and Das[25] conducted tests on diesel engines using blends of mahua and neem oil with diesel. From the results Mathur and Das[25] concluded that neem oil could be substituted at 35 percent with a marginal reduction in efficiency and power output. Mahua oil with diesel had exhaust characteristics similar to those of diesel. The performance of three transesterified vegetable oils and diesel are shown in Fig. 8.2. It is observed that 100 percent methyl ester of Jatropha, karanja, and putranjiva oil gives higher performance in brake-specific fuel consumption (BSFC) and lower in brake thermal efficiency (η_{bt}) than diesel at all loads. Transesterified karanja oil (100 percent) is more efficient than putranjiva and Jatropha, whereas Jatropha is better in BSFC. At high loads the performance of the three biodiesels is almost the same but is different from diesel.

8.1.6 Comparison of the Effect of Load on Biodiesel Emission

The NO_x of biodiesel is lower than diesel, which is due to biodiesel's low CN.[26,27] The NO_x of biodiesel is increased slightly with load, whereas diesel fuel shows a steady increase throughout the load. With increasing load, the fuel consumption rate is increased and more heat is released during burning. NO_x emission is increased with increasing temperature of combustion chamber. Biodiesel has lower calorific value than neat diesel, and therefore the rate of increase of NO_x is low with load than with diesel.[27]

Walker[28] observed that NO_x started at 6.2 g/mile for diesel and decreased to around 5.6 g/mile with 100 percent ester (biodiesel), with a slightly more reduction with rapeseed ethyl ester rather than rapeseed methyl ester.

It is observed that with increasing of load NO_x emission of three transesterified oils (karanja, Jatropha, and putranjiva) is increased (Fig. 8.3). With increasing the percentage of biodiesel in blends (Table 8.2) NO_x emission is decreased due to a decrease of the calorific value of the blend and thus a lower exhaust gas temperature. Jatropha blend shows the lowest NO_x emission than the others. Hydrocarbon emission of biodiesel (Fig. 8.4) is lower than diesel due to better combustion of biodiesel. A higher percentage of biodiesel in blend gives fewer hydrocarbon emissions. Smoke, CO, and particulates (Fig. 8.5) of the three oils are lower than diesel, indicating a good impact on the environment and living beings. Higher concentrations of biodiesel in blend (Table 8.2) results in lower CO emission because of higher ignition temperatures and better combustion of biodiesel than diesel, and fewer exhaust emissions.

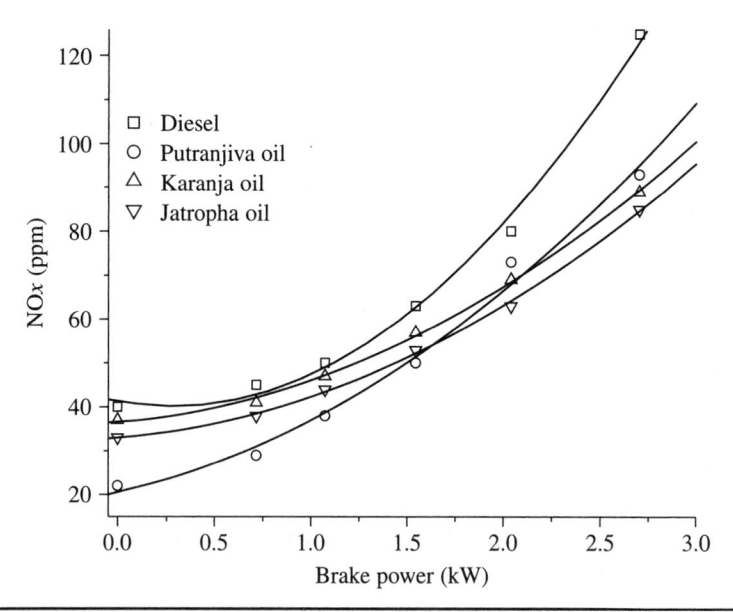

FIGURE 8.3 Nitrogen oxide versus brake power of diesel fuel, 100 percent biodiesel of Jatropha, karanja, and putranjiva at 1200 rpm, 45° angle bTDC, and 20 compression ratios.

Biodiesel	Percentage of Blends	BSFC (gm/kW/hr)	η_{bt} (%)	HC (ppm)	CO (%)	NO_x (ppm)	Smoke (Hu)	Particulate (mg/m^3)	Exhaust Temp (°C)
Putranjiva	10	268	31.7	111	0.037	118	8.1	18.8	210
	30	314	28.5	102	0.034	108	8.3	18.1	206
	50	332	26.5	89	0.032	91	6.7	14.4	201
	70	345	25.9	78	0.031	83	5.9	12.5	197
Karanja	10	311	28.4	116	0.023	99	6.4	18.9	222
	30	320	28.2	104	0.018	86	4.9	18.3	207
	50	325	28.4	95	0.017	78	4.2	18.6	202
	70	328	28.8	84	0.014	72	4.0	18.1	186
Jatropha	10	282	30.2	105	0.019	76	4.4	15.4	170
	30	286	30.3	93	0.017	71	4.1	14.8	167
	50	295	29.9	87	0.014	59	3.8	14.0	166
	70	302	29.68	71	0.011	56	3.2	12.6	162
Diesel	100	272	30.9	120	0.058	80	10.5	22.7	234

TABLE 8.2 Performance and Emissions of Blends of Three Biodiesels and Diesel Fuel at a Load of 2.04 kW

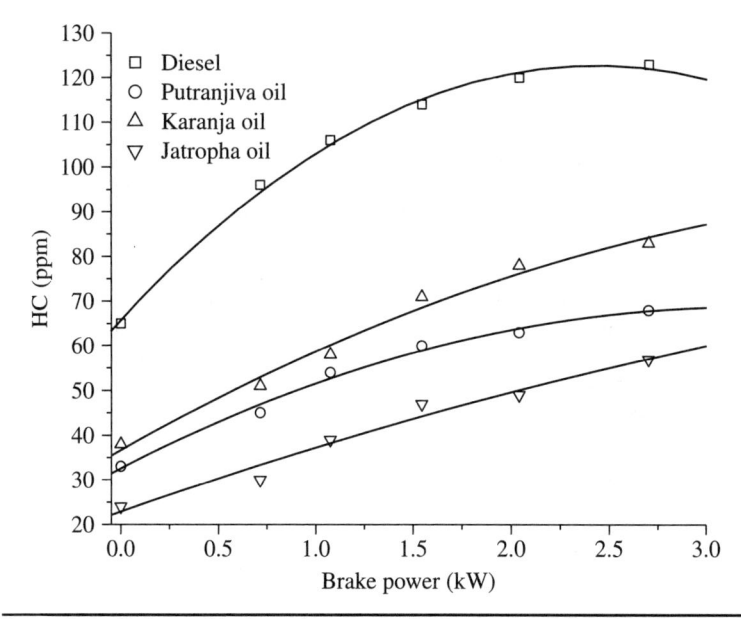

Figure 8.4 Unburned hydrocarbon versus brake power of diesel fuel, 100 percent biodiesel of Jatropha, karanja, and putranjiva at 1200 rpm, 45° angle bTDC, and 20 compression ratios.

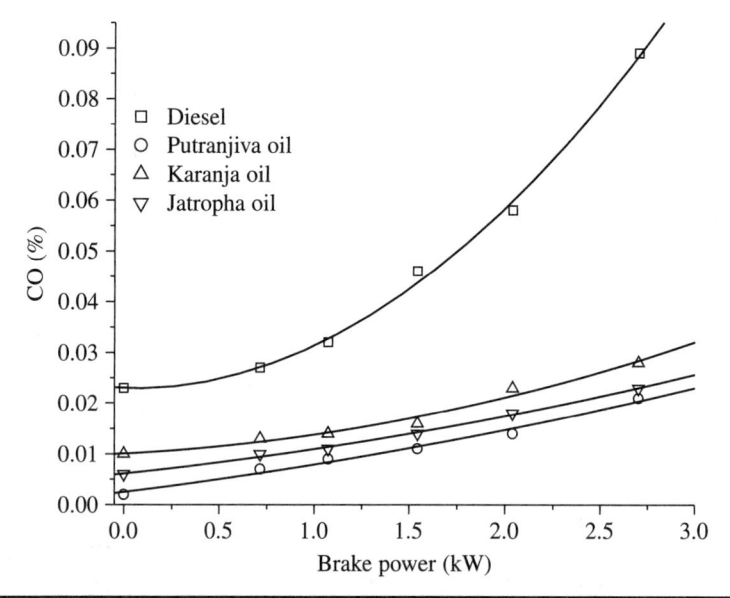

Figure 8.5 Carbon monoxide versus brake power of diesel fuel, 100 percent biodiesel of Jatropha, karanja, and putranjiva at 1200 rpm, 45° angle bTDC, and 20 compression ratios.

8.1.7 Exhaust Gas Temperature

Exhaust gas temperatures of different biodiesels at various loads are compared with that of diesel fuel (Fig. 8.6). The exhaust gas temperatures of 100 percent biodiesel and its blends are always lower than that of diesel. The exhaust gas temperature of a blend is decreased with increasing amount of biodiesel (Table 8.2). This is due to the less calorific value of biodiesel than diesel. The exhaust gas temperature of 100 percent biodiesel from putranjiva is higher than that of 100 percent biodiesel from karanja and Jatropha. From Table 8.2, we can observe that blends of karanja result in higher exhaust gas temperature than Jatropha and putranjiva. This is due to the variation in combustion of blends.

It has been observed that the blends of degummed putranjiva oil and diesel reduce emissions such as CO, NO_x, smoke, particulates, etc. Brake thermal efficiency of blends up to 30 percent shows emissions less than diesel (Fig. 8.7). For blends above 30 percent brake, thermal efficiency and brake-specific fuel consumption shows inferior quality. Thus, we can conclude that up to 30 percent blends of degummed putranjiva oil with diesel can replace diesel in running a diesel engine satisfactorily with a reduction of emissions that are very favorable to the environment.

Transestrified oils from karanja, Jatropha, and putranjiva or that blended with diesel fuel, have shown very satisfactory results as an

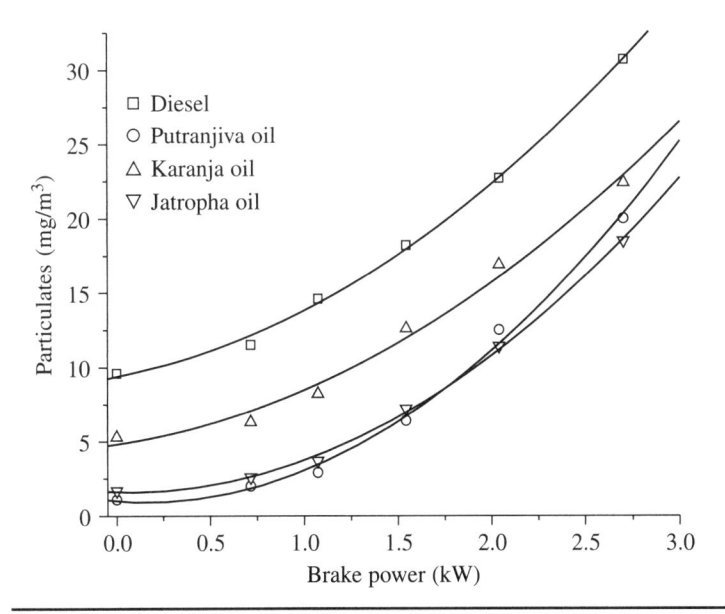

FIGURE 8.6 Particulates versus brake power of diesel fuel, 100 percent biodiesel of Jatropha, karanja, and putranjiva at 1200 rpm, 45° angle bTDC, and 20 compression ratios.

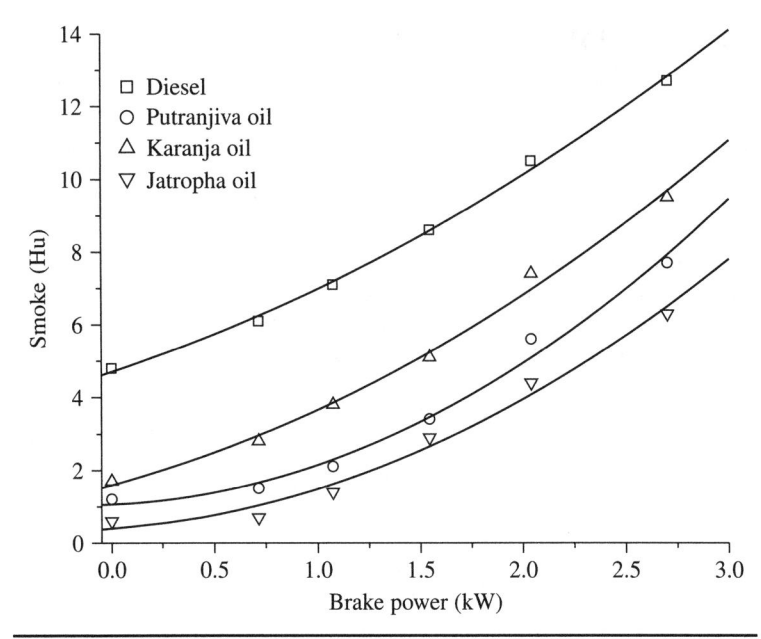

FIGURE 8.7 Smoke versus brake power of diesel fuel, 100 percent biodiesel of Jatropha, karanja, and putranjiva at 1200 rpm, 45° angle bTDC, and 20 compression ratios.

alternative fuel for diesel engines. Out of the three oils, Jatropha promises to yield good performance and lower emissions at each load in all respects. Considering the above-mentioned points, we can conclude that a diesel engine can run very satisfactorily using 100 percent fatty acid methyl ester (pure biodiesel) at 45° angle bTDC timing, and a compression ratio of 20. Any diesel engine can be operated with 100 percent biodiesel as a prime mover without any modifications to the engine.

8.2 Ethanol as an Alternative Fuel

Ethanol as an alternative liquid fuel has recently attracted considerable attention, particularly from the agricultural sector. However, farmers are limited in their ability to use ethanol because most field equipment today are equipped with diesel engines and fuel would be most critically needed for field operations during a shortage. Ethanol may be used to replace, supplement, or extend the use of diesel fuel. The use of alcohol in spark–ignition engines (SI) started in 1954 in countries like the United States, Germany, and France. During World Wars I and II when a gasoline shortage occurred in France and Germany, alcohol was used in all types of vehicles including military planes.

Property	Petrol	Diesel	Ethanol
Specific gravity (at 15°C)	0.73	0.82	0.79
Boiling point (°C)	30–225	190–280	78.3
Specific heat (MJ/kg)	43.5	43.0	28.0
Heat of vaporization (kJ/kg)	400	600	900
Octane no. (research)	91–100	N.A.	N.A.
Cetane no.	Below 15	40–60	Below 15

N.A. = not applicable

TABLE 8.3 Comparative Properties of Petrol, Diesel, and Ethanol

Today, it is used with gasoline (a mixture) in the United States and has become a major fuel in Brazil.

The comparative properties of ethanol with petrol and diesel are listed in Table 8.3.

8.2.1 Production of Ethanol

Alcohol is derived not directly from sugarcane but from molasses–sugarcane by-products. All starch-rich plants like maize, tapioca, and potato can be used to produce alcohol as well as cellulosic waste materials. The production of ethanol from biomass involves fermentation and distillation of crops. The following methods are used to produce alcohol.

Sugarcane There are three stages for obtaining alcohol for energy purposes: (1) extracting the juice from sugarcane, (2) fermenting the juice, and (3) distilling the juice into 90 to 95 percent alcohol.

Molasses The black residue from the purified sugar extracting from sugarcane is molasses. It contains mostly invert sugars and some sucrose. This sucrose also undergoes hydrolysis to produce invert sugar by catalytic action of acids in molasses:

$$C_{12}H_{22}O_{11} + H_2O = C_6H_{12}O_6 \text{ (D-glucose)} + C_6H_{12}O_6 \text{ (D-fructose)}$$

This mixture product is not crystallizable. Yeast organisms in the presence of oxygen oxidize sugars into CO_2 and H_2O, and yeast converts sucrose mostly into ethyl alcohol:

$$C_6H_{12}O_6 = 2\ C_2H_5OH + 2\ CO_2$$

Starch In this process, starchy materials are first converted into fermentable sugars. This is done by enzymatic conversion (using the malt process) or by acid hydrolysis. Then, conversion of sugar to alcohol is completed with yeast:

$$\text{Starch} \quad C_{12}H_{22}O_{11} + C_6H_{12}O_6$$

$$\text{(Maltose)} \quad \text{(Dextrose)}$$

Tapioca Material Tapioca is available in plentiful supply in Asia, the United States, Middle Europe, and Africa. Its production can be increased through modern cultivation techniques. The process consists in converting the tapioca flour into fermentation sugars with enzymes prior to fermentation with yeast.

Modern technology uses α-amyl glycosidase, one of the two enzymes required in the process and then saccharifies the material into alcohol by yeast.

Cellulose Material

1. *From wood:* Cellulose wood is hydrolyzed into simple sugars by means of diluted acid at high temperatures or by concentrated acid at low temperatures. Similarly, cellulosic agricultural waste and straws can be used in place of wood.

2. *Sulphite waste liquor from paper manufacture:* Waste liquor contains 2 to 3.5 percent of sugar, out of which 65 percent is fermentable into alcohol. Before fermentation, all acids in the liquor are removed by adding calcium. Then, sugar fermentation is carried out using yeast. Generally, 1 percent of liquor is converted into alcohol.

Hydrocarbon Gases

1. *Hydration of ethylene:* Conversion of ethylene to ethyl alcohol can be carried out in high yield by first treating ethylene with H_2SO_4, forming ethyl sulphuric acid and diethyl sulphate as given by following reactions:

$$C_2H_4 + H_2SO_4 = C_2H_5HSO_4$$

$$C_2H_5H.SO_4 = (C_2H_5)_2SO_4$$

When treated with water, ethylsulfuric acid and diethyl sulfate become ethanol per the following reactions:

$$C_2H_5.HSO_4 + H_2O = C_2H_5OH + H_2SO_4$$

$$(C_2H_5)_2SO_4 + 2H_2O = 2C_2H_5OH + H_2SO_4$$

2. *By direct hydration:* Is formed per the following chemical reaction:

$$C_2H_4 + H_2O = C_2H_5OH$$

This type of conversion is not found very much, as the reaction is exothermic. This method of production is not suitable for mass production and contains 90 percent gasoline and 10 percent ethanol.

3. *Anhydrous alcohol from vegetable wastes:*

The Philippines carries out an alcogas program to produce its own anhydrous alcohol from local vegetable wastes for blending with petrol. This program is currently based on sugarcane juice and molasses, but it plans to diversify by using other raw materials. In the basic process, cellulose conversion begins with pretreatment of raw materials, which may include coffee hulls, rice straw, and grass-to-sawmill wastes. Enzymes then take over by converting the feedstock into a sugary liquid that is fermented and finally distilled into anhydrous alcohol. After distillation, waste residues can be evaporated into syrup to feed animals, whereas unconverted cellulose is used as the primary fuel for the plant. If the Philippines could engineer a breakthrough in this area, it's agricultural and forestry wastes could supply the energy equivalent to 9.720 million liters of oil annually. In the years to come, this new energy source could make a significant economic impact on a country that depends on imports of crude oil for 95 percent of its energy.

Ethanol has been more expensive than petroleum but can become competitive when petroleum prices rise; for example, the Organization of Petroleum Exporting Countries is able to control supplies to raise crude oil prices.

Ethanol has a very high-octane rating, and it is a good spark–ignition engine fuel. Ethanol is already in widespread use as an additive to gasoline for use in SI engines. Because cetane and octane ratings vary inversely, however, ethanol necessarily has a low-cetane rating and will not self-ignite reliably in most compression ignition engines.

Ecklund et al.[29] state that fumigation is a method by which alcohol is introduced into the engine by vaporizing the alcohol in the intake airstream. This requires adding a carburetor along with a separate fuel tank, lines, and controls. Alcohol delivery must be reduced at low loads to prevent flame quenching and misfire and at high loads to prevent preignition and engine knock. With these methods, fumigation of ethanol offers several advantages, including the following:

1. Fumigation requires a minimum modification to the engine.
2. Fumigation can reduce NO_x emissions and smoke level.

3. Diesel fuel substitution up to 50 percent by the mass of ethanol is possible.

4. Ethanol fumigation system is separate from the diesel fuel injection system, and the engine can be quickly converted back to diesel operation.

Chaplin and Janius[30] studied the effect on engine performance by ethanol fumigation. A Ford 801, four-cylinder direct-injection diesel tractor engine was used for the study. Fumigation of ethanol was accomplished by using a cone jet nozzle mounted in an inlet manifold extension pipe. Tests were conducted at 18, 23, and 30° angle bTDC. Ethanol mass flow rates were set at 0, 25, 32, and 45 g/min, respectively. The mass flow rate of ethanol was not modulated with the engine speed. Cruz et al.[31] tested a turbocharged 650-kW diesel tractor engine with the spray addition method. For this approach, a mixture of water and ethanol was sprayed into the airstream by means of pressure from the turbocharged. Goering et al.[32] installed a multipoint fumigation system on a Case IH 52882 six-cylinder turbocharged 128-kW diesel farm tractor engine. Holes were drilled in the cylinder head to permit the installation of six Bosch model electronic fuel injection nozzles to spray fumigated ethanol on the tops of the intake valve. Denatured anhydrous ethanol was fumigated at a rate of 12 kg/h. Hayes et al.[33] investigated the effect of fumigation of different ethanol proofs on a turbocharged diesel engine. Ethanol was injected directly into the intake ports to ensure even distribution among the cylinders. It was found that the ethanol proof had no apparent effect on the thermal efficiency at any load.

The experiments have been carried out on the Perkins P4 diesel engine (Simpson Electric Co., Wisconsin) of four-stroke direct-injection four-cylinder vertical, in-line 91.4-mm diameter. Bore × 127-mm stroke, capacity 3500 cc, compression ratio 18.5:1, overhead valves, 29° angle bTDC injection timing. The engine is coupled to a hydraulic dynamometer for load measurements. The fumigation nozzle was a Pilot spray gun–type 64M (Mumbai, India) used as a fumigation nozzle in the experiment. It was connected to a manometer and a reciprocating compressor. The nozzle is used to fumigate the ethanol into the intake manifold of the engine. A wide, vertical or horizontal, fan spray or a round concentrated spray is produced by turning the air cap. The air cap-locking nut must be screwed up to grip the air cap firmly and yet allow the operator to turn the air cap by hand. The fumigation rate is controlled by the regulator on the air transformer. The smoke meter of model ED 1949, type ST-00N (Nissan Motor Sales Co. Ltd., Nissalco, Japan) is used to measure the smoke density. The exhaust gases such as CO, NO_x, and gas temperature were analyzed with a gas-analyzing instrument TMT-16-S (Technovation Analytical Instruments Pvt. Ltd., Veravan Beveren, Germany). The engine speed

Biodiesel and Ethanol in Engines 271

is measured by a hand tachometer. A stopwatch is used to measure the consumption of diesel and the fumigation of ethanol.

The performance and emission of the Perkins four-cylinder diesel engine under various brake powers (0, 1.58, 3.12, 4.74, and 6.31 kW), different speeds (800, 900, 1000, and 1100 rpm), and dehydrated ethanol (200 proof) fumigation rates (0, 1.06, 1.45, and 2.06 kg/h) are determined in terms of brake thermal efficiency, brake-specific fuel consumption, diesel substitution, NO_x emission level, CO emission level, smoke density, and exhaust gas temperature.

In order to achieve the objectives of the investigations, 80 experiments were carried out for diesel fuel and dual fuel (diesel + fumigated ethanol) operations.

8.2.2 Effect of Brake Power on Brake Thermal Efficiency

We can observe from Fig. 8.8 that at constant speed the brake thermal efficiency increases with an increase in brake power. The reason may be that lower cycle temperature and pressure at low brake power is not sufficient to cause complete combustion. Again, at constant speed and load, brake thermal efficiency decreases with an increase in ethanol fumigation rate. It may be due to low calorific value of ethanol (65 percent diesel). The long ignition delay resulting from the charge cooling of the vaporizing ethanol causes burning into the expansion

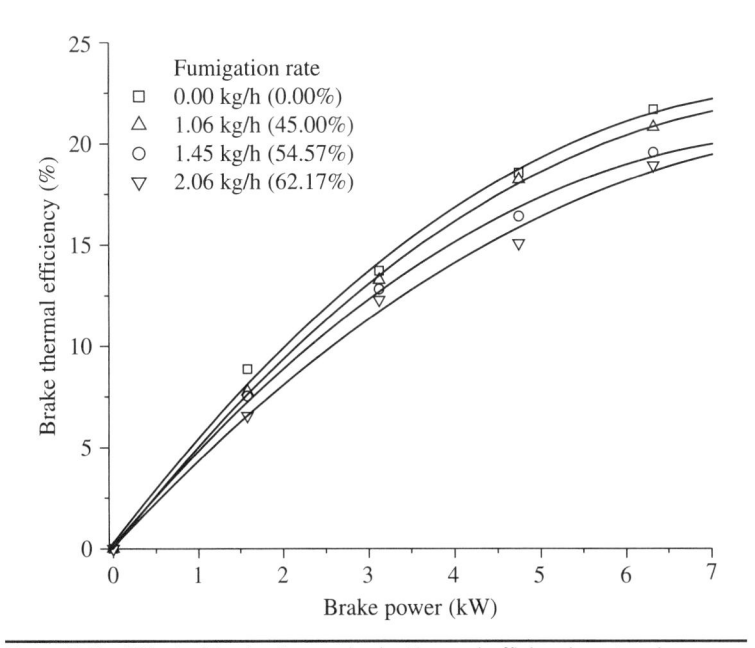

Figure 8.8 Effect of fumigation on brake thermal efficiencies at various brake powers at a speed of 800 rpm.

stroke, resulting in a reduction in brake thermal efficiency. At an engine speed of 800 rpm and at a 1.58-kW load, brake thermal efficiency is decreased by 1.11, 1.33, and 2.30 percent at ethanol fumigation rates of 1.06, 1.45, and 2.06 kg/h (45, 54.57, and 62.17 percent diesel substitution), respectively, when compared to the brake thermal efficiency under diesel operation. At ethanol fumigation rate of 1.06 kg/h (45 percent diesel substitution) and at low load (1.58 kW) the brake thermal efficiency is reduced by 1.11, 2.49, 0.96, and 1.53 percent at engine speed of 800, 900, 1000, and 1100 rpm, respectively, when compared to the efficiency under diesel fuel operation.

8.2.3 Effect of Brake Power on Brake-Specific Diesel Fuel Consumption

Figure 8.9 shows that the brake-specific diesel fuel consumption decreases with increases in brake power. It is observed that the brake-specific diesel fuel consumption (BSDFC) curve rises sharply with decrease in load because the percent decrease in diesel fuel required to operate the engine is less than the percent decrease in brake power.[34] The percent heat loss at low loads and constant speed operation is greater than at the higher loads, which also affects the shape of the BSDFC curve and contributes toward its rise at low loads. It is observed that BSDFC decreases with increase in ethanol fumigation rate at constant speed. This is due to reduced diesel fuel consumption by the

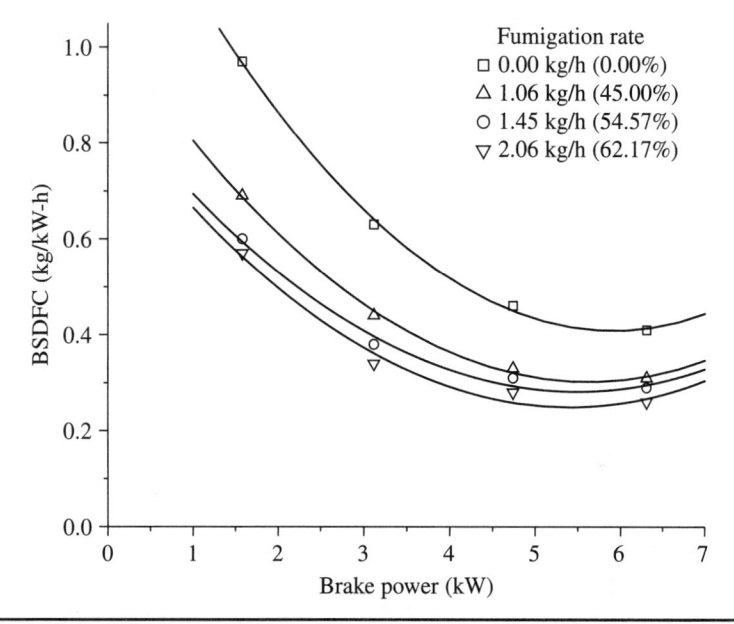

FIGURE 8.9 Effect of fumigation on brake-specific diesel fuel consumptions at various brake powers at a speed of 800 rpm.

engine while the ethanol fumigation rate is increased. At engine speed of 800 rpm and load of 1.58 kW, BSDFC is decreased by 28.86, 41.23, and 38.14 percent at ethanol fumigation rates of 1.06, 1.45, and 2.06 kg/h (45, 54.57, and 62.17 percent diesel substitution), respectively, when compared to the BSDFC under diesel fuel operation. At a high load of 6.31 kW, the BSDFC is decreased by 29.26, 24.39, and 36.58 percent at ethanol fumigation rates of 1.06, 1.45, and 2.06 kg/h, respectively, when compared to the BSDFC under diesel fuel operation. At an ethanol fumigation rate of 1.06 kg/h and at a low load of 1.58 kW, the BSDFC is decreased by 28.86, 9.52, 19.79, and 13.63 percent at engine speeds of 800, 900, 1000, and 1100 rpm, respectively.

8.2.4 Effect of Brake Power on Diesel Substitutions at Various Ethanol Fumigation Rates

Figure 8.10 shows that the diesel substitution decreases with increase in brake power at constant speed. This is due to the increased diesel consumption by the engine while brake power is increased at a constant ethanol fumigation rate. Figure 8.10 also shows diesel substitution increases with increase in ethanol fumigation rate at constant speed. This is due to reduced diesel consumption by the engine while the ethanol fumigation rate is increased. It is observed that at an engine speed of 800 rpm and at no load, diesel substitutions are 59.55, 65.90, and 73.77 percent at ethanol fumigation rates of 1.06, 1.45,

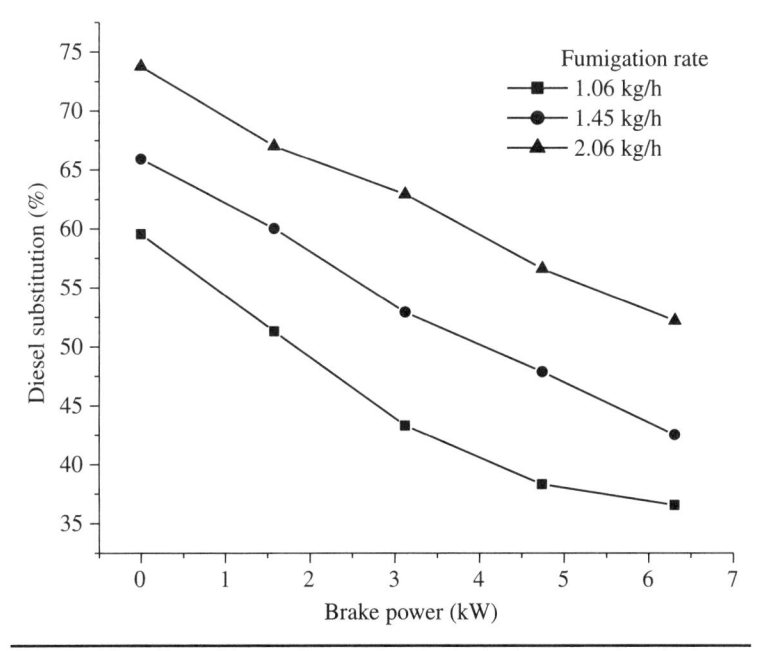

FIGURE 8.10 Effect of fumigation on diesel substitutions at various brake powers at a speed of 800 rpm.

and 2.06 kg/h, respectively. At a high load of 6.31 kW, diesel substitutions are 36.55, 42.52, and 52.17 percent at ethanol fumigation rates of 1.06, 1.45, and 2.06 kg/h respectively. It is observed that at ethanol fumigation rate of 1.06 kg/h and at no load, diesel substitutions are 59.55, 56.88, 51.45, and 48.18 percent at engine speeds of 800, 900, 1000, and 1100 rpm, respectively.

8.2.5 Effect of Brake Power on NO_x Emission Levels at Various Ethanol Fumigation Rates

It is observed that the NO_x emission level increases with increase in brake power at constant speed (Fig. 8.11). This is due to an increase in combustion temperature while increasing brake power. It is also observed that the NO_x emission level decreases with increase in ethanol fumigation rate at constant speed. The reason may be that the latent heat of vaporization of the ethanol cools the intake charge sufficiently to reduce peak combustion temperature, which in turn reduces NO_x formation. At a high ethanol fumigation rate, peak combustion temperature is highly reduced, which results in a reduction of NO_x. It is observed that at an engine speed of 800 rpm and at no load, NO_x emission level is decreased by 100 percent at ethanol fumigation rates of 1.06, 1.45, and 2.06 kg/h (45, 54.57, and 62.17 percent diesel substitution, respectively), when compared to the NO_x emission level under diesel fuel operation. At a high load of 6.31 kW, the

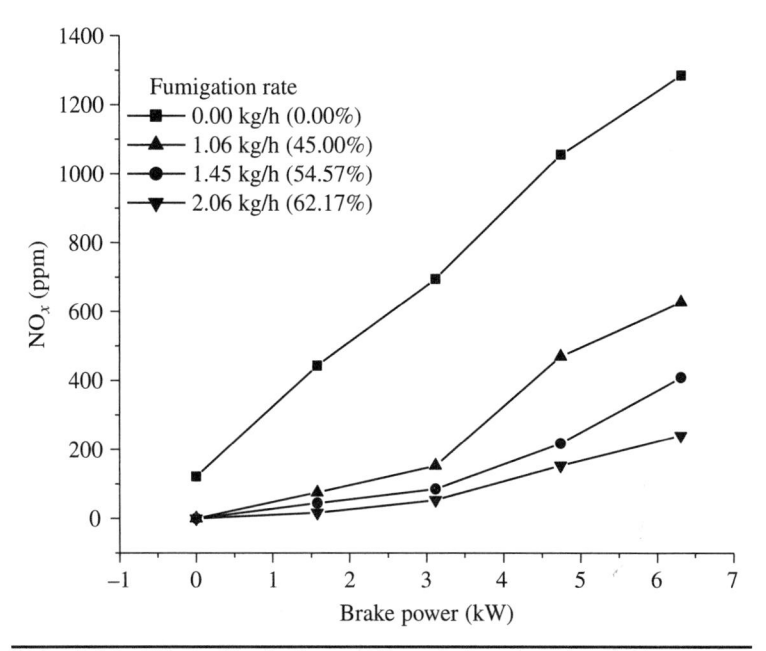

FIGURE 8.11 Effect of fumigation on NO_x emissions at various brake powers at a speed of 800 rpm.

NO_x emission level is decreased by 51.12, 68.72, and 81.24 percent at ethanol fumigation rates of 1.06, 1.45, and 2.06 kg/h, respectively, when compared to the NO_x level under diesel fuel operation. It is observed that at an ethanol fumigation rate of 1.06 kg/h and at no load, the NO_x emission level is reduced by 100, 80, 82.5, and 74.46 percent at engine speeds of 800, 900, 1000, and 1100 rpm, respectively.

8.2.6 Effect of Brake Power on CO Emissions at Various Ethanol Fumigation Rates

Figure 8.12 shows that at constant speed, the CO emission level increases with an increase in brake power and also increases with an increase in ethanol fumigation rate at constant speed.

It is observed that at an engine speed of 800 rpm and at no load, the CO emission level is increased by 0.104, 0.16, and 0.038 percent at ethanol fumigation rates of 1.06, 1.45, and 2.06 kg/h (45, 54.57, and 62.17 percent diesel substitution), respectively, when compared to the CO emission level under diesel fuel operation.

CO emission levels are always higher when coming from a fumigated engine than when the engine operated on diesel fuel only.[35] High CO emissions are usually due to rich combustion and because diesel engines generally operate with a lean overall equivalent ratio. It is possible with ethanol fumigation that one or two cylinders could be

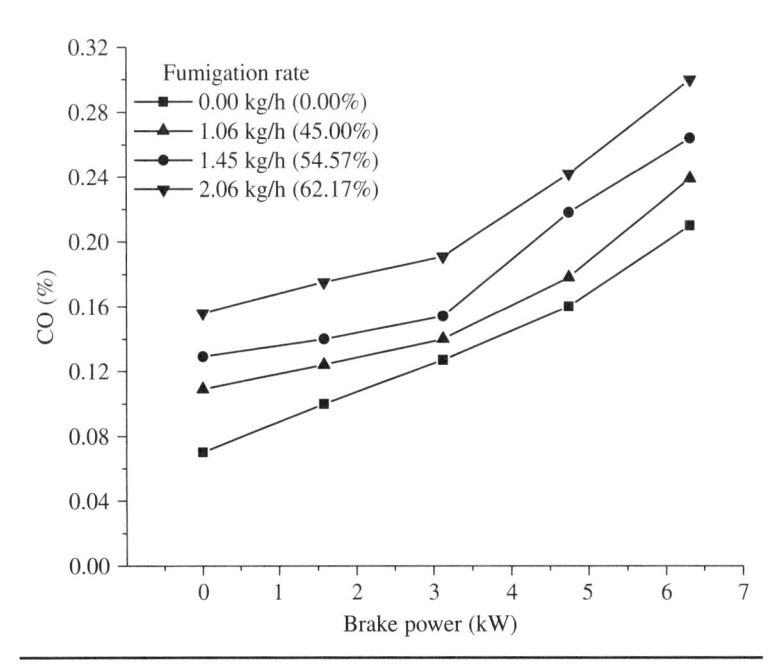

FIGURE 8.12 Effect of fumigation on CO emissions at various brake powers at a speed of 800 rpm.

receiving an especially rich charge while others are lean. Because of poor cylinder-to-cylinder distribution, this one cylinder might be producing high CO emissions while others are low. It is likely that at least a portion of the fumigated ethanol enters the cylinder as a liquid, and because the intake manifold of the diesel engine is not generally designed for good distribution on liquid fuels, poor cylinder-to-cylinder distribution and a high spatial variation in the fuel-to-air ratio within the cylinder are strong possibilities for an increased CO emission level. Therefore, increases in CO emission levels with increasing fumigation rates are a result of incomplete combustion of an air–ethanol mixture. It is observed that at an ethanol fumigation rate of 1.06 kg/h and at no load, the CO emission level is increased by 0.104, 0.03, 0.012, and 0.028 percent at engine speeds of 800, 900, 1000, and 1100 rpm, respectively, when compared to the emission under diesel fuel operation.

8.2.7 Effect of Brake Power on Smoke Number at Various Ethanol Fumigation Rates

Observe in Fig. 8.13 that the smoke level rises from no load to full load. The smoke level is more or less constant as there is always excess air present. However, in a higher load range there is an abrupt rise in smoke level due to decreased available oxygen. A rich fuel-air mixture

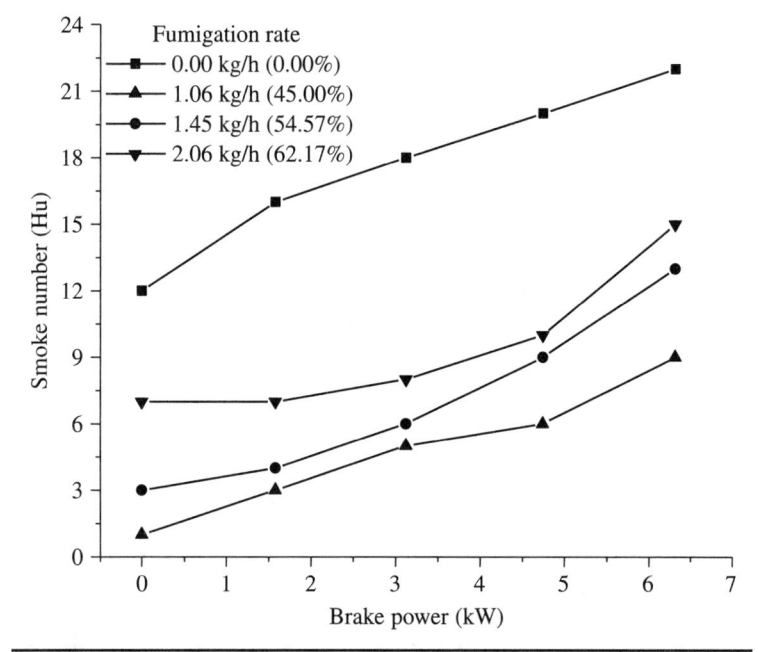

Figure 8.13 Effect of fumigation on smoke emission at various brake powers at a speed of 800 rpm.

results in higher smoke because there is a lesser amount of oxygen available. Hence, overloading the engine will result in a very black smoke. It is observed that smoke level decreases with ethanol fumigation. Fumigation of ethanol starts precombustion reactions before and during a compression stroke, resulting in reduced chemical delay because the intermediate products such as peroxides and aldehydes react more rapidly with oxygen than with original hydrocarbons. This shortening of the delay period curbs thermal cracking, which is responsible for soot formation that results in a reduction of smoke. It is observed that at an engine speed 800 rpm and at no load, the smoke level is decreased by 91.66, 75, and 41.59 percent at ethanol fumigation rates of 1.06, 1.45, and 2.06 kg/h (45.00, 54.57, and 62.17 percent diesel substitution), respectively, when compared to the smoke level under diesel fuel operation. At a high load of 6.31 kW, the smoke level is decreased by 68.18, 50.10, and 40.90 percent at ethanol fumigation rates of 1.06, 1.45, and 2.06 kg/h, respectively, when compared to the smoke level under diesel fuel operation.

8.2.8 Effect of Brake Power on Exhaust Gas Temperatures at Various Ethanol Fumigation Rates

We can see in Fig. 8.14 that exhaust gas temperature increases with an increase in brake power at constant speed. The temperature inside the combustion chamber increases while brake power is increased,

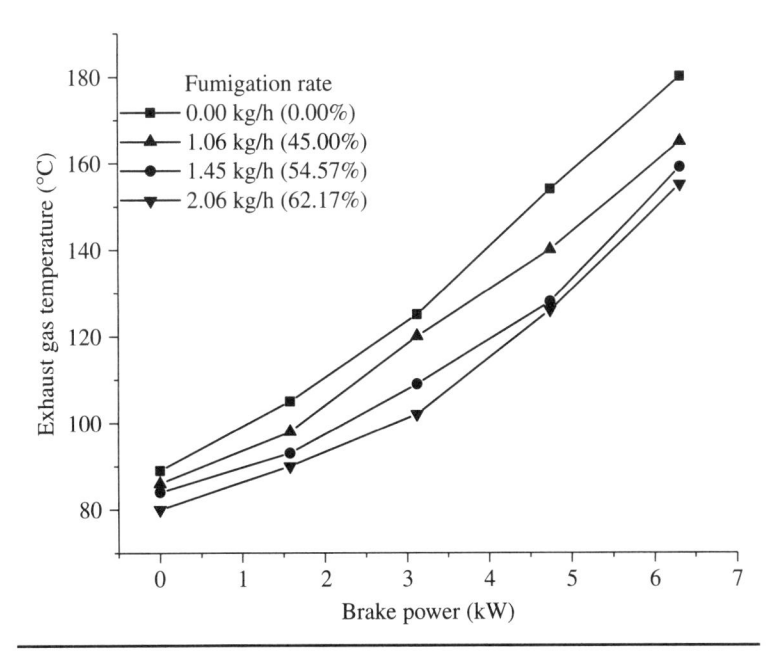

FIGURE 8.14 Effect of fumigation on exhaust gas temperatures at various brake powers at a speed of 800 rpm.

which results in an increase in the exhaust gas temperature. Again, the exhaust gas temperature decreases with an increase in the ethanol fumigation rate at constant engine speed. The latent heat of vaporization of ethanol cools the intake charge sufficiently to reduce the peak combustion temperature, which reduces the exhaust gas temperature. It is observed that at an engine speed of 800 rpm and at no load, the exhaust gas temperature is reduced by 3.37, 4.50, and 12.35 percent at ethanol fumigation rates of 1.06, 1.45, and 2.06 kg/h (45, 54.57, and 62.17 percent diesel substitution), respectively, when compared to the exhaust gas temperature under diesel fuel operation. At a high load of 6.31 kW, the exhaust gas temperature is reduced by 11.09, 14.01, and 16.67 percent at ethanol fumigation rates of 1.06, 1.45, and 2.06 kg/h, respectively, when compared to the exhaust gas temperature under diesel fuel operation. It is observed that at ethanol fumigation rate of 1.06 kg/h engine performs satisfactorily over a range of speeds such as 800, 900, 1000, and 1100 rpm. Hence, the fumigation rate of 1.06 kg/h (45 percent diesel substitution) is selected as the optimum.

Because ethanol is not a compression ignition engine fuel, ethanol is fumigated into the airstream in the intake manifold to improve the ignition quality of the fuel. The brake thermal efficiency, BSFC, NO_x emission level, exhaust gas temperature, and smoke level decreases with ethanol fumigation at constant engine speed. Diesel substitution, energy replacement, and CO emission level increase with an increase in the ethanol fumigation rate at constant engine speed. The fumigation rate of 1.06 kg/h (45 percent diesel substitution) is found to be optimum for good engine performance. Ethanol fumigation in diesels has the advantages for the NO_x and smoke level reduction, which affect good environmental impact. The increase in CO emission level with ethanol fumigation is in a safe range.

References

1. Energy Efficiency and Renewable Energy, U.S. Department of Energy. 2005. Multi Year Program Plan 2007–2012, Office of the Biomass Program. See also: *Biofuels Refining and Performances* by A. Nag; McGraw Hill Publishers, USA, 70–73, 2008.
2. Zhengming, Z. 1999. Renewable energy development in China: The potential and the challenges. Beijing, China: China Sustainable Energy Program, Center for Resource Solutions. See also: http://www.worldbiofuelssymposium.com/Renewable-Energy-Dev-in-China.pdfS. Accessed May 23, 2008.
3. Oko-Institut e.V. 2004. Bioenergy, new growth for Germany. Berlin, Germany: Institute for Applied Ecology-Federal Ministry for the Environment, Nature Conservation and Nuclear Safety.
4. Worgettera, M. 2006. Bioenergy in Austria: Potential, strategies, success stories. Austrian Biomass Association (BLT). See also: http://www.blt.bmlf.gv.at/vero/veroeff/0783_Bioenergy_in_Austria_e.pdfS. Accessed May 23, 2008.
5. Leppiman, A. 2005. Current status of bioenergy and future prospects in Baltic Sea Area. BASREC Bioenergy Working Group 2003–2005, Nordic Bioenergy Conference, Trondheim, Norway.

6. Kumar, A., Bhattacharya, S. C., and Pham, H. L. 2005. Greenhouse gas mitigation potential of biomass energy technologies in Vietnam using the long range energy alternative planning system model. *Energy* 28(7):627–654.
7. Islasa, J., Manzinia, F., and Masera, O. 2008. A prospective study of bioenergy use in Mexico. *Energy* 32:2306–2320.
8. Dorado, M. P., Ballesteros, E., Gomez, J. M., and Lopez, F. J. 2003. Exhaust emissions from a diesel engine fueled with transesterified waste olive oil. *Fuel* 82:1311–1315.
9. Senda, J., Okui, N., Tsukamoto, T., and Fujimoto, H. 2004. On-board measurement of engine performance and emissions in diesel vehicle operated with bio-diesel fuel. Paper published by the Society of Automotive Engineers International, 2004-01-0083.
10. Dorado, M. P., Ballesteros, E., Arnal, J. M., Gomez, J., and Gimenez, F. J. L. 2003. Testing waste olive oil methyl ester as a fuel in a diesel engine. *Energy Fuels* 17:1560–1565.
11. Lang, X., Dalai, A. K., Bakhshi, N. N., Reaney, M. J., Hertz, P. B. 2001. Preparation and characterization of bio-diesels from various bio-oils. *Bioresource Technology* 80:53–62.
12. Antolin, G., Tinaut, F. V., Briceno, Y., Castano, V., Perez, C., and Ramirez, A. L. 2002. Optimization of biodiesel production by sunflower oil transesterification. *Bioresource Technology* 83:111–4.
13. Vicente, G., Martinez, M., and Aracil, J. 2004. Integrated biodiesel production: A comparison of different homogeneous catalysis systems. *Bioresource Technology* 92:297–305.
14. Ghadge, S. V. and Raheman, H. 2005. Biodiesel production from mahua (*Madhuca indica*) oil having high free fatty acids. *Biomass Bioenergy* 28:601–605.
15. Meher, L. C., Dharmagadda, V. S. S., and Naik, S. N. 2006. Optimization of alkali catalyzed transesterification of *Pongamia pinnata* oil for production of biodiesel. *Bioresource Technology* 97:1392–1397.
16. Sarin, R., Sharma, M., Sinharay, S., and Malhotra, R. K. 2008. Jatropha-palm biodiesel blends: an optimum mix for Asia. *Fuel* 86:1365–1371.
17. Demirbas, A. 2006. Biodiesel production via non-catalytic SCF method and biodiesel fuel characteristics. *Energy Conservation Management* 47:2271–2282.
18. Nag, A. 2007. *Biofuels Refining and Performance*. New York: McGraw-Hill.
19. Goering, C. E., Schwab, A. W., Campion, R. M., and Pryde, E. H. 1982. Evaluation of soybean oil-aqueous ethanol microemulsion for diesel engines. *Proceedings of the International Conference on Plant and Vegetable Oils as Fuels*, ASABE, Fargo, ND, Aug. 2–4, pp. 279–286.
20. Knothe, G., Matheaus, A. C., and Ryan III, T. W. I. 2003. Cetane numbers of branched and straight-chain fatty esters determined in an ignition quality tester. *Fuel* 82:971–975.
21. Nag, A., Haldar, S., and Ghosh, B. B. 2008. Utilization of Unattended *Putranjiva roxburghii* nonedible oil as fuel in diesel engine. *Renewable Energy* 34:343–347.
22. Rakopoulos, C. D., Antonopoulos, K. A., Rakopoulos, D. C., Hountalas. D. T., and Giakoumis. E. G. 2006. Comparative performance and emissions study of a direct injection diesel engine using blends of diesel fuel with vegetable oils or bio-diesels of various origins. *Energy Conversion Management* 47:3272–3287.
23. Ziejewski, M. and Kaufman, K. R. 1984. Dynamometer evaluation and engine wear characteristics of palm oil diesel emulsions *Journal of the American Oil Chemistry Society* 61(10):1620.
24. Raheman, H. and Phadatare, A. G. 2004. Diesel engine emissions and performance from blends of karanja methyl ester and diesel. *Biomass and Bioenergy* 27:393–397.
25. Mathur, H. B. and Das, L. M. 1985. Utilization of nonedible wild oil as diesel engine fuel. *Proceedings of the Bioenergy Society*, Hyderabad, India, November 23–26, 1982.
26. Szybist, J. P., Boehman, A. L., Taylor, J. D., and McCormick, R. L. 2005. Evaluation of formulation strategies to eliminate the biodiesel NO$_x$ effect. *Fuel Processing Technology* 86:1109–1126.

27. Ban-Weiss, A. G., Chen, J. Y., Buchholz, B. A., and Dibble, R. W. 2008. A numerical investigation into the anomalous slight NO_x increase when burning biodiesel: A new (old) theory. *Fuel Processing Technology* 88:659–667.
28. Walker, K. 1994. Biodiesel from rapeseed. *Journal of the Royal Agricultural Society of England* 155:43–44.
29. Ecklund, E. E., Bechtold, R. L., Timbario, T. J., and Mc Callum, P. W. 1984. State-of-the art report on the use of alcohols in diesel engines. SAE Paper Series 840118, Warrendale, PA.
30. Chaplin, J. and Janius, R. B. 1987. Ethanol fumigation of a compression ignition engine using advanced injection of diesel fuel. *Transactions of the ASABE* 30(3): 610–614.
31. Cruz, J., Alan, R. C., and Watson, H. D. 1988. Dual fuelling turbo charged diesel with ethanol. *Transactions of the ASABE* 25(5):1163–1168.
32. Goering, C. E., Crowell, T. J., Griffith, D. R., and Savage, L. D. 1992. Compression ignition flexible fuel engine. *Transactions of the ASABE* 25(2):423–428.
33. Hayes, T. K., Savage, L. D., and White, R. A. 1982. The effect of fumigation of different ethanol proofs on a turbo charged diesel engine. SAE Paper Series 880497, Warrendale, PA.
34. Shropshire, G. J. and Goering, C. E. 1982. Ethanol injection into a diesel engine. *Transactions of the ASABE* 25(3):570–575.
35. Jiang, Q., Ottikkutti, P., VanGerpen, J., and VanMeter, D. 1990. The effect of alcohol fumigation on diesel flame temperature and emissions. SAE Paper Series 900386, Warrendale, PA.

CHAPTER 9

Bioseparation Processes

Ahindra Nag
Department of Chemistry
Indian Institute of Technology
Karagpur, India

9.1 Introduction

Bioseparation processes involve recovery, isolation, purification, and polishing of products synthesized by biotechnological processes. Bioseparation processes are necessary for the following reasons:

1. Reduction in bulk and removal of specific impurities
2. Enrichment of target product
3. Enhancement of product stability
4. Achievement of product specifications
5. Prevention of product degradation
6. Prevention of catalysis other than the type desired and also prevention of catalyst poisoning

Bioseparation is largely based on *chemical separation processes* and is different from chemical separation as follows:

1. Bioseparation involves large numbers of low-volume products than chemical processes. For example, monoclonal antibodies are typically present in concentrations of approximately 0.1 mg/mL in mammalian cell culture supernatants.
    Hence, large volumes of dilute product streams have to be processed for obtaining even modest amounts of pure product.
2. Product from bioseparation process should be highly pure as for prophylactic, diagnostic, and therapeutic purposes, both

in terms of active product content as well as in terms of the absence of specific impurities. Injectable therapeutic products should be free from endotoxins and pyrogens.

3. Bioseparation processes involved in biotechnology industries tend to depend on off-the-shelf equipment. Bioseparation is frequently based on multitechnique separation.

4. Many biological products are thermolabile, and hence many bioseparation techniques are usually carried out at subambient temperatures and avoid extremes of physicochemical conditions such as pH and ionic strengths, hydrodynamic conditions such as high shear rates, and exposure to gas–liquid interfaces. Organic solvents, which are widely used in chemical separations, have relatively limited usage in bioseparation because of their tendency to promote degradation of many biological products.

The main disadvantages of bioseparation processes are as follows:

1. High capital cost
2. High operations cost
3. Lower product recovery
4. Requires highly technical scientists (knowledge of both biology and engineering)
5. Requires special safety protection

9.2 Different Stages of Bioseparation Process

Bioseparation processes depend on the following stages:

1. Reduction in bulk or concentration enrichment → Evaporation or distillation in vacuum or extraction
2. Remove insoluble impurities → Filtration or centrifugation
3. Isolation of the product → Extraction or adsorption
4. Purify the product → Chromatographic technique
5. Polish → Drying or crystallization designed to remove solvent, such as water or trace amount of impurities

After final polishing (step 5), required steps are effluent treatment produced in the different stages of bioseparation process.

Upstream Processing → Bio reaction → Downstream Processing → Bio products / Impurities

A good bioseparation process should have the following properties:

1. Desired purity of the product
2. Stability of the product
3. Low cost
4. Reproducibility and scalability
5. Meet regulatory guidelines

Bioseparation processes have in common many familiar chemical engineering unit operations (see Table 9.1). For example, aerobic fermentations involve mixing three heterogeneous phases—microorganism, medium, and air. In the manufacture of antibiotics, mass transfer (e.g., extraction, adsorption, and drying), heat transfer (such as evaporation, drying, and crystallization), and other mechanical operations (e.g., cell rupture, settling thickening, filtration, and centrifuging) all play a vital role. The investment on these operations is often claimed to be about 4 times greater than that for the ferment or vessels and their auxiliary equipment. Often, as much as 60 percent of the fixed costs of fermentation are attributable to the recovery stage in organic acid and amino acid production.

Product	Nature of Bioseparation Required
Alcoholic beverages: beer, wine, spirits	Clarification, distillation
Vitamins: vitamin C, vitamin B_{12}, riboflavin	Precipitation, filtration, adsorption, solvent extraction
Amino acids: lysine, glycine, phenylalanine	Precipitation, filtration, adsorption, solvent extraction
Antibiotics: penicillin, neomycin, bacitracin	Precipitation, filtration, adsorption, solvent extraction
Proteins: food and food additives, nutraceuticals, industrial enzymes, hormones, pharmaceutical enzymes, plasma-derived products, monoclonal antibodies, growth factors Clotting factors: thrombolytics rDNA-derived proteins Diagnostic proteins Vaccines	Filtration, precipitation, centrifugation, adsorption, chromatography, membrane-based separations

TABLE 9.1 Some Methods to Separate a Product by Bioseparation

9.3 Unit Operations in Bioseparation

Various unit operations are involved in bioseparation: (1) disruption of cells, (2) centrifuging, (3) thickening, (4) flocculation, (5) filtration, (6) evaporation, (7) drying and crystallization, (8) chromatographic techniques, and (9) membrane techniques.

9.3.1 Disruption of Cells

The content of microbial cells has high osmotic pressure and is constrained by a fragile, semipermeable membrane, which is protected from rupture by a strong, rigid outer cell wall. Microbial cells are far harder to break than most animal or plant cells. Many cell-breaking methods have been developed, but the cheapest and most effective ones should be selected. Cell rupture can be done using the following methods: (1) mechanical, (2) liquid homogenization, (3) sonication, and (4) other methods.

Mechanical Methods

Breakage by Impact and Shear Machines that can be used for cell rupture include a stamp mill, a mortar and pestle (with powdered glass, sand, or alumina), a colloid mill (Fig. 9.1), and a homogenizer, for example. The colloid mill and pan crusher are described as follows.

Colloid mills are used when there is to be very little breakdown of individual particles and when the task is to disrupt lightly bonded clusters or agglomerates. Dispersion of the material might be important at times, as in emulsion formation, which is basically a two-fluid system (e.g., syrups, milk, ointments, and creams). A special class of mill is used for dispersion and colloidal operations, which work on the principle of high-speed fluid shear (Fig. 9.1). Although colloid

FIGURE 9.1
A colloid mill.

mills are classified as grinders, they do not do much actual grinding. Their value lies in ensuring a breakdown of agglomerates or, in the case of emulsions, shearing the fluid phases to produce droplets of fine size, approximately 1 µm.

Colloid mills, which are used for dispersion or emulsification, fall into four main groups: hammer or turbine, smooth–surface disk, rough-surface type, and valve orifice device. The principle of operation lies in creating a high-velocity fluid stream with very great shear forces existing within the fluid, which acts to disrupt the particles. Chemical aid in the form of dispersing agents is sometimes used. Because of heat generation resulting from friction, a cooling water jacket is provided in many systems.

A *pan crusher* is used not only for crushing but also for mixing liquids and solids. A pan crusher consists of one or more grinding wheels or mullers revolving in a pan; the pan may be driven, whereas the mullers revolve by friction. Mullers are made of tough alloys such as hard nickel, iron scrapers, or plows. The material is properly fed under the mullers. Chambers of dry pan use air cylinders to regulate the grinding pressure under each of the muller tires from 7500 to 20,000 lb. The bottom of the pan rotates and has a central solid-crushing ring as well as an outer ring of screen plates with openings from 0.16 to 1.3 cm. In some instances, a solid pan bottom is used in place of a perforated screen bottom, and the ground material is discharged through a slot in the rim (known as a rim-discharge grinder). It will give greater throughput with wetter materials but will require a greater screening area and a higher circulating load.

A dry pan is useful for crushing medium-hard or soft materials such as clays, shales, cinders, and soft minerals (e.g., barite). The feed size should be 7.5 cm or less and should be a product of a particular feed size. Finer products can be obtained by operating the pan in a closed circuit (a circulating load of 75 percent is common) with a vibrating screen. A high reduction ratio with low power and low maintenance are some of the features of pan crushers (Fig. 9.2). Typical dimensions of a dry pan are as follows:

Pan diameter: 1.8 to 3 m
Muller diameter: 0.71 to 1.6 m
Muller width: 13 to 46 cm
Power: 15 to 100 hp or 1 to 5 hp/ton of product

The production rate varies from 1 to 6 tons/h depending on the pan size and hardness of the material as well as fineness of the feed product. A wet pan is used for developing plasticity or molding qualities in ceramic feed materials. The abrasive and kneading action of the mullers blend finer particles with coarser ones during the crushing operation.

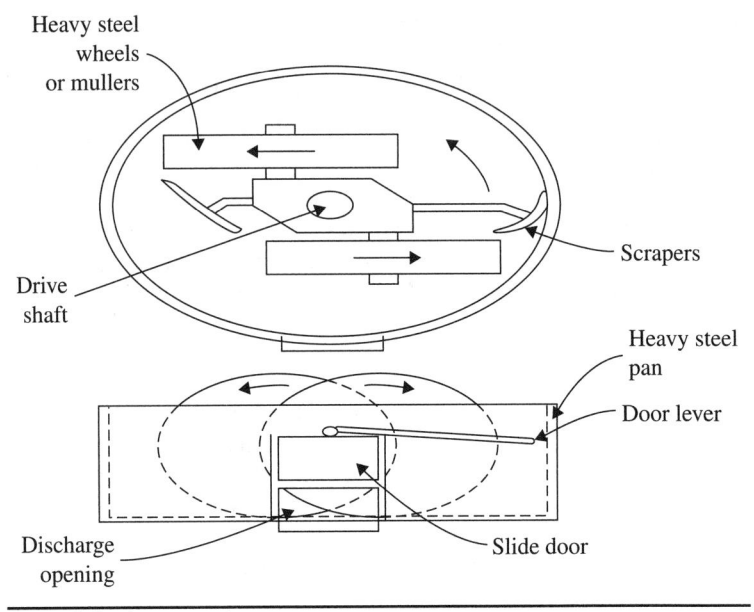

FIGURE 9.2 A pan crusher.

Cell Rupture by Pressure There are several varieties of laboratory presses to disrupt cells; the Hughes press and the French press are especially successful. The original Hughes press consisted of a split block with a half cylinder hollowed in each face. The frozen cell paste (with or without an abrasive) was placed in the hollow. The block was clamped together, and a plunger driven by a fly press forced the frozen paste from the cylinder into channels cut in the block. The frozen disrupted cells were then scraped from the block. With the French press, a hollow cylinder in a stainless steel block is filled with cell paste and subjected to high pressure. The cylinder has a needle valve at the base, and the cells burst as they are extruded through the valve to atmospheric pressure.

Two continuous presses have been scaled up by Duerre and Ribi, which are similar in principle to the French press. Duerre and Ribi found that with *Escherichia coli*, most cells were ruptured at 1000 kg/cm^2 and cells were broken at 1700 kg/cm^2 with little loss of enzyme activity. However, at 2400 to 3800 kg/cm^2, the stability of some enzymes was affected; arginine, decarboxylase, and formic hydrogen lyase were destroyed. Formic oxidase was partially inactivated, whereas lysine decarboxylase and glucose oxidase remained active.

Another mechanical press developed by Lars Edibo, Sweden, in which frozen cells are forced to and fro through a small hole in a disc between two cylinders at low temperature and pressure. Disruption of the cells is probably caused by deformation of organisms embedded in the ice. The ice undergoes transformations in its crystal structure when it passes through the hole because of the temperature and pressure change. Heavy pressures (spring-loaded piston), which can apply pressure above 2100 kg/cm^2, are needed to disrupt heavy cell pastes.

Liquid Homogenization

Liquid-based homogenization is the most widely used cell-disruption technique for small volumes and cultured cells. Cells are lysed by forcing the cell or tissue suspension through a narrow space, thereby shearing the cell membranes. Three different types of homogenizers are in common use:

1. The Dounce homogenizer consists of a round glass pestle that is manually driven into a glass tube.
2. The Potter–Elvehjem homogenizer consists of a manually or mechanically driven Teflon pestle shaped to fit a rounded or conical vessel. The number of strokes and the speed at which the strokes are administered influences the effectiveness of Dounce and Potter–Elvehjem homogenization methods. Both homogenizers can be obtained in a variety of sizes to accommodate a range of volumes.
3. A French press consists of a piston that applies high pressure to a sample volume of 40 to 250 mL, forcing it through a tiny hole in the press. Only two passes are required for efficient lysis due to the high pressures used with this process. The equipment is expensive, but the French press is often the method of choice for breaking bacterial cells mechanically.

Ultrasonic Vibration

Ultrasonic waves of about 20 kg/s that are applied to bacterial suspensions disrupt both the cell wall and the cell membrane; rod-shaped are more readily broken than cocci, and gram-negative cells disintegrate more readily than gram-positive ones. This method is not as successful for breaking fungal cells.

Ultrasonic waves in a liquid cause fluctuation in pressure, forming bubbles, which grow to about 10 µm, begin to oscillate, and then implode violently generating shock waves of several thousand atmospheres and localized high temperatures. Free radicals are also generated in the solution. It is believed that rapid oscillations of the bubbles are responsible for cell rupture, rather than damage from shock waves or free radicals. Nonetheless, free radicals are important because they damage enzymes released into solution. An electronic generator is used to produce ultrasonic waves of 20 kc/s. A transducer converts the waves to a titanium probe immersed in the cell suspension, cooled in ice. This method seems to have possible industrial applications, but, in practice, there are difficulties in dissipating the heat generated;

the free radicals tend to denature sensitive enzymes, and the cell debris is extensively fragmented.

Other Methods
The bacterial cell membrane can be broken by physical stress such as violent depression, osmotic shock, or rupture by freezing. Lytic agents (cationic and anionic detergents, alkalis, bile salts, and solvents) also have restricted use.

The freeze/thaw method is commonly used to lyse bacterial and mammalian cells. This technique involves freezing a cell suspension in a dry ice/ethanol bath or freezer and then thawing the material at room temperature, or 37°C. This method of lysis causes cells to swell and ultimately break because ice crystals form during the freezing process and contract during thawing. Multiple cycles are necessary for efficient lysis, and the process can be quite lengthy. However, freeze/thaw has been shown to effectively release recombinant proteins located in the cytoplasm of bacteria and is recommended for the lysis of mammalian cells in some protocols. Bacterial cells can also be destroyed by enzymes.

9.3.2 Centrifugal Separation

Solids that form a porous cake can be separated from liquids in a centrifuge. A centrifuge uses centrifugal force to separate liquids from solids. It is, essentially, a development of gravity filtering wherein the force acting on the liquid is enormously increased by the use of centrifugal force instead of being restricted to gravity. The slurry is fed to a rotating basket having a slotted or perforated wall covered with a filter medium such as canvas or metal cloth. Pressure, resulting from the centrifugal action, forces the liquid through the filter medium leaving the solids behind. If the feed to the basket is then shut off and the cake of the solids are spun for a short time, much of the residual liquid in the cake drains off the particles, leaving the solids much drier then those from a filter press or vacuum filter.

The major types of filtering centrifugals are suspended batch machines, which are discontinuous in their operation, automatic short-cycle batch machines, and continuous conveyor centrifuges. In suspended centrifugals, the filter media are canvas or other fabric or woven metal cloth. The baskets range (see Fig. 9.3) from 30 to 48 in in diameter and are 18 to 30 in deep, operating at speeds between 600 and 1800 rpm. It may be underdriven or top driven. The solids form a cake 2 to 6 in thick inside the basket.

Immiscible liquids, like oil–water emulsions, can also be separated by special types of centrifuge such as the Sharples Super Centrifuge (Separation Equipment Sales, Inc., Albertson, New York), which operates at about 30,000 rpm.

Bioseparation Processes 289

FIGURE 9.3 Basket-style centrifuge.

Principle of Separation

The centrifugal force acting on a mass of w is as follows:

$$F = 0.000341 W r N^2 \qquad (9.1)$$

where W = weight of basket
 r = inside radius of basket
 N = rpm of basket

The power required by a centrifugal machine is

$$hp = T N_m / 5250 \qquad (9.2)$$

where T is the torque (ft-lb) and N_m is the maximum speed of basket (rpm).

The torque T is equal to

$$T = W R^2 N / 307 t \qquad (9.3)$$

where R = radius of gyration of the basket, which may be taken as 77 percent of the basket radius
 N = 90 percent of the maximum running speed, rpm
 T = shell thickness

The shape of the liquid surface is parabolic in nature and given by

$$y = \frac{\omega^2}{2g} x^2 + y_0 \qquad (9.4)$$

where ω is angular velocity (rad/s) and x, y, and y_0 are characteristics of a point on the parabola.

The volume of a liquid in a basket is

$$V' = \Pi \left(r^2 H \frac{\omega^2 b_0^4}{4g} \right) \qquad (9.5)$$

where H = depth of liquid
b_0 = radius of the liner surface of the liquid at the top of the basket
g = gravitational force

The fluid pressure varies at different points along the wall. However, the centrifugal pressure on the wall of the basket is

$$P_c = \frac{1}{2}S_1\omega^2(r^2 - x^2) \tag{9.6}$$

where S_1 is specific gravity of liquid and x is any distance from the axis.

The stress on the wall of the basket is given by

$$F = \frac{r}{T}(P_c + S_2T, r\omega^2) \tag{9.7}$$

where P_c is fluid pressure (lb/ft^2) and S_2 is density of material of which the basket is made.

9.3.3 Thickening

In thickening, the emphasis is on the solids as the end product of separation—the compacting of suspended solids into dense slurry or sludge in order to facilitate subsequent processing or disposal of solids.

Various settlers and thickeners are used in industrial operations, such as Dorr thickener (batch as well as continuous settling characteristics of the sludge) (Fig. 9.4). Laboratory tests are usually conducted, and the data obtained can be processed to determine the diameter of the thickener. Steps in the design are listed as follows:

1. Determination of compression point
2. Compression depth
3. Classifier area (from overflow data and zone tests)
4. Classifier depth

Total depth would comprise the clarification zone, the compression zone, and ineffective depth due to sloping of rakes and feed well. Normally, mild steel, concrete, or wood is used. For corrosive solutions, stainless, lead-lined, and rubber–covered steels may be used. Considerations may be given to the following:

1. Floor space available
2. Conservation of heat
3. Amount of underflow solids
4. Periodic overloads and corrosive material

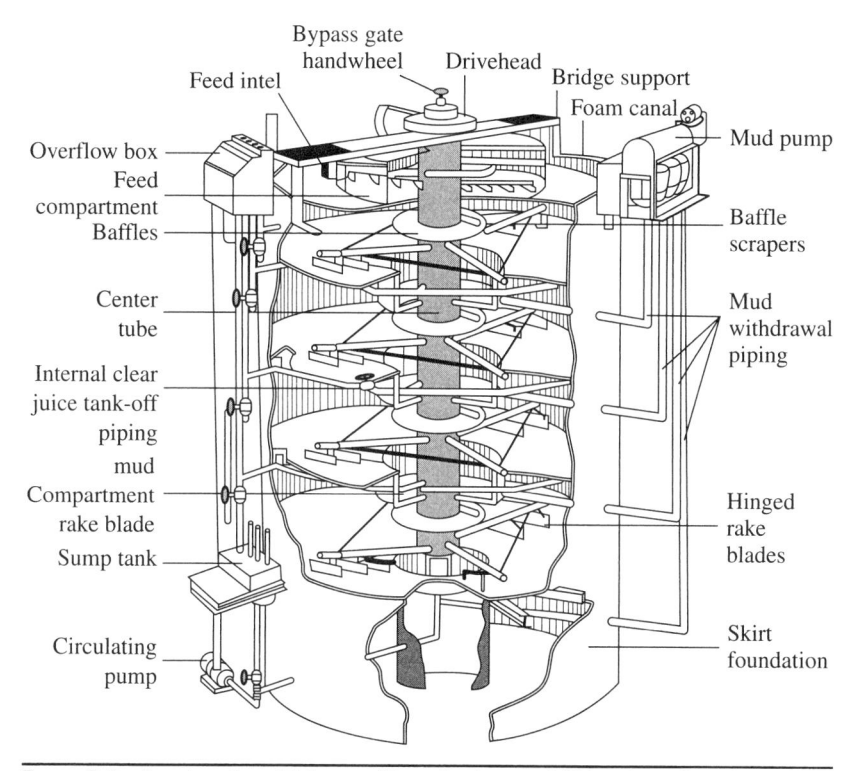

FIGURE 9.4 Four-tray Dorr thickener. (*From Coulson and Richardson*, Chemical Engineering.)

Ancillary materials are as follows:

1. Pump, diaphragm
2. Density control
3. Underflow line features

Other costs include cost of materials, erection, motor horsepower, and rake speed (power cost).

Evidently, we need the densities and sizes of microbial cells to calculate the settling velocity. To recover from suspensions, flocculation of the cells is advisable as a pretreatment to increase the rate of separation.

9.3.4 Flocculation

Colloidal particles and slimes have very low settling velocities and large-diameter thickeners. Also, jamming of the thickeners can cause serious problems. For this purpose, selective aggregation is often practiced (flocculation). Several flocculating agents, among them inorganic (aluminum, iron, and silicon derivatives), organic (such as polymers), and natural products (such as starch and tannin) are used in industry (e.g., clarification of municipal water, processing of uranium ore, and black liquor treatment in paper industry).

9.3.5 Filtration

Filtration is the removal of solid particles from fluid by passing the fluid through a filtering medium, or septum, on which the solids are deposited. Advantages and disadvantages of filter press systems are compared with other dewatering processes in table 9.2.

Advantages	Disadvantages
High solids content cake	Large quantities of inorganic conditioning chemicals are commonly used
Can dewater hard-to-dewater sludges	Very high chemical conditioning dosages or thermal conditioning may be required for hard-to-dewater sludge
Very high solids capture	High capital cost, especially for variable-volume filter presses
Only mechanical device capable of producing a cake dry enough to meet landfill requirements in some locations	Labor cost may be high if sludge is poorly conditioned and if press is not automatic
	Replacement of the media is both expensive and time consuming
	Noise levels caused by feed pumps can be very high
	Requires grinder or prescreening equipment on the feed
	Acid washing requirements to remove calcified deposits caused by lime conditioning may be frequent and time consuming
	Batch discharge after each cycle requires detailed consideration of ways of receiving and storing cake or of converting it to a continuous stream for delivery to an ultimate disposal method

Source: EPA 1987: EPA (1987, pp. 114 through 117). Summarizes performance results and operating and maintenance problems from a survey of 50 filter press wastewater applications.

TABLE 9.2 Advantages and Disadvantages of Filter Press Systems Compared with Other Dewatering Processes

Bioseparation Processes

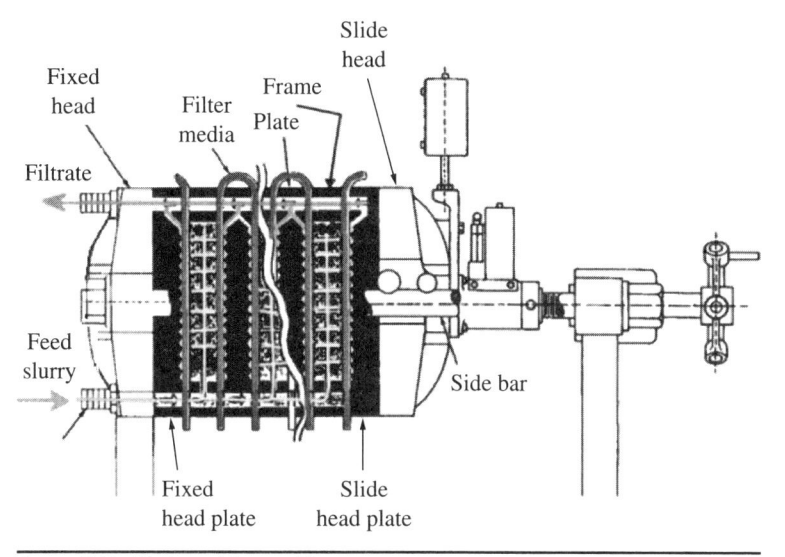

FIGURE 9.5 Plate and frame filter.

Industrial filtrations range from simple straining to highly complex separation. The solid particles may be coarse or fine, rigid or plastic, round or elongated, separate individuals or aggregates. The feed suspension may carry a heavy load of solids or almost none. The fluid may be very hot or very cold and can operate in a vacuum or under high pressure. Sometimes, the fluid is the valuable phase, the solid, or sometimes both. As a result, many filters have been developed to meet specific problems. Some important pressure filters are the plate-and frame-filter presses (Fig. 9.5), the Kelly filter, the Sweetland filter, and the Vallez filter. There are mainly two types of suction—the disc filter and the drum filter. Pressure filters have flexibility in operation and can give a very clear filtrate. The suction filter, on the other hand, gives a muddy filtrate and is also limited by the high vacuum that can be created. The major advantage of a suction filter is that it is continuous in operation.

Theory of Filtration

Filtration is primarily a problem of fluid flow. The generally accepted view is that under conditions of filtration, the flow of filtrate through the porous filter cake and filter medium is laminar. Therefore, Poiseuille's law holds well, the filtration rate is directly proportional to the applied pressure. Assuming laminar flow through filter channels, the filtration rate equation can be written as

$$U = \frac{dV}{Ad\theta} = \frac{Pg_c}{\mu(R_c + R_f)} \tag{9.8}$$

where A = filtration area, ft^2
G_c = conversion factor, 32.17 (lb mass) (ft)/(lb force) (s^2)
P = filtration pressure (i.e., pressure drop of filtrate through the filter), lb force/ft^2
R_c = filter cake resistance, L/ft
R_f = initial filter stance (resistance of filter medium and filter channels), L/ft
U = the filtration rate, ft/s
$dV/d\theta$ = filtration rate, ft^3/s filtrate flow
U = viscosity of filtrate (lb mass)/(ft) (s)
e = filtration time, s

The filtrate cake resistance can be written as

$$R_c = \frac{\alpha w v}{A} \tag{9.9}$$

where w is pounds of solids deposited per cubic foot of filtrate and A is proportionality constant, also known as specific resistance [ft/(lb mass)].

Substituting the value of R_c in Eq. (9.8), we can write

$$\frac{dV}{d\theta} = \frac{A^2 P g_c}{\mu(\alpha w v + A R_f)} \tag{9.10}$$

Considering R_f as being equivalent to the resistance of a fictitious layer of filter cake of equal resistance, Eq. (9.10) can be written as

$$\frac{dV}{d\theta} = \frac{A^2 \rho g_c}{\alpha \mu w (V + V_0)} \tag{9.11}$$

where V_0 is the volume of filtrate required to form a filter cake of resistance equal to the initial filter resistance R_f. The time required to filter V_0 ft^3 of filtrate will be θ_0.

Filtration at Constant Pressure
During constant pressure filtration, the filtrate flow rate goes on decreasing with time. Integrating Eq. (9.10), we can write

$$\frac{\theta}{V} = \frac{\alpha \mu w}{2A^2 P G_c} + \frac{\mu R_f}{A P G_c} = K_1 V + C_1 \tag{9.12}$$

Similarly, by integrating Eq. (9.11), we get

$$(v + V_0)^2 \frac{2 A^2 P G_c}{\alpha \mu w}(\theta + \theta_0) = K(\theta + \theta_0) \tag{9.13}$$

Equation (9.13) is the same as Eq. (9.12). Expanding Eq. (9.13) we get

$$V^2 + 2VV_0 + V_0^2 = K\theta + K\theta_0 \quad (9.14)$$

Dividing by KV, and noting that for $V = 0$, $V_0^2 = K\theta_0$, and $K = 1/K_1$, we obtain Eq. (9.12).

Filtration at a Constant Rate

During constant rate filtration, pressure increases with an increase in cake thickness. Therefore, filtration variables are pressure, filtrate volume, or pressure and filtration time, and so on. The relationships are as follows [arranging Eq. (9.10)]:

$$P = \left[\frac{\alpha\mu w}{A^2 G_c}\left(\frac{dV}{d\theta}\right)\right]v + \frac{\mu R_f}{AG_c}\left(\frac{dV}{d\theta}\right) = K_2 V \quad (9.15)$$

and, in terms of P and θ as variables,

$$P = \left[\frac{\alpha\mu w}{A^2 G_c}\left(\frac{dV^2}{d\theta}\right)\right]K_3\theta + \frac{\mu R_f}{aG_c}\left(\frac{dV}{d\theta}\right) = K_3\theta \quad (9.16)$$

Washing Rate

In most industrial operations, it is required to wash the cake (which may have water, acid, or alkali in it). Because of its configuration, the washing rate in a plate-and-frame filter press becomes one-fourth of the filtration rate. However, in a leaf filter, the washing rate equals the filtration rate.

Filtration Cycle

Industrial filters, especially if they are discontinuous in operation, require some time for dumping plates and frames, cleaning the cake, and reassembling. Consequently, a filtration cycle should consist of filtration time, washing time (if necessary), and time required for cleaning and dumping. If C is filtration cycle, where as θ_F is filtration time, θ_W is the watching time, θ_d is cleaning and dumping, and the volume of filtrate collected in a given cycle is V, then

$$C = \frac{V}{\theta_F + \theta_w + \theta_d} \quad (9.17)$$

It can be shown that for a nonwashing plate-and-frame filter press, the optimum time of filtration equals the time required for cleaning and dumping.

Pretreating Cells to Alleviate Filtration Resistance

It is well-known that broth cultures of actinomycetes and streptomyces, for example, exhibit tremendous resistance to filtration. Many attempts have been made to overcome the difficulty of separating the mycelium from broth on an industrial scale. In the recovery of streptomycin, resistance of the mycelium to filtration has been reduced by heating the fermentation broth. Coagulating mycelia protein with heat to accelerate filtration may have wide application in the fermentation industry. For this purpose, a filter of 110-cm^2 area was used. The filter aid was diatomaceous earth, and pressure was applied constantly at 2 kg/cm^2. The sample of broth used was taken from a batch fermenter (60 m^3), which had been operated for 3 to 4 days. The original culture medium consisted of glucose and soybean powder, supplemented by inorganic salts and dried yeast. In the above experiments, coagulation of mycelia protein was apparently achieved after 30 to 40 minutes at 100°C, but a longer exposure of the broth to heat adversely affected the filtration rate, presumably due to disintegration of the coagulated protein.

To summarize, industrial filtration of streptomyces for the recovery of streptomycin via that of the cells was successful with the pH at 3.7 to 4.3 and the temperature at 80 to 90°C for 30 to 60 minutes to raise the broth to the desired temperature using 2 to 3 percent of the filter aid.

9.3.6 Evaporation

The objective of evaporation is to concentrate a solution containing a nonvolatile solute and a volatile solvent. In the overwhelming majority of evaporations, the solvent is water. Evaporation is conducted by vaporizing a portion of solvent to produce a concentrated solution, or thick liquor. The operation falls under the category of heat transfer to boiling liquids.

In evaporation, consideration should be given to the physical properties of the liquid to be concentrated, such as specific gravity, specific heat, thermal conductivity, and the like, solid concentration (initial and final), heat sensitivity of the material (very important in pharmaceuticals manufacture, preserving foodstuffs, etc.), and scale formation, and the like. Proper boiling temperature should be arrived, taking into account the possible elevation of the boiling point of the solution. The material of construction is an important criterion.

Evaporator performance is dependent on various parameters. Thus, the capacity is influenced by the area of heating surface, temperature drop, overall heat transfer coefficient, and heating load. The economy is governed by load, heats of vaporization of steam and liquid, heat of dilution of liquid, amount of superheat in steam, heat loss, and so on. The multiple effects can be operated in various ways: forward feed, backward feed, parallel feed, and mixed feed. Each of these feeds has its own distinctive features. For example, forward feed is very simple in operation and capital cost is less, but heat transfer in the mixed feed is extremely poor. In backward feed, this difficulty can be overcome, no doubt, but more pumps are required for the transfer of solutions.

In modern systems, vapor compression (mechanical as well as thermal) techniques are used wherein energy use is extremely good.

9.3.7 Drying and Crystallization

The term *drying* usually infers the removal of relatively small amounts of water from a solid or nearly solid material to reduce the content of residual liquid to an acceptably low value. Drying is different from evaporation in that the latter is limited to the removal of a large amount of water from solution. In drying processes, the major emphasis is usually on the solid product. In most cases, drying involves the removal of water at temperature below its boiling point, whereas evaporation means removal of water by boiling the material. Another distinction is that in evaporation, water is removed from the material as practically pure water vapors, mixed with other gases only because of unavoidable leaks. In drying, on the other hand, water is usually removed by circulating air or some other gas over the material to carry away the water vapors; in some drying processes, however, on-carrier gas is used.

Methods of Removing Liquid

Liquids may be removed in the following ways:

1. Deposition of water or liquid as ice
2. Decomposition of water, that is, liquids can be drawn off by means of NaCl, $CaCl_2$, and the likes
3. Precipitation as hydrates, for example, $CaCl_2$, P_2O_5, and $CuSO_4$
4. Absorption technique
5. Adsorption, such as drying of air by silica gel
6. Mechanical separation, for example, removing moisture by squeezing and centrifuging
7. Vaporization, air drying, vacuum drying, and so on

Classifying Dryers

Classification of dryers is based on the form of the material being handled. Thus, for

1. *Materials in sheets (or masses) carried through on conveyors or trays*: Examples are atmospheric compartment and vacuum tray (batch) tunnel dryer (continuous).
2. *Granular or loose materials*: Examples are standard rotary, roto-louver (rotary dryers), turbodryers and conveyor dryers, or filter dryer combination.
3. *Material in continuous sheets*: Examples are cylinder dryers and festoon dryers.

4. *Pastes and sludge or caking crystals*: Examples are atmospheric and vacuum dryers (agitator dryer).
5. *Materials in solution*: Examples are atmospheric and vacuum dryers and spray dryers.
6. *Special methods*: Modern techniques are used, such as infrared dryer, fielectric dryer, greeze dryer, pneumatic dryer, fluidized dryer, and the like.

Drying Conditions
Drying conditions can be constant or variable. Constant drying conditions relate to the constancy of temperature, humidity, air velocity, method of exposure of the material, and so on.

Drying Mechanism
The moisture (or solvent) in wet material occurs in two types, bound and unbound moisture. During the drying of material at a specified drying condition, the free moisture gets vaporized and reaches an equilibrium value. To get a bone-dry material, the temperature has to be raised considerably. Such products may absorb equilibrium water again as soon as the material is taken out of the oven, however.

The rate of the drying curve can show different trends. For some materials, the rate is constant throughout. In some cases, there is an initial constant-rate period, which is followed by a falling (the point of intersection gives critical moisture content) rate. Again, there are a few materials for which there are two falling rates (linear and nonlinear). When the surface of the material is sufficiently wet, a constant rate of drying takes place. When diffusion of liquid from internal pores to the surface becomes predominant, heat flows to the insides of the solid mass when evaporation and subsequent diffusion take place.

The drying rate increases with increasing temperature and velocity of air if the slab thickness increases—the drying rate falls off. If the air is sufficiently humid, there is a less concentrated gradient, and so the drying rate can fall off sharply.

Internal Mechanism of Liquid Flow
Internal liquid flow can occur due to the following:

1. Diffusion in continuous homogeneous solids
2. Capillary flow in granular and porous solid
3. Flow caused by shrinkage and pressure gradient
4. Flow caused by gravity
5. Flow caused by the vaporization–condensation process

Crystallization
Crystallization is the formation solid particles within a homogeneous phase. It may occur as the formation of solid particles in a vapor, as in

snow; as solidification from a liquid melt, as in the freezing of water to ice or the manufacture of large single crystals; or as crystallization from liquid solutions. Good yield and high purity are two important objectives in crystallization. Reasonable size uniformity is desirable for washing, filtering, reacting with other chemicals, transporting, and storing the crystals.

There are seven different classes of crystals depending on the angles and lengths of the axes: triclinic, monoclinic, orthorhombic, tetragonal, trigonal, hexagonal, and cubic.

In the formation of a crystal, two steps are required: (1) the birth of a new particle (nucleation) and (2) the crystal's growth to macroscopic size. In a crystallizer, crystal-size distribution is determined by the interaction of the rates of nucleation and growth; the overall process is complicated kinetically. The driving potential for both rates are supersaturated, and neither nucleation nor growth can occur in a saturated or unsaturated solution.

Crystal Formation

Supersaturation may be generated by one or more of the following three methods: (1) if the solubility of the solute increases strongly with an increase in temperature, as in the case with many common inorganic salts and organic substances; (2) if a saturated solution becomes supersaturated by the relative independence of temperature, as in the case of common salt; supersaturation can be done by cooling or evaporation of the solution; (3) if nearly complete precipitation of the product is required, a new solute may be created chemically by adding a third component that will react with the original solute and from an insoluble substance (precipitation).

Types of Crystallizers

There are six types of crystallizers:

1. Tank
2. Scraped surface
3. Double pipe–scraped surface
4. Circulating liquid evaporator
5. Circulating magma vacuum
6. Continuous oscillatory baffle

Tank Crystallizers In this crystallizer, saturated solutions are allowed to cool in open tanks. After a period of time, the mother liquid is drained and the crystals are removed. Nucleation and size of crystals are difficult to control. This process results in very high labor costs.

Scraped Surface Crystallizers In this type of crystallizer, an open trough consists of a slow-speed agitator and a cooling jacket outside.

The blades pass close to the wall and break off any deposits of crystals on the cooled wall. The product generally has wide crystal-size distribution.

Double Pipe-Scraped Surface Crystallizer This type of crystallizer consists of an internal agitator fitted with spring-loaded scrapers that wipe the wall and provide good heat transfer coefficients. Cooling water passes in the annular space. This crystallizer is used in crystallizing ice cream and in plasticizing margarine.

Circulating Liquid Evaporator Crystallizer In this type of crystallizer, the solution is supersaturated by evaporation. The circulating liquid is drawn down by the screw pump inside the tube side of the condensing stream heater. The heated liquid then flows into the vapor space where flash evaporation occurs, resulting in supersaturation. The vapor leaving is condensed. The supersaturated liquid flows down the downflow tube and then up through the bed of fluidized and agitated crystals, which are growing in size. The leaving saturated liquid then goes back as a recycle stream to the heater, where it is joined by the entering fluid. The larger crystals settle out, and a slurry of crystals and mother liquid is withdrawn as a product.

Circulating Magma Vacuum Crystallizer In this type of crystallizer, a steam-jet ejector provides a vacuum. The suspension of crystals is circulated out of the main body through a circulating pipe by a pump. The suspension flows though a heater, where along with the heated liquor it then mix with body slurry. This causes supersaturation in the swirling liquid near the surface, which is deposited in the swirling suspended crystals until they leave again via the circulating pipe. Vapors leave through the top.

Continuous Oscillatory Baffled Crystallizer This crystallizer is a tubular baffled crystallizer that offers plug flow with a superior heat transfer coefficient and also allows controlled cooling profiles. This crystallizer has much better control over crystal size and consistent crystal products.

9.3.8 Chromatographic Techniques

Chromatography techniques are based on a combination of chemical principles of hydrophobic and hydrophilic interactions, partition behavior, affinity, adsorption, and elution of molecules. In this technique, there are two phases: the stationary phase, where components of a mixture are selectively adsorbed, and the mobile phase, which helps to dislodge the components from the stationery phase at different rates. Pure compounds are thus obtained.

Various types of chromatographic methods are developed depending on the physical states of the different phases. The stationary phase can be solid or liquid, whereas the mobile phase can be liquid or gas. Various types of chromatographic methods are used to separate biological compounds:

1. Column chromatography
2. Paper chromatography
3. Ion-exchange chromatography
4. Gel filtration chromatography
5. Affinity chromatography

Column Chromatography

Column chromatography is an adsorption chromatography where compounds are adsorbed onto silica or alumina in a column. An equilibrium is set up between molecules bound to the column and those free in solution (Fig. 9.6). The charge, van der Waals forces, dipole interactions, hydrogen bonding stearic factors, and structure of the compounds govern the extent of binding. Thus, the choice of correct adsorbent and solvent is, to a large extent, dependent on the polarity of the compounds. It is also essential that the mixture be soluble in the solvent; otherwise, it will remain permanently adsorbed in the adsorbent.

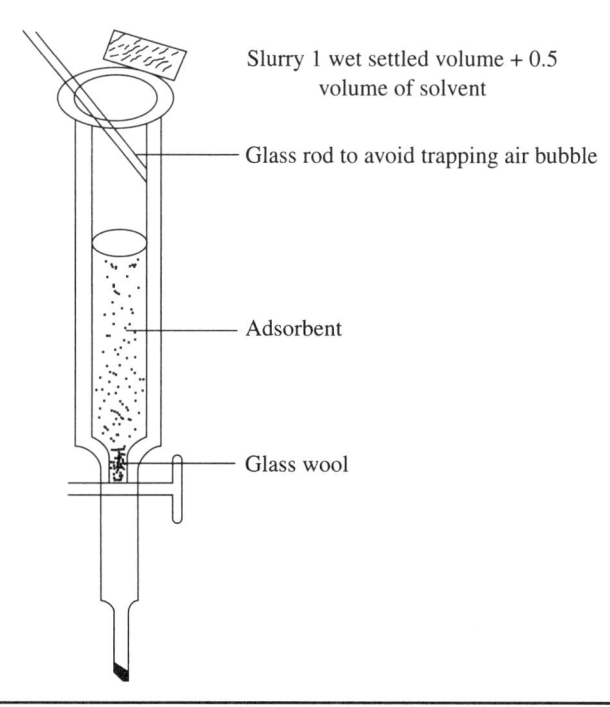

FIGURE 9.6 Chromatographic column.

The eluting powers of various solvents, that is, their ability to move a given component down a column is given below:

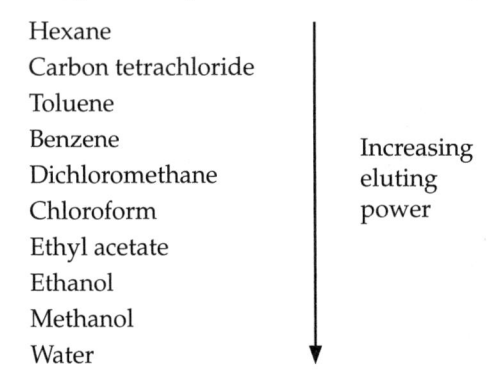

Hexane
Carbon tetrachloride
Toluene
Benzene
Dichloromethane   Increasing
Chloroform   eluting
Ethyl acetate   power
Ethanol
Methanol
Water

Paper Chromatography

In paper chromatography, paper sheets make an ideal support medium where water is adsorbed between the cellulose fibers and forms a stationary hydrophilic phase.

The paper strip containing the mixture is suspended in a jar in such a way that the end of a spotted paper strip is immersed in the developing solvent. The sample is separated into individual spots as the solvent ascends the paper (Fig. 9.7). In this case, distribution takes place between water (adsorbed by the filter paper to an extent of 20 percent) and the mobile solvent. For this reason, it is referred as liquid partition chromatography. The separation of amino acids can be done by paper chromatography.

Ion-Exchange Chromatography

Ion-exchange chromatography is the method used to separate the mixture on the basis of adsorption (cation or anion groups on the matrix) in a column. Generally, the matrix contains charged groups, and the type of groups defines the nature and strength of ion exchange. These groups may be either cationic or anionic, according to the nature of their affinity for either negative or positive ions. There are two different types of ion-exchange materials: strongly ionized groups, such as SO_3H, $-NR_3$, and the weakly ionized groups, such as $-COOH$, $-OH$, and $-NH_2$. In the strong ion exchange, the resins are completely ionized and exist in charged form except at extreme pH values. The weak ion-exchange materials contain a group whose ionization is dependent on the pH and

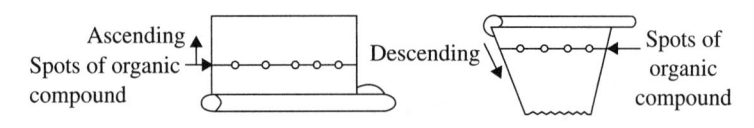

FIGURE 9.7 Ascending and descending paper chromatography.

they can only be used at maximum capacity over a narrow pH range. The extent of binding the ions on the resin depends on the nature of the resin and temperature, ionic strength, and composition of the solvent. In the case of weakly ionized resins, ion uptake also depends on the degree of ionization of the resin and the materials to be separated. The preparation of ion-exchange material is done by washing the ions with an appropriate solution. For example, the Na^+ form is prepared by washing the resin with sodium hydroxide and then with water until the washings are neutral. Similarly, H^+ forms of a cation-exchange resin are obtained by washing the material with hydrochloric acid and then with water until the washings are neutral. Generally, isolation of immunoglobulin (IgG) can be done by ion-exchange chromatography.

Gel Filtration Chromatography

Gel filtration chromatography is the technique where the molecules can be separated on the basis of their size by passing them down a column containing swollen particles of an uncharged gel. Typically, the gels consist of macromolecules with great affinity for the solvent used. These have been cross-linked to form a three-dimensional network, thereby making them insoluble. Instead of dissolving the liquid, the macromolecules swell, taking up large amounts of liquid. Small molecules can enter the gel, but larger molecules are completely separated, leaving the column first (Fig. 9.8). The separation of proteins with different molecular weights can be done by gel filtration.

Affinity Chromatography

Affinity chromatography is a single-step method that can be applied to purify an enzyme from a very complex mixture. This method is based on a biological property of some enzymes, that is, their capacity for specific binding another molecule called a ligand. In the enzyme, the ligand attached to the matrix is usually a powerful inhibitor with a high affinity constant that will bind only one enzyme in a complex mixture of proteins. Choice of the proper matrix is very important parameter

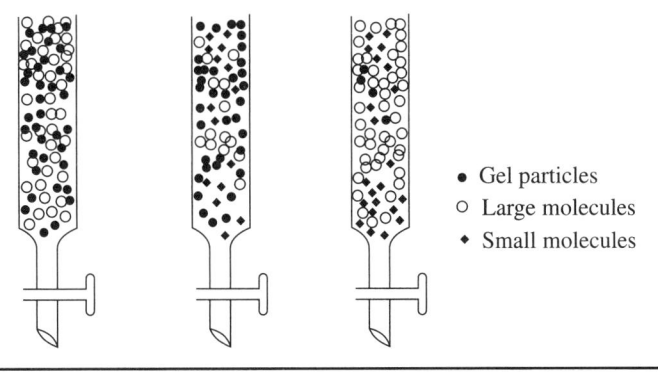

FIGURE 9.8 Gel filtration chromatography.

in the affinity chromatographic process. A good matrix for this method should have the following properties:

1. *Hydrophilic:* Reduce the nonspecific interactions.
2. *Large pores:* Allow all areas of the matrix to be available to most of the molecules in the mixture. Some matrices allow binding only to the outer surface and some are useful for separating very large molecules, cells, or viruses.
3. *Rigid:* The matrix must withstand the pressures of packing and solvent flow during elution or washing.
4. *Inert:* The matrix should not contribute to the separation.
5. *Chemical stability:* The matrix must be stable to all solvents in the separation.

The ligand is covalently attached to supporting medium so that the chromatographic material can be designed for a specific purification task. A spacer arm is used between the matrix and the ligand so that the active site of the ligand is available to the sample. Generally, the spacer arm should be optimum; 2 to 13 carbon atoms have been found to be optimal (Table 9.3). If the spacer arm is too short, steric hindrance can occur, whereas if the arm is too long there is an increased risk of nonspecific adsorption.

The solvent system chosen for separating the enzyme from the ligand is also a critical factor to a good separation. The elution buffer should release the sample safely and rapidly. Again, the buffer should neither denature the sample nor cause any change in the specific activity. Purifying acetyl cholinesterase can be done by affinity chromatography.

Linkage	Ligand Group	Spacer Length	Active pH	Specificity
CNBr	NH$_2$	Equivalent to about 3 carbons	8–10	Proteins, peptides
Thiopropyl	SH		9–11	Sulphydryls
Thio	SH		9–13	Sulphydryls, peptides
Epoxy	NH$_2$	Equivalent to about 11 carbons	9	Carbohydrates
	OH		10	Sulphydryls
	SH		11	Peptides
Tresyl	NH$_2$	6 carbons	8–10	Proteins, aminoacids
Aminohexyl	COOH	6 carbons	8	Proteins, carboxylic acid
Carbothexyl	NH$_2$	6 carbons	8	Acids

TABLE 9.3 Coupling Chemistry

9.3.9 Membrane Technology

In membrane separation, the membrane used may be a solid matrix. A liquid may also act as a separation barrier between two phases or mediums where transport occurs using the solution–diffusion mechanism. The advantages of the process are as follows:

1. Operation of the process is easy.
2. It replaces the conventional process and provides clean technology.
3. The process can recover high-value product with appreciable energy savings.
4. There is a possibility of modifying the design at any time.
5. This process is more easier to separate without heat treatment, for example, separation of alcohol from water than chemical process.
6. Separation of amino acids can be done due to like electrostatic charge, hydrophobicity, molecular size, or solubility.
7. Membrane technology can be utilized for the immobilization of the enzymes in enzyme emulsions. This can be done by immobilizing the enzyme systems in liquid surfactant membrane emulsions. Enzyme emulsions have a broad field of applications in cell-free fermentation broth, to preselect the desired substrate for enzymatic reaction in the complex mixture.
8. Membrane technology specially liquid membrane can be used in gas-separation techniques. High selectivity is also achieved due to extraordinary solubility of certain gases in liquids.

Disadvantages of the process are as follows:

1. The membrane process is expensive due to the high cost of the membrane.
2. Membrane fouling is a common problem in the separation process.
3. There is an upper solid limit in the membrane separation process.

Ultrafiltration Membrane

This is a pressure-driven process (Fig. 9.9) in which water and low-molecular-weight components can pass through the membrane, whereas particles, colloids, and macromolecules are retained. Ultrafiltration pore ratings range approximately from 1000 to 500,000 Du. Thus, filtration makes the membrane more permeable than nanofiltration. In ultrafiltrated milk, lactose and minerals pass in a 50 percent separation ratio; for example, in the retention there will be

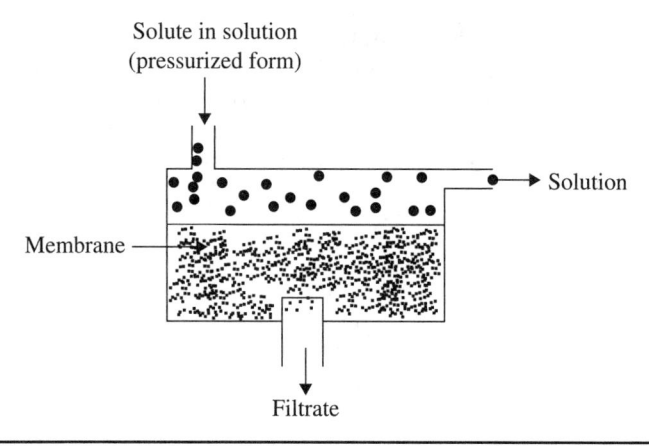

FIGURE 9.9 Ultrafiltration process.

100 percent fat, 100 percent protein, 50 percent lactose, and 50 percent free minerals.

Liquid Membrane

A liquid membrane is made of liquid that is highly selective. With the use of carriers for the transport mechanism, specific molecular recognition can be achieved.

There are, in fact, two basic types of liquid membranes: an emulsion liquid membrane (ELM), and an immobilized liquid membrane (ILM), also called a supported liquid membrane

ELM An emulsion liquid membrane may be either oil-in-water (O/W) or water-in-oil (W/O). W/O emulsion is such that the dispersed phase or the internal phase is water, and the continuous phase or dispersion medium is oil. However, O/W emulsion is such that the dispersed phase or internal phase is oil, and the continuous phase or dispersion medium is an aqueous base. The liquid membrane separates two miscible liquid phases across it.

An emulsion liquid membrane is made first by making a primary emulsion (either O/W or W/O) and then dispersing the emulsion into another phase—either oil or water—this phase is also referred to as the third or external phase giving rise to a multiple emulsion of type O/W/O or W/O/W. Multiple emulsions are actually emulsions within emulsions. The W/O/W emulsion is prepared by incorporating the emulsion W/O in an aqueous phase.

ELM creates a huge surface area for mass transfer than in other techniques, and extraction and stripping are achieved simultaneously in a single operation, unlike conventional solvent extraction. The disadvantage of this process, however, is its stability and its tendency to undergo swelling due to a rapid increase in volume of the internal phase that finally retards the efficiency of separation.

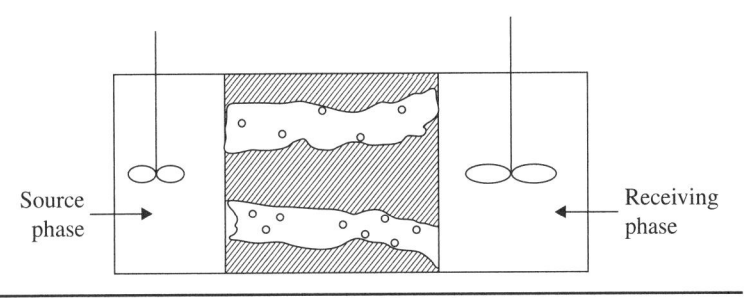

FIGURE 9.10 Thin sheet–supported liquid membrane.

Thin-Supported Liquid Membrane

A thin sheet–supported liquid membrane can be used for a laboratory scale, but it cannot be scaled up for industrial use (see Fig. 9.10). Essentially, this is just a porous polymer membrane whose pores are filled with an organic liquid and carrier, set between the source phase and the receiving phase, which are being gently stirred.

Hollow Fiber-Supported Liquid Membrane

Supported liquid membrane (SLM) extraction, with flat membrane or hollow fiber membrane as supporting material for an organic solvent to form an aqueous–organic solvent–aqueous three-phase system, has been applied to sampling of various compounds, including weak acids and bases, as well as metal ions. It is a new versatile tool for speciation of trace metals, in particular Cu, Pb, and Cd based on a hollow fiber–supported liquid membrane. Low plutonium content acidic waste is generated in nuclear chemical facilities. In this technique for quantitative separation and recovery of plutonium (Pu) from such wastes using tri-*n*-butyl phosphate (TBP) in dodecane as carrier was successful. Citric acid was successfully separated from an aqueous solution using this membrane. Tri-*n*-octylamine diluted in various organic solvents was used as a complexing agent for facilitated transport of citric acid from an aqueous solution. Pure water and aqueous sodium hydroxide were used as stripping agents.

In the liquid membrane (Fig. 9.11) the outer shell in made up of nonporous material restricting the transport of materials inside. The thin fibers run along the length of the shell inside. The source phase is piped through the system from top to bottom, and the pores in the fibers are filled with the organic phase.

The waste liquid containing the carrier was transported across the receiving phase (may be organic or aqueous), and the receiving phase is forced out through the sides of the shell. There is rapid transportation due to surface area and membrane thickness. There are no problems with leakage and contamination, but the system is highly expensive and requires intermittent cleaning to remove aqueous and contaminant buildup.

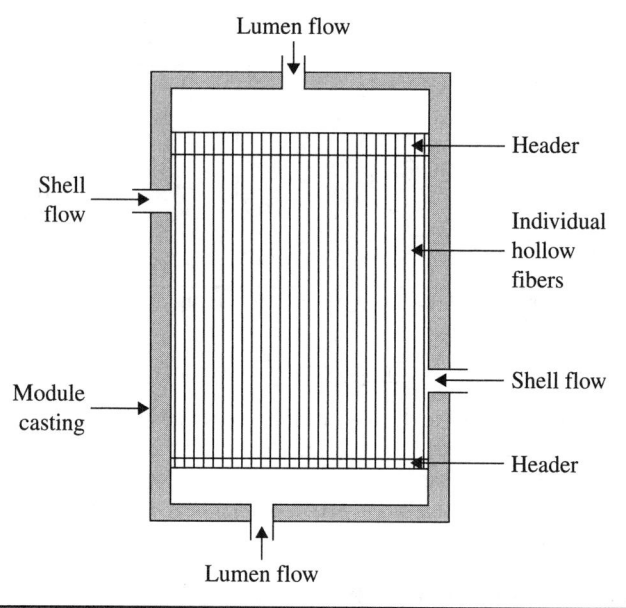

FIGURE 9.11 Hollow fiber–supported liquid membrane.

Further Reading

Foust, A. S., Wenzel, L. A., Clump, C. W., Maus, L., and Andersen, L. B. 1960. *Principles of Unit Operations*. New York: Wiley.

John A. Duerre and Edgar Ribi. *National Institute of Allergy and Infectious Diseases*. Rocky Mountain Laboratory, U.S. Public Health Service, USA.

McCabe, W. L., Smith, J., and Harriott, P. 1997. *Unit Operations of Chemical Engineering*. 7th ed. New York: McGraw-Hill.

Nag, A. 2007. *Analytical Techniques in Agriculture, Biotechnology and Environmental Engineering*. New Delhi, India: Prentice Hall of India.

Nath, K. 2008. *Membrane Separation Processes*. New Delhi, India: Prentice Hall of India.

CHAPTER 10
Food Safety Management

Enda Cummins

UCD School of Agriculture, Food Science and Veterinary Medicine
Agriculture and Food Science Centre
Belfield, Dublin, Ireland

10.1 Introduction

Food safety is the assurance that food will not cause harm to the consumer when it is prepared and/or eaten according to its intended use and generally refers to its chemical and microbiological content (FAO/WHO 2001). Food safety management refers to the process of ensuring and controlling food safety through regulation or other policy mechanisms for the health and well-being of consumers. Food safety management is of prime importance in maintaining consumer confidence and is critical for a healthy population and economy, directly influencing consumer sentiment, national and international trade, and tourism. Food safety is among the top 11 organizational priorities identified by the World Health Organization (FAO/WHO 2001; Toyofuku 2006). A food contamination scare can have devastating economic consequences; for example, the bovine spongiform encephalopathy (BSE) crisis is estimated to have cost the United Kingdom 6 billion U.S. dollars (Burlingame and Pineiro 2007). Thus, it is of prime importance for national and international regulatory agencies to facilitate international trade in food and food products while ensuring the protection of human health. Recent health scares, including the BSE crisis, growth hormones in meat, dioxin, Sudan Red, avian flu, and influenza A (H1N1) scares, have made consumers more wary about the origin, traceability, and safety of the food they eat. It has been reported that foodborne microbiological hazards may be responsible for as many as 45.4 million cases of food poisoning within the European Union (EU) every year (Cummins 2008). Food safety worldwide faces increased pressures and challenges arising from the globalization of food trade, intensive production systems, and changing consumer preferences (Burlingame and Pineiro 2007). This in turn has prompted the need to harmonize and integrate food safety policy.

In 2005, zoonotic diseases reportedly affected over 387,000 people in the EU. Although these diseases can be acquired directly through animal contact, majority are acquired through ingestion of contaminated food. Given the potential severe consequences on human health and negative publicity from such contamination events, there is a requirement to control foodborne pathogens from farm to fork. As the global food supply widens and evolves, responsibility for food safety is seen as the responsibility of all stakeholders. This chapter looks at zoonotic foodborne hazards and assesses current food safety management strategies and targets used worldwide, and within the EU, in controlling and managing zoonotic pathogens along the farm-to-fork continuum.

10.2 Zoonotic Foodborne Hazards

Campylobacter and *Salmonella* are responsible for the greatest number of zoonotic infections in humans (EFSA 2006). Meat and meat products appear to be the primary sources of foodborne outbreaks of salmonellosis and campylobacteriosis in the EU. An incidence of 51.6 per 100,000 population makes campylobacteriosis the most frequently reported zoonotic disease in the EU (EFSA 2006). Chicken meat is the most significant sources of human exposure to *Campylobacter*. EU member states reported a total of 176,395 cases of human salmonellosis in 2005 (38.2 cases per 100,000 population; EFSA 2006). Eggs and egg products are a significant source of human infection with *Salmonella*. Although the incidences of other infections (e.g., verocytotoxigenic *Escherichia coli* (VTEC) and *Listeria monocytogenes*) may be lower, the consequences of exposure can be more severe, particularly among children and immunocompromized individuals. Outbreaks of verocytotoxigenic *E. coli* VTEC O157 have been associated with minced meat and minced meat products, including hamburgers, sausages, and patties (Cummins et al. 2008). Ready-to-eat (RTE) meat products are commonly involved in outbreaks of human listeriosis. Human listeriosis is rare, but the disease can be severe with over 1439 cases of listeriosis reported in the EU in 2005 (Nørrung and Buncic 2008). *Listeria monocytogenes* are able to grow at refrigeration temperatures, which make it a particular challenge for the food industry. Yersiniosis caused by *Yersinia enterocolitica* has a lower incidence (2.6 per 100,000 population) in the EU and affects mainly young children. Pork and beef are the main sources of food contamination, although human pathogenic *Yersinia* serotypes have been shown to come mainly from pork (Nørrung and Buncic 2008).

In addition to the normal strains of these zoonotic diseases, drug-resistant strains of bacteria (such as *E. coli* and *Listeria*) are emerging and causing health scares. Centralization of the food industry means that a single contaminated product may appear in many different foods and many different forms and infect a considerable number of people before it is identified. Monitoring, control, and management of such hazards are central to any food safety strategy. Food safety management within the EU and worldwide is governed by food safety policy decisions and effective implementation of these policies in member states through food safety regulation. Food safety has become a priority research area worldwide as the global food supply evolves (Tent 1999).

10.3 Food Safety Measures

A firmer grip on food safety issues has been aided by the introduction of pertinent regulatory responses on the international stage. This has enabled a closer analysis of possible risk factors and risk-reduction measures for existing and emerging pathogens. Food regulation engenders fair trade in food and food products within and between nations while protecting the consumer's health and increasing economic viability (Aruoma 2006).

10.3.1 World Trade

The World Trade Organization (WTO) was formed in 1995 following the Uruguay round of multilateral trade negotiations and is currently the only global international organization dealing with the rules of trade between nations. WTO negotiations and agreements are therefore agreed on by trading nations (Aruoma 2006). WTO sanitary and phytosanitary (SPS) agreements, and the codes of practices issued by the CODEX Alimentarius Commission (CAC), represent current best practices that are required to ensure trade of safe food in a world market (Jukes 1993, 2000; Boutrif 2003). All WTO members are required to ensure that their SPS measures are based on the assessment of risks to human, animal, or plant life or health (Klapwijk et al. 2000). WHO also works closely with the Food and Agriculture Organization (FAO) of the United Nations to address food safety issues worldwide and to ensure safe food production and trade that is based on risk assessment.

CODEX was established in the 1960s by the FAO and WHO to ensure safe and fair trade in food worldwide and to establish international standards to ensure this. As stipulated under the SPS agreement reached at the Uruguay round of the General Agreement on Tariffs and Trade (GATT) negotiations, member states must justify food regulations implemented outside internationally recognized food standards (Klapwijk et al. 2000). This justification must be complemented by a thorough and comprehensive risk analysis.

The CODEX Alimentarius, as administered by FAO and WHO, sets out key principles for the global harmonization of food safety standards, including codes of practice on labeling requirements, risk assessment, and food analysis. This in turn is the basis for many national food standards (Burlingame and Pineiro 2007). FAO outlines the importance of an integrated food chain approach, placing the responsibility of safe food squarely on the shoulders of all involved from primary production to food preparation and consumption (FAO 2003). The WTO SPS agreements and the codes of practices issued by the CAC represent a benchmark for international harmonization of food safety management that tries to ensure the global trade of safe food (Aruoma 2006).

10.3.2 European Trade

Within the EU, a number of safety management systems are in place and harmonized by European food safety regulations. Hazard Analysis Critical Control Points (HACCP) was established as an essential food safety management system. HACCP is a risk management tool, which identifies, prevents, controls, and then monitors the most vulnerable points in a food production system (Voysey and Brown 2000). HACCP is a system used for food safety where the process of control is placed at critical points in the production chain, thus reversing reliance on end-product testing as a means of detecting contaminants (McMeekin et al. 2006). Concepts for HACCP originated in the 1960s with attempts to minimize the risk of foodborne illness to astronauts in space. This was followed by the implementation of HACCP as a general food safety measure throughout the food production chain. Because of the implementation of the European Communities Hygiene of Foodstuffs Regulations in 2000, HACCP is a legal and customer requirement for all food businesses (Mayes 1998).

CODEX has endorsed HACCP through its Hazard Analysis and Critical Control Point System and Guidelines for its Application (CODEX 1997) as a cost-effective system for ensuring food safety. This coupled with prerequisite systems such as good manufacturing and good hygiene practice; ensure that food professionals are equipped with the tools to certify product quality and safety (van Schothorst 2005). Thus, as part of a food safety management system, HACCP complements general hygiene requirements for food premises (Sun and Ockerman 2005).

The new EU food hygiene regulations harmonizes food safety policy within the EU with emphasis on the need to protect public health in a way that is effective, proportionate, and based on risk analysis. The farm-to-fork approach recognizes that food safety is the responsibility of all those involved along the food chain from primary production to final preparation and final human consumption. The main EU food safety regulation, Regulation EC 178/2002, resulted in the establishment of the European Food Safety Authority and established the common basis for food law in EU member states. The regulation also includes common definitions, general provisions, and specific requirements for food safety. The regulation aims to provide guiding principles and objectives for the food/feed industry for human health protection while ensuring safe and effective trading between member states.

Regulation 178 also has specific requirements for risk analysis, feed and food importation, public consultation, traceability, and general safety requirements for food and feed. In particular, under the general principles of the food law and Article 6, it states that food law shall be based on risk analysis. As a regulation, the provisions are directly applicable to member states without the need to be transposed into national legislation. The regulation also establishes in EU law the three interrelated components of risk analysis (risk assessment, risk management, and risk communication) as the basis for food law and sets requirements for traceability procedures for food and feedstuffs.

Regulation EC 852/2004 lays down general rules for food business operators on the hygiene of foodstuffs, with particular emphasis on the necessity for establishing microbiological criteria and temperature-control requirements based on a scientific risk assessment. Also, food business operators need to ensure that any imported food is equivalent or better than that produced within the EU. The regulation lays down general hygiene requirements to be respected by food businesses at all stages of the food chain. The regulation must be implemented by all food businesses that have a responsibility to ensure that all the requirements are properly implemented in order to promote food safety.

Regulation EC 2073/2005 lays down microbiological criteria for certain microorganisms and implementation rules to be complied with by food business operators when implementing general and specific hygiene measures. For food businesses dealing with products of animal origin (meat, fish, shellfish, milk, and associated products), there is also the need to comply with Regulation EC 853/2004, which generally applies to wholesale activities.

The new EU food safety regulations ensure that national and international food safety laws and best practices are based on risk analysis. Risk analysis provides a framework from which foodborne hazards can be monitored and the identification of emerging risks can be clarified. WHO/FAO provide a comprehensive framework for food safety risk analysis. Risk assessment (the evaluation component of risk analysis) can be used to assist in identifying hazards, critical control points, and establishing critical limits for the HACCP plan.

Thus, although risk assessment and HACCP are related, they are fundamentally different processes (Voysey and Brown 2000).

10.3.3 Food Safety Targets

Specific targets have been developed in food safety management to monitor overall food safety and to assist in complying with trade regulations. In the WTO SPS agreement, reference is made to members ensuring an appropriate level of protection (ALOP) (WTO 1995). The decision on what this level should be is a matter for risk managers. Risk assessment is a tool that can provide data to assist in making that decision (Gorris 2005), which in turn can help in the formulation of food safety objectives (FSO). Havelaar et al. (2004) define an FSO as being "a limit to the prevalence and the average concentration of a microbial hazard in food, at an appropriate step in the food chain at or near the point of consumption, that provides the appropriate level of protection." The concept of FSOs is important for governmental bodies working on food safety control and legislation. Previous risk assessments have resulted in FSOs being suggested for *Salmonella enteritidis* in eggs, aflatoxins in raw peanuts, and *Salmonella* in powdered milk (Stringer 2005).

Another term of reference is the concept of a performance objective (PO), which is, essentially, a target set earlier in the food chain prior to consumption and one that must be adhered to in order to fulfill a FSO (Fig. 10.1). POs provide guidance on what should be happening along the food chain in order for an FSO to be achieved. Thus, FSOs and POs are food safety targets (van Schothorst, 2005). Control measures (CMs) may be introduced as necessary in order to ensure compliance with POs and FSOs. The CAC (2004) have proposed an additional term, performance criteria (PC), to describe the overall effect that a CM must have in order to reduce hazard during that step so that the resulting hazard level from that step complies with the PO or the FSO (Gorris 2005). The relationship between food safety targets and how they relate to different stages in the food production chain are illustrated in Fig. 10.1.

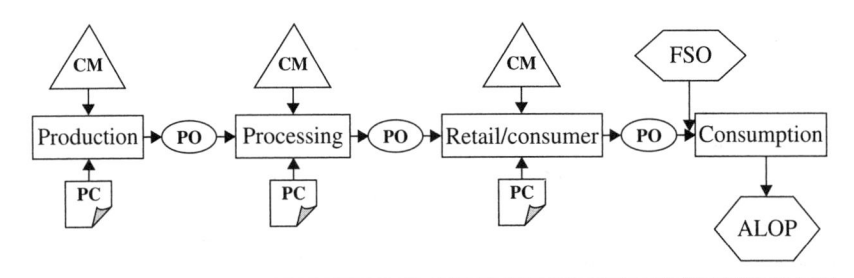

FIGURE 10.1 Food safety targets pertaining to a food safety management system (ALOP = appropriate level of protection; CM = control measure; FSO = food safety objective; PC = performance criteria; PO = performance objective).

10.4 Risk Analysis

It is clear that risk analysis has gained international support as a process to assist in ensuring global trade in safe food. It provides a structured framework for guiding policy makers and assisting in decision making. The CAC (1999) provides a structured approach for risk analysis and describes the process as comprising of three interconnected components: risk management, risk communication, and risk assessment. The integration of the different stages is highlighted in Fig. 10.2.

Risk management is the process of making an informed decision as to whether a risk is acceptable or not, and, if not, what risk reduction measures can be employed to reduce or prevent risk. Risk communication refers to the interaction between risk assessors, risk managers, and stakeholders; this interaction is essential in order to establish clear objectives for risk assessment to answer and to ensure that both sides are clear on the risk question to be addressed in the analysis (Hugas et al. 2007). Risk assessment is defined as "the qualitative or quantitative estimation of the likelihood of adverse effects from exposure to specified health hazards or from the absence of beneficial influences" (Hathaway et al. 1988). The North American Free Trade Agreement (NAFTA) defines risk assessment as "an evaluation of the potential for the introduction, establishment or spread of a pest or disease and associated biological and economic consequences" (NAFTA 1993). By quantifying the possible hazards or risks associated with a product/action, the safety of the product may be ascertained. Risk can be defined as the mathematical expression of a potential harmful event (hazard), measured by the likelihood of its occurrence.

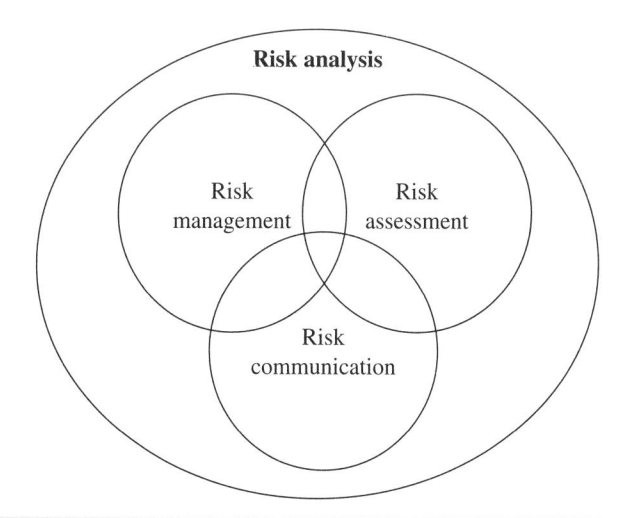

FIGURE 10.2 Components of risk analysis.

Risk assessment of industrial processes and reactions has been widely used by engineers. However, the use of risk assessment techniques in assessing disease risk is relatively recent (Hugas et al. 2007). Risk assessment has been identified as being applicable to food safety issues in two main areas: (1) international trade in animals and animal products, where the possibility of importing infected/contaminated products has long been recognized, and (2) food hygiene, to examine toxic residues and infectious organisms in meat and meat products (Wooldridge et al. 1996). Traditional methods of managing risks, particularly when considering international movement, have tended to rely on zero-risk policies when asking whether imports should be allowed into a disease-free area. Recent developments in international trade mean that this is no longer acceptable, and disease risks must be balanced against trading benefits. This brings into prominence the need for food safety management systems based on risk assessment.

In many cases, given the availability of some basic parameters, risk assessment modeling is often able to create the same results as a pilot study at a lower cost and in a shorter time. In addition, recent developments in information technology have meant that personal computers are now capable of handling enormous amounts of data easily and carrying out complicated calculations quickly. Risk assessment can be a useful tool in improving our understanding of the behavior of foodborne pathogens and enables the evaluation of different control methods. This approach allows an estimation of the relationship between different parameters influencing foodborne pathogens and can lead to an improved understanding of the effect of each factor on the future course of disease occurrence.

There is a clear need for a risk assessment framework within which foodborne pathogens can be understood, interventions planned, and the potential risk to humans quantified. Risk assessment must be undertaken in an independent, objective, and transparent manner based on the best available science (EC 2008). It provides a basis for evidence-based decision making in food safety and helps to improve the transparency of the process. This will result in better risk communication and help to build trust among stakeholders (Hugas et al. 2007).

Risk analysis has formed an important role in the development of food safety management and encourages science-based decision making to improve food safety controls, to identify data gaps, and to assist in devising risk management strategies. International trade agreements, such as GATT and NAFTA, have requirements for risk assessment in their sanitary and SPS agreements (WTO 1995). Risk assessment has been useful as a means of influencing European policy and determining control techniques; for example, risk assessment was essential in determining policies to control the BSE epidemic (Ferguson et al. 1997). Risk assessment has come to the fore as the method of quantifying risks to both the human and animal populations and as a valuable tool for policy makers.

With the emergence of more resilient strains of bacteria, risk assessment plays a large role in ensuring the safety of our food and food products as consumers become increasingly conscious of emerging health problems. Risk assessment models can be used to test scenarios and, by doing so, allow some insight into the underlying dynamics of a pathogen and associated risk. This can lead to further understanding of the pathogen and enable further—and much needed—research and development in the area. Risk assessment modeling of food-manufacturing processes for the agrifood industry will bring to prominence any deficiencies in the process while highlighting areas of potential health risk. It is often the case that modeling exercises bring to light important deficiencies in the available body of knowledge (Martin et al. 1987). Risk assessment models can also be used to demonstrate the importance of the missing data and direct data-collection efforts. Model building can be an important means of generating and formalizing hypothesis (Hurd and Kaneene 1993).

Risk assessment provides a measure of risk to consumers while accounting for uncertainty and variability in data measurements and knowledge. The result of a risk assessment is to estimate the level of risk to risk managers. The purpose of risk assessment is not to establish a safe level of risk; this responsibility lies with risk managers. Should the level of risk be deemed to be unacceptable by the risk managers, then a risk-minimization strategy needs to be implemented to minimize or eliminate the risk (Buchanan 2002).

Risk assessment can be used to assess alternative strategies and to evaluate the impact of policy decisions. One of the advantages of risk assessment is that multiple scenarios can be created and evaluated. Risk managers need to decide which control strategy is appropriate, in addition to a multitude of other factors, including cost, acceptability, and ethical issues. The evaluation of strategies is the responsibility of the risk managers.

The advantage of quantitative risk assessment is that it is able to describe complex biological systems, including the growth and development of pathogens and the resulting human exposure. This results in an improved understanding of pathogen dynamics with an extensive documentation of the pathogenic characteristics and their potential exposure pathways. Ross and McMeekin (2003) discuss the application of predictive microbiology within quantitative microbial food safety risk assessment. Predictive microbiology provides data on the likely growth rate of microbial pathogens within a food substrate (USDA/ARS 1999). Only when microbial growth dynamics are assessed can a clearer picture of possible risks be evaluated (Coleman and Marks 1999). Major food safety systems such as HACCP and risk assessment are now underpinned by predictive microbiology (McMeekin 2007; McMeekin et al. 2002).

Within a risk analysis, the CODEX Alimentarius Commission has developed the concept of quantitative microbiological risk assessment (MRA), which will complement other quantitative food safety targets that are currently under development, such as FSOs and POs (Hugas et al. 2007). MRA has been driven by the requirement for science-based risk assessment, as well as sanitary and SPS agreements. MRA facilitates safe international trade of food and food products (van Schothorst 2002).

10.4.1 Stages in Risk Assessment

Risk assessment is a well-established method of characterizing the impact of foodborne bacteria on the food chain and consists of four main stages: hazard identification, exposure assessment, hazard characterization (or dose-response assessment), and risk characterization (Fig. 10. 3).

The first stage in the process is identifying hazards. This involves identifying the hazard in question and necessitates the collation of all relevant data pertaining to the hazard. This stage will usually highlight factors that contribute to the survival, mode of transmission, and growth of the foodborne pathogen. Hazard identification focuses on what can go wrong and how it might happen.

During exposure assessment, the pathways by which the hazard can reach a host are identified, and the effects of process stages and possible critical control points are assessed. Exposure assessment evaluates the likelihood of these hazards occurring and the resulting implications. Thus, the ultimate goal of exposure assessment is to evaluate the occurrence and levels of a hazard in a food at the time of consumption (Voysey and Brown 2000).

An accurate exposure assessment means that we need accurate information on the following: the presence of the pathogen in raw ingredients and the effect of food processing, distribution, handling,

```
Hazard identification
        ↓
Exposure assessment
        ↓
Hazard characterization
        ↓
Risk characterization
```

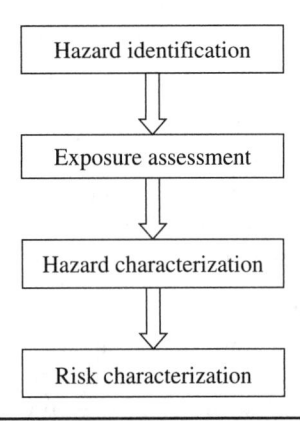

FIGURE 10.3 Stages of risk assessment.

and preparation on the hazard in terms of levels and consumption patterns (e.g., portion size). Model validation is a welcome addition to the process where possible (Whiting and Buchanan 1994).

When we characterize risk, we recognize the fact that pathogenic bacteria may vary widely in terms of their ability to cause illness, depending on their numbers, virulence, and host susceptibility. Thus, a dose response is used to translate the exposure assessment into a response in terms of infected or ill individuals, depending on the nature of the dose–response relationship. The dose–response assessment evaluates the impact of a certain level of exposure on human health in terms of illness or mortality levels. Changes in exposure can be assessed in terms of change in disease incidence or mortality rate. Of course, susceptibility of the host and exposure levels will influence response levels and are critical in formulating a dose–response relationship. Such information can be difficult to ascertain, usually being based on outbreak studies or limited challenge tests; however, some data are available (Teunis et al. 1996).

All the data generated in previous stages are pulled together in the risk-characterization stage of the process. Input uncertainty and variability, including measurement and sampling errors for microbiological methods, need to be accounted for regularly (Nauta 2002). This can be achieved by representing uncertain inputs using probability distributions. This allows us to identify those inputs having the greatest impact on model inputs and allows us to identify risk-reduction strategies. This can also direct future research efforts and provide a scientific basis for policy decisions.

10.4.2 Methodologies

There is no standard methodology for conducting a risk assessment. The type of assessment that is performed will vary depending on the complexity of the system being assessed, the available resources, and the original risk question posed by risk managers (Coleman and Marks 1999). Qualitative assessments and fully quantitative assessments, which incorporate elements of uncertainty and variability, have been used. Qualitative risk assessments are usually conducted initially to assess the risk of a hazard and use descriptors to detail the level of risk. Descriptions of risk can vary from "negligible" to "low," "medium," or "high," resulting in a ranking or categorizing system. In contrast, quantitative assessments generate numerical estimates of risk. Qualitative risk assessments are used to give an initial estimate of risk. They usually require less time and resources than a fully quantitative risk assessment. HACCP is primarily a qualitative approach to food safety. Thus, it does not lend itself to assessing the effects of parameter uncertainty or the effect of control measures (Hoornstra et al. 2001). This generates the need for a quantitative approach that can be used to provide more insight into the effect of control measures.

Two approaches are typically used within quantitative risk assessment. The first is the deterministic or point estimate. In this approach, point estimates are used for each of the input parameters with no real understanding of the uncertainty in the input parameters. A stochastic or probabilistic approach, on the other hand, uses probability density distributions to characterize the uncertainty of the input parameters. Using probability density distributions as inputs result in probability density distributions being generated for the model outputs and, therefore, allows confidence intervals to be generated. The type of model used is dependent on the initial risk question and the availability of data and resources to conduct a risk assessment.

10.5 Perspective

There is a legislative requirement for food safety management, dominated internationally by WTO legislation and by EU hygiene regulations within the EU (Allio et al. 2006). Risk assessment has played a pivotal role in providing guidance for food safety management and control of risks and has become the cornerstone in ensuring safe food. Risk assessment has provided a scientific basis for HACCP systems and allows risk-mitigation strategies to be tested while providing a structured framework for data gathering and analysis. Risk assessment procedures play an important role in establishing import criteria and facilitating international trade by ensuring that scientific controls and food safety practices are in place between trading nations. A recent survey (Walls and Scott 2005) found that scientists foresee a major role for risk assessment in setting regulatory policy and facilitating international trade. Consumer confidence is an important outcome of a successful food safety policy. There is a major role for quantitative risk assessment in establishing food safety targets, including FSOs and POs. This is likely to support its future role in food safety management.

10.6 Conclusions

With increased consumer awareness about food safety and quality, a more preventative food safety management approach is being adopted for international and national trade. No food can be considered to be risk free, and all individuals involved from farm to fork have a role to play in ensuring its safety. Zoonotic diseases pose a considerable threat to human health. The main vector for transmission to humans is *via* food. In particular, meat and meat products are significant sources of these infections. Zoonotic pathogens in food can be managed and controlled through risk assessment strategies.

The importance of different exposure routes for consumers needs to be investigated and risk assessment offers considerable flexibility as an evaluation tool and in identifying and optimizing risk management strategies. Risk assessment can be used to set FSOs and POs. Current world and EU trade legislation highlights the need for food safety and quality to be addressed along the entire food chain. Increased emphasis is placed on international collaboration to ensure safe food across boundaries and on what is now a global food market. International policy advocates risk assessment as a tool to evaluate risk and also to evaluate risk management strategies. Risk assessment is regarded as essential for managing food safety issues and can act as an extension or validation of HACCP. Important elements yet to gain exposure within food safety management are the establishment of FSOs on the basis of ALOP or other public health goals, as well as the possible enforcement of POs (or FSOs) or related PCs, standards, and control measures.

The evolution of food safety management has occurred at an uneven pace, propelled at times by the public and by governmental concern about food safety; it has been restrained in some cases by limitations of technical knowledge, data, and resources. The global harmonization of food safety regulations is necessary to ensure a safe supply of food and to ensure fair and transparent competition among countries in terms of trade. In light of this, risk assessments are being used as tools to help manage risk and to help make policy decisions. These have a significant impact on the importing and exporting of goods to and from a country. Movement of animals and their products has long been recognized as having the potential to spread infection. For the facilitation of trade, international trade agreements such as GATT and NAFTA have specified requirements for risk assessment in their sanitary and SPS agreements. For animal health, SPS measures are the most important, allowing countries to give food safety and animal health priority over trade requirements only if a scientific basis can be demonstrated (ICMSF 1998). Although each country can determine the level of risk appropriate to its own conditions, this must be based on defendable assessments available for objective evaluation. These factors have resulted in increasing numbers of risk assessments being completed for increasing numbers of trade situations. Guidelines and other documents produced by CODEX have become reference standards for international trade. It is clear that all stakeholders from farm to fork play a role in ensuring food safety. Current food safety legislation highlights the importance of food safety controls that are scientific and proportionate to the risks and emphasize the need to protect public health in a way that is effective, proportionate, and based on risk analysis.

References

Allio, L., Ballantine, B., and Meads, R. 2006. Enhancing the role of science in the decision–making of the European Union. *Regulatory Toxicology and Pharmacology* 44(1):4–13.

Aruoma, O. I. 2006. The impact of food regulation on the food supply chain. *Toxicology* 221:119–127.

Boutrif, E. 2003. The new role of Codex Alimentarius in the context of WTO/SPS agreement. *Food Control* 14:81–88.

Buchanan, R. L. 2002. The role of quantitative microbiological risk assessment in risk management options assessment. Background paper for the Joint FAO/WHO Consultation held in Kiel, Germany, on March 18–22, 2002. Available at www.foodrisk.org/exclusives/qmra.pdf. Accessed May 12, 2009.

Burlingame, B. and Pineiro, M. 2007. The essential balance: Risks and benefits in food safety and quality. *Journal of Food Composition and Analysis* 20:139–146.

CAC (CODEX Alimentarius Commission). 1999. Principles and guidelines for the conduct of microbiological risk assessment. Reference Number CAC/GL 30. Available at http://www.codexalimentarius.net/download/standards/357/CXG_030e.pdf. Accessed May 12, 2009.

CAC (CODEX Alimentarius Commission). 2004. Report of the twentieth session of the CODEX Committee on General Principles, Paris, France, 3–7 May 2004, ALINORM 04/27/33A, Appendix II, pp. 37–38. Available at ftp://ftp.fao.org/codex/alinorm04/al0433ae.pdf. Accessed May 12, 2009.

CODEX. 1997. Hazard analysis and critical control point system and guidelines for its application. Alinorm 97/13A, Codex Alimentarius Commission, Rome.

Coleman, M. E. and Marks, H. M. 1999. Qualitative and quantitative risk assessment. *Food Control* 10:289–297.

Cummins, E. 2008. The role of quantitative risk assessment in the management of foodborne biological hazards. *International Journal of Risk Assessment and Management* 8(3):318–330.

Cummins, E., Nally, P., Butler, F., Duffy, G., and O'Brien, S. 2008. Development and validation of a probabilistic second-order exposure assessment model for *Escherichia coli* O157:H7 contamination of beef trimmings from Irish meat plants. *Meat Science* 79(1):139–154.

EC (European Commission). 2008. Food safety—from farm to fork. Available at http://ec.europa.eu/food/food/foodlaw/principles/index_en.htm. Accessed May 12, 2009.

EFSA (European Food Safety Authority). 2006. The community summary report on trends and sources of zoonoses, zoonotic agents and antimicrobial resistance and foodborne outbreaks in the European Union in 2005, vol. 94. *The EFSA Journal*.

FAO. 2003. FAO's strategy for a food chain approach to food safety and quality: A framework document for the development of future strategic direction. Item 5 of the Provisional Agenda, Committee on Agriculture (COAG), 17th session, Rome, 31 March–4 April 2003. Document COAG/2003/5/. Available at http://www.fao.org/DOCREP/MEETING/006/Y8350e.HTM. Accessed May 12, 2009.

FAO/WHO. 2001. CODEX Alimentarius—Recommended International Code of Hygienic Practice—General Principles of Food Hygiene (CAC/RCP 1–1969, Rev. 3 (1997), amended 1999. FAO/WHO, Rome. Available at www.codexalimentarius.net/download/standards/23/cxp_001e.pdf. Accessed May 12, 2009.

Ferguson, N. M., Donnelly, C. A., Woolhouse. M. E. J., and Anderson, R. M. 1997. The epidemiology of BSE in GB cattle herds: II. Model construction and analysis of the transmission dynamics. *Philosophical Transactions of the Royal Society London B* 352:808–838.

Gorris, L. G. M. 2005. Food safety objective: An integral part of food chain management. *Food Control* 16:801–809.

Hathaway, S. C., Pullen, M. M. and McKenzie, A. I. 1988. A model for assessment of organoleptic postmortem inspection procedures for meat and poultry. *Journal of the American Veterinary Medical Association* 192:960–966.

Havelaar, A. H., Nauta, M. J., and Jansen, J. T. 2004. Fine-tuning food safety objectives and risk assessment. *International Journal of Food Microbiology* 93:11–29.

Hoornstra, E., Northolt, M., Notermans, D. S., and Barendsz, A. W. 2001. The use of quantitative risk assessment in HACCP. *Food Control* 12(4):229–234.

Hugas, M., Tsigarida, E., Robinson, T., and Calistri, P. 2007. Risk assessment of biological hazards in the European Union. *International Journal of Food Microbiology* 120:131–135.

Hurd, S. H. and Kaneene, J. B. 1993. The application of simulation models and systems analysis in epidemiology: A review. *Preventive Veterinary Medicine* 15:81–99.

ICMSF (International Commission on Microbiological Specifications for Foods). 1998. Potential application of risk assessment techniques to microbiological issues related to international trade in food and food products. *Journal of Food Protection* 61:1075–1086.

Jukes, D. 1993. Regulation and enforcement of food safety in the UK. *Food Policy* 18:131–142.

Jukes, D. 2000. The role of science in international food standards. *Food Control* 11:181–194.

Klapwijk, P. M., Jouve, J. L., and Stringer, M. F. 2000. Microbiological risk assessment in Europe: The next decade. *International Journal of Food Microbiology* 58:223–230.

Martin, S. W., Meek, A. H., and Willesberg, P. 1987. *Veterinary Epidemiology. Principles and Methods*. Ames, IA: Iowa State University Press.

Mayes, T. 1998. Risk analysis in HACCP: Burden or benefit? *Food Control* 9(2–3):171–176.

McMeekin, T. A. 2007. Predictive microbiology: Quantitative science delivering quantifiable benefits to the meat industry and other food industries. *Meat Science* 77:17–27.

McMeekin, T. A., Baranyi, J., Bowman, J., Dalgaard, P., Kirk, M., Ross, T., Schmid, S. et al. 2006. Information systems in food safety management. *International Journal of Food Microbiology* 112:181–194.

McMeekin, T. A., Olley, J., Ratkowsky, D. A., and Ross, T. 2002. Predictive microbiology: Towards the interface and beyond. *International Journal of Food Microbiology* 73(2–3):395–407.

NAFTA. 1993. North American Free Trade Agreement, Sanitary and Phytosanitary Measures, Chapter 7, Section B.

Nauta, M. 2002. Modeling bacterial growth in quantitative microbial risk assessment. Is it possible? *International Journal of Food Microbiology* 38:45–54.

Nørrung, B. and Buncic, S. 2008. Microbial safety of meat in the European Union. *Meat Science* 78:14–24.

Ross, T. and McMeekin, T. A. 2003. Modeling microbial growth within food safety risk assessments. *Risk Analysis* 23:179–197.

Stringer, M. 2005. Summary report: Food safety objectives—role in microbial food safety management. *Food Control* 77:55–59.

Sun, Y. M. and Ockerman, H. W. 2005. A review of the needs and current applications of hazard analysis and critical control point (HACCP) system in food service areas. *Food Control* 16:325–332.

Tent, H. 1999. Research on food safety in the 21st century. *Food Control* 10:239–241.

Teunis, P. F. M., Van der Heijden, O. G., Van der Giessen, J. W. B., and Havelaar, A. H. 1996. The dose–response relation in human volunteers for gastro-intestinal pathogens, Report number 284550002. National Institute of Public Health and the Environment (RIVM), Bilthoven, Netherlands.

Toyofuku, H. 2006. Harmonization of international risk assessment protocol. *Marine Pollution Bulletin* 53:579–590.

USDA/ARS. 1999. Pathogen Modeling Program. Available at http://pmp.arserrc.gov/PMPOnline.aspx. or from USDA/ARS, National Program Staff, BARC–West, Building 5, Beltsville, MD 20705. Accessed May 12, 2009.

van Schothorst, M. 2002. Microbiological risk assessment of foods in international trade. *Safety Science* 40:359–382.

van Schothorst, M. 2005. A proposed framework for the use of FSOs. *Food Control* 16:811–816.

Voysey, P. A. and Brown, M. 2000. Microbiological risk assessment: A new approach to food safety control. *International Journal of Food Microbiology* 58(3):173–179.

Walls, I. and Scott, V. N. 2005. Use of predictive microbiology in microbial food safety risk assessment. *International Journal of Food Microbiology* 36:97–102.

Whiting, R. C. and Buchanan, R. L. 1994. Microbial modeling. *Food Technology* 48:113–120.

Wooldridge, M., Clifton-Hadey, R., and Richards, M. 1996. I don't want to be told what to do by a mathematical formula. Overcoming adverse perceptions of risk analysis. *Proceedings of the Society for Veterinary Epidemiology and Preventive Medicine*, Glasgow, 27–29 March 1996, Eds. Thrusfield, M.V. Goodall, E.A., 37–47.

WTO (World Trade Organization). 1995. Agreement on the application of sanitary and phytosanitary measures (SPS). World Trade Organization, Geneva. Available at http://www.wto.org/english/docs_e/legal_e/15-sps.pdf. Accessed May 12, 2009.

CHAPTER 11
Food Package Engineering

Shyam S. Sablani

Department of Biological Systems Engineering
Washington State University
Pullman, Washington

11.1 Introduction

Food packaging process requires integration of the principles of materials science, food science, information science, and socioeconomics needs. In the last two decades, significant research and development in packaging materials has led to innovative and environmentally sound packaging to suit product characteristics and consumer needs. Comprehensive textbooks and reference books have been published that cover several aspects of food packaging.[1-5] The modern food industry is a major user of polymer-based packaging. Selected books focusing on technology, properties, and applications of plastic are available.[6-8]

The International Packaging Institute defines packing as the enclosure of products in a pouch, bag, box, cup, tray, can, tube, bottle, or other container form to perform one or more of the following functions: containment, protection and/or preservation, communication, and utility or performance. On the other hand, the U. K. Institute of Packaging considers food packaging to be a coordinated system of preparing foods for transport, distribution, storage, retailing, and end use.[1] The importance of packaging in the food supply chain is fully recognized. The scope of food wrapping is very broad and includes many activities including packaging material properties, package development, package manufacturing, graphic design, machinery design, shelf life testing, food packaging interaction, distribution, and marketing.[2]

This chapter discusses selected aspects of food package engineering. It gives an overview of the most commonly used materials for foodstuff wrappings including their processing and manufacturing, physical and barrier properties of packaging materials, and recent technological advances.

11.2 Packaging Materials

Packaging materials are used to protect a food during storage and distribution. They may provide protection from microbiological, chemical, and physical contamination. The most commonly used materials for packaging of foods are polymer, glass, metal, paper, and paper-based materials. Once dominated by metal, glass, and paper, polymeric materials are now increasingly being used for packaging. Composite materials are also being used in a growing number for packaging of foods due to their desirable characteristics. Such materials are formed by combining two or more materials that have quite different properties. Together, these properties give composite unique properties. The following section describes commonly used packaging materials and package-manufacturing process using these materials.

11.2.1 Packaging Polymers

Polymers include a wide range of natural and synthetic materials; however, only a small portion is used for packaging of foods. Synthetic polymers, commonly known as plastics, are compounds having a very high molecular weight. They are constructed of many repeating units or building blocks, known as monomers, combined together by a chemical reaction. These monomers are gases or liquids at room temperature and pressure, whereas polymers are normally solids under these conditions. Other components in plastic are small amounts of additives (typically <1 percent) such as heat and light stabilizers, antioxidants, plasticizers, and ultraviolet absorbers as well as lubricants, slip agents, and antistatic agents to improve their manufacturability.[1,2,6]

Packaging polymers may be classified into thermoplastic and thermosets depending on their cross-linked structures. Most food-wrapping polymers are thermoplastics that are not cross linked and have either amorphous or semicrystalline structures. Thermosets are hard and stiff materials usually formed at high temperature by cross-linking polymer chains with covalent bonds into a three-dimensional network. The applications of thermosets in food packaging are limited mostly to coatings, adhesives, caps, and trays.[2,6]

The properties of plastics are determined by the chemical and physical nature of the polymers used in their manufacture. The properties of polymers are determined by their molecular structure, molecular weight, degree of crystallinity, and chemical composition. These factors in turn affect the physical properties of the polymers. In terms of chemical composition, polymers can be classified as

Low-density polyethylene (LDPE)	Polyvinyl chloride (PVC)
High-density polyethylene (HDPE)	Polyvinylidene chloride (PVDC)
Polypropylene (PP)	Nylon
Polyethylene terephthalate (PET)	Ethylene vinyl alcohol (EVOH)
Polystyrene (PS)	Polysulfone
Polyvinyl alcohol(PVOH)	Polyimide
Polyphenylene sulfide	Polyvinyl fluoride
Ethylene-vinyl acetate copolymer (EVA)	

TABLE 11.1 Most Commonly Used Plastics for Packaging

homopolymer and copolymer. The former is synthesized using a single type of monomer, for example, polyethylene from an ethylene monomer; the latter is synthesized using more than one type of monomer, for example, ethylene vinyl acetate produced from a mixture of ethylene and vinyl acetate monomers. In addition, two or more polymers (homopolymers or copolymers) are blended to obtain a polymer with desirable properties.[2,6] The most widely used thermoplastic used for packaging are listed in Table 11.1.

Polymer Processing

Extrusion Low-cost synthetic thermoplastics can be converted into various shapes such as films, bags, bottles, trays, and other forms using different polymer-processing techniques. Extrusion is one of the most common methods of plastic processing. This process is used to produce cast films, blown films, multilayer films, coatings, and laminated structures. Extrusion involves melting a thermoplastic resin with an extruder, which consists of screw fitted in an electrically heated metal barrel, a hopper for feeding the resin, a motor for rotating a screw, and a die where the polymer melts exits in certain forms (Fig. 11.1). The heat from the barrel softens the polymeric material and rotating screw melts the resin pellets and forces the melted polymer through the die. The melting of the resin is mainly due to the mechanical energy input to the screw, which compresses and shears the resin pellets and dissipates frictional heat to melt the pellets. The screw is the most important element of the extruder, and different designs are used for extruding different polymers. Extruder screws are characterized by their length-to-diameter (L/D) ratios and their compression ratios, this being the ratio of the volume of one flight of the screw at the feed end to the volume of one flight at the die end. L/D ratios commonly used for single-screw extruders are between about 15:1 and 30:1, whereas the compression ratio varies from 2:1 to 4:1.[1,2]

Extruded thermoplastic is converted into films by two processes: cast film extrusion and blown film extrusion. In cast film extrusion,

328 Chapter Eleven

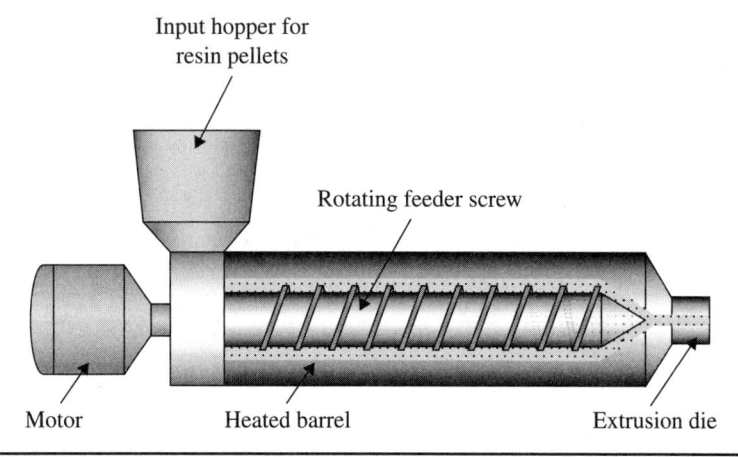

Figure 11.1 Single-screw extruder. *(Adapted from Ref. 2.)*

the polymer melt is pumped through a slit die and film is passed over chilled rollers and wound into rolls. In the blown film process, a thin tube is extruded through an annular slit die in a vertically upward direction, followed by air blown through the die head to inflate into a thin tube. The tube of film is cooled and then passed through nip rolls where the tube is flattened and then wound up (Fig. 11.2). The

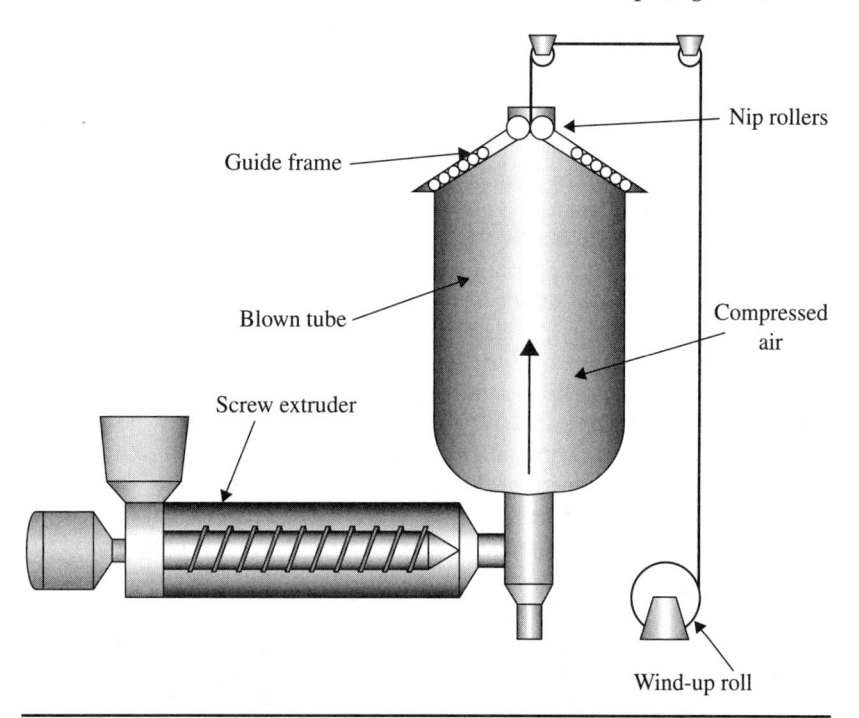

Figure 11.2 Extrusion-blown tubular film. *(Adapted from Ref. 2.)*

mechanical properties of blown film are generally better because molecular orientation is achieved in both machine and transverse directions. The stretching of film is controlled by adjusting the air pressure inside the tube and haul-off speed. The expansion ratio between die and blown film is 1.5:4 times the die diameter. The cost for making a wide tubular film is much lower than for a wide cast film due to the cost of precision grinding long chill rolls. However, blown films may contain such defects as variations in film thicknesses, surface defects, low tensile and impact strengths, and inferior optical properties. Cast film extrusion results in less thickness variation, very high outputs, and superior optical properties. Low-density polyethylene (LDPE) and high-density polyethylene (HDPE) are the most commonly used polymers for making blown films. The multilayer structure is obtained using multiple extruders; this process is known as coextrusion.[2,6]

Injection Molding This process involves softening the thermoplastic in a heated cylinder and injecting it under high pressure into a mold, cooling the part and finally removing the part (Fig. 11.3). Injection molds are expensive, but they allow high throughput, thus short production runs are not economical. Injection molding is widely used for converting thermoplastics into jars, trays, caps, spouts, and dispensers.[2] The most commonly used resins used in injection molding are HDPE, polypropylene (PP), polyethylene terephthalate (PET), and polystyrene (PS).

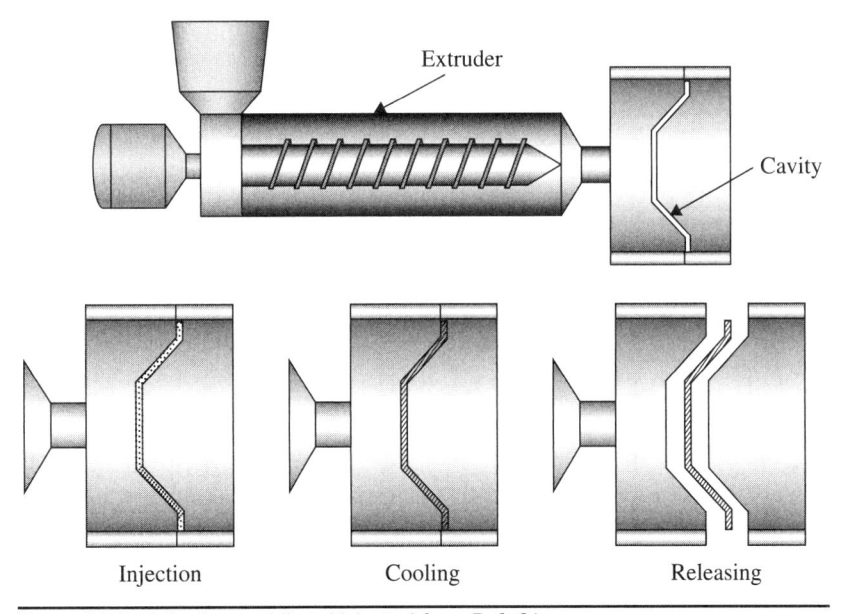

FIGURE 11.3 Injection molding. (*Adapted from Ref. 2.*)

Blow Molding The blow molding process involves blowing into a molten thermoplastic tube with high-pressure air to conform to the shape of a cooled mold. Air or nitrogen gas is introduced into the tube causing the molten mass to expand against the walls of the mold where it solidifies on cooling. The mold is then opened to remove the container. There are three common types of blow molding: extrusion blow molding, injection blow molding, and injection stretch blow molding.[1,2,6]

In extrusion blow molding, the polymer melt is extruded through a die as a hollow tube, known as a *parison*. The chilled split mold arranged around the parison is then closed. Air is blown through a blow pin mounted inside the die head that blows the parison into the final container shape (Fig. 11.4). Extrusion blow molding is used with HDPE, PP, and polyvinyl chloride (PVC). Some of these polymers are coextruded with ethylene vinyl alcohol (EVOH) or nylon to provide desired barrier properties at an economical price. Injection molding is a two-step process for making a container. At first stage, a molten plastic is injection molded into a thick hollow tube known as preform and then preform is transferred to a second mold where blowing with compressed air forms the final shape (Fig. 11.5). This process is virtually scrap free and no further trimming is required of the finished containers. Compared to extrusion blow molding, the tooling cost of this process is higher and is typically limited to the production of relatively small bottles. The most common polymers for injection blow molding are LDPE, HDPE, PP, PS, PVC, and PET. Injection stretch blow molding is an extension of injection blow molding where containers with orientation in both longitudinal and transverse directions are produced. A stretch rod is used to stretch the preform in the

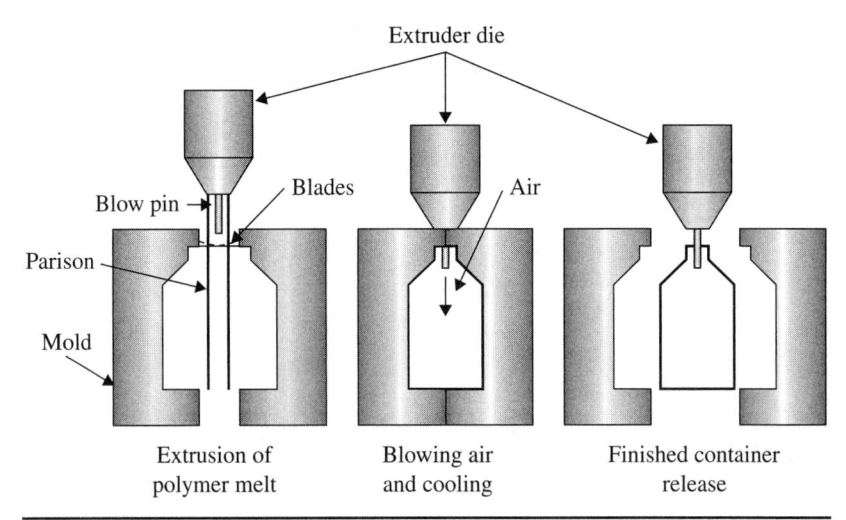

FIGURE 11.4 Extrusion-blown molding. (*Adapted from Ref. 2.*)

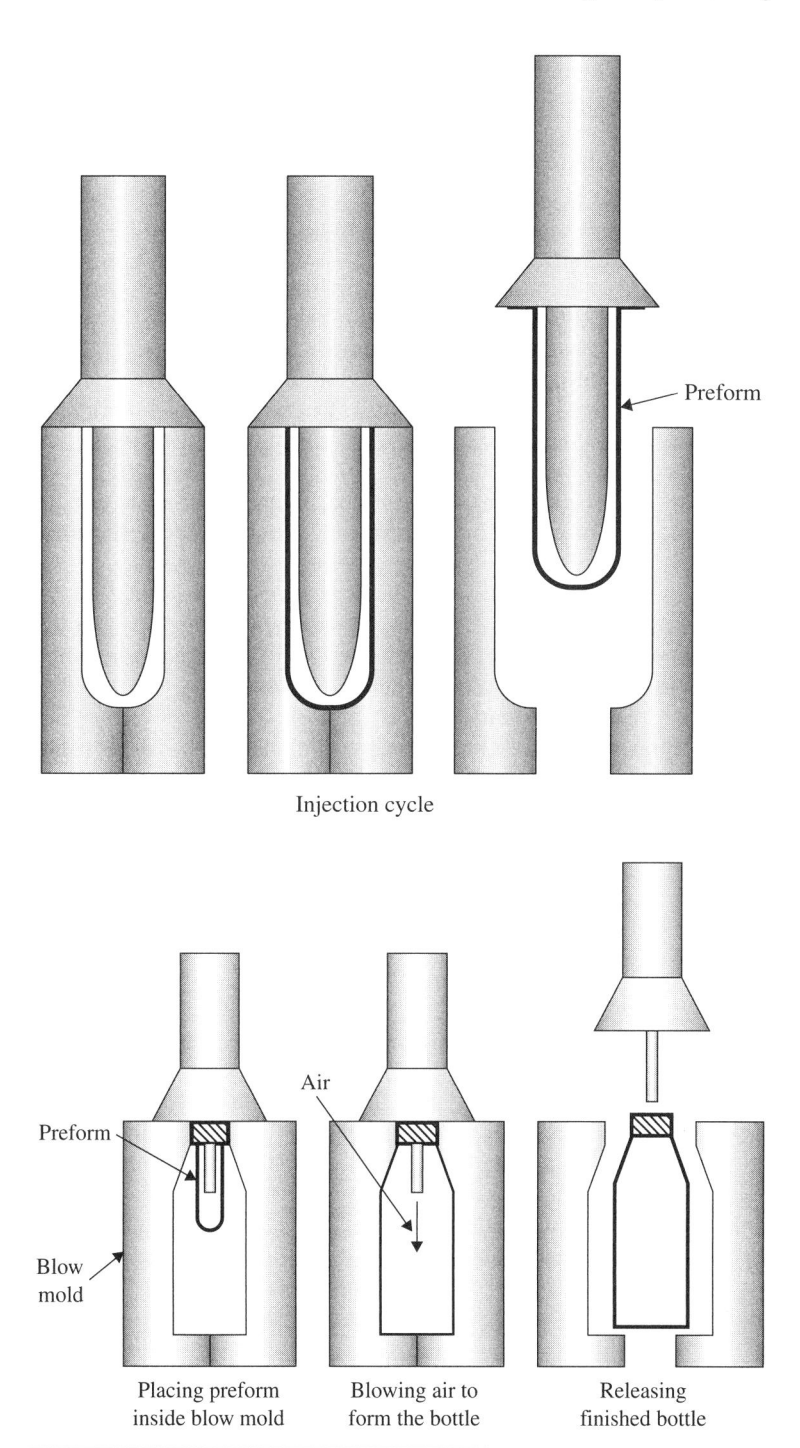

FIGURE 11.5 Injection-blown molding. (*Adapted from Ref. 1.*)

longitudinal direction, whereas orientation in the transverse direction is produced as in normal extrusion blow molding by blowing air. This biaxial orientation produces bottles with superior mechanical, optical, and gas barrier properties.[1,2,6]

Thermoforming In this process, a thermoplastic sheet, 75 to 200 μm thick, is heated to its softening temperature with infrared radiation. Mechanical or pneumatic (vacuum or pressurized air) forces are used to press the soft sheet against the mold contours. The most common thermoplastics used for thermoforming include PS, PVC, PP, PET, LDPE, HDPE, cellulose acetate, and nylon. The sheets may be monolayer, multilayer, coated, or laminated structures. Typical food packaging containers produced using thermoforming are trays, cups, and tubs.

A number of processes as described above are used to process thermoplastics into films, bottles, trays, foamed products, and other packaging forms. Some polymer processes used are orientation, vacuum metallization, lamination, and coatings to improve the function properties of final products. A detailed discussion of several of these processes is found in the literature.[1,2,6,7] The use of polymer-based packaging has increased over the years. The major advantage is their broad spectrum of properties. It is relatively inexpensive, has low density, is easily processed and shaped, and is easy to seal. The density of most plastics is similar to the density of paper but is less than half of the density of glass or aluminum and about one-eighth the density of steel. Plastics do not shatter like glass or buckle like metals.[9] The major disadvantages of polymer-based packages are their permeability to gases and vapors and their interacting properties with food.[10]

11.2.2 Glass Packaging

Molecular units in glass have a disordered arrangement but sufficient cohesion to produce mechanical integrity. Glass is defined as an amorphous, inorganic product of fusion that has been cooled to a rigid condition without crystallizing. Glass is also indicated as a supercooled liquid. Glass containers are considered to be a valuable means of packaging liquids, semisolids, and liquid/solid mixture products such as wines, liqueurs, soft drinks, sauces, dressings, and pickles, among others. Using glass to store foods is reported to be 5000 years old. In terms of the chemical composition of common glasses, silicon is their major component. Silicon does not occur freely in nature, but it is found as silica of silicon dioxide or as silicate.

Glass is hard but fragile, easily susceptible to mechanical failures. Glass is highly inert, impermeable to gases and vapors, and transparent to light. The glass transparency is modified by selecting appropriate ingredients and glass can be given different desired colors. Glass is perfectively transparent to microwave, leading to low-energy dissipation. Because glass is an amorphous material, it does not show a real melting temperature. It progressively softens, achieving a liquid-like state. This change of state occurs over a temperature range known as glass transition temperature range. Glass can be easily recycled. The main disadvantages of glass packaging are its fragility, higher density, and high energy requirement during manufacturing. Glass containers do not withstand a large temperature differential during heating (up to 60°C) or cooling phase (up to 40°C). Recent advances in glass-manufacturing technology have resulted in improved strength, reduced and density, and reduced energy in manufacturing.[1,2]

Manufacturing of Glass Containers

This process involves mixing of silica, the largest constituent (68 to 73 percent) being cullet (scrap or recycled glass). Using cullet is economically desirable because much less energy is required to melt cullet than raw silica. Cullet also reduces the amount of dust and other particulate matter that is often associated with manufacturing process that uses raw silica only. In most cases, boron (B_2O_3) and aluminum oxide (Al_2O_3) are also mixed to the thermal and mechanical performance of the final container. To reduce the viscosity of melts and melting temperature of silica, sodium carbonate and calcium carbonate are also added. A large quantity of carbon dioxide gas is released in glass manufacturing, together with a small amount of sulfur dioxide and water vapor. Ingredients such as, sulfate, nitrate, or sulfite of alkaline ions is added to remove gases from the mass of molten glass. Finally, other ingredients are added to give glass certain physical properties. For example, lead gives clarity and brilliance, although at the expense of softness of the glass; alumina increase hardness and durability. The mixture of all ingredients in a preweighted quantity is then transferred to a glass-melting furnace maintained at a temperature of approximately 1500°C. At this temperature, molten glass is chemically homogeneous and gas free. Then the molten glass is cooled to about 1150°C to form containers.

Glass containers are made using commonly known forming processes: (1) blow and blow and (2) press and blow. In both processes, two different molds are used in two subsequent stages. In the first mold, the lump of molten glass, also known as a *gob*, is transformed into a preform (or a parison) by pressurized air (in blow and blow process) or by a plunger in the gob (in press and blow process). The final shape of the container is achieved in the second mold by pressurized (around 160 to 250 kPa) air. Most bottles or narrow-neck containers are produced using blow and blow process, whereas jars and wide-mouth containers are produced by the press and blow process. However, it is currently possible to produce narrow-neck glass packages using the press and blow process, which produces containers with uniform wall thickness and higher strength. The industrial production rate of glass packages is slower compared to other food packaging containers. It takes about 10 to 12 s from the time that the gob falls in the blank mold to the time the formed package exits.[2]

Postprocessing Treatments The temperature of the glass container is lowered quickly in the second mold, which produces internal stresses in the containers, making them very fragile. The glass containers are subjected to annealing to remove internal stresses. This is achieved by raising the temperature of the containers to about 540°C (about softening temperature), holding them at this temperature for few minutes, and then cooling the containers slowly to well below softening temperature. Glass containers are also subjected to surface treatments to improve their mechanical properties and chemical resistance. The inner surface is treated with SO_2 or fluorocarbon to form sodium sulfate or sodium fluoride, which increases the container's resistance to chemical attack. An outer surface is treated with a vapor containing tin or titanium prior to annealing. This treatment forms a metal oxide layer, which improves adhesion properties of postannealing coating. Before the annealing process ends, the outer surface of the container is sprayed with an aqueous solution of waxe, stearates, silicones, oleic acid, or polyethylene to improve its lubricity.[1]

11.2.3 Metal Packaging

Aluminum and steel are the most commonly used metals for packaging foods. Aluminum is used in the form of an alloy containing magnesium and manganese, whereas steel is coated by tin or chromium oxides. Lead and copper are used for soldering or welding tinplate and chrome-coated steel containers. Aluminum is purified from aluminum oxide (Al_2O_3) using an electrolytic process. Aluminum oxide is obtained from ore, which is mined as bauxite. The energy requirements to produced purified aluminum are extremely high, seven- to ten-fold the energy required to produce the same mass of tinplate or steel. Aluminum is typically added with alloying agents to increase its mechanical properties and corrosion resistance and to improve formability. Steel is produced from pig iron and may contain manganese, chromium, nickel, molybdenum, cooper, tungsten, cobalt, and silicon, depending on the desired properties. Based on its chemical composition, steel is classified into three categories: carbon steels, alloy steels, and stainless steels. Carbon steel is most important for food packaging applications. Tinplate is produced by coating the steel with tin using an electrolytic process. The method of electroplating is very intricate.[1,2] Because tin is an expensive metal, the industry has developed tin-free steel as an alternative to tinplate. Tin-free steel, also known as electrolytic chromium-coated steel, is produced by coating low-carbon steel with chromium or chromium oxide.

Manufacturing Metal Containers

Cylindrical cans are the most frequent types of container made from metal, although other shapes are also manufactured. These cans are manufactured in three- and two-piece units. For manufacturing of three-piece cans, a coil is cut into rectangular sheets, lacquered and decorated as required, and then cut into strips of size based on desired can dimensions. The slit strips are cut into body blanks and corners are notched. The side edges forming the seams are bent to form hooks, and then a strip is rolled to form the can's body and side seam. The seam area is then soldered with a mixture of 2 percent tin and 98 percent lead. These days silver is used in place of lead for soldering due to environmental concerns and lead toxicity. Welding is also done, which offers a stronger side seam of narrow margin needing less metal. A flange is then formed on each end of the can. One end, the top or the bottom, of the formed body is then closed.

Two-piece cans are manufactured from both aluminum and tinplate and have no side seam. A higher rate of production is achieved in manufacturing two-piece cans compared to manufacturing three-piece cans. Two processes—drawn and ironed (D & I) and drawn and redrawn (D & RD)—are used to make two-piece cans (Figs. 11.6 and 11.7). In the D & I process, a circular blank of metals fed into a cupping press that stamps and draws shallow cups. The cup is then transferred to

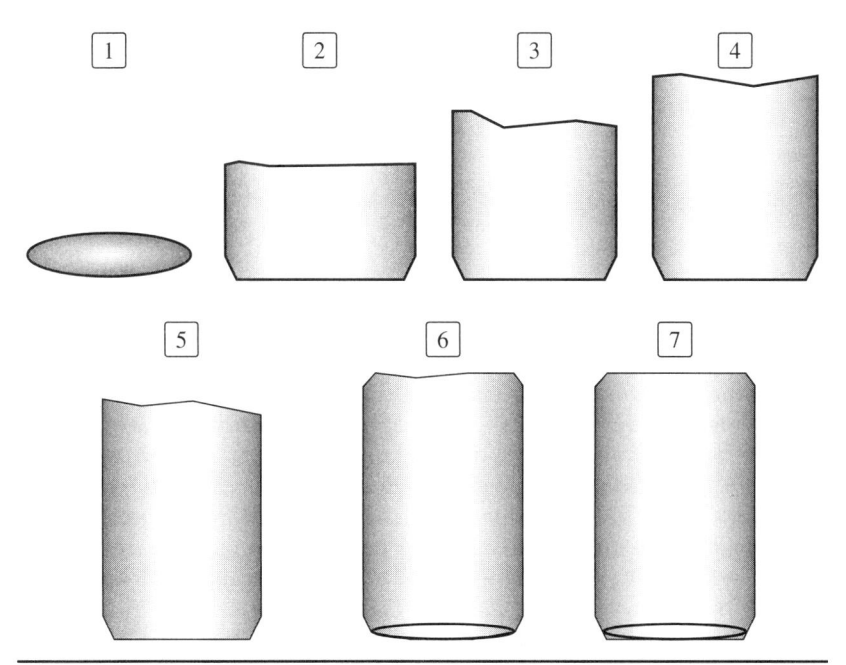

FIGURE 11.6 Different stages of drawn and ironed can production. (1) Body blank, (2 and 3) drawn and redrawn cup, (4 through 6) wall ironing and base formation, (7) trim to required height. (*Adapted from Ref. 1.*)

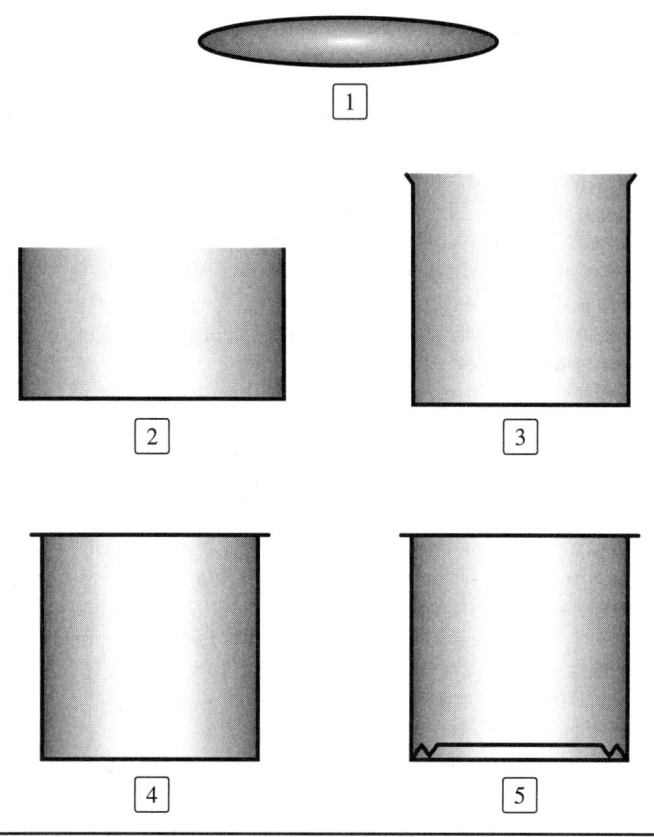

FIGURE 11.7 Different stages of drawn and redrawn (D & RD) can production. (1) Body blank, (2 through 4) drawn and redrawn cup, (5) trimming and profiling base. (*Adapted from Ref. 1.*)

an ironing press where it is pressed through series of dies (Fig. 11.6). In this process, the wall thickness is reduced and the body height is increased. The diameter of the bottom and the thickness of the sheet remain the same during the ironing process. The bottom end is domed to provide added strength, and the top portion is trimmed to the desired height.[1,2] The trimmed cans are chemically cleaned and flanged. Cans intended for beverages are necked. In the D & RD process, the final height and diameter of the container is produced by sequentially drawing cups to a smaller diameter, keeping the thicknesses of the wall and the base as well as the surface area matching the original blank (Fig. 11.7).

The metal containers are strong and provide an effective barrier against gases and light and are resistant against high temperatures during processing. The main disadvantages of metal cans are their heavy mass, higher cost, and tendency to interact with contents and environment (internal and external corrosion). A metal such as aluminum is also used extensively as foils for packaging, caps, lids, and laminates. Aluminum foil is also used in multilayer retort pouches. Other metal containers are trays, collapsible tubes, and aerosol containers. Most metal packaging intended for food packaging is coated on the inside and outside surfaces. The coating on the container surfaces is called lacquer or enamel. An internal coating prevents food contamination by metallic ions. Many types of internal coatings are available for food containers including oleoresins, vinyl, acrylic, phenolic, and epoxy-phenolic. External surfaces of a container are coated to improve corrosion resistance and provide a barrier layer against external abrasion. The coating also improves a container's appearance and its marketability.

11.2.4 Paper and Paper-Based Packaging

Paper and paper-based packaging include bags, wrappings, paperboard, corrugated boards and boxes, folding cartons, composite cans, and fiber drum. The packaging is used in primary, secondary, and tertiary packaging. Pulp, which is obtained from plant fiber, is the raw material used to produce paper-based packaging. Paper and paper-based packaging is based on a renewable resource and is biodegradable. The main source of fiber for making pulp are trees, including spruce, pine, birch, eucalyptus, fir, poplar, and jasmine. Plant fibers consist of cellulose, hemicelluloses, and lignin, and their proportions in fiber may vary depending on the type of tree and different parts of the tree. The characteristics of paper and choice of paper-making technique are determined by the amount of cellulose, hemicelluloses, and lignin present in the fiber.

Manufacturing Paper

The first step in paper making is to prepare a pulp from fibers. Pulp can be produced from wood chips using two processes: chemical and mechanical. In the chemical process, diluted (5 percent) sodium hydroxide at high temperature (170°C) is used to prepare the pulp. In addition, sodium sulfite and a small amount of sodium sulfate solution are used to improve the yield and mechanical resistance of fibers. Mechanical pulp is produced by forcing wood chips against a rotating grindstone. This mechanical process is less expensive due to the high yield obtained.

Mechanical pulp contains 70 to 80 percent fiber bundles, broken fibers, and fines in addition to individual fibers. Most often, the pulp-making process is improved by combining mechanical and chemical processes. In addition, thermal treatment is used to enhance the pulp yield.

The next step is bleaching the pulp to improve fiber brightness. Mechanical pulp is bleached using hydrogen peroxide followed by neutralization and destruction of excess peroxide with SO_2. The bleaching of chemical pulp is achieved in a number of stages using chlorine, hypochlorite, chlorine dioxide, oxygen, and peroxide. The pulp is treated with alkali to dissolve the degradation products. After bleaching, the pulp is passed into a beater where the surface area of fibers is increased for water absorption and interfiber bonding. This process is known as beating or refining, and the extent of beating in the process determines the physical properties of the final paper.[1] A small amount of beating will result in high-tear strength with low tensile and burst strength. Beating time is varied to produce a paper with the desired tensile and burst strength. Several additives are mixed during the beating process to increase the performance and reduce the cost of paper manufacturing. A dilute solution of fibers is then deposited onto a fine screen in the Fourdrinier system (a rotating wire-mesh cylinder connected to a vacuum pump). The sheet of paper is then mechanically pressed and dried using steam-heated rolls to reduce the moisture content below 10 percent.

In many applications, the paper is further treated to improve its characteristics. These treatments include (1) calendering to improve surface smoothness, control texture, and develop a glossy finish, (2) sizing and coating to improve the paper's appearance, (3) adhesive application to increase surface strength, gloss, brightness, opacity, smoothness, ink receptivity, and surface firmness, and (4) barrier coatings to provide water vapor and oxygen barrier properties. The quality of the final paper is measured by its physical properties including thickness, tensile strength, stretch, bursting strength, tearing strength, stiffness, and folding endurance. Paper is generally converted to a variety of other products such as paper boards, paper bags, folding cartons, and corrugated and solid boards.

11.3 Physical Properties of Packaging Materials

During the manufacturing process and packaging fabrication, the material is subjected to various physical treatments. A good knowledge of physical properties of packaging materials is needed to design and optimize physical processes of package manufacturing and fabrication. Physical properties of packaging materials are also important in designing and selecting the appropriate type of packaging for different food products to achieve desirable shelf life. The physical properties in context of food packaging are classified in different categories: mechanical, thermal, optical, and mass transport properties. These properties are very well defined, and appropriate scientific methods are available for their measurement. The following section describes relevant physical properties of packaging materials, referring to their importance in food packaging.

11.3.1 Mechanical Properties

The behavior of a material to mechanical forces is described in these properties. Mechanical properties are important in selecting materials and designs and producing packaging, optimization of transport, and distribution.

Density This is defined as mass per unit volume (kg/m^3). There are several standard methods available for measuring density of packaging materials. Densities of these materials vary from 800 to 8000 kg/m^3. Density of materials is used to compare different physical attributes of similar or dissimilar materials. The density of different foamed plastics can indicate the amount of voids. Papers with different densities can indicate the level of fiber compaction.

Tensile Strength This is the maximum tensile stress that a material can withstand before breaking. During tensile testing of the material, three other parameters measured are elongation, yield strength, and Young's modulus. Elongation is defined as the percent of change in original length of the material before it breaks. A large value of elongation indicates that the material will absorb a large amount of energy before breaking. Yield strength indicates the maximum stress the material can take before it undergoes nonreversible distortion. Both elongation and yield strength is important in the design of an unwinding operation of the polymer film roll. Young's modulus is defined as the ratio of stress to strain over the elastic range (reversible deformation). Young's modulus is also useful for characterizing paper stiffness and polymer-based packaging.

Burst Strength This is measured as the maximum pressure at which the package fails. Pressure is applied at a right angle to the surface of the package at a slowly increasing rate. The bursting test is common for paper and polymer-based packaging. Metal cans and glass bottles are also tested for their resistance for internal pressure.

Impact Strength This is a measure of package strength to withstand a sudden (shock) load. This property is related to the material's toughness and the ability of the material to absorb applied energy. The impact resistance of a package is measured using notched Izod, or dart drop tests. In practice, packages may experience a sudden load as a result of the package falling onto a rigid surface or knocking against hard objects. The impact test is commonly performed for glass, paper, and polymer-based packaging.

Tear Strength Paper and polymer-based packaging are frequently subjected to tests that measure tear strength. The energy required to propagate a tear is measured as the tear strength of the material. The Elmendorf test is one of the most common methods for measuring the tear strength of paper and plastic films. The tear in packages is initiated by sharp edges of products (frozen beans, French fries) or any other physical defects (microholes, microbubbles, sudden thickness changes, cracks, etc.) in a package.

Stiffness This is the resistance of the paper, film, or sheet to distortion or bending. Young's modulus, the cantilever test, and Handle-O-Meter stiffness tester are used to quantify the stiffness of packaging.

Creep Resistance Packages and packaging materials may be subjected to low but constant stress for very long periods of time, for example, packages stacked on a shelf and pressurized containers. Creep resistance is the deformation measured over time under constant stress. Most packaging materials and packages intended for long storage under constant stress are tested for creep analysis.

Crease Resistance Resistance to repeated flexure or creasing of a film or a paper may result in increased permeability to gases, reduced tensile properties, or change in optical properties. Crease resistance is measured as the number of cycles to failure or significant changes of a material's properties.

Cushioning This is an ability of a material to absorb energy of dynamic stress (vibration). Materials with higher energy-absorption ability provide better cushioning for food. The cushioning property of packaging is important in providing protection to packaged foods against dynamic stresses during manufacturing, transport, and distribution.

Coefficient of Friction This is a measure of ease with which the surface of one material will slide over another. This property is important for all packaging materials used for packaging foods. Knowing the coefficient of friction of different materials is valuable for stacking packages over each other, rolling films/paper in printing and laminating equipment, sliding, or moving packages over conveyor belts.

11.3.2 Thermal Properties

Knowledge of thermal properties is required to quantify heat transport in packages during package manufacturing, in-package processing, and storage. Heat transfer occurs in materials during package fabrication and may also occur in packages during processing (e.g., in-container sterilization and pasteurization) and storage (under fluctuating temperatures).

Specific Heat This is defined as the amount of heat required to raise a temperature of a unit mass by one degree. A larger specific heat of a material indicates that is has better thermal insulation properties. In general, amorphous solids have a larger specific heat than crystalline solids. The specific heat of semicrystalline materials depends on the amorphous and crystalline fraction in solids. Metals and glasses have lower specific heat than plastics and cellulose materials.

Thermal Conductivity   This indicates the ability of a material to transport thermal energy within the solids by conducting the mode of heat transfer. Thermal conductivity of solids depends on their atomic structures and intermolecular bonds. Metals have larger thermal conductivities than glass, plastic, or cellulose. Thermal conductivity of plastic and paper-based packaging can be changed significantly by incorporating the air of low thermal conductivity into the packaging. Corrugated board and expanded polystyrene show better insulating properties than solid board and plastics of the same thickness.

Thermal Diffusivity   This is the ratio of thermal conductivity to the product of density and specific heat. The rate of conduction heat transfer in solids is proportional to their thermal diffusivities. Thermal diffusivity of packaging material influences heating or cooling rates during in-container thermal processing or cooling. Metal cans and multilayer aluminum/plastic pouches show a very high rate of heat transfer during thermal processing.

Coefficient of Thermal Expansion   This expresses the relative change in length for a temperature change at constant pressure. Most polymers and alumina have a larger coefficient of thermal expansion than steel, glass, and paper. The linear expansion in materials due to heat may damage the container seal during thermal processing or cooling.

State/Phase Transition Temperature   The melting point of crystalline materials and glass transition temperatures for amorphous materials are important during fabrication of packaging. The mechanical properties of materials are affected dramatically around their state/phase transition temperatures. The amorphous and crystalline materials above their transition temperatures can easily be formed into different shapes. Amorphous packaging materials with very high transition temperatures behave like rigid solids at room temperature. The molecular mobility in materials is restricted to their glassy state, resulting in low diffusion of gases. Thermal and mechanical properties of materials are affected by their transition temperatures.[1,2]

11.3.3   Optical Properties

Metal and paper-based packaging do not allow (visible) light to transit through them. For thermoplastic polymers and glass, a number of optical properties of importance are transmittance, reflectance haze, clarity, gloss, and refractive index. These properties are important for their aesthetic characteristics and viewing the food through the package. Transmittance of ultraviolet light though the packaging material may be important for food protection.

Transmittance This is the ratio of the light passing through to light incident on the material, and reflectance is the ratio of the reflected light to the incident light. Haze indicates a material's "milkiness" or "cloudiness." For transparent material, transmittance is measured as the percentage of transmitted light that deviates more than 2.5 percent from the incident radiation. Surface imperfections and in homogeneities in a material's structure affect the haze of material. Clarity is a measure of degree of distortion of an object when viewed through the material. The ability of a material's surface to simulate a mirror (shiny surface) is defined as (specular) gloss. The refractive index is the measure of refracted light when it passes through a medium of one density to a medium of another density. The refractive index depends on density of the (packaging) material. The higher the index of refraction of a transparent material, the clearer the vision appears.

11.3.4 Properties of Mass Transport

Metal and glass are impermeable to gases and vapors and cellulose and polymer-based packaging are not. The transport of water vapor, oxygen, carbon dioxide, and other gases can influence the quality, and the shelf life, of packaged foods. A decrease in moisture in fresh produce can lead to texture changes and economic loss, whereas an increase in moisture in dehydrated products can cause microbial growth, increase enzyme activity, and change texture. Oxygen transfer in a package may lead to quality deterioration in dehydrated products (e.g., lipid oxidation in fatty foods) and loss of color and nutrients. However, in some cases modification of the atmosphere within packaging gases may be desirable, which is achieved by transfer of gases through the packaging material, for example, modified atmosphere packaging of fresh produce and freshly roasted coffee.[2]

The transport of water vapor and gas through a polymeric material is described by two processes: permeation and pore effect. Permeation involves three steps: adsorption of gas or water vapor onto the high-concentration side of the polymeric surface, diffusion across the film thickness, and desorption of the gas or water vapor from the low-concentration side of the polymeric surface. In the case of pore effect, water vapor and gas flow through microscopic pores, pinholes, and cracks in the materials. Fick's first law of diffusion gases is used to characterize transport of water vapor or gas through a polymeric film [Eq. (11.1)]:

$$J = -D\frac{dc}{dx} \qquad (11.1)$$

where J = flux, mol/cm² s
  D = diffusion coefficient, cm²/s
  c = concentration of water vapor or gas, mol/cm³
  x = film thickness, cm

It is convenient to measure vapor pressure (p) than measure the actual concentration of gas or vapor. Under steady-state conditions, Eq. (11.1) can be integrated across the total thickness of the polymer film and vapor pressure across the film [Eq. (11.2)]:

$$\frac{Q}{t} = \frac{PA\Delta p}{x} \quad (11.2)$$

where Q = amount of gas or water vapor, cm³
  t = time, s
  P = permeability cm³ cm/cm² s atm

Over 30 different units for P are reported in the literature. Some of the common units of permeability are cm³ cm/cm² s Pa, mL mm/m² day kPa, and mL cm/m² day atm. Equation (11.2) assumes (1) steady-state diffusion, (2) a linear concentration gradient across the film, (3) unidirectional diffusion, and (4) permeability independent of concentration.

Unsteady-state transport of water vapor or gas through polymeric film is described using Fick's second law [Eq. (11.3)]:

$$D \frac{\partial^2 c}{dx^2} = \frac{dc}{dt} \quad (11.3)$$

where t is time (s).

The permeability of a polymer is affected by the nature of the polymer, the nature of the gas or water vapor, and ambient conditions. The barrier properties of a polymer's gas and vapor is increased with the increasing polarity of certain types of polymer, regularity of molecular structure, and close chain-to-chain packing in the polymer matrix. Crystalline polymers have lower gas permeability than amorphous polymers.[8] Permeability (diffusivity × solubility) also depends on the molecular size of the gas and its chemical affinity to the polymer. Larger gas molecules generally have lower diffusivity and higher solubility compared to smaller molecules. Temperature and relative humidity also affect a gas's permeability.

Many methods are used for measurement of water vapor and gas permeability of polymer-based packaging.[1,2] One method is based on the permeation rate equation [see Eq. (11.2)]. Quantity of water vapor (Q) transferred through a film is measured using a precision balance, whereas a sensitive detector is used in a permeation cell to quantify the amount of gas transfer. The exposed surface area (A) of the polymer film is accurately defined by the opening in the permeation cell. The pressure difference (Δp) across the film is controlled by the instrument. The film thickness (L) is measured using a micrometer. The choice of detector determines which gas is being measured. Thermal conductivity detectors are used for most gases—flame ionization detectors for volatile organic vapors, coulometric sensors for oxygen, and infrared detectors for water vapor.[2] The permeability values of some of the polymers are presented in Table 11.2.

11.4 Recent Advances in Packaging

Food packaging has become very sophisticated in today's modern society. Consumers and the packaging industry seek enhanced features in packaging beyond its basic functions of containment, protection, convenience, and communication. Some recent advances in food packaging have been in the area of active and intelligent packaging. The former involves physical, chemical, or biological action for altering the interactions between the package, the product, and the package headspace to achieve a desired outcome.[2] Examples of active packaging are absorbing systems such as oxygen, carbon dioxide, moisture, ethylene, and releasing systems including carbon dioxide, ethanol, antimicrobials, antioxidants, enzymes, flavors, and nutraceuticals to protect and enhance food quality. Other packaging systems are self-heating, self-cooling, microwave susceptors, and selective permeable film. Intelligent packaging is defined as a packaging system that is capable of carrying out functions such as detecting, sensing, recording, tracing, communicating, and applying scientific logic to facilitate decision making to extend shelf life, enhance safety, improve quality, provide information, and warn of possible problems.

Increased use of synthetic polymer bases has raised environmental concerns. Packaging constitutes approximately one-third of the municipal solid waste stream by weight in many developed countries. Many different ways are being used to manage packaging waster: reduce, reuse, recycle, compost, incinerate, and landfill. The presence of some polymers such as PVC in incinerators may lead to dioxins, a family of highly toxic chlorinated hydrocarbons. Due to environmental concerns, PVC containers are being replaced by PET or PE. Considerable efforts are being made to produce biodegradable packaging from biobased materials such as polylactic acid, polyhydroxybutyrate valerate, and thermoplastic starch. The use of biobased packaging for food application is limited due to the production cost of biobased polymers and their weak mechanical and poor barrier properties. Research is underway to improve the physical properties of biodegradable plastic by incorporating nanoclay particles and blending with other polymers.[11,12] Some innovative technologies such as photoelectronic scanners and microprocessor-controlled video cameras are being developed for online detection for seal defects or

Food Package Engineering 345

Polymer	O_2 Permeability cm³ mil/100 in² day atm at 25°C	CO_2 Permeability cm³ mil/100 in² day atm at 25°C	Water Vapor Permeability g mil/100 in² day at 38°C, 90% RH
LDPE*	300–600	1200–3000	1–2
HDPE[†]	100–250	350–600	0.3–0.6
PP (unoriented)[†]	150–250	500–800	0.6–0.7
PP (oriented)[†]	100–160	300–540	0.2–0.5
PS[†]	250–350	900–1050	7–10
PET[†]	3–6	15–25	1–2
PVC (unplasticized)[†]	5–15	20–50	2–5
PVC (plasticized)*,[†]	50–1500	200–8000	15–40
PVDC[†]	0.1–2	0.2–0.5	0.02–0.6
EVOH (0% RH)[†]	0.0007–0.1	0.01–0.5	—
EVOH (100% RH)[†]	0.2–3	4–10	—
Ionomer[†]	300–450	—	1.5–2
Nylon 6[†]	2–3	10–12	10–20
PC[†]	180–300	—	10–15
Polysulfone[‡]	230	950	45
PVOH[‡]	0.05–0.2	0.1	—
PPS[‡]	30	75	1.6
Polyimide[‡]	6.5	7.5	1.4
Polyamide Nanocomposite[‡]	0.001	2.5	—
Polyvinyl fluoride[‡]	3	11	3.3
EVA[‡]	460	2800	11.5
Multilayer Films			
Nylon 6-LDPE[‡]	4–5	—	0.5
HDPE-EVA-PVDC-EVA[‡]	1.1	2.8	0.15
EVOH-LDPE[‡]	0.05–0.1	—	—
PP-PVDC-PP[‡]	3.55–4.55	—	—
PP-EVOH-PP[‡]	0.9	—	—
PS/PC-EVOH-PP[‡]	0.7	—	—
HDPE-EVOH-LDPE[‡]	1.5	—	—

—Data not available.
*Value depends on amount of plasticizer.
[†]*Source:* Data from Ref. 2.
[‡]*Source:* Data from Ref. 8.

TABLE 11.2 Permeability Values for Various Polymers

faults on the surface and inside the containers. Intelligent packaging devices such as bar codes memory tags and radio frequency–emitting tags are used in distribution channels to keep track of pallet loads of food products.[2] Consumer demands for convenience, product safety, and need of environmental friendliness, will continue to be major forces of innovation in food packaging.

References

1. Robertson, G. L. 1993. *Food Packaging, Principles and Practice*. New York, NY: Marcel Dekker.
2. Lee, D. S., Yam, K. L., and Piergiovanni, L. 2008. *Food Packaging Science and Technology*. Boca Raton, FL: CRC Press, Taylor and Francis Group.
3. Han, J. H. 2005. *Innovations in Food Packaging*. London, UK: Elsevier Academic Press.
4. Coles, R., McDowell, D., and Kirwan, M. J. 2003. *Food Packaging Technology*. Oxford, UK: Blackwell Publishing Ltd.
5. Ahvenainen, R. 2003. *Novel Food Packaging Techniques*. Boca Raton, FL: CRC Press.
6. Brown, W. E. 1992. *Plastics in Food Packaging*. New York, NY: Marcel Dekker.
7. Osborn, K. R. and Jenkins, W. A. 1992. *Plastic Films: Technology and Packaging Application*. Lancaster, PA: Technomic Publication Company, Inc.
8. Massey, L. K. 2003. *Permeability Properties of Plastic and Elastomers: A Guide to Packaging and Barrier Materials*, 2nd ed. Norwich, NY:William Andrew Publishing.
9. Nordmark, B. 1998. The miracle material from the 20th century: Plastics, the safe barrier. *Food Technology in New Zealand* October:10–13.
10. Miltz, J. 1992. Food packaging. In *Handbook of Food Engineering*, eds. D. R. Heldman, and D. B. Lund. 667–740. New York, NY: Marcel Dekker.
11. Rhim, J. W. 2007. Potential use of biopolymer-based films in food packaging applications. *Food Science and Biotechnology* 16:691–709.
12. Chaudhary, Q., Scotter, M. Blackburn, J., Ross, B., Boxall, A., Castle, L., Aitken, R. et al. 2008. Application and implications of nanotechnologies for the food sector. *Food Additives and Contaminants* 25:241–258.

CHAPTER 12
Evaluation of Transgenic Wood for Wood Productivity and Quality

Siqun Wang
Department of Forestry, Wildlife & Fisheries
University of Tennessee
Knoxville, Tennessee

12.1 Introduction

Wood is a cellular solid characterized by a high degree of anisotropy at all structure levels. It consists of different cell types that are oriented in the axial direction (tracheids in softwood and libriform, tracheid fibers, and vessels in hardwood) or in radial direction (ray cells). The cell walls of wood fibers are built up of layers of different thickness. Each layer has different spiral microfibril angles (MFA) and chemical components. The secondary cell wall layer (S_2) is the predominant factor that determines the mechanical properties of the fiber because its thickness accounts for approximately 80 percent of the thickness of the entire cell wall. The stiffness of wood mostly relies on the semicrystalline cellulose microfibril as found in Z-helix form around the lumen within the cell walls of the wood (Meylan and Butterfield 1978) as well as its microfibril angle (Cave 1968; Tze et al. 2007). As a natural material, wood has been used for homes and other structures, furniture, tools, vehicles, papers, decorative objects, and biopolymers.

There is no doubt that America is experiencing its second major energy crisis in 30 years, and bioenergy from renewable resources will be a very promising solution. Ethanol as an alternative fuel is receiving a great attention as it addresses concerns related to dwindling oil supplies, energy independence, and climate change. The ethanol in the United States is produced mainly from corn starch. The U.S. Congress envisioned a 30 percent replacement of the current U.S. petroleum consumption with biofuels by 2030. Accomplishing this goal would require approximately 1 billion dry tons of biomass feedstock per year. The current production of corn could not meet this anticipated demand, which could mean that less grain will become available for food and feed purposes. A study by Perlack et al. (2005) found over 1.3 billion dry tons per year of biomass potential—enough to produce biofuels to meet more than one-third of the current demand for transportation fuels. This annual potential is based on a more than sevenfold increase in production from the amount of biomass currently consumed for bioenergy and biobased products. About 368 million dry tons of sustainably removable biomass could be produced on forestland, and about 998 million dry tons could come from agricultural lands.

Although wood has been used many years in broad areas, there are a lot of disadvantages and problems during wood process and in use. There is a need to engineer forest trees for specific usage or to be fast growing to provide additional biomass for future bioenergy.

12.2 Overview of Transgenic Woods

12.2.1 Transgenic Trees

In recent years, the rapidly increasing knowledge of plant genomes has raised the possibility of genetically manipulating trees and other woody plants at a rate faster than that afforded by traditional tree-breeding programs. Advancements in gene cloning and genomics technology in forest trees have fostered the introduction of value-added traits for wood quality and for resistance to biotic and abiotic stresses into genotypes, adding a new dimension to forest-tree improvement programs (Koehler and Telewski 2006).

There are many traits being introduced and modified in wood species, such as spruce, English elm, pine, poplar, yellow poplar, and aspen. Poplars were the first forest tree to be transformed genetically (Fillatti et al. 1987) and the first to have a sequenced genome (Brunner et al. 2004). Pines and spruces as coniferous forest species are the most studied transgenic trees, primarily due to their substantial economic importance, particularly in developed countries, for construction lumber as well as for pulp and paper. In the United States, loblolly pine is more widely planted than any other species of tree.

A tree's trunk is the major harvested organ in forest trees, and breeding programs are usually aimed directly at improving trunk performance and wood quality. This includes biochemically modifying wood characteristics and trunk structure, increasing its growth rate, and altering its shape. Additional breeding targets include improving the root system and tree canopy (leaf) performance, pest resistance, and tolerance to abiotic stresses (Tzfira et al. 1998).

Herbicide-resistant transgenic crops are considered to be one of genetic engineering's major successes. Direct herbicide detoxification and reduction of target-enzyme sensitivity to the applied herbicide are the two main approaches that have been used for engineering plants with herbicide resistance (Tzfira et al. 1998). Most tree species are susceptible to the herbicide used for weed control. These chemicals interfere with key metabolic pathways of trees, hampering their normal growth and having great impact on their commercial value in forestry. Thus, introducing the herbicide-resistance trait in trees has great economic potential. It will allow the establishment of tree plantations in weed-infested sites that would otherwise be economically unsuitable to plant (Walter et al. 2002).

Important losses caused by defoliating insects occur in various tree species. The damage often translates into a reduction of tree growth and survival, as well as in alterations in tree shape or fruit quality. In practice, the use of insecticides is rather limited in forestry, in part due to the large forest areas and tree sizes and to the development of resistance by insects and the environmental impact of insecticides. Genetic engineering currently allows the production of plants resistant to a wide range of insects through the induction of transgenes derived from plants, microorganisms, and mammals. These transgenes code for an extensive range of biomolecules that attack the insects' digestive systems using different mechanism (Schuler et al. 1998).

In general, most natural forest-tree species are well adapted to their environment, exhibiting high ecological competence. However, forestation with plantation-improved or imported exotic tree species will probably reveal their sensitivity to several ecological factors. Cold, drought, salinity, and heavy-metal toxicity are the main stresses specifically affecting trees, which are subjected to many annual changes during their life cycle (Tzfira et al. 1998). Genetic engineering for cold tolerance has been evaluated in several transgenic plants expressing an ice-nucleation gene from bacteria, antifreeze genes from fish, and altered lipid composition in their cell membranes (Tzfira et al. 1998). Cold tolerance in trees would enable the use of cold-sensitive species in northern areas, as well as providing better protection of native plants from chilling damage.

Phytoremediation, the use of transgenic plants to remove contaminants from soil or water, has a potential impact on environmental pollution and, in the long term, the preservation of natural forestry (Herschbach and Kopriva 2002). Rugh et al. (1998) found that overexpression of the bacterial mercuric reductase in yellow poplar resulted in transgenetic plants that were resistant to toxic levels of mercuric ions and were able to release elemental mercury.

12.2.2 Challenges for Evaluation and Safety

Because the ability to transform forest-tree species is becoming a reality, the release and use of transgenic wood has not been exempt from criticism, generally due to activist groups. Unlike annual crops, a forest tree has a long perennial life, needs a large field size, and requires stability of the transgenes over several years. A forest tree must adapt to seasonal climatic changes; has resistance to pests, diseases or herbicides, and enhanced stress tolerance. All of these raised a fear of the generation of "superplants" that could displace the natural ecosystem or the occurrence of pollen-mediated transfer of undesired transgenic traits into unexpected hosts.

James et al. (1998) have described potential impacts from the escape of transgenic forest trees, outlined some potential avenues to decrease the possibility for such escapes, and encouraged increased field testing and monitoring. Many of their suggestions were incorporated into a position statement on the benefits and risks of transgenic tree plantations adopted by the IUFRO Working Party on Molecular Biology of Forest Trees (2.04.06) in September 1999 (Strauss et al. 1999; Merkle and Dean 2000). There are many options to mitigate their impacts. Strauss et al. (1999) listed the following priority research areas: (1) isolation, modification, and testing of additional genes, and systems for gene regulation, to impart traits without undesired effects on tree development or ecosystem function; (2) studies to support resistance-management programs for use of pest-resistant trees; (3) efficient transformation methods so that genetic diversity is not unintentionally impaired as result of the inability to produce large numbers of useful transgenic genotypes; and (4) methods to modify flowering to allow reliable containment of transgenes within plantations when ecologically prudent. Field trials are crucial for all of these research objectives and can be done with a high degree of environmental safety.

Rendering transgenic trees sexually sterile is one means of minimizing, if not avoiding, the presence of foreign genes to natural populations (Dinus et al. 2001). Genetic constructs for sterility in poplars should be available in 5 to 10 years. Useful side effects may also accrue, such as eliminating reproductive structures to channel more energy, water, and nutrients into wood production. Concern might be eased further by using genetic markers to monitor the potential gene flow. Confidence and acceptance can also be built by responsible, transparent testing along with clear, consistent communication of both risks and benefits.

Agencies primarily responsible for regulating biotechnology in the United States are the U.S. Department of Agriculture (USDA), the Environmental Protection Agency (EPA), and the Food and Drug Administration (FDA) (Tang and Newton 2003). Products from transgenic organisms are regulated according to their intended use, with some products falling under the purview of more than one agency. In the case of transgenic trees, the agency responsible is the USDA; the EPA is also involved when modified traits may have an environmental impact. The USDA regulates novel plant release through its Animal Plant Health Inspection Agency (APHIS). APHIS permits for release into the environment are usually issued or denied within 120 days, and during that time state officials will inspect facilities to determine their security and operating conditions. Permits for field trails are renewed annually.

12.3 Productivity and Quality of Transgenic Wood

Wood quantity is the total amount of wood and can be measured in diameter, length, volume, weight, or any combination. Breeding for fast growth may lead to poor wood quality. Wood quality is defined in relation to a norm, which depends entirely on the end product, for example, structural lumber for construction requires a high wood density, whereas low-density trees (aspen and pine) are more desirable for oriented strandboards and paper manufacturing. There is no absolute measure for wood quality. Quality assessment is multifaceted and depends on the intended application. Quantity and quality may not be treated as independent factors. Quality control should form an integral part of most breeding programs.

12.3.1 Growth Rate

Magnussen and Keith (1990) developed six selection strategies aimed at genetically improving volume production and wood quality factors such as density, heartwood content, and stem taper. Selection indices were computed under various assumptions about economic values of the traits under selection and with constraints on the magnitude and direction of expected genetic gain. Stem taper, wood density, and heartwood content were under strong genetic control; however, the low phenotypic variation of wood density limits its potential for genetic improvement. Heartwood content emerged as a trait amenable for rapid genetic improvement.

A strong correlation between wood density and growth rate is often found. There are a number of reports of a negative relationship between growth rate and wood density in several genera such as spruce (Zobel and Jett 1995). Factors affecting growth rate are environment, age, and heredity. The evolution from juvenile to mature wood is the expression of age effects. Foresters can influence growth rate with silvicultural practices, such as initial spacing or thinning. Most studies on the influence of initial spacing or thinning intensity conclude that these factors have little effect on the wood density of Sitka and Norway spruce. Stand fertility seems to have more influence on wood density through growth rate than initial spacing

(Rozenberg and Cahalan 1997). The negative phenotypic and genetic relationships between diameter growth and wood density seem very general within the genus *Picea*. They are moderate for Sitka spruce and other North American spruces, and most of the time, stronger for Norway spruce.

In several cases, bacterial genes have been used to alter plant form and performance. The use of the *iaaM* and *iaaH* auxin-biosynthetic genes from *Agrobacterium tumefaciens* significantly affected several wood characteristics in transgenic hybrid aspen: the transgenic trees were generally smaller than the controls, exhibiting reduced growth rate, leaf size, and stem diameter (Tuominen et al. 1995). Although these changes seem undesirable for the improvement of wood quality, other changes, such as a reduction in the number of side shoots following decapitation and changes in the xylem structure and composition, show some potential for transgenic trees to express these genes. Transgenic hybrid aspen plants overexpressing a peroxidase gene showed an increase of approximately 17 percent of stem length compared with untransformed control trees. Although little is known about the precise action of peroxidase family genes, their use for improving plant performance seems possible, as they showed a similar accelerated-growth phenotype in tobacco (Tzfira et al. 1998).

12.3.2 Physical and Mechanical Properties

Density and Mechanical Properties

Among traits in wood quality, wood density is the most widely used, as it is relatively easy to determine and is well correlated to many other physical and mechanical properties of wood. For example, the compression strength of wood increases linearly with increasing density and the correlation, which is based on 111 representative wood species of both hardwood and softwood in North America, is strong (Fig. 12.1). Density is often used as a wood quality indicator, (i.e., it is related to the suitability of wood to different end uses). Structural

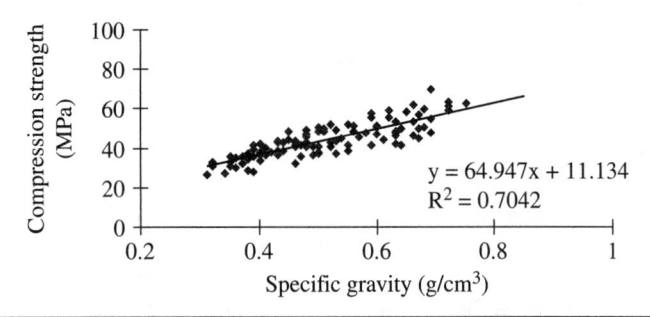

FIGURE 12.1 Relationship between compression strength and specific density of 111 wood species (Zhang 2008).

timber needs a high density and strength. Low-density wood is more suitable for pulp, paper, and composite production.

The mass density of a substance is its mass per unit volume. A wood's structure can be simplified into solid materials (cell walls) and voids (cell lumens). High-density wood contains thicker cell walls and smaller lumens than lower-density wood. Our recent research (Wu et al. 2009) investigated the bulk wood and cell wall mechanical properties of ten hardwoods with densities varying from 0.41 to 1.18 g/cm^3 by nanoindentation and SilviScan as affected by wood species (density) and microfibril angle (MFA). The elastic moduli of bulk wood and the cell walls of wood were both significantly different, whereas the hardness of the cell wall was not significantly different among the ten species. The SilviScan elastic modulus increased with wood density and decreased with microfibril angle. At the cell wall level, the elastic modulus and hardness obtained by nanoindentation were more related to the properties of natural libriform fibers. However, there was no significant trend found for the hardness of cell wall as affected by either wood bulk density or microfibril angle.

Research by Park et al. (2004) showed that the 54 independent transgenic poplars produced longer internodes and a larger amount of cellulose in the secondary xylem than the wild type. The cellulose content in the tenth internode was also assessed by measuring specific gravity, which showed relative larger values in the transgenic than in the wild type. Specific gravity was increased in the transgenic probably by increasing the density of cellulose, which could result in an increase in cellulose content. By measuring the tensile Young's modulus of the stem via an automatic material-testing machine, Park et al. (2004) found that the transgenic stem had a lower modulus in the second internode compared to the wild type, but in the remaining internodes the modulus was increased gradually from the fourth to the tenth internodes with the increased length of the internode.

Differences in the location of cell wall fracture after tangential splitting were observed among three genetic groups of radiate pine (Donaldson 1995). In a control radiate pine, cell wall fracture occurs predominantly at the middle lamella/S1 boundary or within the S1 layer, producing large numbers of fines on the fracture surface. On the other hand, in two genetically select groups of radiate pine, cell wall fracture occurs predominantly at the S1/S2 boundary, producing fewer fines on the surface. The differences appear to be related to reduced lignifications in the inner S1 layer of the genetically select trees. Observed differences in the type of fracture and its location were unrelated to cell dimensions. The observed changes in fracture behavior may be related to differences in the thermomechanical pulping properties observed among these three groups of trees and may provide a useful nondestructive screening method for selecting clones with advantageous pulping properties.

MFA

The cellulose microfibril angle of the S2 layer is a critical factor in determining the mechanical properties of wood. The microfibril angle is defined as the angle between the most probable cellulose microfibril orientation and the long axis of the cell. Figure 12.2 shows an average 4.17° microfibril angle in the cell wall of kwila wood. The microfibril angle can be determined by a number of techniques. X-ray diffraction has become a popular method in recent years.

The microfibril angle varies considerably between trees and within the trunk of a tree with a large angle near pit and small angles in the outerwood. Our group (Tze et al. 2007) measured the mechanical properties of individual, native wood fibers for five different annual rings of a loblolly pine, with microfibril angles (MFA) between 14 and 36° using continuous stiffness measurement techniques. Results showed that a predictable pattern of stiffness values was found as a function of MFA, and stiffness can at least be considered to be a relative measure of the longitudinal stiffness properties of wood cell walls (Fig. 12.3). To determine hardness values, a dependence on orientation was also

FIGURE 12.2 Radial section of kwila wood showing microfibrils aligning in the direction of the fiber axis (average MFA 4.17°).

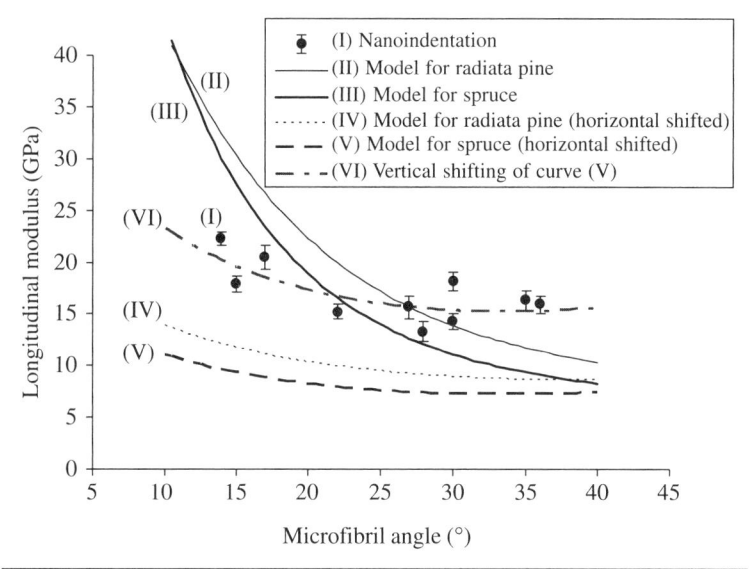

FIGURE 12.3 The S-2 cell wall longitudinal stiffness as a function of microfibril angles: experimental and modeled values (Tze et al. 2007).

observed, and there is a preliminary indication that the dependence could be affected by cell wall extractives. In cell wall modification studies, it is desirable to minimize any unintended variations by using samples that are from the same growth ring, so that any treatment-induced changes in cell wall hardness can be identified.

12.3.3 Chemical Composition

Wood tissue is composed of a group of polymeric components, which make up the bulk of wood cell walls. All polymeric components are mixed in the wall and determine the physical and mechanical properties of wood. A typical wood cell includes a primary wall, a secondary wall (S1, S2, S3), and a lumen.

There are two chemical components in a wood cell wall. The primary components, which establish the chemical and physical nature of the cell wall and constitute the bulk of materials in the wood cell wall, include cellulose (40 to 50 percent), hemicellulose (20 to 35 percent), and lignin (15 to 35 percent). Cellulose is the most important single component in the cell wall and has a primary relationship to the physical and mechanical behavior of the wood. Hemicellulose and lignin are also important in the wall because they act as matrixes in the natural polymer composites. The secondary components in the wall include ash (<1 percent) and extractives, such as tannins, volatile oils and resins, gums, latex, alkaloids, and other compounds. The secondary components may not significantly affect the structure of the cell but may give the bulk of wood many specific characteristics, such as resistance to natural degradation, color, and odor.

Lignin

Lignin is the most abundant organic compound in the earth after cellulose and makes up 15 to 35 percent of the dry weight of trees. It is an important component in all vascular plants and occurs in the secondary cell walls of fibers, xylem vessels, and tracheids, providing them with mechanical support and helping in the plant's defense against pathogens (Boudet et al. 1995). There is a very limited market for lignin. The majority of lignin is used in pulp mills as energy sources. Lignin is an obstacle to efficient pulp and paper production because the lignin must be removed in order to extract the cellulose from the wood. This process is energy consuming and requires the use of polluting chemicals. It is of great interest to engineer trees to have a lower lignin component or a lignin type that is easily extracted without reducing tree growth rates or bole form (Pilate et al. 2002).

An extensive review by Baucher et al. (2003) provided detailed information regarding lignin genetic engineering and the impact on pulping. The lignin polymer is produced by the dehydrogenative polymerization of essentially three different cinnamyl alcohols (*p*-coumaryl, coniferyl, and sinapyl alcohol) that differ in the degree of methoxylation at the C3 and C5 positions of the aromatic ring. The terms *H-type*, *S-type*, and *G-type* lignin refer to lignin containing hydroxyphenyl (H), guaiacyl (G), and syringyl (S) units, correspondingly. Determining the amount, structure, or monomeric composition of lignin in a plant is extremely difficult because of the heterogeneity of the polymer and the high proportion of covalent bonds linking different monomers. Moreover, during isolation, lignin undergoes secondary modifications, such as condensation, oxidation, addition, or substitution. Therefore, a combination of several methods has to be used to obtain reliable information on lignin structure. Pyrolysis gas chromatography-mass spectrometry (pyrolysis GC-MS), Fourier transform infrared (FTIR) spectroscopy, ultraviolet and IR spectroscopy, and nuclear magnetic resonance spectroscopy have been used to investigate lignin content and composition. Many different transgenic plants and a few mutants are now available with altered lignin content, altered lignin composition/structure, or both.

Introducing antisense O-methyltransferase into a hybrid poplar reduced the syringyl:guaiacyl ratio (a direct result of a reduction in the level of syringyl and an increase in the level of guaiacyl groups) (Van Doorsselaere et al. 1995). Furthermore, a novel lignin component (a 5-hydroxyguaiacyl residue) was identified. Both the increase in guaiacyl units and the production of 5-hydroxyguaiacyl were hypothesized to result from an alternative pathway in the synthesis of guaiacyl units. Although no reduction in total lignin content was observed, the reduction of syringyl units is considered to be beneficial to the kraft-pulping process because G-type lignin is more easily extractable using chemical pulping techniques.

Cellulose and Hemicellulose

The perceivable shortage of petroleum resources demands novel techniques to produce fuels and products from biorenewable resources. The success of future bioenergy production will be dependent on whether or not we can develop economically and environmentally viable processes to convert biomass to biofuels. Celluloses are the most abundant natural macromolecules that are promising for production of biofuels and bioproducts. Current technologies of converting lignocellulose into liquid fuels mainly convert cellulose and hemicellulose into ethanol.

Cellulose is a high-molecular-weight linear homopolymer, consisting of repeating β-D-glucopyranosyl units joined by (1-4) glycosidic linkages in a variety of arrangements. The hydrogen bonds formed between the ring oxygen atom of one glycosyl unit and the hydrogen atom of the C-3 hydroxyl group of the preceding ring hinder the free rotation of the rings on their linking glycosidic bonds, resulting in a stiffening of the chain. The adjacent cellulose chains fit closely together in an ordered crystalline region, so that high strength can be observed in plants and some cellulose-constituted animals; cellulose is insoluble in common solvents. In nature, cellulose chains have a degree of polymerization of approximately 10,000 glucopyranose units in wood cellulose and 15,000 in native cellulose cotton (Sjöström 1993).

Hemicelluloses are heteroglycans, which consist of various sugar units arranged in different proportions and with different substituents. Most hemicelluloses comprise two carbohydrate polymers: a xylose- and a mannose-containing polysaccharide. The molecular chains of these hemicelluloses are much shorter than that of cellulose. Hemicelluloses have a side group and a branch group. Hemicelluloses are more associated with lignin than cellulose because of its amorphous state. Hemicelluloses are relatively low-molecular-weight polysaccharides. Xylan is the predominant type of hemicellulose in hardwoods. Xylan is a polyose with a homopolymer backbone of xylose units. It is linked by β-(1-4)-glycosidic bonds.

Our understanding of cellulose synthesis and deposition is still limited; progress in this field has been slow due to our inability to isolate active and intact cellulose synthase complexes and the intricacy of the mechanistic processes involved in the wood cellulose development (Joshi et al. 2004). It appears that hemicelluloses are instrumental in the formation of cellulose fibril networks, and genetic alterations of hemicelluloses can both disrupt or enhance cellulose formation. Through expressed *Aspergillus* xyloglucanase constitutively in *Populus alba*, the expression increased the length of the stem even in the presence of sucrose. Increased stem growth was accomplished by a decrease in Young's elastic modulus in the growth zone but an increased elasticity in mature tissue (Park et al. 2004). Increased (9 to 15 percent) cellulose content and alteration of hemicellulose composition was observed in antisense 4CL transgenic poplars as the result of a drastic reduction (45 percent) of lignin content (Hu et al. 1999). This research also showed an increased growth rate.

Research on genetic variations and control of hemicellulose content are quite rare. Analyses of trees from a 6-year-old *Eucalyptus* species/provenance trial in South Africa showed that variation in pentosan, cellulose, and lignin contents among species was statistically significant, as was that in pentosan and cellulose contents among provenances within species (Clarke and Wessels 1995).

12.3.4 Durability

Lumber from most wood species usually lacks natural durability, and this is of great concern for the building industry. Pine-dimensional lumber is frequently treated with various chemicals for durability, including copper–chrome–arsenate (CCA) pressure treatment for outdoor and foundation application. Although this method is very efficient, CCA is becoming less acceptable, even banned in some applications, due to environmental concerns. Alternatives could be found as genes that provide durability in other tree species may be used for genetic engineering for higher durability.

The extraction of nonstructural hemicellulose sugars from wood strands, prior to the production of oriented strandboard (OSB) is expected to have a great impact on the stiffness and durability of OSB. Hemicelluloses sugars are valuable resources for making other by-products such as ethanol and polymers. Our recent study was to extract hemicelluloses from southern yellow pine strands and using the preextracted strands to produce enhanced OSB (Hosseinaei et al. 2008). A total of 14 OSB panels with a 42-lb/ft^3 target density were made by the same hot-press parameters (200°C and 4 min). Four percent of phenol–formaldehyde resin and 1 percent wax were used. The results from this study indicated that weight loss of wood strands and recovered sugars can be easily controlled by extraction conditions, and all hemicelluloses can be removed from a strand without damaging the strands. OSB panels made from extracted strands have much better dimension stability and bending properties as compared to control panels. Water absorption and thickness swelling of panels could be reduced to 51.8 and 37.8 percent, respectively. All panels met the internal bonding requirements of Canadian Standard Association standard. The mold susceptibility of the panels made from preextracted strands was significantly reduced as compared to control panels (Taylor et al. 2008). This suggests that reduction and/or alteration of hemicelluloses through genetic modification on trees could improve wood durability, especially mold resistance.

12.3.5 Advanced Analysis Tools

To efficiently monitor and control the properties of transgenic wood, rapid and small-scale analytical tools are required. These tools will be more important when a tree that is just a few months old needs to be investigated. Several advanced analysis tools, which have been used or will be used to analyze transgenic wood, will be discussed as follows.

Tree-Ring Machine/SilviScan

In recent years, several developments have taken place in biotechnology, including genetic tree improvement. Characterizing wood quality quickly and reliably becomes more important for evaluating if the genetic application is successful or not. The recent development of SilviScan-2 (Evans 1997, 1999) has provided a tool to conduct the rapid scanning of increment cores for tree improvement programs and large-scale resource assessments. SilviScan provides pith-to-bark measurements of fiber width, wood density, and microfibril angle. From these data, fiber-wall thickness, coarseness, and wood stiffness can be estimated.

SilviScan is an automated wood-microstructure analyzer developed for the rapid assessment of wood properties, including wood density and MFA by a combination of x-ray diffractometry, x-ray densitometry, and digital microscopy (Evans 1997; Evans and Ilic 2001). Diffraction patterns collected on SilviScan can be integrated over a specified radial section; thus, pith-to-bark profiles of wood density and MFA variation can be obtained at different spatial resolutions. Density was scanned at a resolution of 25 µm, whereas MFA had a resolution of 1 mm.

The elastic modulus of wood specimens can be calculated from the SilviScan wood density and MFA using the following equation:

$$E = A(DI_{CV})^B \quad (12.1)$$

where A and B are constants calibrated using sonic resonance data. D (g/cm^3) is the scanning wood bulk density, and I_{CV} is the coefficient of variation of the 002 azimuthal diffraction profile determined by MFA (Evans 1999).

The Quintek Measurement Systems also developed an instrument to measure increment core tree ring samples. A complete printout listing ring counts, ring years from core date, radial growth by year, early wood and late wood percentage, early wood and late wood density, ring density, and average core density is generated from single thin strips. Figure 12.4 shows a density profile of southern pine measured by the x-ray method. The profile shows six tree rings with detailed earlywood and latewood information.

360 Chapter Twelve

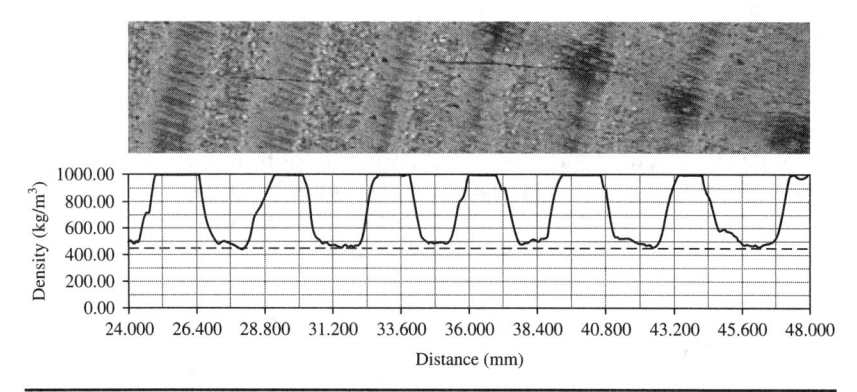

FIGURE 12.4 Density profile of southern pine. (*Courtesy of Chuck Dawson, Quintek Measurement Systems, Inc.*)

Nanoindentation

Nanoindentation testing is a technique that determines the mechanical properties of a material in the micron or submicron scale. The test involves penetrating a sample material using an indenter, whereas the penetration depth and load are recorded so that the stiffness and hardness of the indented location can be subsequently calculated. The indenter head can have a radius of 100 nm (in the case of the Berkovich indenter), and penetration can be up to 1 or 2 µm deep, with the resulting indent having a linear dimension on the order of micrometers. This dimension is in the same order of magnitude of the thickness of the wood cell wall. The wood cell walls were reported to be 5 to 6 µm and 9 to 13 µm thick, respectively, for earlywood and latewood of loblolly pine (Barefoot et al. 1965). Therefore, the local mechanical properties of wood cell walls can be probed using nanoindentation tests. More specifically, the test detects the mechanical properties of the cell wall S2 layer, which constitutes about 80 percent of the total cell wall thickness and is the major contributor to the mechanical properties of wood cell walls. Figure 12.5 shows two indent marks on the cell wall of the Keranji hardwood (*Dialium* spp.).

The micron spatial (lateral) resolution in nanoindentation tests renders the foregoing technique very useful in the investigation of the wood cell-wall level as a result of growth processes or utilization operations. To date, a few studies have used nanoindentation to investigate the effects of seasonal growth response (earlywood versus latewood) (Wimmer et al. 1997), cell wall lignification (Gindl et al. 2002), growth ring (Tze et al. 2007), and wood species (Wu et al. 2009) on the mechanical properties of single cell wall. An attractive feature demonstrated in these studies is that the measurements were made without requiring chemical or mechanical modifications to isolate individual wood fibers as required in single-fiber tensile tests. These chemical and mechanical modifications change the mechanical properties of the wood fibers in poorly defined ways.

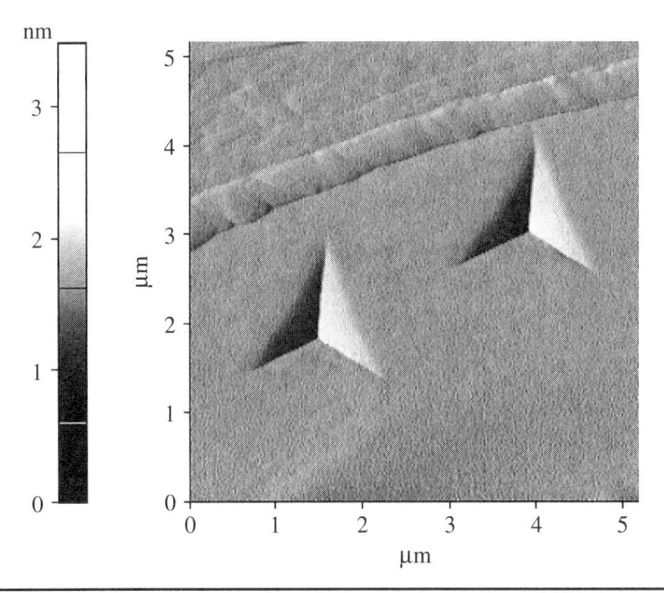

FIGURE 12.5 AFM topography of two indent marks on Keranji (*Dialium* spp.) cell wall after nanoindentation.

By using the nanoindentation technique, Gindl et al. (2002) examined possible differences in the behavior of developing and fully lignified spruce tracheid cell walls under axial compressive loading, which could not be investigated by other techniques. The average lignin content of developing tracheids was 0.10 g.g^{-1}, as compared with 0.21 g.g^{-1} in mature tracheids. The modulus of elasticity of developing cells was, on average, 22 percent lower than that measured in mature, fully lignified cells. For longitudinal hardness, a large difference of 26 percent was observed.

Atomic Force Microscopy

Atomic force microscopy (AFM) has been used increasingly to characterize biological samples. AFM has a high resolution and measurement accuracy, far surpassing the capabilities of both contact and tapping modes (www.parkafm.com). The resolution of the tapping mode is not as high as that of the noncontact AFM because the very sharp end of the tip is extremely fragile and is blunted instantaneously when it forcefully establishes contact with the sample. This problem is more severe in the case of imaging soft materials such as thermoplastic polymers. On the other hand, a constant tip–sample distance of a few nanometers is maintained in the true noncontact AFM without damaging the tip or sample to obtain a high resolution.

One obvious advantage to using AFM is that it is possible to obtain sample topography (height image), elasticity (phase image),

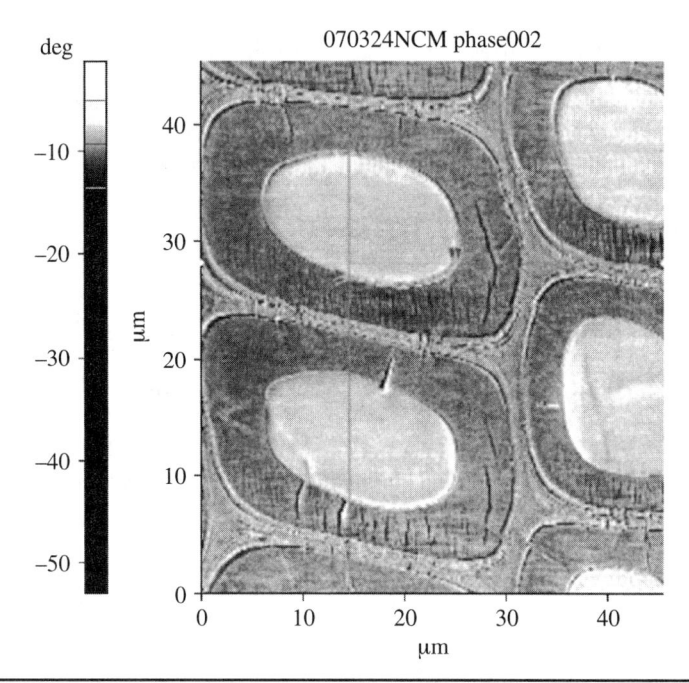

FIGURE 12.6 Phase images of the sixteenth growth ring of loblolly pine.

thermal conductivity contrast imaging (Lee et al. 2009), and even acoustic imaging (Nair et al. 2008) simultaneously. These measurements can be made in a nearly in vivo physiological environment (in air or under fluid), which is essential to eliminate structure modification resulting from a sample preparation. Direct visualizations of the cellulose crystal, microfibril, and surface of the plant cell wall have been achieved using AFM (Kirby et al. 1996; Baker et al. 1997).

Noncontact tapping mode AFM imaging was used to characterize the cell wall surface of loblolly pine. Phase image (Fig. 12.6) distinct layers were obtained. A sharp phase difference was obtained between the different layers of the cell wall and the image that was produced showed a very clear S3 layer. The width of the S3 layer was around 0.6 µm.

Near-Infrared Spectroscopy

Near-infrared spectroscopy (NIR) is known as a powerful tool that provides quantitative information on the chemical and physical properties of woody materials. It is nondestructive, rapid, and inexpensive and requires a little sample preparation. NIR has been used to study lignin content, cellulose content, extractive content, fiber length, wood density, stiffness, and microfibril angle. Transmittance NIR spectroscopy has been used as a rapid analysis tool for determining lignin content, syringyl-to-guaiacyl (S/G) ratio, cellulose, and xylose content of transgenic aspen trees (Yamada et al. 2006). Pellets were prepared from 75 mg of wood meal and directly scanned using transmittance NIR spectroscopy. Very strong correlations were obtained between the NIR data and conventional wet-chemistry results for the lignin content, S/G ratio, cellulose, and xylose content. The results indicate that transmittance NIR is a powerful tool for determining and screening the chemical properties of transgenic trees.

Fourier transform infrared (FT-IR) microimaging spectroscopy and pyrolysis molecular beam mass spectrometry (py-MBMS) were used as rapid-analysis tools to evaluate differences in the chemical composite of 1-year-old transgenic aspens (Labbé et al. 2005). In their work, multivariate analysis of the spectroscopic data sets was used to compare the cell wall composition of a nontransformed control to transgenic aspen plants with the GRP-*iaa*M gene and with the GRP-*iaa*M/35SACCase gene. Principal component analysis (PCA) was applied to both the FT-IR and py-MBMS spectra, which revealed sample groupings dues to differences in chemical composition. The analysis showed that changes in the composition of the xylem that occur over one annual growth ring can be monitored with FT-IR microimaging. The py-MBMS provided detailed and specific information on the chemical composition of the samples and is particularly valuable for its ability to quickly, and unambiguously differentiate syringyl and guaiacyl lignins and C6 and C5 sugars.

12.3.6 Impacts on Process and Utilization

Although much of the research on genetic modification was motivated by a desire to improve wood for a more efficient and environmentally friendly process, such as papermaking, very few transgenic trees have yet been tested for understanding its impact on wood process and utilization. Kraft pulping experiments on the wood of 2-year-old greenhouse-grown poplars resulted in a pulp with a lower kappa number and an increase in brightness. This genetic improvement may increase pulp throughput by 60 percent, concomitantly with a decreased consumption of pulping chemicals (Huntley et al. 2003).

The fracturing properties of wood are important in relation to wood processing including log sawing, thermomechanical pulping, production of fiber or wood chips for use in wood composites, and size reduction for bioenergy production. After comparing a control radiate pine with genetically improved radiate (control pollinated progeny from a single tree of the 850 family known as NZ850-55 or clone 55), and New Zealand–grown spruce, Corson et al. (1989) found that energy consumption for pulping to a specific freeness was significantly lower in the genetically improved radiate pine when compared to control-radiate pine.

12.4 Conclusions

In increasing demand for wood products, bioenergy, and reduction of available harvestable forests has recently led to the development of genetic engineering into forest-tree research and improvement. Due to the significant genetic transformation progress, future forest trees will be more tolerant to abiotic and biotic stresses, will express genes for accelerated growth rate, and will have a modified wood structure and composition. Selective genetic alterations to wood structure and/or composition clarify the mechanical role of particular wood features. Several advanced tools, especially nanotechnology-based tools, have the potential to accelerate the characterization of transgenic trees.

References

Baker, A. A., Helbert, W., Sugiyama, J., and Miles, M. 1997. High–resolution atomic force microscopy of native *valonia* cellulose I microcrystals. *Journal of Structural Biology* 119(2):129–138.

Barefoot, A. C., Hitchings, R. G., and Ellwood, E. L. 1965. Wood characteristics of kraft paper properties of four selected loblolly pines. III. Effect of fiber morphology on pulp. *Tappi* 49:137–147.

Baucher, M., Halpin, C., Petit-Conil, M., and Boerjan, W. 2003. Lignin genetic engineering and impact on pulping. *Critical Reviews in Biochemistry and Molecular Biology* 38(4):305–350.

Boudet, A. M., Lapierre, C., and Grima-Petternati, J. 1995. Biochemistry and molecular biology of lignification. *New Phytology* 129:203–236.

Brunner, A. M., Busov, V. B., and Strauss, S. H. 2004. Poplar genome sequence: Functional genomics in an ecologically dominant plant species. *Trends in Plant Science* 9:49–56.

Cave, I. D. 1968. The anisotropic elasticity of the plant cell wall. *Wood Science and Technology* 24:268–278.

Clarke, C. R. E. and Wessels, A. E. 1995. Variation and measurement of pulp properties in Eucalypts. I. *Proceedings of the CRC–IUFRO Conference: Eucalypt Plantations: Improving Fibre Yield and Quality*, eds. B. M. Potts, N. M. G. Borralho, J. B. Reid, R. N. Cromer, W. N. Tibbits, and C. A. Raymond, 19–24 Feb., 93–100. Hobart, Australia: CRC for Temperate Hardwood Forestry.

Corson, S. R., Foster, R. S., and Richardson, J. D. 1989. New Zealand grown spruce and radiate pine can have similar TMP properties. *Appita* 42:345–349.

Dinus, R. J., Payne, P., Sewell, N. M., Chiang, V. L., and Tuskan, G. A. 2001. Genetic modification of short rotation popular wood: Properties for ethanol fuel and fiber productions. *Critical Reviews in Plant Sciences* 20(1):51–69.

Donaldson, L. A. 1995. Cell wall fracture properties in relation to lignin distribution and cell dimensions among three genetic groups of radiate pine. *Wood Science and Technology* 29:51–63.

Evans, R. 1997. Rapid scanning of microfibril angle in increment cores by X–ray diffractometry. In *Microfibril Angle in Wood*, ed. B. G. Butterfield, 116–139. *Proceedings of the IAWA/IUFRO International Workshop on the Significance of Microfibril Angle to Wood Quality*. Westport, New Zealand: University of Canterbury Press.

Evans, R. 1999. A variance approach to the X-ray diffractometric estimation of microfibril angle in wood. *Appita Journal* 51:53–57.

Evans, R. and Ilic, J. 2001. Rapid prediction of wood stiffness from microfibril angle and density. *Forest Products Journal* 51(3):53–57.

Fillatti, J. J., Sellmer, J., McCown, B., Haissig, B., and Comai, L. 1987. Agrobacterium-mediated transformation and regeneration of Populus. *Molecular and General Genetics* 206:192–199.

Gindl, W., Gupta, H. S., and Grünwald, C. 2002. Lignification of spruce tracheid secondary cell walls related to longitudinal hardness and modulus of elasticity using nano-indentation. *Canadian Journal of Botany* 80:1029–1033.

Herschbach, C. and Kopriva, S. 2002. Transgenic trees as tools in tree and plant physiology. *Trees* 16:250–261.

Hosseinaei, O., Wang, S., Xing, C., Rials, T. G., Kelley, S. S., and Enayati, A. A. 2008. Production of enhanced OSB using hemicellulose extracted strands. Abstract in *Biographies and Abstracts*. Forest Products Society 62nd International Convention, St. Louis, Missouri, June 22–24, p. 20.

Hu, W. J., Harding, S. A., Jacqueline, J. L., Popko, L., Ralph, J., Stokke, D. D., Tsai, C.-J., et al. 1999. Repression of lignin biosynthesis promotes cellulose accumulation and growth in transgenic trees. *Nature Biotechnology* 17:808–812.

Huntley, S. K., Ellis, D., Gilbert, M., Chapple, C., and Mansfield, S. D. 2003. Significant increases in pulping efficiency in C4H–F5H transformed poplar. Abstract presented at the Tree Biotechnology Symposium, Umeå, Sweden, June 7–12 (S10.5).

James, R. R., Difazio, S. T., Brunner, A. M., and Strauss, S.H. 1998. Environmental effects of genetically engineered woody biomass crops. *Biomass and Bioenergy* 14(4):403–414.

Joshi, P. C., Bhandari, P., Ranjan, K., and Kalluli, U. C. 2004. Genomics of cellulose biosynthesis in poplars. *New Phytologist* 164:53–61.

Kirby, A. R., Gunning, A. P., Waldron, K. W., Morris, V. J., and Ng, A. 1996. Visualization of plant cell wall by atomic force microscopy. *Biophysics* 70(3):1138–1143.

Koehler, L. and Telewski, F. W. 2006. Biomechanics and transgenic wood. *American Journal of Botany* 93(10):1433–1438.

Labbé, N., Rials, T. G., Kelley, S. S., Cheng, Z. M., Kim, J. Y., and Li, Y. 2005. FT–IR imaging and pyrolysis–molecular beam mass spectrometry: New tools to investigate wood tissues. *Wood Science and Technology* 39(1):61–77.

Lee, S., Wang, S., Endo, T., and Kim, N. H. 2009. Visualization of interfacial zones in lyocell fiber-reinforced polypropylene composite by AFM contrast imaging based on phase and thermal conductivity measurements. *Holzforschung* 63(2):240–247.

Magnussen, S. and Keith, C. T. 1990. Genetic improvement of volume and wood properties of jack pine: Selection strategies. *The Forestry Chronicle* 66(3): 281–286.

Merkle, S. A. and Dean, J. F. D. 2000. Forest tree biotechnology. *Current Opinion in Biotechnology* 11(3):298–302.

Meylan, B. A. and Butterfield, B. G. 1978. Helical orientation of the microfibril in tracheids, fibres and vessels. *Wood Science and Technology* 12:219–222.

Nair, S. S., Wang, S., and Hurley, D. 2008. Evaluation of interphase properties in fiber reinforced polymer composite using atomic force acoustic microscopy (AFAM). Paper presented at the SWST Annual Meeting, University of Bio–Bio, Concepción, Chile, Nov. 10–12.

Park, Y. W., Baba, K., Furuta, Y., Lida, I., Sameshima, K., Arai, M., and Hayashi, T. 2004. enhancedment of growth and cellulose accumulation by overexpression of xyloglucanase in poplar. *FEBS Letters* 564:183–187.

Perlack, R. D., Wright, L. L., Turhollow, A. F., Graham, R. L., Stokes, B. J., and Erbach, D. C. 2005. Biomass as feedstock for a bioenergy and bioproducts industry: The technical feasibility of a billion-ton annual supply. Technical Report A357634, U.S. Department of Energy, Oak Ridge National Laboratory, Oak Ridge, Tenn. 78 pp.

Pilate, G., Guiney, E., Holt, K., Petti-Conil, M., Lapierre, C., Leple, J. C., Pollet, B. et al. 2002. Field and pulping performances of transgenic trees with altered lignification. *Nature Biotechnology* 20:607–612.

Rozenberg, P. H. and Cahalan, C. H. 1997. Spruce and wood quality: Genetic aspects (a review). *Silvae Genetica* 46(5):270–279.

Rugh, C. L., Senecoff, J. F., Meagher, R. B., and Merkle, S. A. 1998. Development of transgenetic yellow poplar for mercury phytoremediation. *Nature Biotechnology* 16:925–928.

Schuler, T. H., Poppy, G. M., Kerry, B. N., and Denholm, I. 1998. Insect-resistant transgenic plants. *Trends in Biotechnology* 16:168–175.

Sjöström, E. 1993. *Wood Chemistry Fundamentals and Applications*. New York: Academic Press.

Strauss, S. H., Boerjan, W., Cairney, J., Campbell, M. M., Dean, J., Ellis, D., Jouanin, L., et al. 1999. Forest biotechnology makes its position known. *Nature Biotechnology* 17:1145.

Tang, W. and Newton, R. J. 2003. Genetic transformation of conifers and its application in forest biotechnology. *Plant Cell Reports* 22(1):1–15.

Taylor, A. M., Hosseinaei, O., and Wang, S. 2008. Mold susceptibility of oriented strandboard made with extracted flakes. Paper presented at the 39th Annual Meeting of the International Research Group on Wood Preservation, Istanbul, Turkey, 25–29 May. IRG/WP 08–40402, p. 8.

Tuominen, H., Sitbon, F., Jacobsson, C., Sandberg, G., Olsson, O., and Sundberg, B. 1995. Altered growth and wood characteristics in transgenic hybrid aspen expressing agrobacterium tumefaciens T–DNA indoleacetic acid-biosynthetic genes. *Plant Physiology* 109:1179–1189.

Tze, W. T. Y., Wang, S., Rials, T. G., Pharr, G. M, and Kelley, S. S. 2007. Nanoindentation of wood cell walls: Continuous stiffness and hardness measurements. *Composites: Part A* 38:945–953.

Tzfira, T., Zuker, A., and Altman, A. 1998. Forest-tree biotechnology: Genetic transformation and its application to future forests. *Trends in Biotechnology* 16:439–446.

Van Doorsselaere, J., Baucher, M., Chognot, E., Chabbert, B., Tollier, M. T., Petit-Conil, M., Leplé, J. C. et al. 1995. A novel lignin in poplar trees with a reduced caffeic acid/5–hydroxyferulic acid O–methyltrasferase activity. *Plant Journal* 8:855–865.

Walter, C., Charity, J., Grace, L., Höfig, K., Möller, R., and Wagner, A. 2002. Gene technologies in *Pinus radita* and *Picea abies*: Tools for conifer biotechnology in 21st century. *Plant Cell Tissue and Organ Culture* 70:3–12.

Wimmer, R., Lucas, B. N., Tsui, T. Y., and Oliver, W. C. 1997. Longitudinal hardness and Young's modulus of spruce tracheid secondary walls using nanoindentation technique. *Wood Science and Technology* 31:131–141.

Wu, Y., Wang, S., Zhou, D., Xing, C., Zhang, Y., and Pharr, G. M. 2009. Use of nanoindentation and SilviScan to determine mechanical properties of 10 hardwood species. *Wood and Fiber Science* 41(1):64–73.

Yamada, T., Yeh, T. F., Chang, H. M., Li, L., Kadla, J. F., and Chiang, V. L. 2006. Rapid analysis of transgenic trees using transmittance near-infrared spectroscopy (NIR). *Holzforschung* 60(1):24–28.

Zhang, X. 2008. Characterizing strength and fracture of wood micropillars under uniaxial compression. MS thesis, University of Tennessee, Knoxville, Tenn.

Zobel, B. J. and Jett, B. J. 1995. *Genetics of Wood Production*. Berlin: Springer-Verlag.

CHAPTER 13
Extraction, Refining, and Stabilization of Edible Oils

Michel Eskin
Department of Human Nutritional Science
University of Manitoba
Manitoba, Canada

Ahindra Nag
Department of Chemistry
Indian Institute of Technology
Karagpur, India

13.1 Introduction

Oils and fats are composed of different mixtures of glycerides of various fatty acids. *Oil* is an ester containing more unsaturated fatty acids and having a lower melting point. The more saturated ester is a *fat* and is generally solid.

The properties of fats depend on the type of fatty acids present, whether they are saturated or unsaturated. Fatty acids are mostly straight-chain aliphatic carboxylic acids. In certain fats, the straight chain may be linked with cyclic or aromatic groups. With some exceptions, fatty acids are of even number of carbon atoms and have more than 12 carbon atoms.

Saturated Fatty Acids Saturated fatty acids that are present in naturally occurring fats are given below. The general formula is $C_nH_{2n+1}COOH$, *n*-butyric (C_4), isovaleric (C_5), *n*-caproic (C_6), *n*-caprylic (C_8), *n*-capric (C_{10}), *n*-lauric (C_{12}), *n*-myristic (C_{14}), *n*-palmitic (C_{16}), *n*-stearic (C_{18}), and a few others.

Above caprylic acid, fatty acids are solid at room temperature at 25°C. The lower fatty acids up to capric acid are steam volatile. Butyric acid is soluble in water and others are insoluble.

Unsaturated Fatty Acids Unsaturated fatty acids present in fats make them industrially important. Actually, the drying power of the oils depends on the presence of unsaturated fatty acids in oil. The use of oil in the foundry industry depends on the presence of unsaturated fatty acids (e.g., oleic acid and linoleic acid). These acids contain one, two, or three double bonds that may be conjugated or nonconjugated. The various types of unsaturated acids present in fats are as follows:

Acids containing one double bond: The general formula is $C_nH_{2n-1}COOH$. These fatty acids are 9,10-decanoic acid ($C_{10}H_{18}O_2$), present in milk fat; dodecanoic acid ($C_{12}H_{22}O_2$), and 9,10-hexadecanoic acid ($C_{16}H_{30}O$), present in marine animals. By far, the most important monoethylenic unsaturated fatty acid is octadecanoic acid, or oleic acid, ($C_{17}H_{33}COOH$) or ($CH_3-(CH_2)_7CH = CH(CH_2)_7COOH$. This acid is important and very common in most fats and oils.

Acids containing more than one double bond: Linoleic acid and linolenic acids are widely present in drying and semidrying vegetable oils. Linolelic acid is dienoic acid and the double bonds are present at the 9,10 and 12,13 carbon positions. For linolenic acid the double bonds are present at the 9-10, 12-13, and 15-16 positions.

$$\overset{18}{} \quad \overset{13}{} \quad \overset{12}{} \quad \overset{11}{} \overset{10}{} \overset{9}{} \quad \overset{1}{}$$
$$CH_3(CH_2)_4 - CH = CH_2 - CH_2CH = CH(CH_2)_7 COOH$$

Linoleic or 9,12-octadecadienoic acid ($C_{18}H_{32}O_2$)

$$CH_3CH_2CH = CHCH_2CH = CHCH_2CH$$
$$= CH(CH_2)_7 COOH (C_{18}H_{30}O_2)$$

Linolenic or 9,12,15-octadecatrienoic acid: An acid isomeric with linolenic acid is the conjugated acid eleostearic acid. It is 9,11,13-octadecatrienoic acid. It is present in Chinese wood oil (tung oil) and has the following structure:

$$\overset{18}{} \quad \overset{17}{} \quad \overset{16}{} \quad \overset{15}{} \quad \overset{14}{} \quad \overset{13}{} \quad \overset{12}{} \quad \overset{11}{} \quad \overset{10}{}$$
$$CH_3 - CH_2 - CH_2 - CH_2 - CH = CH - CH = CH - CH$$
$$\overset{9}{} \quad \overset{1}{}$$
$$= CH - (CH_2)_7 COOH$$

Tung oil: Due to the presence of conjugated double-bond fatty acids in its glyceride form, polymerize very rapidly to a gelatinous form.

Extraction, Refining, and Stabilization of Edible Oils

The 4-ketoeleostearic acid is present in the glycerides of oiticica oil. Examples of fatty acids with double bond in a ring are chaulmoogric and hydnocarpic acids. The glycerides of these acids are found in chaulmoogra and similar oils of considerable therapeutic values. These oils have antimetabolite properties and are effectively used to combat leprosy and other external infections.

$$HOOC(CH_2)_{10} - CH_2 - CH_2 - CH = CH_2$$

Hydnocarpic acid ($C_{16}H_{28}O_2$)

$$HOOC(CH_2)_{12} - CH_2 - CH_2 - CH_2 - CH = CH_2$$

Chaulmoogra acid ($C_{18}H_{32}O_2$)

Glycerides of very highly unsaturated fatty acids containing 20 and 22 carbon atoms and having up to 6 double bonds have been found to be present in marine animal oils; 1.0 percent is found in human fat, especially in the adrenal gland. It has the following structure:

$$\overset{20}{CH_3}(CH_2)_4 \overset{15}{C}H = \overset{14}{C}HCH_2\overset{12}{C}H = \overset{11}{C}HC H_2\overset{9}{C}H = \overset{8}{C}HC\overset{7}{H}_2\overset{6}{C}H$$

$$= \overset{5}{C}H - (CH_2)_3 \overset{1}{C}OOH$$

Arachidonic acid ($C_{19}H_{32}O_2$)

Only one hydroxy unsaturated acid ricinoleic acid (12-hydroxyoleic acid) has been reported in natural fat. It is found in castor oil, which is widely used for medicinal purposes (purgative) and for the production of surfactants. It can also be converted to drying oil.

Oils are either edible or nonedible. Edible oils such as groundnut and sunflower among others, are used for cooking purposes and salad dressings.[1] Nonedible oils are used in soaps, paints, and varnishes, politicizes, and similar products. About 1000 different fats have been identified and characterized. One hundred fats are used regularly and 12 of them constitute more than 90 percent of the world's vegetable oil production.

13.2 Determining Oil Content in Oil Seeds and Its Extraction

Determining the oil content in oil seeds have been traditionally accomplished using the soxhlet extraction method with different organic solvents, as discussed by Lumley and Colwell.[2] Improvements in apparatus, sample preparation, and methodology for determining the oil content of seeds have continued to appear as alternative methods,[3] such as pulsed nuclear magnetic resonance and near-infrared spectroscopy, but solvent extraction remains the most commonly used method.

13.2.1 Extracting Oil from Seeds

The oil seed is first freed from impurities by screening and winnowing. Once completed the following processes are adopted.

Decortications

In this process, the seed kernels are separated from the hulls. Decortications are done by passing the seeds between rollers with sharp edges or grooves to break the shells without damaging the kernel inside.

Milling

To break down the oil-containing cell, the kernel is crushed. The crushing machine consists of a number of chilled iron rolls placed one above the other. Seeds pass through successive rollers under increased pressure, crushing the kernel.

Cooking

The crushed kernel is placed in a steam-jacketed pan fitted with a stirrer and heated to about 80°C by live steam to coagulate the albuminous matter and to lower the viscosity of the oil.

Extraction

Oil is extracted by the following processes.

Pressing There are two types of pressing machines, the open press and the enclosed or cage press.

In the open-type press, oil-seed meal is fed into bags of filter cloth resting between horizontal plates; the oil escapes through the horizontal plates. However, a cage press consists of a barrel-shaped box with perforated walls where the meal is fed directly. Pressure is applied in two stages by a hydraulic press. The oil content of the cake is reduced to 5 to 6 percent. The oil escapes through the perforation, whereas solid matter is retained in the box.

Expelling The oil screw press carries out expeller extraction of oil continuously. The press consists of a warm shaft in a perforated cylinder. The oil meal thus receives a steadily increasing pressure as the revolution of the shaft pushes it forward against the small outlet at the end of the cylinder. Oil is discharged through the perforations and the cake through the cylinder outlet. It can be used for any type of seed but is mostly used for groundnut and tung seeds. Two types of expellers—a low-capacity hand-operated oil expeller and a high-capacity expeller—are shown in Figs. 13.1 and 13.2.

Extraction, Refining, and Stabilization of Edible Oils 371

FIGURE 13.1 Hand-operated mechanical press and oil expeller of low-capacity oil extraction.

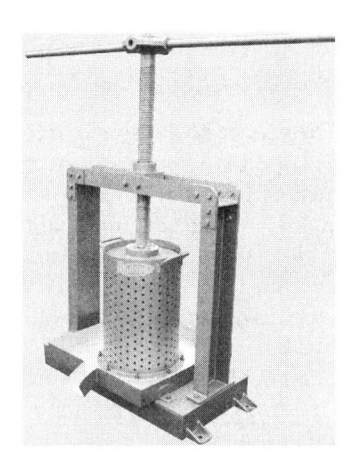

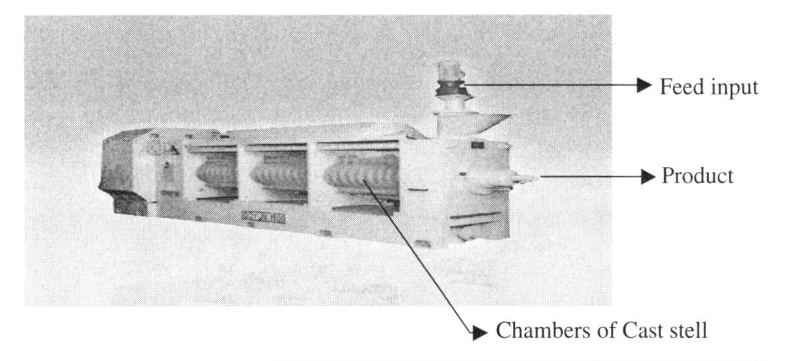

FIGURE 13.2 High-capacity (approx. 1500 tons) oil expeller. This expeller has three cast-steel chambers, vertically hinged. Each chamber is 36 in long. Cake thickness can be adjusted on a running machine.

Solvent Extraction Solvents that are used for extracting oil from the oil meal are hexane and petroleum ether. Hexane and petroleum ether have low boiling points and can be easily volatilized to recover oil.

Oil extraction is carried out in a series of large vessels by the countercurrent process. In this process, free solvent is added to the last vessel and fresh meal is fed into the first vessel. The meal and the solvent move in the opposite directions. In this way, fresh meal containing the maximum amount of oil comes in contact with the solvent containing oil, whereas a nearly exhausted meal is subjected to pure solvent. The oil solution is then redistilled to drive off the solvent. This is recovered for cyclic reuse, and the oil is collected in a storage tank. The residual cake, which is practically free of oil, is steam heated to remove the solvent, partially dried and packed. Low oil content seeds or oil cakes having residual oil (below 10 percent) are subjected to the solvent-extraction process. This process yields more oil than other processes. However, the process is expensive because solvent cannot be recovered completely.

Oil Recovery: A Novel Azeotropic Mixture for Extracting Solvents from Edible Oils Edible oil extracted by a conventional mechanical press bears many impurities such as free fatty acids, colored and other gummy materials, which are known to be detrimental to oil flavor and stability.[2] Hexane is generally used as a solvent to extract oil, but a question arises about its safety[3] due to the presence of solvent[2] in the oil as well as solvent vapor, which is a hazardous air pollutant. Ethanol, methanol, and acetone have been recommended as solvents for for extracting vegetable oils. Nag et al.[4] carried out oil extraction from flax and bahera seeds with isopropyl alcohol, ethanol, and ethyl acetate and have reached the following conclusions:

1. Isopropyl alcohol, ethanol, and ethyl acetate are equally effective for extracting oil compared with n-hexane (Table 13.1).
2. The aqueous solvent and the respective azeotropes are also effective, but an added advantage of a higher content of aqueous solvent in azeotrope solvent is that it reduces the amount of organic solvent.
3. The azeotrope solvent reduces organic gummy materials in solvent-extracted oil.

The advantage of a higher aqueous content solvent is the lower amount of organic solvent required; however, more energy would be required to remove the water from the solvent. However, amendments to the U.S. Clean Air Act—1990, Lusas et al.[5] suggested for alternative solvents to avoid the formation of hazardous air pollutants.

Solvent	\multicolumn{8}{c	}{Temperature (°C)}
	30	40	50	60	70	80	90	100
n-Hexane	M							
Ethanol	4.4	21	41	74	M			
Isopropanol (IPA)	67	83	M					
Ethyl acetate	M							
Ethanol + water	2.5	12	25	40	55	65	M	
Isopropanol (IPA) + water	35	50	60	65	M			
Ethyl acetate + water	35	50	60	70	75	M		

IPA = isopropyl alcohol; M = miscible.

TABLE 13.1 The Solubility of Oils in Different Absolute Solvents and Mixture Solvents at Different Temperatures (°C)

Supercritical Fluid Extraction An alternative technique to the organic solvent–based extraction methods is supercritical fluid extraction (SFE). Extraction with a solvent at temperatures and pressure above its critical point is known as *supercritical fluid extraction* with carbon dioxide as the extraction solvent. This has been tried as an alternative deacidification process for high free fatty acid (FFA) containing oils. The efficiency of this technique for extracting oil from a variety of seeds has been demonstrated by different investigators.[2,3]

However, the emphasis of these prior researches has been to develop an alternative option for large-scale processing of oil seeds, rather than to use SFE as an alternative analytical technique to determine oil content in seeds. King et al.[6] performed the SFE experiment in the following way.

In this experiment, 10,000 psig (680 bar) at 80°C was applied to the soybeans seeds in a 316 stainless steel extraction cell measuring 6 in in length with an internal diameter of 5.8 in. The flow rate and the amount of CO_2 used in each extraction were measured with a dry test meter under ambient conditions. The flow rate was maintained at 5 L/min, and the amount of CO_2 ranged from 200 to 600 L (expanded volume) until no further oil was extracted. The collected oil was transferred to 250-mL round-bottom flasks. After SFE, the oil was subjected to rotary evaporation for 75 min at 50°C to remove any residual CO_2 or water from collected oil. King et al.[6] also found that results obtained with SFE were in excellent agreement with those obtained by a conventional soxhlet technique with organic solvent.

Rao[7] has shown the percentage contents of oil and protein in the seeds of *Brachychiton rupestris, Brachychiton australis, Brachychiton acuminatus,* and *Brachychiton gregori* were 28.7, 29.6, 32.5, and 29.8 percent and 23.6, 23.5, 22.0, and 22.0 percent, respectively. The major fatty acid was oleic acid in *B. acuminatus* and linoleic acid in the other three seed oils. Cyclopropane fatty acids, malvalic and sterculic, were present in appreciable concentrations (6.6 to 10.6 percent and 0.5 to 2.2 percent, respectively). Cyclopropane fatty acids were identified and estimated using a combination of spectroscopic, chemical, and chromatographic analyses.

Pearl millet (*Pennisetum typhodeum* L.) grain is widely cultivated and consumed throughout the world and is rated as the world's sixth most important crop. It can be grown in regions with comparatively low rainfall (arid zone) and can also withstand the vagaries of nature to a great extent. Nag et al.[8] have discussed the processing oil from pearl millet. Studies revealed that each layer of the grain contains oil. Bran with 16 percent pearling contains about one-third of the total oil available in the kernel. The oil cake is rich in protein and can be converted to useful feed or fodder.

Nag et al.[1] observed that *Terminalia belerica Roxb* (*Combretaceae*), known as bahera, is found abundant in tropical Asia, is a source of new edible oil (37 percent by dry weight of kernel), biofertilizer, tannin, and antioxidant. The oil cake contains a high amount of nitrogen (8.34 percent). Regarding the biochemical evaluation from the oil cake, it is evident that about 60 percent NaCl extractable protein is digestible, which can be converted into biofertilizer or other useful fodder. The extractable high quality of tannin present in fruit pulp is used in the leather industry and in herbal medicines. The maximum tannin was extracted at 135°C over 12 h with shaking. The seed coat contains a high amount of gallic acid (3.2 mg/mL), which showed good antioxidant properties in different vegetable oils.

13.3 Refining Edible Oil

The refining process is generally associated with the removal of trace metals and various organic impurities known to be detrimental to oil flavor and stability. Production costs depend on the amount of oil lost in the process of removeing these impurities from the oil and the cost of operating and maintaining the equipment used in the refining process. A refinery running at 0.6 percent adsorbent with a production rate of 1000 tons/day with 28 percent oil retention would have an annual oil loss of 613 tons. The main object of each stage is to obtain the best oil quality at the lowest cost.

Two schemes are generally employed: physical refining and chemical refining. Physical processes are generally preferred over chemical techniques.

Physical Refining The oil is heated to about 260°C to coagulate albuminoidal matter, which absorbs with it self much of the coloring matter. When cooled, the albuminoidal matter settles down. The clear oil can be decanted off from the top or by filtration. Sometimes, physical refining is done by distillation, which is steam stripping the oil at an elevated temperature under reduced pressure. A new degumming protocol has been developed that employs electrolyte solutions to remove nonhydratable gums from soybean, rice bran, and mustard oils. This protocol removes nonhydratable gums, mostly phosphatidic acid (PA) and phosphatidyl ethanolamine (PE), which is left over after water degumming and cannot be removed without using phosphoric acid or citric acid. Acid degumming is associated with oil loss (~10 percent) due to its emulsifying nature and subsequent washings.[9]

Chemical Refining Chemical refining is done to remove the free fatty acids and leave a neutral fat. The oil is churned with a solution of caustic alkali to form soap with a free fatty acid. The soap settles down along with the break and coloring matter. The clear oil is drawn off, washed with water to remove the excess alkali, dried, and filtered. This process results in a very pale-colored oil. Sometimes sodium carbonate is used to remove the free fatty acids. Alternative processes such as liquid–liquid extraction with 98 percent methanol and without the aid of stripping steam[3] have been used to a limited extent to deacidify expensive specialty oils with high initial acidity.

13.4 Bleaching and Deodorization

Bleaching Edible Oil Refined oil can be further bleached to an almost colorless state using activated earth. Fuller's earth is prepared by treating natural earth with HCl, which consists of a large number of small pores where the coloring matter of oil is absorbed, and then dried. The oil is mixed with activated earth in a tank fitted with agitators and heated to a moderate temperature. The oil is then clarified by filtration. In addition to fuller's earth, activated carbon, kieselguhr, or diatomaceous earth may be used. The colored compound is also removed in certain cases either by oxidation or reduction to colorless derivatives or by decomposition with concentrated H_2SO_4.

The quality of edible oils depends on the removal of trace metals, free fatty acids, gummy materials, coloring, and various organic matter that promote oxidation and produce objectionable compounds that are detrimental to oil flavor and stability. Different types of adsorbents[9,10] are used to remove coloring materials and other impurities in oils.

The term "bleaching" refers to the process of removing color-producing substances and for further purifying the fat or oil. Normally, bleaching is accomplished after the oil has been refined.

Chlorophyll is readily extracted along with the oil from oil seeds. Chlorophyll imparts a greenish color to crude oil. In conventional refining techniques, the chlorophyll is converted to the pigment pheophytin, which makes oil dark and contributes to an off flavor. It may also promote oxidation of the oil and thus reduce its storage stability.

Removal of chlorophyll is essential; however, removal of other colored pigments is not particularly important because the pigments are effectively removed during either hydrogenation or deodorization due to the heat-bleaching effect during those two processes.[11] There are indications that chlorophyll and its derivatives may be present in the oil—not in true solution but as a suspension of very small particles, probably colloidal in size. During the last decade, new bleaching agents have been developed that are more effective in removing chlorophyll than conventional bleaching clays. The industry is carefully evaluating whether the increased costs of these agents is offset by the savings in clay requirements and reduced oil losses.[12]

The usual method of bleaching is by adsorption of the color-producing substances on an adsorbent material. Acid-activated bleaching earth or clay as bentonite[22] or alumina-pillared (Al-pillared) acid-activated clay[12] is the adsorbent material that has been used most extensively. This substance consists primarily of hydrated aluminum silicate. Anhydrous silica gel and activated carbon are also used as bleaching adsorbents to a limited extent. The desired color of the oil can be obtained by adjusting the amount and type of earth used and the physical conditions under which the bleaching operation is carried out. Synthetic adsorbents and attapulgites[13] have also been used.

Because the necessity of using fuller's earth involves not merely an expense but also loss of oil, and because the costs of decolorizing rise with the amount of earth used, the price paid for oil, such as cottonseed oil in the United States, among other factors, also depends on the ease with which the oil may be decolorized or bleached. Furthermore, if too much fuller's earth has to be used, the oil acquires an earthy flavor. The American trade has, therefore, established two classes of cottonseed oil: bleachable and unbleachable. A bleachable oil is one that may be reduced to a very definite color standard when treated in a specifically prescribed way.[14]

Nag et al.[10] have studied the effect of temperature and cost required to bleach oil. Kinetic studies on bleaching edible oil reported using a low-cost adsorbent (e.g., charred sawdust) at different temperatures. The changes in energy, enthalpy, and entropy during this process have been calculated as follows.

Nag et al.[11] have calculated the change in free energy (ΔG) due to the adsorption of inorganic salts from the solution to the surface of the solid [Eq. (13.1)]:

$$\Delta G = RT/M \int_0^{C_p} \Gamma_p / C_p \, dC_p + RT/M \ln C_p \tag{13.1}$$

The first term on the right-hand side of Eq. (13.1) is based on the Gibbs adsorption equation and represents the change in free energy (KJ/mole) due to the transfer of Γ (mg/m^2) of the adsorbate to an interface when the solute concentration is changed from zero to C_p. The value of this term can be obtained by graphical integration using the experimental data. The value of Γ_p/C_p has been plotted for various values of C_p within the concentration range from zero to a fixed value of adsorbate concentration, and the graphical value of the area has been estimated. Here, M is the molecular weight of the oils, and the second term on the right-hand side represents the free-energy change due to hypothetical dilution of the bulk solution from the concentration range C_p to one molal value, keeping Γ_p (mg/m^2) hypothetically fixed during the dilution process. ΔG for different values of C_p has been estimated from the experimental data. For a diluted solution, the difference between molarity and molality has been neglected.[13] ΔG is the free-energy change of the crude oil per gram of the adsorbent, that is, the saturated surface area of charred saw dust referred as Φ, where Φ = surface area = 1. From kinetic studies, it is observed that adsorption values will decrease, which show that colorants and impurities in the oils are adsorbed by charred sawdust.

Extraction, Refining, and Stabilization of Edible Oils

Crude oil becomes lighter in color during the course of experiment. According to Beer–Lambert's law, adsorption is proportional to the concentration, which is as follows:[15]

$$\ln A/A_0 = -kt \quad (13.2)$$

where A is the absorbance of the colorants at time t and A_0 is the absorbance of the colorants at $t = 0$. If we plot $\ln A$ versus t according to Eq. (13.2), the slope of the curves gives the value of k.

The activation energy can be determined by the Arrhenius equation[15]

$$k = Ae^{-E_a/RT} \quad (13.3)$$

where k = rate constant
A = frequency factor or the Arrhenius constant
E_a = activation energy
R = universal gas constant (8.314 $JK^{-1}mol^{-1}$)
T = absolute temperature

When $\ln k$ is plotted against the reciprocal of the temperature $(1/T)$, $-E_a/R$ represents the slope and the intercept gives $\ln A$. The thermodynamic parameters are assigned to activated complex as follows:

$$\Delta H = E_a - RT \quad (13.4)$$

Here, ΔH represents the activation enthalpy change of the oils; these values are represented in Table 13.2. The negative value of ΔH indicates that energy is released during the adsorption process (i.e., this adsorption process is an exothermic one). ΔS represents the activation entropy change; its values are calculated at different temperatures [Eq. (13.5)] as follows:

$$A = RT/Nhe^{\Delta S/R} \quad (13.5)$$

Name of Oil	Temperature (°C)	ΔH (KJ/mole)	ΔS (KJ^{-1}/mole)
Bahera oil	30	−2.503	−22.80
	60	−2.752	−19.96
	75	−2.877	−18.74
Mustard oil	30	−2.506	−16.81
	60	−2.755	−14.55
	75	−2.880	−13.56
Sesame oil	30	−2.505	−14.17
	60	−2.754	−12.10
	75	−2.879	−11.27

TABLE 13.2 Measurement of ΔH, ΔS at Different Temperatures of Bahera, Mustard, and Sesame Oil

Chapter Thirteen

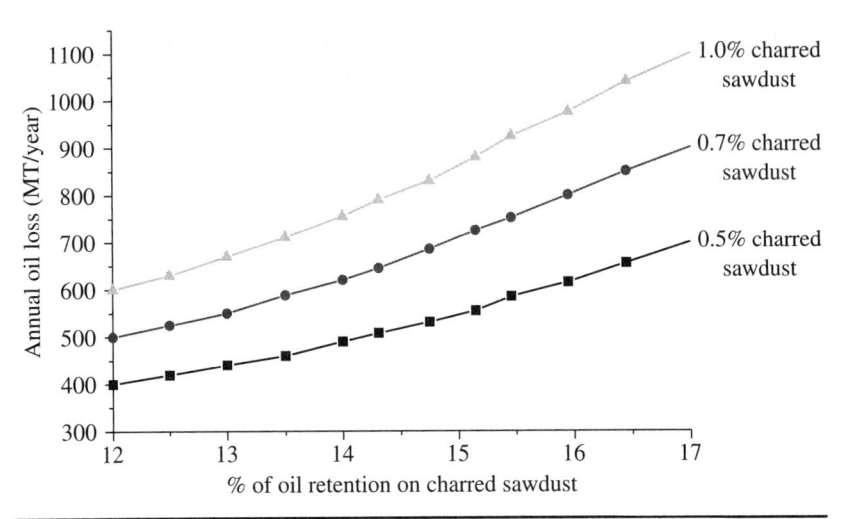

FIGURE 13.3 Graphical representation of oil retention (i.e., annual oil loss) during the bleaching process of edible oil using charred sawdust as an adsorbent.

where N is Avogadro's constant (6.023×10^{23}), and h is Planck's constant (6.626×10^{-31} JS).

The negative value of ΔS indicates there is less disorder, that is, the crude oil contains more colorant and impurities before the bleaching process, and these are decreased during the adsorption process because disorder is directly proportional to entropy.

Generally, loss of oil in the bleaching process results from oil lost to the filter cake and oil lost to the changes in the free fatty acid content. From David D. Brooks's statistical data[16] we observe that for bleaching edible oil in different refinery industries, oil loss varies from 24 to 30 percent. A refinery running at 0.6 percent adsorbent with a production rate of 1000 tons per day has a 28 percent oil loss of 613 tons. In our system with 1 percent charred sawdust, oil loss is only 17 percent (Fig. 13.3), which is very economical for bleaching edible oil.

Vacuum bleaching,[12] batch or continuous, is somewhat more effective than atmospheric bleaching. It can use less earth, emphasize quantity not quality, operate at lower top temperatures, and minimize oxidation by reducing exposure to air and providing an opportunity to cool the oil before returning it to atmospheric conditions. Although vacuum bleaching practices are preferred, atmospheric bleaching can and does produce high-quality bleached oils. Some refiners prefer batch vacuum or atmospheric bleaching because it provides absolute control, whereas continuous bleaching gives an average treatment because it is not a plug–flow system.

Deodorization Zosel[17] described the deodorization of edible oil with CO_2 at a temperature between 50 and 55°C and pressures of 10 to 25 MPa

Extraction, Refining, and Stabilization of Edible Oils

in a continuous countercurrent device. However, there has been the possibility of negative effects at high temperatures during deodorization. Ziegler et al.[18] worked on deodorizing and deacidifying peanut oil using CO_2. The efficient deodorization and deacidification of an actual crude oil pressed from roasted peanuts and accomplished by extraction with CO_2 was done at 47°C and 20 MPa.

Extraction with deacidification is expensive, especially with oils with high initial acidity or where the quality and purity of the extracted components are important. Good manufacturing practice also implies the following: use of stainless steel equipment, careful deaeration at <100°C before heating to the final stripping temperature, use of oxygen-free steam, and strict feedstock specifications (typically Fe 0.1, Cu 0.01, P S, bleaching earth 5 mg/kg oil maximum).

13.5 Stability of Edible Oils and Antioxidants

Lipid oxidation is primarily responsible for the deterioration of the sensory and nutritional quality of foods. This process, known as rancidity, is responsible for economic losses associated with the deterioration of edible oils. When this deterioration occurs, the oil or the oil-containing food becomes unacceptable to consumers because of off-flavors and unpleasant odors associated with the formation of oxidized products. In addition, many of these oxidized products are considered unhealthy and harmful. One way of preventing lipid oxidation and enhancing the shelf-life of oils and oil-containing foods is by adding antioxidants. These antioxidants may be indigenous to the food product or added after processing. Because of health concerns associated with the use of synthetic antioxidants, there has been an increasing interest in natural antioxidants. This chapter briefly reviews lipid oxidation and discusses the efficacy of natural antioxidants in minimizing or preventing oxidative changes from occurring.

Food components such as lipids are very susceptible to oxidation, which results in detrimental changes to the color, odor, and nutritive value of the affected food products. Antioxidants prevent such changes by retarding or slowing down the process of oxidation or rancidity.

Fat deterioration can be divided into four types[19]

1. *Hydrolysis*—leads triacylglycerols to form free fatty acids and glycerol, often characterized by a "soapy flavor."

2. *Rancidity*—a term widely used in the food industry that normally covers a large number of objectionable off-flavor volatile components generated from the auto-oxidation of polyunsaturated fatty acids.

3. *Reversion*—a type of flavor and odor degradation usually associated with the oxidation of certain highly unsaturated vegetable oils (e.g., soybean oil) and fish oils. This flavor degradation is attributed to oxidation of linolenic acid, and other omega-3 fatty acids.

4. *Polymerization*—a term usually used to describe the cross-linking of unsaturated fats between two carbon atoms. Polymers are also formed with oxygen between two fatty acid chains at an unsaturated site with both types of polymers containing cyclic structures.

13.5.1 Mechanism of Antioxidants

Generally, the oxidation of fats or oils involves a free-radical mechanism. This process can be induced catalytically by light, temperature, enzymes, metals, metalloproteins, and microorganisms, with the reaction involving free radicals or active oxygen species.[20]

The initiation step, the formation of a fatty acid free radical, occurs when hydrogen departs from the α-methylenic carbon in the unsaturated fatty acid group of the fat molecule. The resultant radicals become very susceptible to attack by oxygen to form peroxide free radicals. These free radicals themselves further serve as strong initiators and promoters (catalysts) of oxidation by extracting hydrogen from another fatty acid molecule, which triggers propagation. In the final stage of fatty acid oxidation, the primary oxidation products of rancidity formed are hydroperoxides. These hydroperoxides are quite unstable and subsequently degrade into smaller-chain organic compounds such as aldehydes, ketones, alcohols, and acids—the secondary oxidation products of rancidity. It is the latter compounds that actually render fats and oils rancid and unacceptable and unsuitable for food use.

Fat or oil molecule where R = fatty acid group

Double-bond linkage and methylenic carbon

Propagation stage

$$\underset{H}{\overset{H}{\underset{|}{C}}}-\overset{H}{\underset{|}{C}}=\overset{H}{\underset{|}{C}}- \xrightarrow{-\dot{H}} -\overset{H}{\underset{|}{\dot{C}}}-\overset{H}{\underset{}{C}}=\overset{H}{\underset{}{C}}- \xrightarrow{+O_2}$$

Fat free radical

$$-\overset{H}{\underset{|}{C}}-\overset{H}{\underset{}{C}}=\overset{H}{\underset{}{C}}- \xrightarrow{+\dot{H}} -\overset{H}{\underset{|}{C}}-\overset{H}{\underset{}{C}}=\overset{H}{\underset{}{C}}-$$
$$\underset{\dot{O}}{\overset{O}{|}} \qquad \underset{H}{\overset{O}{|}}$$

Peroxide free   Hydroperoxide

Triplet oxygen lipid oxidation, a free-radical process, has been extensively studied during the past 70 years. However, triplet oxygen lipid oxidation does not fully explain the initiation step of lipid oxidation. Singlet oxygen is involved in the initiation of triplet oxygen lipid oxidation because singlet oxygen can react directly with double bonds without the formation of free radicals. During the last 30 years, attention has been given to singlet oxygen oxidation of foods because (1) the rate of singlet oxygen oxidation is much greater than that of the triplet oxygen oxidation, and (2) singlet oxygen oxidation produces compounds absent in triplet oxygen oxidation due to the different reaction mechanisms. Interaction with light, sensitizers, and oxygen is mainly responsible for singlet oxygen formation in food. The reaction rates of singlet oxygen with oleic, linoleic, and linolenic acids were 0.74, 1.3, and $1.9 \times 10^5 \, M^{-1}s^{-1}$, respectively.[19] These values were relatively proportional to the number of double bonds in the molecules. On the other hand, triplet oxygen reacts with unsaturated fatty acids by abstracting allylic hydrogen. Once hydrogen is removed, a pentadienyl radical intermediate is formed.[20] The energy required for hydrogen removal varies with different carbon atoms. The relative reaction ratio of triplet oxygen with oleic, linoleic, and linolenic acid for hydroperoxide formation is 1:12:25, which is dependent on the relative difficulty for the formation of the free radical in the molecule.[21] The reaction rate of the triplet oxygen with linolenic acid is approximately twice as fast as that of linoleic acid because linolenic acid has two pentadienyl groups in the molecule, whereas linoleic acid has one.[20] The classical mechanism for free-radical oxidation of unsaturated fatty acids involves hydrogen abstraction at the allylic carbon to produce delocalized three-carbon allylic radicals.

Termination Phenolic substances (antioxidants) function as free-radical acceptors and could terminate free radicals at the initiation stage. Hindered phenolics such as BHA, BHT, TBHQ, EQ, and tocopherols, as well as polyhydroxy phenolics like propyl gallate, are primary antioxidants that delay or inhibit the initiation step by reacting with a lipid free radical or by inhibiting the propagation step by reacting with the peroxy or alkoxy radicals. Thymol, a phenolic compound obtained from an extract of the spice ajowan, also actively prevents further oxidation in the glycerides, which was suggested by Nag et al.[22]

Some compounds contribute to antioxidant defense by chelating transition metals and preventing them from catalyzing the production of free radicals in the cell. Particularly important is the ability to sequester iron, which is the function of iron-binding proteins such as transferrin and ferritin.[20] Selenium and zinc are commonly referred to as antioxidant nutrients, but these chemical elements have no antioxidant action themselves and instead are required for the activity of some antioxidant enzymes.[20]

13.5.2 Functions of Antioxidants

Antioxidants mixed with food materials have the potential to

1. Control damaging free radicals.
2. Minimize the effects of aging.
3. Help control fatty deposits in the arteries and maintaining elasticity of the veins and good circulation.
4. Helps to relieve localized oxidative stress and also promote healthy joint cells.
5. Although antioxidants may not cure cancer, much research has been done, and is still continuing, regarding the role of antioxidants in the prevention and alleviation of cancer.
6. It can help boost the immune system and clear the blood.
7. Promotes tissue growth and helps prevent destruction of DNA. Antioxidants are also indicated to assist with memory loss.
8. Oxidation processes can hamper wound healing, and because antioxidants will help control this, better healing should be promoted and the immune system boosted in general, which will help to prevent infections.
9. Assist in free-radical scavenging.

The wide array of solid antioxidants and liquid antioxidant blends available in the market necessitates consideration of the following factors before using a particular antioxidant in food:

1. The type of food to be stabilized.
2. The need to "carry through" from the oil to the cooked food product.
3. The ease of antioxidant solubility and disposition into the fat phase of the product.
4. The presence of metal ions and the possibility of discoloration.
5. The relative severity of processing (e.g., baking versus frying) and the legality of using a specific antioxidant for a specific antioxidant application.

13.5.3 Different Types of Antioxidants

There are two types of antioxidants: primary and synergetic.

Primary Antioxidants Different synthetic antioxidants, such as tertiary butyl hydroquinone, butylated hydroxytoluene, hydroxyanisole, and tocopherols, for example, have been used in foods to prevent deterioration during storage, transportation, and discoloration from oxidation. The primary antioxidants act as follows:

1. They can terminate the free-radical chain by donating hydrogen to free radicals and converting them to more stable products.
2. They can form lipid–antioxidant complexes by reacting with lipid radicals.
3. They may either delay or inhibit the initiation step by reacting with a lipid free radical or inhibit the propagation step by reacting with peroxy or alkoxy radicals.

Synergetic Antioxidants These antioxidants decompose the hydrogen peroxide formed during the reaction of primary antioxidant with free radicals and are used synergically with primary antioxidants. Examples of synergistic antioxidants include phosphates, sulphites, and thioesters.

Both natural and synthetic antioxidants are used in the food industry. Recently, there has been an increased demand for natural antioxidants. When compared to synthetic antioxidants, natural antioxidants have the following advantages:

1. They are readily acceptable by consumers.
2. They are considered to be safe.

3. These natural antioxidants (not synthetic chemical antioxidants) are identical to the food, which people ate for hundreds of years.

4. These antioxidants not only stabilize food but also add nutritional value to foods, such as ginger rhizome, β-carotene, ascorbic acid from lemons, tocopherol, ajowan extracts, and cereal extracts.

There is a continuing need for widely usable natural antioxidants. In addition to having the same safety and technological requisites of synthetic compounds, the natural antioxidant additive should also originate from an abundant and inexpensive vegetable matrix, be produced by an economic and compatible technology, and be effective.

Most natural antioxidants are phenolic compounds (with the exception of the tocopherols) that contain *ortho*-substituted active groups, whereas synthetic antioxidants (with the exception of gallates) are *para* substituted.[20] The most potent sources of natural antioxidants—spices and herbs—have been used for hundreds of years not only for flavoring but also for their food-preserving abilities.

Nag et al.[22] observed that an azoan extract contained thymol as an active ingredient, which acted as a natural antioxidant suitable for the stabilization of oil (Fig. 13.4).

Hu et al.[23] determined the antioxidant properties of the whole leaves of aloe vera (*Aloe barbadensis* Miller) compared to BHT and α-tocopherol using the DPPH radical scavenging method. The antioxidant activity of aloe vera extracts and reference compounds follows this order: 3-year-old aloe vera > BHT > 4-year-old aloe vera > α-tocopherol > 2-year-old aloe vera. The 3-year-old extract exhibited the strongest radical scavenging activity of 72.19 percent higher than that of BHT at 70.52 percent and α-tocopherol at 65.20 percent. These data suggest that the stage of development of aloe vera dramatically affects its composition and antioxidant activity. Nag[24] observed that the antioxidant properties of capsicum in flaxseed oil are due to the presence of compound N-(4-hydroxy-3-methoxy-benzyl)-alkylamide.

Phenolics in vegetable oils may generally exert a beneficial effect on their oxidative stability. As an example, sesame oil, which contains sesamine, sesaminal, and sesamole, is known to be highly stable.[25] Similarly, olive oil is a rich source of tyrosols and hydroxytyrosol, and these compounds are known to possess good antioxidant activity.[26] Nag et al.[26] recovered natural antioxidants from olive oil mill wastewater and studied the oxidative stabilization of lard.

Studies on ginger extracts were also found to have higher antioxidant activity than α-tocopherol.[27] In Nigeria, ginger is one of the few commercial spices produced for local consumption that is used in bread, oil, essences, pastries, and biscuits. In addition to spices and herbs, teas, vegetables, and fruits are also good sources of

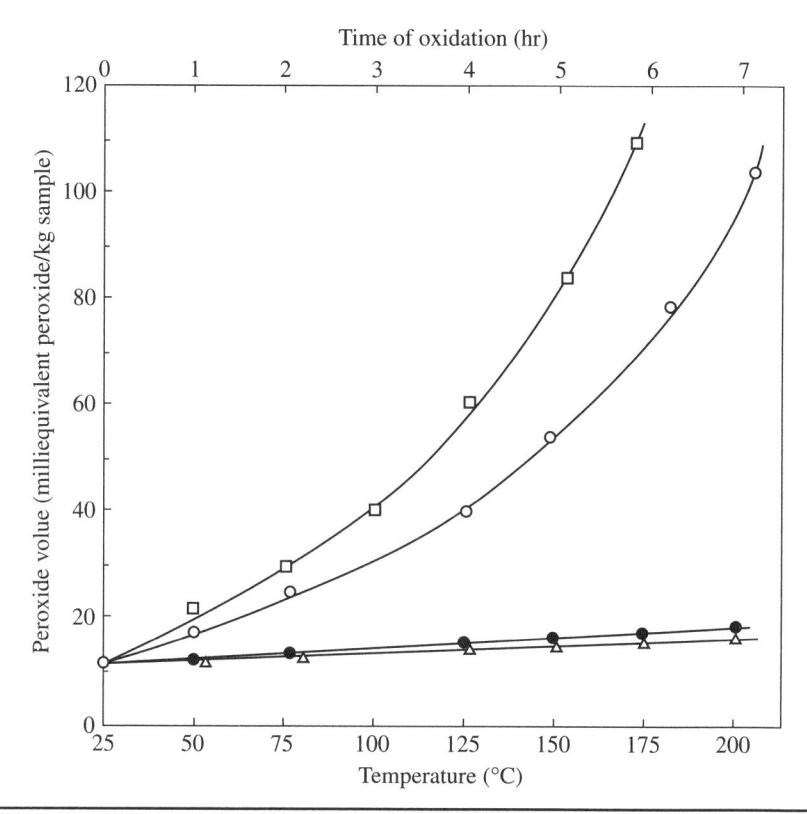

Extraction, Refining, and Stabilization of Edible Oils 385

FIGURE 13.4 Peroxide values of flaxseed oil at different temperatures in the presence of air: without antioxidant (○); with antioxidant (△). Peroxide values of flaxseed oil at different times in the presence of air at 30°C: without antioxidant (□); with antioxidant (●).

antioxidants, such as, tocopherols, ascorbic acid, and β-carotenes.[28] Black tea and green tea also have very good antioxidant potential.[29] Extracts from grapes were also shown to exert good antioxidant activities, which proved beneficial in reducing the incidence of cardiovascular disease.[28]

13.6 Measuring Antioxidant Activities

The efficacy of synthetic and natural antioxidants is measured by monitoring the oxidative stability of oil. After the sample is oxidized under standard conditions, the extent of oxidation is measured by chemical, instrumental, or sensory methods.

A variety of methods used to assess antioxidant activities including the following.

Peroxide Value Method A 250-mL standard joint round-bottom flask containing 100 g of oil was placed over a thermostatic bath. Air from an air cylinder (pressure of the cylinder 1.7 kg/cm^2) was passed through the oil (2 mL/min). At different intervals oil was withdrawn to determine peroxide values (PV) as follows. Weigh 0.1 to 0.2 g of oil in a 250-mL iodine flask and add 15 cc of chloroform or carbon tetrachloride to dissolve it. Along with this, carry out a blank experiment. Add 25 cc of Wij's solution (7.9 gm of pure iodine trichloride in 100 mL of glacial acetic acid in a beaker by warming on a water bath. In another flask containing dissolve 8.7 gm of iodine in a second portion of warm 100 cc glacial acetic acid. Mix the two solutions in a 1000-mL measuring flask and fill the solution to the mark with glacial acetic acid. Store the Wij's solution in a freezer) to the flasks containing the fat and the blank. Allow the flasks to sit in the dark for about 30 min after adding 25 mL of a 10 percent solution of potassium iodide to the neck of the flask. Then dilute the mixture with 50 to 60 mL of water. Titrate it immediately against a 0.1-N sodium thiosulphate solution. Shake vigorously and titrate again until the yellow color almost disappears. Add a starch solution and titrate again until the blue color disappears.

Let V_b and V_w be the number of cubic centimeters of $N/10$ sodium thiosulphate required for W g of fat. The amount of thiosulphate that is the equivalent of iodine that reacts with W g of fat is calculated using the following formula:

$$\frac{(V_b - V_w) \times 10^{-3}}{10 \times W}$$

From this, we get the amount of iodine in grams that react with W g of fat:

$$\frac{(V_b - V_w) \times 10^{-3}}{10 \times W} \times 127$$

where 127 is the equivalent weight of the iodine.

Therefore, the peroxide value (meqv of O_2/kg of lipid) and the formula is meqv O_2/kg: [V (mL of thiosulphate) $\times N$ (thiosulphate)] $\times 1000/W$ (g of fat sample):

$$\frac{(V_b - V_w) \times 10^{-3} \times 127}{10 \times W} \times 100 = \frac{(V_b - V_w) \times 1.27}{W}$$

Thiobarbituric Acid Test The thiobarbituric acid (TBA) test was based on the color reaction of TBA with malondialdehyde (mg/kg) in the sample.

Mix 10 g of fatty food with 50 mL water for 2 min and pour into a distillation flask with 47.5 mL water. Add 2.5 mL of 4-N hydrochloric acid to bring the pH to 1.5. Heat the flask with an electric mantle so that the 50-mL distillate is collected in 10 min from the time boiling commences. Pipette the 5-mL distillate into a glass-stoppered tube, add 5 mL TBA reagent (0.2883 g/100 mL of 90 percent glacial acetic acid), stopper, shake, and heat in boiling water for 35 min. Prepare a blank similarly using 5-mL water with 5 mL of reagent. Then cool the tubes in water for 10 min and measure the optical density (D) against the blank at 538 µm using 1-cm cells. The TBA number (as mg malonaldehyde/kg sample) is $7.8D$.

Measuring the Oxidation Stability of Fats Using the Rancimat Method
Oxidation processes that slowly take place in fats at ambient temperature are known as autoxidation. They start with the reaction of the fatty acids. In multistage processes, this leads to a variety of decomposition products such as peroxides, alcohols, aldehydes, and different carboxylic acids. In the Rancimat method, the sample is exposed to a particular high temperature (50 to 200°C), and the volatile oxidation products are transferred to the measuring vessel by air stream and are absorbed in the measuring water solution. When the conductivity of the measuring solution is recorded continuously, an oxidation curve is obtained whose point of inflection is known as induction time, which provides characteristic values of the oxidation stability.

Apparatus Required (Rancimat Apparatus Attached to a Computer)—Methods

1. Check initially that the Rancimat apparatus is attached to a computer correctly and then switch both machines.
2. Sample vessel and measuring vessels should be cleaned, otherwise reaction conditions are hampered.
3. Before starting measurement, program the required temperature and flow of air.
4. Place the weighed sample (approximately 2.0 to 2.5 g) and cover the vessel with caps fitted with screws and tubes.
5. Place the measuring vessel containing 60 mL of distilled water in the Rancimat apparatus (Fig. 13.5).
6. Attach the outlet tube of the sampling vessel in the measuring vessel to absorb the oxidized products in water.
7. Place the exhaust tube of the measuring vessel in a hood because oxidation products of are very irritating to the eyes.
8. After the required time as programmed on the computer, take the induction result with the time from the computer.

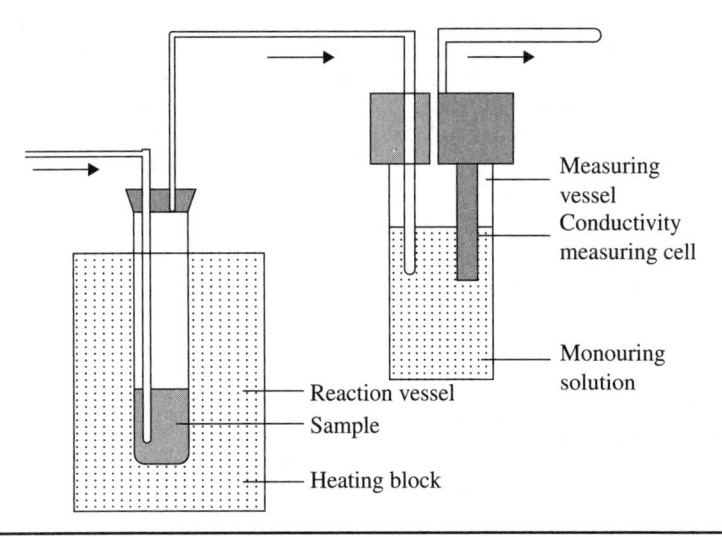

FIGURE 13.5 Rancimat apparatus.

Rapid Method for Determining the Oxidation of Fatty Acids   Samples were dissolved in isooctane for gravimetric transfer into 5-mL high-pressure headspace vials. The isooctane was evaporated under a stream of nitrogen followed by equilibration of the vials in air for 10 min. The vials were then sealed with Teflon tape before heating.

Capillary Gas Chromatography   Sample vials were heated at 85°C for 0, 5, 35, 65, and 95 min. With a gas-tight syringe, 1 mL of headspace volatile was injected directly into a Hewlett Packard gas chromatograph (GC) with a flame ionization detector, integrator, and IBM computer for data handling and storage. Helium was the carrier gas in the fused silica capillary, 30 m × 0.32 mm, the injector was set at 250°C and the detector at 270°C; the temperature was programmed at 50 to 110°C (at 4°C/min) and from 110 to 220°C (at 20°C/min) with a 1-min hold. Volatiles were identified by comparing retention times to reference standards.[30]

GC/MS Analysis   Confirmation of volatile identities was accomplished on a HP GC/MS (mass spectrometer) containing a similar capillary column; ionization energy was 70 eV, and scan range was 38 to 260 m/z. After the respective fatty acids (i.e., Lo/DHA), were heated (80°C/100 min), 100 μL of the headspace volatile was injected into a DB5 silica capillary column 30 m × 1.0 mm. Conditions remained as described earlier. After the ionization pattern was recorded from GC/MS system, and the relative retention times were generated using the GC technique, most of the volatile compounds could be identified.

Fatty Acid Analysis Loss of fatty acids due to heating was determined from the fatty acid methyl esters of Lo/EPA/DHA. Prior to methylation, C19:0 acid was added as an internal standard to each sample. A HP GC/DB-225 silica capillary column (30 m × 0.25 mm) used at 180 to 230°C (2°C/min), and values compared with initial and final fatty acids remaining gave the fatty acid loss for various periods of heating at 80°C.

Triplet oxygen lipid oxidation, a free-radical process, has been extensively studied during the past 70 years. However, triplet oxygen lipid oxidation does not fully explain the initiation step of lipid oxidation. Singlet oxygen is involved in the initiation of triplet oxygen lipid oxidation because singlet oxygen can react directly with double bonds without forming free radicals.

Termination Antioxidants are classified into two groups—primary and chain-breaking antioxidants and secondary antioxidants—depending on their mechanisms of action. The former react with lipid peroxy radicals to convert the antioxidants to stable products. These groups include free-radical inhibitors. Secondary antioxidants such as oxygen scavengers reduce the rate of chain initiation.

$$AH + -\overset{H}{\underset{\cdot}{C}} - \overset{H}{C} = \overset{H}{C} - \longrightarrow A^+ \quad + -\overset{H}{\underset{H}{C}} - \overset{H}{C} = \overset{H}{C} -$$

Antioxidant fat　　　　　　　　　　Original fat molecule

References

1. Nag, A., Bera, D., and De Leonardis, A. 2007. Biotechnological application of new edible oil, biofertilizer and antioxidant from new resource. *Journal of Food Engineering* (UK) 81(4):678–672.
2. Lumley, I. D. and Colwell, K. K. 1991. *Analysis of Oilseeds, Fats and Foods*, eds. J. B. Rossell and, J. L. R., Pritchard. London, UK: Elsevier.
3. Gunstone, F. 2004. *The Chemistry of Oils and Fats: Sources, Composition, Properties and Uses*. Oxford, UK: Blackwell Publishing.
4. Bera, D., Lahiri, D., De Leonardis, A., De, K. B., and Nag, A. 2006. A novel azeotropic mixture for solvent extraction of edible oils. *CIGR European Journal* 24.
5. Firestone, D., ed. 1990. *Official Methods and Recommended Practices of the American Oil Chemists Society*. 4th ed. Champaign, IL: American Oil Chemists' Society.
6. King, J. W., Taylor, S. L., Snyder, J. M., and Holliday, R. L. 1993. Total fatty acid analysis of vegetable oil soapstocks by supercritical fluid extraction/reaction. *Journal of the American Oil and Chemistry Society* 71(10):1291–1294.
7. Bera, D. 2009. Physico-Chemical Studies in the Extraction, Purification and Stabilization of Bahera oil and Application of its by-products, PhD thesis, Vidyasagar University, West Bengal, India.
8. Nag, A., Jain, R. K., and Bal, S. 1996. Processing of pearl millet oil and digestibility characteristics of protein cake. *Indian Chemical Engineering A* 38:33–35.

9. Nasirullah. 2005. Physical refining: Electrolyte degumming of nonhydratable gums from selected vegetable oils. *Journal of Food Lipids* 12(2):103–111.
10. Bera, D., Lahiri, D., and Nag, A. 2004. Kinetic studies on bleaching of edible oil using charred sawdust as a new adsorbent. *Journal of Food Engineering* 65: 33–36.
11. Woerfel, J. B. 1981. Processing and utilization of by-products from soy oil processing. *Journal of the American Oil and Chemistry Society* 58:188–190.
12. Christidis, G. E. and Kosiari, S. 2003. Decolorization of vegetable oils: A study of the mechanism of adsorption of β-carotene by an acid activated bentonite from Cyprus. *Clays and Clay Minerals* 51:327–333.
13. Falaras, P., Lezou, F., Seiragakis, G., and Petrakis, D. 2000. Bleaching properties of alumina–pillared acid activated montmorillonite. *Clays and Clay Minerals* 48:549–556.
14. Erickson, D. R., Pryde, E. H., Brekke, O. L., Mounts, T. L., and Falb, R. A. 1980. *Handbook of Soy Oil Processing and Utilization*. St. Louis, MO and Champaign, IL: American Soybean Association and the American Oil Chemists' Society.
15. Laidler, J. K. 1997. *Chemical Kinetics*, 3rd ed. New York, NY: Harper & Row. 40–47.
16. Brooks, D. D. *Proceedings of PORIM International Palm Oil Changes*. Palm Oil Research Institute of Malaysia, Malaysia 1–6 February, pp. 234–238.
17. Zosel, K. 1976. Process for the simultaneous dehydrogenation and deodorisation of fats and oils. U.S. Patent 3969382, filed August 29, 1974.
18. Ziegler, G. R. and Liaw, Y. J. 1993. Deodorization and deacidification of edible oils with dense carbon dioxide. *Journal of the American Oil and Chemistry Society* 70:947–953.
19. Nakatam, N., Achiban, T. Y., and Kituzaka, K. 1988. Biochemical and chemical aspects of free radicals. *Proceedings of the 4th Biennial General Meeting of the Society for Free Research*, Kyoto, Japan, April, pp. 453–466.
20. Addis, P. B. and Warner, G. L. 1991. The potential aspects of lipid oxidation products in food. In *Free Radicals and Food Additives*, eds. O. I. Aruoma and B. Halliwell. London, UK: Taylor and Francis Ltd. 140–150.
21. Newton, I. S. 1996. Food enrichment with long chain n–3 PUFA. *INFORM* 7:169–177.
22. Nag, A., Bera, D., and Lahri, D. 2005. Studies on a natural antioxidant for stabilization of edible oil and comparison with synthetic antioxidants. *Journal of Food Engineering* (UK) 32:76.
23. Hu, Y., Xu, J., and Hu, Q. 2003. Evalution of antioxidant potential of Aloe vera (*Aloe barbadensis Miller*) extracts. *Journal of Agriculture and Food Chemistry* 51: 7788–7791.
24. Nag, A. 2000. Stabilization of flaxseed oil with capsicum antioxidant. *Journal of the American Oil and Chemistry Society* 77:799–800.
25. Kikugawa, K., Arai, M., and Kurechi, T. 1983. Participation of sesamol in stability of sesame oil. *Journal of the American Oil Chemists' Society* 60:1528–1533.
26. Nag, A., De Leonardis, A., Macciola, V., Lembo, G., and Aretini, A. 2006. Studies on oxidative stabilization of lard by natural antioxidants recovered from olive oil mill wastewater. *Journal of Food Chemistry* 6:309–320.
27. Kikuzaki, H. and Nakatani, N. 1993. Antioxidant effects of some ginger constituents. *Journal of Food Science* 58:1407–1410.
28. Loligert, J. 1991. The use of antioxidants in food. In *Free Radicals and Food Additives*, eds. O. I. Aruoma and B. Halliwell. London, UK: Taylor and Francis Ltd. 121–130.
29. Cao, G., Sofic, E., and Prior, R. L. 1996. Antioxidant capacity of tea and common vegetables. *Journal of Agriculture and Food Chemistry* 44(11):3426–3431.
30. Haumann, B. F. 1997. Nutritional aspects of n-3 fatty acids. *INFORM* 8: 428–447.

CHAPTER 14
Phenolic Substances from Olives (*Olea europeae*) and Related Olive Mill Products

Antonella De Leonardis and Vincenzo Macciola

Department of Agricultural, Food, Environmental and Microbiological Science and Technologies (DiSTAAM)
University of Molise, Campobasso, Italy

Ahindra Nag

Department of Chemistry
Indian Institute of Technology
Karagpur, India

14.1 Introduction

Epidemiological evidence shows that the Mediterranean diet is associated with a lower incidence of coronary heart disease and some cancers. Virgin olive oil is considered to be one of the best edible lipids, and it is the primary fat used in Mediterranean countries.

Several researchers demonstrate that the beneficial effect of eating olive oil is due to favorable fatty acid composition and also to the presence of minor components. Among the minor components, very important are the polyphenols that prevent lipid autoxidation and give unique flavor and taste to extra-virgin olive oil.

Several component of *Olea europaea* plant contain bioactive polyphenols. Typical phenols of *O. europaea* are oleuropein and hydroxytyrosol; both these phenols show antioxidant properties not only during olive oil oxidation but also in cellular and animal models.

Recently, the feasibility of isolating, extracting, and producing olive oil phenols as pure compounds (to use them as antioxidant and for other biological properties) has been investigated extensively.

An olive oil industry, produces also high amounts of organic by-products besides the oil. A considerable quantity of native olive polyphenols is lost in the above-said by-products and so these have been considered a source of bioactive olive polyphenols.

14.2 Olives and Relative Olive Mill Products

The olive tree (*O. europaeae*) is a plant cultivated typically in Mediterranean countries and the olive fruits have been used to produce edible oil since ancient times.

Olives are drupes, consisting of four parts: epicarp (skin or peel), mesocarp (pulp or flesh), endocarp, and kernel. A representative chemical composition of olive fruit is showed in Table 14.1 (Niaounakis and Halvadakis 2004).

Most of the oil is in the pulp with content variable in function to climate, season, variety, and ripeness of the fruit.

Mediterranean countries produce 95 percent of the total world production of olive oil, estimated to be 2.5 million tons per year and European Union countries produce 75 to 80 percent of the world's olive oil. The largest European producers are Spain, Italy, and Greece, plus Portugal and, to a lesser extent, France, Cyprus, and Croatia (IMPEL 2003).

Components	Olive Pulp	Stone	Seed
Water	50–60	9.1	30.0
Oil	15–30	0.7	27.3
N total	2–3	3.4	10.2
Sugar	3.0–7.5	41	26.6
Cellulose	3–6	38	1.9
Minerals	1–2	4.1	1.5
Polyphenols	2.3–3.0	0.1	0.1–1
Other compounds	—	3.4	2.4

TABLE 14.1 A Representative Chemical Composition of the Olive Fruit (%)

"Virgin olive oil" is the oil extracted exclusively with mechanical techniques; when one virgin olive oil has specific chemical-analytical and good organoleptic characteristics, it can be sold as *extra-virgin*, which indicates that it is of superior quality (UE Regulation 2003).

The olive oil industry also produces a large amount of by-products. It is estimated that for every 100 kg of olives worked, 35 kg of solid waste (olive cake or pomace) and variable quantities of liquid wastewaters are produced. The volume of wastewaters could range from 40 to 100 L for every 100 kg of olives, depending on the oil-extraction process used.

Pomace contains residue oil and so, it can be processed further to extract "pomace oil" with husks or kernel wood as additional by-products.

Other olive mill by-products are the leaves, which represent 5 percent of the weight of olives worked.

More diffused mechanical systems of olive oil extraction, shown in Fig. 14.1, are the following: traditional press, two-phase, or three-phase systems.

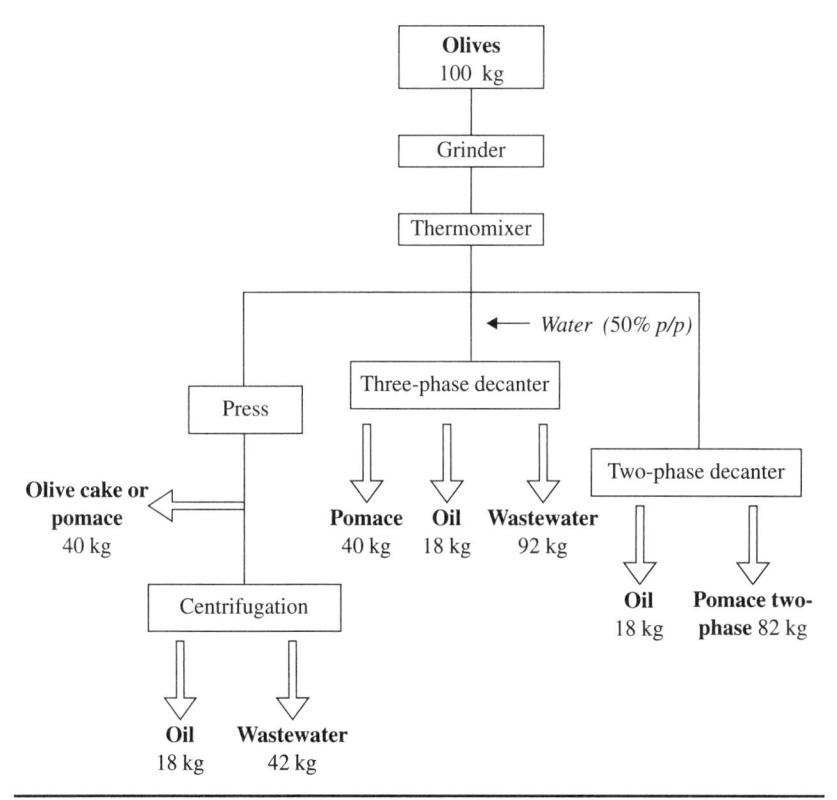

Figure 14.1 Mechanical extraction of olive oil: comparison among traditional press, two-phase decanter, and three-phase decanter.

Olive oil's yield and commercial quality are substantially equivalent among the three different extraction systems, whereas characteristics and quantities of the by-products change significantly.

In the traditional system, the effluent is essentially constituted by the water deriving from the olives, also called "vegetation water."

In the three-phase system, to facilitate the separation of the oil by decanter, potable water (in quantities equal to around 50 percent of the original olive weight) is added to the olive paste before thermomixing. For this reason, three-phase systems produce the largest volumes of wastewaters.

The two-phase system appeared on the market in the early 1990s. It is based on a technological improvement of the decanter that directly separates the oil form olives without the addition of water. This system practically does not produce liquid effluent but only one liquid–solid mixed residue remains after industrial process. Nevertheless, the two-phase pomace is more difficult to handle due higher moisture.

Environmental impact of olive oil production is considerable; actually, about 5 to 7 million tons of olive mill effluent is produced yearly. Olive mill wastewaters have a very high polluting level because they are rich in organic substances. In fact, their chemical oxygen demand (COD) and biological oxygen demand (BOD5) are 100 to 200 gL^{-1} and 20 to 100 gL^{-1}, respectively (Belice et al. 1990).

On the other hand, olive mill by-products are a rich source of bioactive olive polyphenols. In fact, the wastewaters are characterized by a high content of polyphenols, ranging from 0.5 to 24 gL^{-1} (De Leonardis et al. 2009). Polyphenol level of the olive pomaces is very variable, according to industrial procedures of oil extraction. In particular, phenolic compounds are around 1.2 percent for press process, 0.5 percent for the three-phase process, and 2.4 percent for the two-phase process.

Finally, also olive leaves contain polyphenols in quantities variable from 1.4 to 6.4 g 100 g^{-1} of dry matter (Ryan et al. 2001; De Leonardis et al. 2007).

14.3 Nature of Olive Polyphenols

At least 30 different phenolic substances are identified in olives and in their derived products (Montedoro et al. 1993). Although olive polyphenols have been studied extensively, their qualitative and quantitative compositions still vary too much because the data are strongly affected by the analytical techniques applied and by the solvent used to extract the phenols. At present, there is no worldwide accepted official analytical method. Generally, in virgin olive oil, the total phenols, extracted by different percentage concentrations of methanol and water solution, are measured using the Folin–Ciocalteu spectrophotometric method and are expressed as caffeic acid or gallic acid in milligrams per kilogram of oil (ppm) .

Phenolic Substances from Olives & Olive Mill Products

Acid phenols
- caffeic
- cinnamic
- p-coumaric
- o-coumaric
- 3, 4-di-hydroxyphenylacetic
- ferulic
- gallic
- p-hydroxyphenylacetic
- 4-hydroxybenzoic
- protocatechuic
- vanillic

Alcohol phenols
- tyrosol
- hydroxytyrosol

Flavonoids
- apigenin
- cyanidin flavone
- anthocyanin
- luteolin
- luteolin-7-glucoside
- quercetin

Secoiridoids
- oleuropein
- demethyloleuropein
- verbascoside
- ligustroside

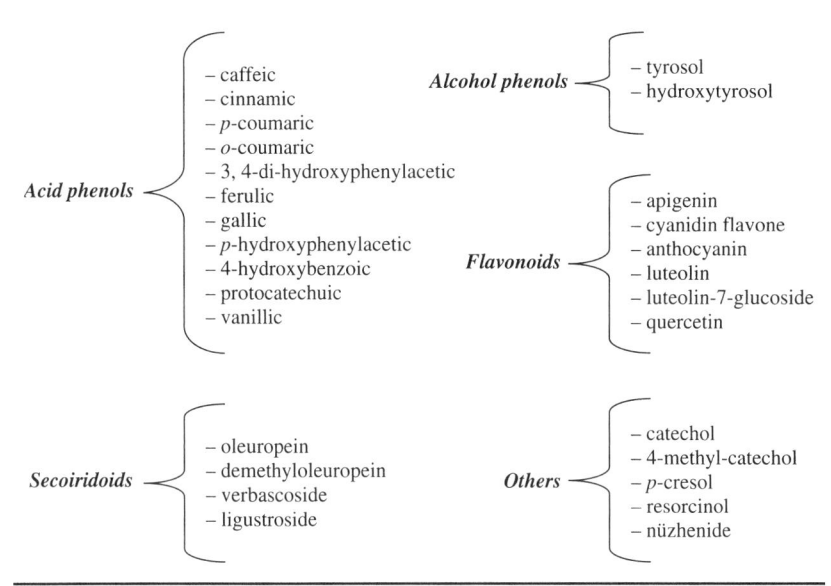

Others
- catechol
- 4-methyl-catechol
- p-cresol
- resorcinol
- nüzhenide

FIGURE 14.2 Major phenol substances present in the olives and relative olive mill products.

Polyphenols present in the olives are generally water soluble. Due to their chemical characteristics and partition coefficients, higher quantities of polyphenols are found in the aqueous phase than in the oil. In addition, olive phenol substances are numerous and variable and can be free or bound to other substance to form complex molecules.

The major olive polyphenols are listed in Fig. 14.2, and their formulas appear in Figs. 14.3 and 14.4.

It is known from long time that the principal phenolic compound present in the green parts of the olive tree, including the fruit and the leaves, is oleuropein (Panizzi et al. 1960).

Oleuropein belongs to a specific group of complex molecules called secoiridoids.

Olive tree secoiridoids are typically composed of a glycoside, phenol, and elenolic acid. Similar oleuropein secoiridoids are abundant also in other *Oleaceaes*, *Gentianales* and *Cornales*, and in many other plants. In the plant, generally, secoiridoids are produced from the secondary metabolism of terpenes, as a precursor of various alkaloids (Soler-Rivas et al. 2000).

The glycoside bound to the elenolic acid forms the oleoside, while the elenolic acid bound with a phenol forms the aglycone. Oleuropein is an ester of the 3′,4′-dihydroxyphenyl-ethanol (hydroxytyrosol) with the oleoside formed by elenolic acid and glucose.

Other important secoiridoids are demethyloleuropein and ligustroside, containing tyrosol, and verbascoside, containing hydroxytyrosol and caffeic acid (Ragazzi et al. 1973; Bianchi and Pozzi 1994). Recently, a new complex phenolic compound, called nüzhenide, was

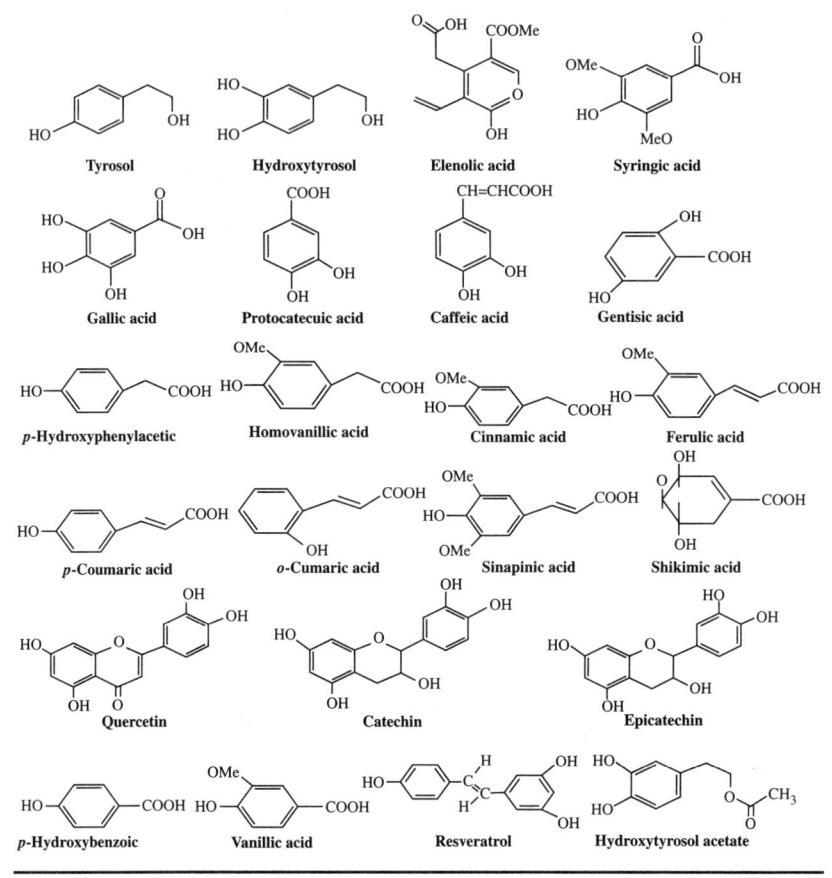

FIGURE 14.3 Simple phenol substances present in the olives and relative derivate products.

recognized in the olive seed (Servili et al. 1999a). Oleuropein is present also in high amounts (60 to 90 mg g^{-1} dry weight) in the olive tree leaves (Le Tutour and Guedon 1992; De Leonardis et al. 2007).

In olive fruits, acid phenol, flavonoids, and secoiridoids represent approximately 1 to 3 percent of dry matter (Brenes et al. 1993). Oleuropein represents approximately 80 percent of polyphenols present in unripe olives; it is very bitter and toxic, and therefore its presence makes unripe olives inedible. However, a small quantity of oleuropein improves organoleptic characteristics not only of ripe olives or properly treated green olives, but also of virgin olive oil.

Polyphenols accumulate in the fruit pulp during growth; in the first phases of ripeness polyphenols are generally bound with glycosides and have lypophilic and hydrophilic properties (Damtoft et al. 1993; Rovellini and Cortesi 1998).

During fruit ripening, oleuropein significantly decreases and its derivatives are formed. A native olive enzyme, α-glycosidase, separates

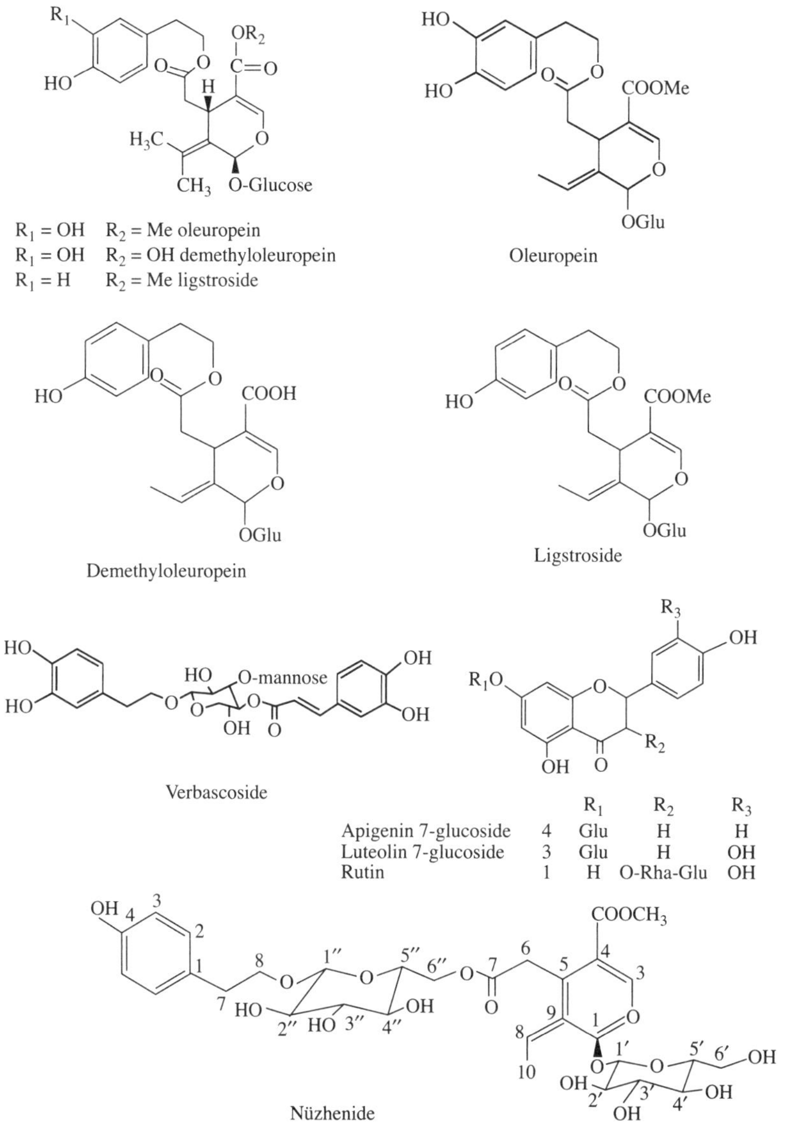

Phenolic Substances from Olives & Olive Mill Products

FIGURE 14.4 Complex phenol substances present in the olives and relative derivate products.

the oleuropein aglycone from the glucose (Gariboldi et al. 1986). Finally, in ripe olives, the principal phenols are hydroxytyrosol and tyrosol, demethyloleuropein, verbascoside, hydroxytyrosol-4-β-D-glucoside, and elenolic acid. Demethyloleuropein is the major constituent of black olives and is probably formed by the action of an esterase present in the fruit (Amiot et al. 1989).

Hydroxytyrosol is the olive phenol most studied because it is characterized by major bioantioxidant activity. In ripe olives, it is present in quantities ranging from 1.0 and 2.9 g per 100 g^{-1} of dry matter (Amiot et al. 1986; Romero et al. 2002).

In the virgin olive oil, total phenol ranged from 50 to 500 mg kg^{-1} of oil; the final content is the result of several factors including cultivar and quality of the olives, degree of maturation, climate, time and condition of olive storage, and oil-extraction system (Montedoro and Garofalo 1984; Solinas 1987; Bruni et al. 1994; Caponio et al. 1999). Actually, only one or two of the total phenols present in olives are found in virgin oil because the majority of them remain in the wastewaters and in the pomace (Rodis et al. 2002; Di Giovacchino et al. 1993).

During the first phase of the olive oil–extraction process, during the initial grinding and the thermomixing of the olive paste, chemical and enzymatic reactions are carried out that degrade at least 80 percent of the olive's complex polyphenols. In the three-phase system, it is estimated that the major part of olive polyphenols, especially hydroxytyrosol, relocate themselves in the wastewater (Niaounakis and Halvadakis, 2004). Hydroxytyrosol is very abundant in olive mill wastewaters in the free form or as a component of oleuropein, demethyloleuropein, verbascoside, and hydroxytyrosol-4-β-D-glucoside (Vásquez et al. 1987; Gutfinger 1998; Servili et al. 1999b; Capasso et al. 1999; Visioli and Galli 1999).

On the contrary, in virgin olive oil, simple phenols are lower than complex phenols; principal simple phenols are hydroxytyrosol, tyrosol and caffeic acid, *p*-coumaric, syringic and vanillic acids (Montedoro et al. 1992a; Montedoro et al. 1992b; Limiroli et al. 1995).

When virgin oil is inedible, the refining process is necessary where almost all phenols are destroyed (Tiscornia et al. 1982). For the loss of natural antioxidants, refined olive oil is of lower quality than extra virgin olive oil.

14.4 Antioxidant and Nutraceutical Properties

In the best storage conditions, virgin olive oil has a protracted shelf life, up to almost 2 years. Accordingly, olive oil's higher stability has been attributed to its phenolic content (De Leonardis and Macciola 1998; Psomiadou and Tsimidou 2002).

It was repeatedly found that among olive polyphenols, *o*-diphenols, especially hydroxytyrosol and caffeic acid, are the most effective substances (Papadopoulos and Boskou 1991; Tsimidou et al. 1992; Baldioli et al. 1996). However, the monophenol tyrosol and its derivatives and *p*-hydroxybenzoic, *o*-coumaric, and *p*-coumaric acids have insufficient or no antioxidant effect (Litridou et al. 1997).

Hydroxytyrosol is a good antioxidant that is more effective than tocopherol, butylated hydroxytoluene (BHT), and other synthetic antioxidants permitted for use in foods (De Leonardis and Macciola 2008).

Few studies have been done on the phenol's evolution during storage and cooking of virgin olive oil. During frying, olive antioxidants degrade with different ratios: tocopherols degrade before the hydroxytyrosol, which degrades before tyrosol (Pellegrini et al. 2001). In optimal storage conditions, in a dark and cool room, total phenols and hydroxytyrosol decrease significantly in relationship with the time (Cinquanta et al. 1997).

Olive polyphenols, especially oleuropein and hydroxytyrosol, have been intensively studied regarding their effect on human health and their medicinal potential. Although most studies have been done in vitro, some statistical studies on human populations confirm the benefit of olive oil (Hill and Giacosa 1992). Higher virgin olive oil consumption is associated with a lower incidence of coronary heart disease and the risk of breast, prostate, and colon cancers (Tuck and Hayball 2002).

Finally, potential nutraceutical properties of olive polyphenols are synthesized in Table 14.2.

Cytostatic and cytotoxic activity against tumoral cells	Owen et al. 2000; Saenz et al. 1998
Hypocholesterolaemic effect	De Pasquale et al. 1991
Increase resistance of LDLs to oxidation	Auroma et al. 1998; Visioli et al. 1995; Visioli et al. 2001; Salami et al. 1995; Fito et al. 2000
Scavenging peroxy radicals, hydroxy radicals and superoxide anions	Visioli et al. 1998; Visioli and Galli 1999
Hypoglycaemic effect and antidiabetic activity	Trovato et al. 1993
Inhibition of platelet aggregation	Petroni et al. 1995
Reduction of formation of proinflammatory molecules thromboxane B_2 and leukotriene B_4 by activated human leukocytes	De La Puerta et al. 1999; Martinez-Domingues et al. 2001
Antioxidant effects in cellular and animal models	Manna et al. 1997; Speroni et al. 1998
Prevention of passive smoking oxidative stress	Visioli et al. 2000
Antimicrobial activity against many pathogenic microorganisms, including bacteria, fungi, viruses, and parasitics, and protozoans	Juven and Henis 1970; Fleming et al. 1973; Bisognano et al. 1999

TABLE 14.2 Some Potential Nutritional Properties Ascribed to Phenolic Substances Present in Olives and Related Olive Mill Products

14.5 Future Biotechnological Applications

The use of olive polyphenols in different sectors, such as pharmaceutical, cosmetics, food industry, and medical, is being actively considered.

In the recent years, interest in olive mill by-products is increasing because they could be used as a good source of bioactive olive polyphenols (Visioli et al. 1999; Fernández-Bolaños et al. 2002).

At the moment, recovering olive phenols from olive mill wastewater is the most promisings as a new biotechnology application. Under this point of view, the wastewaters, generally considered only as a polluting effluent, actually could furnish organic natural substances with considerable commercial value.

Possible uses of olive polyphenols recovered from olive mill by-products are (1) as integrator and antioxidant in food, (2) as a pesticide in agriculture, (3) for preventing and treating human diseases, and (4) in cosmetics.

Interest on recovering and applying move after olive mill by-products phenols is ongoing and includes remarkable biotechnological perspectives.

Several methods exist to recover olive polyphenols in olive mill by-products, including solvent extraction, chromatography on resin, selective concentration by ultrafiltration and inverse osmosis, and solid–liquid or liquid–liquid extraction (Fernández-Bolaños et al. 2002).

The Table 14.3 lists recent patents relative to olive polyphenols.

Patent No.	Purpose
EP811678 (1977)	A process of extracting antioxidants from olives. The final product contains hydroxytyrosol, tyrosol, other acid phenols, and oleuropein
WO0145514 (2001)	A method for extracting antioxidants from olives, olive paste, olive oil, and olive mill wastewater
WO02064537 (2002)	A method for recovering hydroxytyrosol from olive mill by-products
JP000319161 (2000)	Using olive mill wastewater extract to produce cosmetics
WO03079794 (2003)	Using phenol olive leaves as antimicrobial agents against aerobic and anaerobic germs
WO03080006 (2003)	Using olive polyphenols as antidandruff agents
WO0004794 (2000) WO0218310 (2002)	Process of recovering oleuropein from olive vegetation water. Using phenol extract as a food additive and an integrator tablet
CS2004A000001	Using oleuropein as an insecticide

TABLE 14.3 Recent Patents Relative to Olive Polyphenols

Finally, in the few last years, numerous dietary supplement containing olive polyphenols have appeared in the market.

References

Amiot, M. T., Fleuriet, A., and Macheix, J. T. 1986. Importance and evolution of phenolic compounds in olive during growth and maturation. *Journal of Agriculture and Food Chemistry* 34:823–826.

Amiot, M. T., Fleuriet, A., and Macheix, J. T. 1989. Accumulation of oleuropein derivatives during olive maturation. *Phytochemistry* 28:67–69.

Auroma, O. I., Deiane, M., Jenner, A., Halliwell, B., Kaur, M., Banni, S., Corongiu, F. P., et al. 1998. Effects of hydroxytyrosol found in extra virgin olive oil on oxidative DNA damage and on low-density lipoprotein oxidation. *Journal of Agriculture and Food Chemistry* 46:5181–5187.

Baldioli, M., Servili, M., Perretti, G., and Montedoro, G. F. 1996. Antioxidant activity of tocopherols and phenolic compounds in virgin olive oil. *Journal of the American Oil Chemists' Society* 73:1589–1593.

Belice, V., Carrier, C., and Cera, O. 1990. Caratteristiche analitiche delle acque di vegetazione. *Rivista Italiana Delle Sostanze Grasse* 67:9–16.

Bianchi, A. and Pozzi, N. 1994. Dihydroxyphenylglycol: a major C6–C2 phenols in *Olea europaeae* fruits. *Phytochemistry* 35:1335–1337.

Bisognano, A., Tomaino, A., Lo Cascio, R., Crisafi, G., Uccella, N., and Saija, A. 1999. On the in-vitro antimicrobial activity of oloeuropein end hydroxytyrosol. *Journal of Pharmacy and Pharmacology* 51:971–292.

Brenes, M., Garcia, P., Duran, M. C., and Garrido, A. 1993. Concentration of phenolic compounds change in storage brines of ripe olives. *Journal of Food Science* 58:347–350.

Bruni, U., Fiorino, N., and Cortesi, N. 1994. Influence of agricultural techniques: cultivars and area of origin on characteristics of virgin olive oil and on levels of some of its minor components. *Olivae* 53:28–34.

Capasso, R., Evidente, A., Avolio, S., and Solla, F. 1999. A highly convenient synthesis of hydroxytyrosol and its recovery from agricultural waste waters. *Journal of Agriculture and Food Chemistry* 47:1745–1748.

Caponio, F., Alloggio, V., and Gomes, T. 1999. Phenolic compounds of virgin olive oil: Influence of paste preparation techniques. *Food Chemistry* 64:203–209.

Cinquanta, L., Esti, M., and La Notte, E. 1997. Evolution of phenolic compounds in virgin olive oil during storage. *Journal of the American Oil Chemists' Society* 74:1259–1264.

Damtoft, S., Franzyk, H., and Jensen, S. R. 1993. Biosynthesis of secoiridoids glycosides. *Oleaceae*: 34:1291–1299.

De La Puerta, R., Ruiz-Gutierrez, V., and Hoult, J. R. 1999. Inhibition of leukocyte 5-lipoxygenase by phenols from virgin olive oil. *Biochemistry and Pharmacology* 57:445–449.

De Leonardis, A. and Macciola, V. 1998. Evaluation of the shelf-life of virgin olive oils. *Rivista Italiana Delle Sostanze Grasse* 75:391–396.

De Leonardis, A., Aretini, A., Alfano, G., Macciola, V., and Ranalli, G. 2007. Isolation of a hydroxytyrosol-rich extract from olive leaves (*Olea europaea L.*) and evaluation of its antioxidant properties and bioactivity. *European Food Research and Technology* 226:653–659.

De Leonardis, A., and Macciola, V., 2008. The hydroxytyrosol recovered from oil mill by-products as a possible food antioxidant. *Journal of Environmental, Agricultural and Food Chemistry* 7:3310–3314.

De Leonardis, A., Macciola, V., and Nag, A. 2009. Antioxidant activity of various phenol extracts of olive-oil mill wastewaters. *Acta Alimentaria* 38:77–86.

De Pasquale, R., Monforte, M. T., Trozzi, A., Raccuia A., Tommasini, S., and Ragusa, S. 1991. Effects of leaves and shoots of *Olea europeaea L.* and oleuropein on experimental hypercholesterolemia in rat. *Plants in Medical Phytotherapy* 25:1–34.

Di Giovacchino, L. M., Solinas, M., and Miccoli, M. 1993. Effect of extraction systems on the quality of virgin olive oil. *Journal of the American Oil Chemists' Society* 71:1189–1194.

EC Commission Regulation n. 1989/2003 on the characteristics of olive oil and olive-pomace oil and on the relevant methods of analysis.

Fernández-Bolaños, J., Rodríguez, G., Rodríguez, R., Heredia, A., Guillén R., and Jiménez, A. 2002. Production in large quantities of highly purified hydroxytyrosol from liquid-solid waste of two-phase olive oil processing or "Alperujio." *Journal of Agriculture and Food Chemistry* 50:6804–6811.

Fito, M., Covas, M. I., Lamuela-Raventos, M. R., Villa, J., Torrents, J., De la Torre, C., and Marrugat, J. 2000. Protective effect of olive oil and its phenolic compounds against low-density lipoprotein oxidation. *Lipids* 35:633–638.

Fleming, H. P., Walter, W. M., and Etchells, J. L. 1973. Antimicrobial properties of oleuropein and products of its hydrolysis from green olives. *Applied Microbiology* 26:777–782.

Gariboldi, P., Jommi, G., and Verrotta, L. 1986. Secoiridoids from *Olea Europeae*. *Phytochemistry* 25:865–869.

Gutfinger, T. 1998. Phenols in olive oil. *Journal of the American Chemists' Society* 68:966–968.

Hill, M. J. and Giacosa, A. 1992. The Mediterranean diet (editorial). *European Journal of Cancer Prevention*:1:339–340.

IMPEL. 2003. The European Union Network for the Implementation and Enforcement of Environmental Law (IMPEL). http://ec.europa.eu/environment/impel/pdf/olive_oil_project.pdf

Juven, B. and Henis, Y. 1970. Studies on the antimicrobial activity of olive phenolic compounds. *Applied Bacteriology* 33:721–732.

Le Tutour, B. and Guedon, D. 1992. Antioxidative activities of *Olea European* leaves and related phenolic compounds. *Phytochemistry* 31:1173–1178.

Limiroli, R., Consogni, R., Ottolina, G., Marsilio, V., Bianchi, and G., Zetta, L. 1995. 1H and 13C NMR characterization of new oleuropein aglycones. *Journal of the Chemical Society: Perkin Transactions* 1:1519–1523.

Litridou, M., Linssen, J., Schols, H., Bergmans, M., Posthumus, M., Tsimodou, M., and Boskou, D. 1997. Phenolic compounds in virgin olive oils: fraction by solid phase extraction and antioxidant activity assessment. *Journal of the Science of Food Agriculture* 74:169–174.

Manna, C., Galletti, P., Cucciolla, V., Moltedo, O., Leone, A., and Zappia, V. 1997. The protective effect of the olive oil polyphenol (3-4-dyhydroxyphenyl)-ethanol counteracts reactive oxygen metabolite induced cytoxicity in caco-2-cells. *Journal of Nutrition* 127:286–292.

Martinez-Domingues, E., De la Puerta, R., and Ruiz-Gutierrez, V. 2001. Protective effects upon experimental inflammation models of a polyphenol-supplemented virgin olive diet. *Inflammation Research* 50:102–106.

Montedoro, G. F. and Garofalo, L. 1984. The qualitative characteristics of virgin olive oils. The influence of variable such as variety, environment, preservation, extraction, conditioning of the finish product. *Rivista Italiana Delle Sostanze Grasse* 61:157–168.

Montedoro, G. F., Servili, M., Baldioli, M., and Miniati, E. 1992a. Simple and hydrolysable phenolic compounds in virgin olive oil. 1. Their extraction, separation, and quantitative and semiquantitative evaluation by HPLC. *Journal of Agriculture and Food Chemistry* 40:1571–1576.

Montedoro, G. F. , Servili, M., Baldioli, M., and Miniati, E. 1992b. Simple and hydrolysable phenolic compounds in virgin olive oil. 2. Initial characterization of the hydrolysable fraction. *Journal of Agriculture and Food Chemistry* 40:1577–1580.

Montedoro, G. F., Servili, M., Baldioli, M., Selvaggini, R., Miniati, E., and Macchioni, P. 1993. Simple and hydrolysable compounds in virgin olive oil. 3. Spectroscopic characterizations of secoiridoids derivatives. *Journal of Agriculture and Food Chemistry* 41:2228–2234.

Niaounakis, M. and Halvadakis, C. P. 2004. In *Olive Mill Waste Management. Literature Review and Patent Survey*, ed. T.-G. Dardanos. Amsterdam: Elsevier.

Owen, R. W., Giacosa, A., Hull, W. E., Haubner, R., Spiegelhalder, B., and Bartsch, H. 2000. The antioxidant/anticancer potential of phenolic compounds isolated from olive oil. *European Journal of Cancer* 36:1235–1247.

Panizzi, L., Scarpati, M. L., and Oriente, E.G. 1960. Structure of oleuropein bitter glycoside with hypotensive action of olive oil. Note II. *Gazzetta Chimica Italiana* 90:1449–1485.

Papadopoulos, G. and Boskou, D. 1991. Antioxidant effect of natural phenols on olive oil. *Journal of the American Oil Chemists' Society* 68:669–671.

Pellegrini, N., Visioli, F., Buratti, S., and Brighenti, F. 2001. Direct analysis of total antioxidant activity of olive oil and studies on the influence of heating. *Journal of Agriculture and Food Chemistry* 49:2532–2538.

Petroni, A., Blasevich, M., Salami, M., Papini, N., Montedoro, G. F., and Galli, C. 1995. Inhibition of platelet aggregation end eicosanoid production by phenolic component of olive oil. *Thrombosis Research* 78:151–160.

Psomiadou, E. and Tsimidou, A. 2002. Stability of virgin olive oil. 1. Autooxidation studies. *Journal of Agriculture and Food Chemistry* 50:716–721.

Ragazzi, E., Veronese, G., and Guiotto, A. 1973. Demethyloleuropein: A new glycoside isolated from ripe olives. *Annali di Chimica* 63:13–20.

Rodis, P. S., Karathanos, V. T., and Mantzavinou, A. 2002. Partitioning of olive oil antioxidant between oil and water phase. *Journal of Agriculture and Food Chemistry* 50:596–601.

Romero, C., Brenes, M., Garcia, P., and Garrido, A. 2002. Hydroxytyrosol-4-β-D-glucoside:an important phenolic compound in olive fruits and derived products. *Journal of Agriculture and Food Chemistry* 50:3835–3839.

Rovellini, P. and Cortesi, N. 1998. Identification of three intermediates compounds in the biosynthesis of secoiridoids as oleoside-type in the different anatomic parts of *Olea europeae L.* by HPLC-ESMS. *Rivista Italiana Delle Sostanze Grasse* 75:543–550.

Ryan, D., Lawrence, H., Prenzler, P. D., Antolovich, M., and Robards, K. 2001. Recovery of phenolic compounds from *Olea europeae*. *Analytica Chimica Acta* 445:67–77.

Saenz, M. T., Garcia, M. D., Ahumada, M. C., and Ruiz, V. 1998. Cytostatic activity of some compounds from virgin olive oil. *Il Farmaco* 53:448–449.

Salami, M., Galli, C., De Angelis, L., and Visioli F. 1995. Formation of F2-isoprostanes in oxidized low density lipoprotein: Inhibitory effect of hydroxytyrosol. *Pharmacy Research* 31:275–279.

Servili, M., Baldioli, M., Selvaggini, R., Macchioni, A., and Montedoro, G. F. 1999a. Phenolic compounds of olive fruits one- and two-dimensional nuclear magnetic resonance characterization of nüzhenide and its distribution in the constitutive parts of fruits. *Journal of Agriculture and Food Chemistry* 47:12–18.

Servili, M., Baldioli, M., Selvaggini, R., Miniati, E., Maccioni, A., and Montedoro, G. 1999b. HPLC evaluation of phenols in olive fruit:virgin olive oil:vegetation waters and pomace:and 1D– and 2D-NMR characterization. *Journal of the American Oil Chemists' Society* 76:873–882.

Speroni, E., Guerra, M. C., Minghetti, A., Crespi-Perellino, N., Pasini, P., Piazza F., and Roda, A. 1998. Oleuropein evaluated in vitro and in vivo as an antioxidant. *Phytotherapy Research* 12:98–100.

Soler-Rivas, C., Espín, J. C., and Wichers, H. J. 2000. Oleuropein and related compounds. *Journal of Science and Food Agriculture* 80:1013–1023.

Solinas, M. 1987. HRGC analysis of phenolic components in virgin olive oils in relation to the ripening and variety of olive. *Rivista Italiana Delle Sostanze Grasse* 64:255–261.

Tiscornia, E., Forina, M., and Evangelisti, F. 1982. Composizione chimica dell'olio di oliva e sue variazioni indotte dal processo di rettificazione. *Rivista Italiana Delle Sostanze Grasse* 54:519–575.

Trovato, A., Forestieri, A. M., Lauk, L., Barbera, R., Monforte, M. T., and Galati, E. M. 1993. Hypoglycemic activity of different extracts of *Olea europeae L.* in the rats. *Plants in Medical Phytotherapy* 26:300–308.

Tsimidou, M., Papadopoulos, G., and Boskou, D. 1992. Phenolic compounds and stability of virgin olive oil. *Food Chemistry* 45:141–144.

Tuck, K. L. and Hayball, P. J. 2002. Major phenolic compounds in olive oil: Metabolism and health effects. *Journal of Nutrition and Biochemistry* 13:636–644.

Vásquez, R., Maestro Duran, R., and Graciani, C. 1987. Phenols components of olive oil mill waste water. *Grasas y Aceites* 25:341–345.

Visioli, F., Bellomo, G., and Galli, C. 1998. Free radical-scavenging properties of olive oil polyphenols. *Biochemistry and Biophysics Research Communications* 247:60–64.

Visioli, F., Bellomo, G., Montedoro, G. F., and Galli C. 1995. Low density lipoprotein oxidation is inhibited in vitro by olive oil constituents. *Atherosclerosis* 117:25–32.

Visioli, F., Caruso, D., and Plasmati, E. 2001. Hydroxytyrosol:as a component of olive:mill waste water:is dose-dependently absorbed and increase the antioxidant capacity of rat plasma. *Free Radical Research* 34:301–305.

Visioli, F. and Galli, C. 1999. Free radical scavenging of olive oil phenolics. *Lipids* 34:S315.

Visioli, F., Galli, C., Plasmati, E., Hernandez, A., Colombo, C., and Sala, A. 2000. Olive phenol hydroxytyrosol prevents passive smoking-induced oxidative stress. *Circulation* 102:2169–2171.

Visioli, F., Romani, A., Mulinacci, N., Zarini, S., Conte, D.Vincieri, F. F., and Galli C. 1999. Antioxidant and other biological activities in olive oil mill waste water. *Journal of Agriculture and Food Chemistry* 47:3397–3401.

CHAPTER 15
Effect of Exogenous Bioregulators on the Mineral Composition and Storability of Fruits

Alina Basak
Plant Bioregulators Department
Research Institute of Pomology and Floriculture
Skierniewice, Poland

15.1 Introduction

One of the characteristics determining the quality of fruits is their mineral composition. Of special importance here is the element calcium. It is the most important factor in delaying the aging process and thus the ripening of the fruit. Calcium deficiency in fruits, as well as incorrect ratios of the calcium content to that of other elements, mainly K, Mg, and N, are the cause of physiological diseases of fruit in storage.

Calcium uptake and its transport within a plant is a complex process. Various environmental and agrotechnical factors have an effect on that process. Bioregulators (BRs) are one such factor.[55] The mineral composition of plants depends both on exogenous BRs (delivered to the plant from outside) and endogenous BRs (i.e., hormones produced naturally by the plant). At the same time, the activity of endogenous BRs is controlled by the mineral elements delivered from outside.[67]

BRs are known to have an effect on a number of physiological processes taking place during the acquisition of mineral nutrients (e.g., on the proton pump mechanism, permeability of cell walls, and metabolism of the ions taken up). Faust and Miller[26] divided exogenous BRs into ones that affect these processes through changes in the metabolism of IAA (NAA, IBA, PB) and gibberellins (GA_3, GA_{4+7}, PB), and the ones whose mode of action is complex and so far not specified definitively (daminozide).

BRs are believed to affect the calcium content of fruit directly, through changes in the transport of the auxin IAA (indole acetic acid), and indirectly, as a result of changes in fruit size, limiting competition among shoots, or changes in root weight.

According to Faust and Miller,[26] BRs, when acting directly, usually cause a drop in the calcium content of fruit. It is generally difficult to determine when BRs act indirectly and when their direct effect begins. The problem is quite complicated, particularly in the case of trees that are already bearing fruit. It is easier to explain it using young plants that have not yet started bearing fruit because it is possible to do experiments on such plants under controlled conditions. However, the results obtained for nonfruiting seedlings do not fully reflect the mineral nutritional status of mature, fruit-bearing trees.

15.1.1 Auxins and the Mineral Status of Fruits

Calcium migrates upward within a plant.[2,4,29] The direct effect of BRs on Ca uptake is related to the transport of IAA in the plant, and the upward transport of calcium depends on the downward transport of the auxin IAA. Auxin transport is modified by BRs, proved by Banson and Stahly.[14] These researchers, using TIBA (2,3,5-triiodobenzoic acid), an auxin transport inhibitor 2 to 6 weeks after flowering, found a marked decrease in the calcium content of apples. TIBA caused the greatest reduction in calcium levels when it was applied 2 weeks after flowering. Because the treatment was delayed, the preparation had a progressively weaker effect on calcium transport. The TIBA-treated apples, however, always contained less calcium than the control apples, even when apples of the same size were compared. Particularly large differences in calcium levels between the TIBA-treated and TIBA-untreated apples were found 7 weeks after flowering. As a result of spraying with TIBA, a large number of apples were affected by a bitter pit. Other studies by Stahly and Benson[66,67] indicated that TIBA, while reducing calcium levels, increased the levels of potassium at the same time, but not until the second half of the summer. This caused an increase in the potassium-to-calcium ratio, which meant a reduction in the storability of apples. TIBA did not affect the levels of such elements as Mg, P, or B in apples but did cause a higher nitrogen content.

Because the reduction in auxin transport decreased the supply of calcium to the fruit, it could be expected that increased levels of auxins in a tree would cause greater accumulation of calcium in its fruits and, consequently, an improvement in their storability. This was proved for the first time by Stuivenberg and Pouwer in 1950.[69] By spraying apple trees with the cultivar "Notaris" with IAA, they brought about a decrease in the number of apples with bitter pit. It is interesting to note that this effect of IAA was achieved only when the substance was applied between the end of June and the middle of July (i.e., just after the natural fruit drop in June). Twenty years later, in studies by Sharples,[61] IAA delivered to the core of apples with the cultivar "Cox's Orange Pippin" was found to reduce the incidence of core browning. IAA thus improved the keeping quality of apples, as in the experiments by Stuivenberg and Pouwer.[69]

Similar experiments were carried out by Looney,[42] who used IBA (indole butyric acid) on the apple trees cultivar "Spartan." Within plants, this auxin gets converted to IAA. In one of three experiments, the auxin IBA caused the amount of Ca in the skin to increase by 13 percent, and in the core by 28 percent. This reduced the decomposition rate of apples. The effect of IBA was greatest when the apples were treated with calcium chloride at the same time.

The auxin IAA delivered onto apples in a synthetic form does not always cause an increase in their calcium content. According to Bangerth,[2] this happens in the case of apples that have a large number of seeds. Apples of this kind can themselves produce auxins in amounts necessary for Ca transport. This means that the additional amount of auxins can have a toxic effect, destroying the seeds. This was confirmed by the results of the experiments conducted by Wills and Scott in 1974[77] in which injections of IAA into the seed core speeded up the breakdown of apples. This effect was also shown by the results of our experiments in which IAA was applied directly to apples followed by an examination of the uptake and transport of a Ca isotope through the parts of the tree consisting of fruits and shoots.[7] It was found that IAA deposited on the apples of the cultivar "Double Red McIntosh," which had a large number of seeds, decreased the uptake of the Ca isotope by whole parts of the trees, but the apples treated with IAA in general maintained higher levels of calcium than the untreated ones.

The role of seeds in the acquisition of calcium by fruits was demonstrated by the low levels of the element in parthenocarpic (seedless) fruits found following the use of GA_3 or GA_{4+7}. Bangerth's experiments[2] indicated that seedless fruits—both apples and pears—contained less Ca than those with seeds. Griggs et al.[36] demonstrated that seedless fruits respond to the delivery of exogenous auxins in a different way than seed-containing fruits do. In seedless fruits, the exogenous auxin, particularly when used in combination with gibberellins, can replace the endogenous auxin produced by the seeds. This can result in an increase in the Ca content of the fruits and an improvement in their storability.

Higher levels of Ca in fruits and better storability are also caused by other auxins, such as 2,4,5-T (trichlorophenoxyacetic acid), which, in the experiments by Sharples,[71] caused a reduction in the incidence of low temperature breakdown and core browning in apples of cultivar "Cox's Orange Pippin," and 2,4,5-TP (trichlorophenoxypropionic acid), which in Stahly's experiments,[66,67] when used at 20 mg/L 24 days after full bloom, produced higher levels of Ca, and also of K and Mg, in "Golden Delicious" apples.

The most commonly used auxin in orchard production is NAA (α-naphthalene acetic acid). In one of three experiments carried out by Martin et al.,[50] NAA reduced the incidence of bitter pits and the decomposition rate in "Merton" apples. These apples are also richer in calcium and many other elements. NAA was found to enhance the effectiveness of $CaCl_2$ in preventing the breakdown of Cox's apples.[48] NAA also reduced the severity of bitter pit in apples sprayed with daminozide and ethephon.[53] NAA, used in our experiments for fruitlet thinning, increased the level of calcium in apples or had no effect, probably in relation to the degree of crop reduction and the range of the increase in apple size.[12] In Link's experiments,[40] NAAm (amide of naphthalene acetic acid), used for fruitlet thinning, affected the mineral composition of apples, just as NAA did in our experiments, but it sometimes increased the incidence of bitter pit. NAA is commonly used to prevent preharvest fruit drop. According to Wertheim,[74] NAA used in this way does not affect the storability of apples provided, however, that the harvest date is right. In contrast, in the experiments by Wills and Scott,[77] NAA injected into the seed core at a concentration of 20 µmol increased apple breakdown.

15.1.2 Exogenous Use of Gibberellins

The effect of these substances on the mineral composition and storability of fruits depend on the time of their application and their concentration.

For example, parthenocarpic fruit obtained in Bangerth's experiments[2] following the use of GA_3 and GA_{4+7} at flowering time contained less Ca. In Stahly's experiments,[66] GA_3 used later (i.e., 24 days after full bloom), did not change the Ca content of "Spartan" apples.

In studies by Looney,[42] where both gibberellin A_3 and auxin IBA were used, most of the apples affected by breakdown were found after both these substances were used in combination. The best-keeping apples were those treated with the auxin only. This proves that the gibberellin eliminated the beneficial effect of the auxin and lowered the level of calcium in apples treated with the auxin, although it did not cause definite changes in the mineral content and storability of apples by itself.

According to Wills and Scott,[77] gibberellins can, under certain conditions, enhance the process of attracting calcium by the fruit, for example, when other parts of the tree are less active or after repeated applications of gibberellins at high concentrations over a long period resulting in a holding back of the ripening process. It was probably for that reason that GA_3 reduced the incidence of internal breakdown in "Jonathan" apples in Clijsters's[19] experiments. It has repeatedly been shown that GA_3 used just before or just after the harvest improves the keeping quality of apples. Among other things, GA_3 reduced the incidence of breakdown and core browning in studies by Scott and Wills[60] and Sharples.[61] When used at those times, the gibberellin modified the calcium distribution in fruit during storage, making the fruit more resistant to physiological diseases.

In orchards, the gibberellins GA_{4+7} are often used on their own or in combination with benzyladenine in order to reduce russeting and to improve the shape and size of apples. In Looney's experiments,[43] these substances, used only once or twice in the period from petal fall to 5 weeks after flowering, always decreased the level of calcium in apple flesh, on average by 25 percent, and increased the incidence of fruit breakdown in storage. In our experiments,[6] GA_{4+7} reduced the keeping quality of apples of the cultivar "Starkrimson." Particularly badly affected was the keeping quality of apples treated with these substances at high concentrations. Similar effects were shown by the gibberellins GA_{4+7} in the experiments by Greene et al.[34] These researchers demonstrated that as the concentration of the gibberellins was increased, the calcium content of apples decreased. This could have been caused by the reduced number of seeds as well as by the large increase in apple size. The gibberellins A_{4+7} used in combination with benzyladenine reduced calcium uptake by apples treated with daminozide and adversely affected their storability. Similar effects of GA_{4+7} and BA were observed in our experiments, in which they were used on daminozide-treated "McIntosh" apple trees (unpublished data).

15.1.3 Retardants Are Also Bioregulators

Use of retardants was very popular in the 1980s. They were used in orchards very often. There is, therefore, a lot of information on the effects of retardants on the storability and mineral composition of fruits. According to Faust and Miller,[26] retardants affect calcium uptake by fruits as a result of a direct influence on the transport of IAA, or indirectly by reducing fruit size, limiting competition among shoots, or changing root weight.

Retardants are antigibberellins. They reduce the levels of gibberellins in shoot tips. They can also impair the diffusion of auxinlike substances from shoot tips. As a result, the transport of Ca to fruits can be weakened. The direct effect of retardants on Ca uptake was demonstrated by experiments in which these preparations caused a drop in calcium levels in fruits and deterioration in their storability.

For example, in the experiments by Schumacher et al.,[59] apples of the cultivar "Grafsztynek" whose trees had been sprayed with daminozide had 20 percent less Ca and were affected by bitter pit more often than apples from control trees of similar size and degree of ripeness. In Poland,[12] daminozide (SADH) often increased the incidence of core browning in McIntosh apples. This also provides evidence for calcium deficiency in those fruits or incorrect ratios of Ca to other elements. Spraying such apples with or dipping them in calcium chloride reduced the incidence of the disease. Also in Kallai's experiments,[39] paclobutrazol (PB) decreased the calcium content of "Jonathan" apples by as much as 28 percent, already 1 month after full bloom. As harvest time approached, the differences in the calcium content between the treated and control apples became smaller. However, even at harvest the calcium content amounted to 11 percent.

15.1.4 Effect of Retardants on the Mineral Composition of Fruits

The indirect effect of retardants on the mineral composition of apples is related to a reduction in fruit size, among other things. A negative correlation is known to exist between the Ca content of apples and the apples' diameter. It has repeatedly been shown that retardants [SADH (daminozide; 1-N-(dimethylamino)succinamic acid), as well as PB (paclobutrazol; 1-(4-chlorophenyl)-2-(1H-1,2,4-triazol-1-yl)-penten-3-d)] decreased apple size, particularly after retardants had been used at high concentrations. The decrease in size was often followed by an increase in the calcium content of apples, which improved their storability. In experiments by Schumacher et al.,[59] in which daminozide reduced apple size by various degrees depending on how it was used, the largest number of healthy apples without bitter pit symptoms were found among the smallest apples. All the daminozide-treated apples had a better keeping quality than the untreated control. The trees sprayed with daminozide only for 1 year produced 20 percent less fruit with bitter pit symptoms. The use of daminozide for 2 years, however, reduced the number of apples with symptoms of this disease by 34 percent, whereas the use of the retardant over three consecutive years resulted in as much as a 75 percent reduction. A small, 5 percent, reduction in apple size was found on the trees where only the crown's periphery was sprayed with daminozide. Such a slight decrease in fruit size had the smallest effect on the incidence of bitter pit. In this experiment, SADH was not very good at retarding shoot growth.

The most frequent reason for changes in calcium content of fruit following the use of retardants was the restricted growth of shoots and, at the same time, reduced competition between them for mineral nutrients. As it is known, large amounts of calcium are transported in the spring when shoots grow very intensively. Young shoots are supplied with calcium and other nutrients first. The calcium content of fruits depends on the ratio of fruit growth to shoot growth. Under the conditions of greatly restricted shoot growth and a small reduction in fruit size, the Ca content of fruits increases. In experiments by Miller and Swietlik,[51] the effect of PB on the incidence of bitter pit and the Ca content of fruit was found to decrease as the retarding effect on shoot growth became smaller. The usually heavily cropping, but not vigorously growing, trees supply sufficient amounts of nutrients to growing fruits. The small number of leaves in that case restricts fruit size and improves their calcium supply. Fruits from trees like that usually keep well. As the axial shoots grow longer (e.g., from 12 to 55 cm), the incidence of bitter pit increases—from 0 to 60 percent in experiments by Terblanche et al.[70]

The most convincing evidence of the competition between shoots and fruits for calcium was provided by Naumann.[52] This researcher followed the uptake of a calcium isotope by the fruits and shoots of Cox's Orange Pippin apple trees growing in pots under controlled conditions. Calcium was supplied to the soil. In the case of untreated control trees, Naumann[52] found that most of the calcium was taken up by the leaves of young shoots and much less was taken by the fruits. The fruits were, to a large extent, affected by bitter pit. After retardants were applied, shoot growth became weaker. At the same time, the fruits took up more calcium and kept much better. Of a particularly good keeping quality were the apples treated with SADH. However, 3-year-long experiments of Ludders and Fischer-Bolukbasi[45] indicated that the mineral content of apples treated with SADH depended on the fruiting intensity of the apple trees; levels of Ca, Mg, and N were higher in the fruits of abundantly fruiting trees.

In studies by Greene,[30,33] paclobutrazol deposited on the leaves slowed down shoot growth, reduced apple size, and, at the same time, increased the amounts of calcium in fruit flesh and limited the incidence of diseases, even in the second year after treatment. However, the total calcium content per fruit remained unchanged. Therefore, the increase in Ca content was related to the reduction in fruit size. At the same time, the total calcium content per tree increased. This could have been caused by the shoots attracting less calcium. PB was found to produce similar changes in the experiments by Elfving et al.,[24] but only in the year of application. The preparation retarded shoot growth and decreased fruit size in the next 3 years. Greene[31] tried to explain the divergent results with different environmental conditions, cultivar-related differences, and different application rates and times at which retardants were used. Only retardants applied at the stage of intensive growth had an effect on the uptake of mineral nutrients. When shoot growth was slowed down, the retardants ceased to work.

The effect of PB on calcium acquisition by fruit depends on the way the preparation is applied. The very same retardant, PB, applied to the soil in the experiments by Steffens et al.[68] caused the calcium content to increase only in the leaves of long shoots. It did not, however, have any effect on changing the amounts of Ca in fruits and the leaves of dwarf shoots. This could have been caused by a marked reduction in shoot growth, whereas fruit size decreased only slightly. According to these researchers, in order to bring about physiological changes in fruits, PB should be supplied directly to the fruits or the dwarf shoots. Only foliar applications of the retardant produce significant effects on the postharvest physiology of fruits, but the retardation effect is then weaker compared with applications to the soil.

Studies carried out in Korea[18] indicate that the retardants PB and daminozide, as well as NAA, intensify Ca transport to the plant mainly under conditions of a low calcium content of the soil.

15.1.5 Effect of Retardants on Calcium Uptake

Retardants affect the uptake of calcium by plants as a result of an increase in root weight. The supply of carbohydrates to the roots has a significant function in Ca uptake by fruits[26] because it stimulates root growth. Carbohydrates increase the weight of young roots whose nonwoody growth apices take up more calcium. The acquisition of calcium is confined only to the root tips.

As was shown by Steffens et al.,[68] retardants cause an increased efflux of carbohydrates to the trunk and roots, and this favors the development of numerous young roots, actively taking up calcium from the soil. In experiments by Curry and Williams,[21] the weight of roots following the use of PB was found to have doubled. However, other retardants in the same conditions had only a slight effect on root weight.

Retardants can increase the calcium content of above-ground plant parts[26] because by retarding growth, they increase the proportion of roots to top growth.

In orchard production, various applications have been found by products in which ethephon is the active ingredient. For example, these preparations are used in fruit growing to improve color development in apples. Information on the fact that apples treated with ethephon have a lower keeping quality and contain less calcium has repeatedly been published. For that reason, during the period when daminozide was available, the storability of apples treated with ethephon was enhanced by using daminozide on the same trees. As a result of the combined application of the two substances, apples stored better and contained more calcium.[5]

More and more often, synthetic cytokinins are used in fruit production. One of them is benzyladenine (BA), the latest compound used for thinning apple fruitlets. According to Greene et al.,[35] this compound can affect the calcium content of apples indirectly by changing their size. It is probably for this reason that in the experiments by Ben,[13] Jonagold apples treated with BA 3 weeks after flowering were affected by flesh browning. Also, in experiments by Basak,[9] BA that was applied after flowering to thin fruitlets sometimes reduced the keeping quality of apples.

The results of using BRs depend on many environmental and agrotechnical factors, and often also on the concentration, time, and method of application, as well as tree age, species, and even cultivar. For that reason, the literature often contains contradictory information on the effects of the same BRs on the mineral composition of fruits.

To synthesize all this information is extremely difficult because, in many experiments, the mineral composition of fruits was regarded as an additional observation. Moreover, most of the studies concerned the effect of BRs on the mineral composition of leaves. Many experiments concerning the effect of BRs on the uptake of mineral nutrients were carried out on young, not-yet-fruiting plants, or their parts, growing under controlled conditions. The results of such experiments do not fully reflect the mineral nutritional status of plants already bearing fruit and growing in the field.

Taking into consideration the present state of knowledge, it can be stated that BRs can improve the mineral composition and storability of fruits, but it is difficult to predict the results that using these substances can produce because they depend on both environmental conditions and tree physiology. Thus, it does not seem likely that BRs can be used for controlling Ca nutrition of fruits.

15.2 Fruit Quality

Quality of fruit such as an apple is closely related with the apple's mineral composition; among them calcium has a special meaning. Ca content and its relationship with other elements decide about apple resistance to physiological disorders in storage.[17,70] The uptake of mineral nutrients, calcium in particular, was the subject of many investigations. The process is complicated and dependent on the physiological state of trees.[26] BRs and other preparations influence uptake and distribution of minerals, among them calcium.[7,11,26,47,72]

Apple trees take the highest amount of Ca in spring at the time of intensive shoot growth. In condition of weak shoot growth, the fruit supply with calcium improves,[59] but at strong shoot growth it worsens.[70] There is much evidence that retardants that are natural growth inhibitors favor the accumulation of Ca in fruit.[21,51,59]

Calcium moves to fruits and shoots mainly acropetally; however transport is related to the basipetal movement of auxins.[29] Auxins are mainly delivered by seeds, young leaves, and other intensively growing parts of plants. It is presumed that quantitative changes in fruits are due to BRs controlling auxin levels and their transports in plants.[2,4,43]

This work aimed to trace the uptake of mineral nutrients in apple trees in the condition of shoot growth weakened by retardants. It was assumed that retardant uptake by shoots would decline, too. Thus, shoot competition for these elements would be lessened, favoring fruits. There were also trials to attract nutrient accumulation in fruits through the application of synthetic auxin. To weaken shoot growth, two retardants (daminozide and paclobutrazol) were used. Both have different modes of action. These retardants influence auxin metabolism in the tips of shoots. To assess direct and indirect action of a retardant in accumulating mineral nutrients in fruits, daminozide was applied to whole plants or only to the shoots.[11]

The research was done in three experiments, each one in a different year, on 17- to 20-year-old "Double Red McIntosh" apple trees grafted onto Antonovka seedlings. To weaken shoot growth, two retardants—daminozide (Alar, Uniroyal Chemical Co., Middlebury, Connecticut) and paclobutrazol (Cultar, AkzoNobel, Slough, UK)—were used. Both preparations were applied 3 weeks after blooming at 0.2 percent concentration in the form of one spray. Daminozide was given to shoots only in other trees. To protect the fruit, the trees were covered with aluminium bags that were removed after the preparation had dried. Chemically pure IAA (Koch-Light, UK) was used as the auxin. The auxin was given to fruit about 2 weeks after bloom, 3 days after spraying with retardants (first term) or 2 weeks later, that is, about 4 weeks after bloom (second term). This compound was given 3 times in three consecutive days: first and second as submersion of fruit in a solution containing 50 mg/L with addition of Tween 20 at 0.1 percent concentration and the third one as an injection of 0.1 percent IAA solution containing 50 mg/L of active ingredient without a wetting agent.

Untreated trees and trees treated with retardant or auxin only served as controls. Also, trees with 1-year shoots, together with undeveloped leaves, tipped off at the time the retardant was applied to restrain their growth were taken as controls. Another group of control trees with 1-year shoots were sprayed at the same time with a solution of gibberellins (Gibrescol, Polfa, Kutno, Poland) containing 500 mg/L of active ingredient with glycerol as a wetting agent, amounting 5 mL/L to achieve maximum stimulation of shoot growth. For each treatment, five frame branches were chosen, one per tree (replicate) on parts of the tree situated in the same cardinal point. Control branches and branches treated solely with retardants as well as with auxin (one branch per treatment) have been on the same trees. Just after bloom, fruitlets were thinned with 1 per spur left.

Analytical material consisted of apples collected at harvest. Fruits from one branch (about 50 apples) served as the control. Chemical analysis was made separately on the apples' flesh and seeds. Because of limited seed volume from the replicates, chemical analyzes were executed on mean sample from all replicates, leaves from the middle of terminal shoots longer than 10 cm were sampled in the middle of August. Each sample consisted of about 100 leaves taken from the same branch (five samples per treatment).

The content of minerals Ca, K, Mg, and N (in percent of dry weight) was assessed in fruit, seeds, and leaves. The collected samples (five per treatment) were air dried at 60°C and ground down. Nitrogen content was assessed using the Kjeldahl method. Before analysis, other elements were mineralized in a mixture of HNO_3, H_2SO_4, and $HClO_4$. Calcium, potassium, and magnesium were assessed using the atomic absorption method with a Pye Unicam SP 09 spectrophotometer (Thermo Scientific, Cambridge, UK).

The following measurements were also taken: mean fruit weight, number of seeds per fruit, mean length of 1-year terminal shoots and mineral composition of the soil. Soil was sampled according to generally accepted rules (data are not shown here).

The results concerning fruit and leaf content were analyzed statistically with analyzes of variance. The significance of means differences was assessed with Duncan's *t* test at a 95 percent probability level. The results of the mineral content of the seeds were not analyzed statistically because chemical analyzes was made on a mixed sample of five replicates.

From changes in calcium content (Fig. 15.1), the soil in the orchard showed pH from 4.8 to 5.3, which indicates relatively low calcium content. The Ca supply of leaves and fruits differed in years. Also, the influence of bioregulators on Ca content was inconsistent in separate experiments.

Ca in Fruits In experiment 2, apples from control trees, untreated with growth BRs, contained almost double the amount of calcium as compared to the calcium in experiments 1 and 3. Retardants did not affect the Ca volume in experiment 2 when fruits were rich in this element. In the remaining experiments, however, retardants had slightly increased its contents of Ca. However, the significance of the differences was proven only in the case of fruits and shoots after treatment with daminozide in experiment 1 and paclobutrazol in experiment 3. Auxin did not affect calcium content in fruits as compared to plants treated with retardant only.

Ca in Seeds The seeds contained over 2 to 3 times more calcium than did fruit flesh. Retardants caused a slight increase of this element in seeds. Auxin influenced a further increase of Ca in seeds from plants treated with daminozide to shoots only in experiment 1 and from plants treated with paclobutrazol applied to shoots and fruits in experiment 3. In experiment 2, however, seeds from trees from which shoots and fruits were treated with daminozide, after application

416 Chapter Fifteen

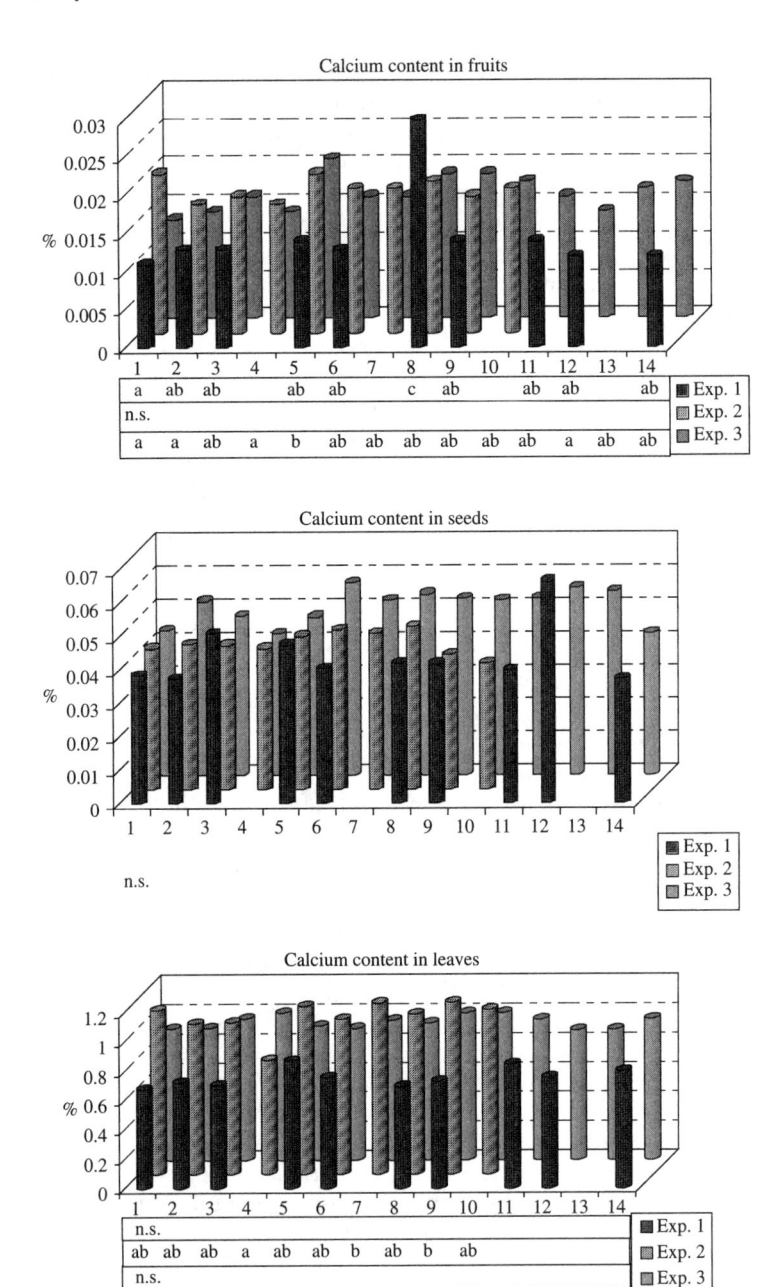

FIGURE 15.1 Effect of treatment with retardants (SADH, PB) and auxins (IAA) on calcium content in fruits, seeds, and leaves of Double Red McIntosh apples.

Figure concerning on experiments 1, 2, 3 signed with the same letter don't differ significantly at $P = 0.05$ t Duncan's test; n.s. means no significant difference between treatments.

of auxin, contained less calcium than trees treated with the retardant only.

Ca in Leaves Both retardants, alone or with auxin, did not change Ca content significantly during the entire experimentation period. However, in four cases (paclobutrazol applied to shoots and fruits in experiments 1 and 2), there was a tendency to find a decreased amount of calcium in leaves after application of auxin and treatment with retardant.

From changes of potassium content (Fig. 15.2) in fruits, it was observed that the control fruit contained the least amount of this element in experiment 3 and the most in experiment 2. The greatest changes in the content were caused by BRs in experiment 1. Paclobutrazol evoked a significant decrease of potassium in fruits. But the effect of daminozide on K fruit content varied. The auxin influenced the increase of the element both in the control (unsprayed) fruits and in fruits treated with daminozide (especially when given to leaves only). The reverse effect (i.e., decreased potassium content) was caused by auxin in experiments 1 and 3 after treating trees previously sprayed with paclobutrazol and in experiment 2, after covering leaves only with daminozide. Regarding the low potassium content in fruits (experiment 3), both daminozide and auxin did not change the amount of this element in fruit. In all experiments, the greatest accumulation of this element in fruit was recorded after treatment with GA_3 and the lowest accumulation was in fruits with shoots tipped off earlier.

K in Seeds The results presented here do not unanimously describe the influence of retardants on the potassium content in seeds. But auxin, in many cases, caused an increase of this element in seeds from trees treated with a retardant, especially when used on shoots and leaves.

K in Leaves According to the results presented here, the influence of retardant and auxin on potassium content in leaves was inconsistent.

From changes in magnesium content (Fig. 15.3) in fruits in spite of differences in magnesium content, no matter what the type and method of treatment, the retardants did not affect its accumulation in fruits. However, auxin caused a Mg decrease in apples after treatment with daminozide, both in shoots and fruits in experiment 3.

Mg in Leaves It was observed that in case of low content of magnesium in plants not treated with retardants, both retardants and auxin generally induced a small (insignificant) increase of Mg in leaves. The reverse changes were observed after applying auxin following treatment with a retardant.

With changes in the content of nitrogen, it can be explained that simultaneous treatment with retardants and auxin decreased the amount of nitrogen in apple leaves. This treatment did not influence

418 Chapter Fifteen

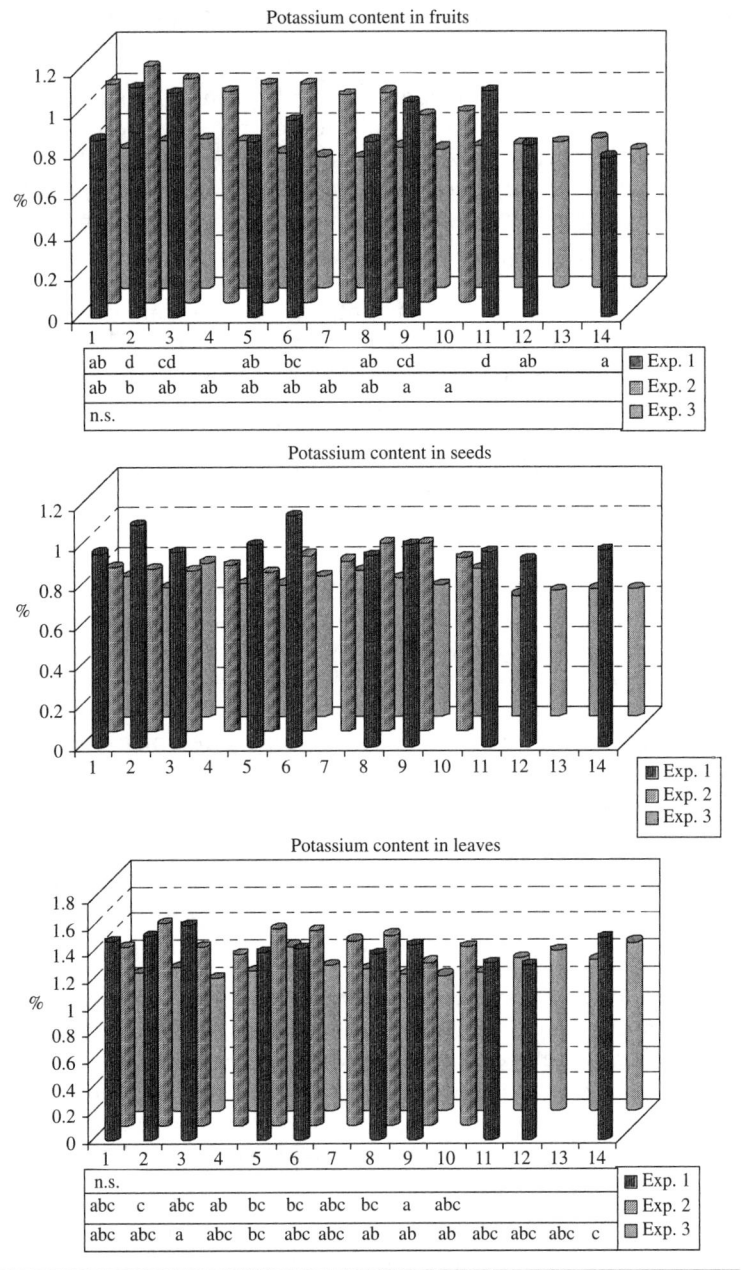

FIGURE 15.2 Effect of treatment with retardants (SADH, PB) and auxin (IAA) on potassium content in fruits, seed, and leaves of Double Red McIntosh apples.

Figure concerning on experiments 1, 2, 3 signed with the same letter don't differ significantly at $P = 0.05$ t Duncan's test; n.s. means no significant difference between treatments.

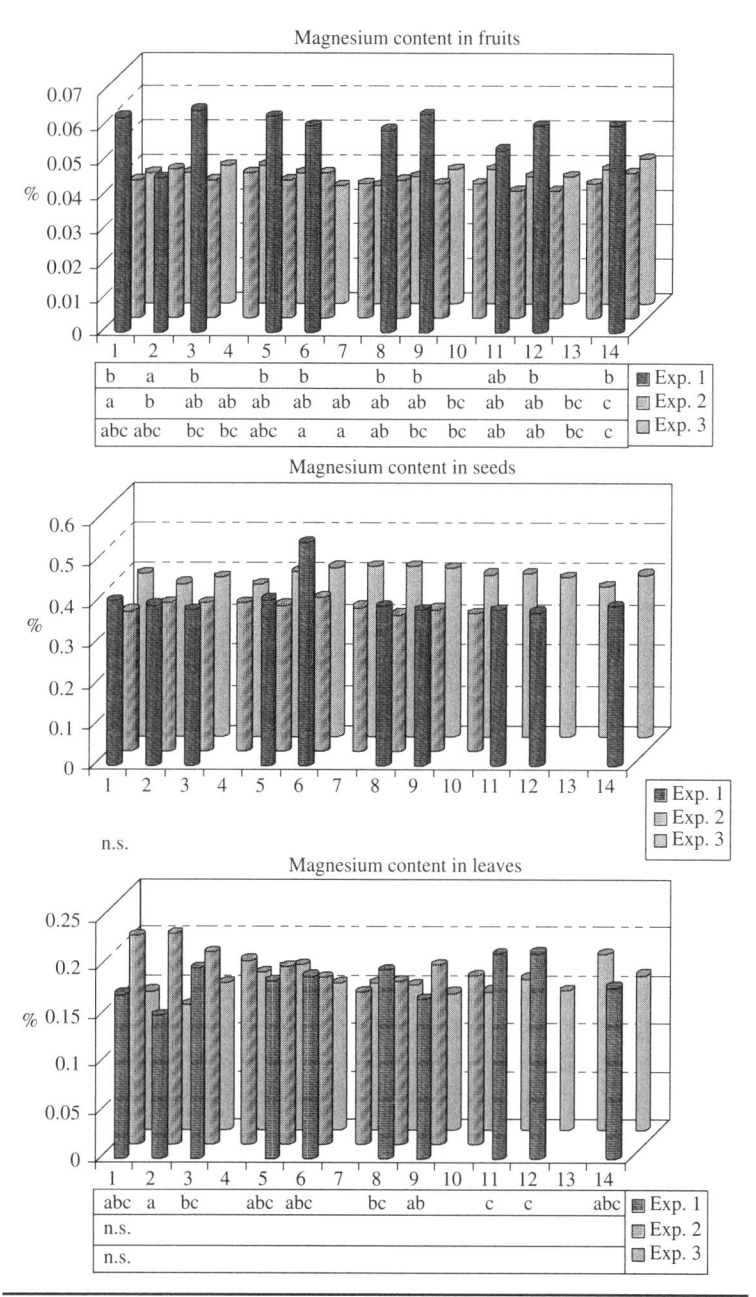

FIGURE 15.3 Effect of treatment with retardants (SADH, PB) and auxin (IAA) on magnesium content in fruits, seeds, and leaves of Double Red McIntosh apples.

Figure concerning on experiments 1, 2, 3 signed with the same letter don't differ significantly at $P = 0.05$ t Duncan's test; n.s. means no significant difference between treatments.

No.	Treatments	Week after Blooming
1	Control, untreated	—
2	GA$_3$	2
3	IAA	2
4	IAA	4
5	PB (shoots and fruits)	2
6	PB (shoots and fruits) + IAA	2
7	PB (shoots and fruits) + IAA	4
8	SADH (shoots and fruits)	2
9	SADH (shoots and fruits) + IAA	2
10	SADH (shoots and fruits) + IAA	4
11	SADH (shoots only)	2
12	SADH (shoots only) + IAA	2
13	SADH (shoots only) + IAA	4
14	Shoot thinning	2

Description of treatments to Figures 15.1–15.3.

the amount of nitrogen in fruit and seeds (data are not shown here). The auxin alone, as well as gibberellin and shoot tipping, did not change N content.

The results presented here suggest that retardants can improve fruit supply in calcium. However, this concerns several factors.[26] Daminozide did not affect Ca uptake in calcium-rich fruits. Daminozide did not change its calcium uptake power in fruits, although it diminished the fruits' competition for Ca through weakened shoot growth. At that time, fruits were small and supposedly seeds produced a sufficient amount of auxin to enable transportation of an adequate volume of calcium to the fruits.

There is much evidence that diaminozide reduces the growth of shoots and fruits. Thus, diaminozide could influence the flow of calcium to these parts of the plant.[26,59] The results presented here show that direct contact of daminozide with fruits has often changed calcium uptake by fruit and leaves. Usually, leaf and seed uptake has been changed insignificantly. In fruits, calcium uptake was slightly increased, and decreased or did not change the amount of calcium, depending on their physiological stage during treatment with retardant. Presumably, in one experiment at the time of treatment, fruitlets were larger, although they were more resistant to retardant than in other experiments. In small fruits that are rich in calcium, the retardant did not affect the calcium content in fruit, regardless of its mode of application.

Retardants (paclobutrazol and daminozide) that have been tested influenced differently the auxin transportation to tips of shoots. In this test, daminozide evoked larger changes in the amount of calcium than paclobutrazol. Some authors report increased Ca uptake by fruit after applying auxins.[1,47] Also, our results confirmed that auxin can increase Ca uptake by fruit and leaves treated with retardant in shoots. However, the amount of Ca in fruit did not change. However, when retardant was applied to shoots and fruit, auxin acted in a reverse way (i.e., it slightly decreased calcium uptake by fruits and leaves). It seems that applying a retardant to shoots only did not change the auxin's influence on calcium uptake by leaves; it weakened the transport to the leaves. When a retardant was also applied to fruit, its influence on calcium uptake caused by auxin was greater and concerned the leaves, too.

Retardants did not affect unanimously potassium content in analyzed plant organs. However, paclobutrazol acted more efficiently than daminozide, although auxin has a distinct influence on potassium uptake, particularly by fruits and seeds. Control fruit not treated with auxin accumulated the highest amount of potassium. Auxin did not affect potassium uptake unanimously. It was probably due to the effectiveness of the retardant. However, in many cases, retardant and auxin caused an increased K content in seeds.

A joint treatment with retardant and auxin had no great effect on magnesium content in fruit. A slight increase of Mg in seeds and a decline of Mg in leaves were observed.

Storage ability assessed on relationship between potassium and calcium (also potassium plus magnesium and calcium) worsened after treatment with auxin (Table 15.1). Apples treated with retardant and not treated reacted similarly to auxin application. In conclusion, in case of apples with seeds producing endogenous auxins, exogenous auxin depreciates storage ability. It might be caused by too high an auxin level in fruit. Auxins at high concentrations can be toxic, limiting the development of fruit and diminishing their size. It was proven in the presented work. According to Marcelle et al.[47] auxin stimulates Ca transport to fruit but only at low, limited levels. Exogenous auxin can intensify calcium transport to partenocarpic fruit,[1] thus these apples store better. According to our results, auxin affected storage ability similar to giberellin A_3.

Retardants increased the storage ability of apples. A similar effect was produced in conditions of extreme growth inhibition through shoot tipping. In fruits with seeds that produce endogenous auxin, their supply of Ca depends mainly on the sucking power of the shoots. Retardants diminish the level of auxin-like substances in shoots and weakens their competition with fruits. As a result, shoots facilitate a flow of a greater volume of mineral nutrients. Investigations on the uptake and distribution of calcium using a Ca isotope confirm this.[7]

Treatment	Time of Treatment (weeks after bloom)	Potassium:Calcium Exp. 1	Potassium:Calcium Exp. 2	Potassium:Calcium Exp. 3	(Potassium + Magnesium):Calcium Exp. 1	(Potassium + Magnesium):Calcium Exp. 2	(Potassium + Magnesium):Calcium Exp. 3
Control, untreated	—	74.9 ab	52.1 ab	52.8 b	85.1 b	53.8 a	55.7 b
GA_3	2	86.0 b	69.4 b	54.1 b	89.6 b	71.6 b	57.0 b
IAA	2	86.7 b	62.0 ab	47.3 ab	91.7 b	64.3 ab	47.2 ab
IAA	4	—	59.5 ab	53.0 b	—	60.0 ab	55.9 b
PB (shoots and fruits)	2	79.6 ab	—	43.5 ab	83.4 b	—	45.8 ab
PB (shoots and fruits) + IAA	2 + 2	68.0 ab	—	49.7 ab	73.8 ab	—	52.3 ab
PB (shoots and fruits) + IAA	2 + 4	—	—	51.1 b	—	—	53.8 b
SADH (shoots and fruits)	2	63.2 a	51.0 a	32.6 a	67.8 ab	52.9 a	34.5 a
SADH (shoots and fruits) + IAA	2 + 2	76.1 ab	56.6 ab	39.9 ab	80.7 ab	59.4 ab	42.0 ab
SADH (shoots and fruits) + IAA	2 + 4	—	56.1 ab	40.4 ab	—	58.3 ab	42.6 ab
SADH (shoots only)	2	57.4 a	51.0 a	37.4 ab	57.2 a	53.1 a	39.4 ab
SADH (shoots only) + IAA	2 + 2	77.9 ab	56.2 ab	38.7 ab	76.6 ab	53.8 a	41.9 ab
SADH (shoots only) + IAA	2 + 4	—	50.6 a	45.1 ab	—	52.7 a	47.6 ab
Shoot pinching	2	64.9 ab	—	38.6 ab	70.0 ab	—	1.1 ab

The means in columns with the same letter do not differ significantly at $P = .05$, Duncan's t test.

TABLE 15.1 Effect of Treatment with Retardants (SADH: daminozide, PB: paclobutrazol) and Auxin (IAA) in Relation to Potassium, Magnesium, and Calcium in Double Red McIntosh Apples

The results presented here are a further proof of the complicated influence of growth regulators on plant mineral composition.[11] However, the collected information can not be used to control apple tree nutrition in practice, but they have broadened our knowledge of the influence of growth regulators on mineral uptake and distribution. The retardants tested—daminozide and paclobutrazol—are solely the examples of preparations used for plant growth retardation. Application of the obtained results should be facilitated so that we may learn more about other new retardants.

15.3 Influence of Retardant and Auxin Treatment of Apple Shoots and Fruits on Calcium Uptake and Distribution

Calcium plays an important role in fruit physiology, especially in fruit ripening and in preventing physiological diseases in storage.[1,17,25,27,62,64] These phenomena are controlled by endogenous hormones[23,58] to a great extent and are also modified by the application of growth regulators.[41]

Hormonal regulation in a plant is connected at many points with calcium physiology of the cell.[16,71] In addition, typical acropetal transport of calcium in a plant is dependent on the polar basipetal translocation of the auxin.[2]

Retardants may influence calcium uptake in a fruit probably by changing the activity of physiological sinks. Suppressing vegetative growth with a retardant may diminish the sink activity of the long shoot apex and young leaves. Retardants may change some features of the fruit and, therefore, may also influence the sink activity of calcium in the fruit itself.[26]

This recent investigation analyzed calcium uptake and distribution in apple shoots and fruits as affected by the growth retardant daminozide, which is supposed to diminish the competition between young fruits and their vegetative parts. In separate treatment, auxin was also applied to the fruits, in untreated shoots and in retardant treated ones, to augment their sink activity (as previously supposed).

The investigations were carried out from 1987 to 1988, on 19- and 20-year-old Double Red McIntosh apple trees grafted onto Antonovka seedlings. The trees were from the Pomological Orchard of the Institute of Pomology and Floriculture in Skierniewice, Poland, and had round, almost natural, crowns.

To diminish the growth of current-year shoots, daminozide was applied 2 weeks after flowering. The compound was used at a concentration of 0.2 percent, and the chosen branches were sprayed once in two ways: whole branches (including shoots and fruits) or only shoots (the fruits being protected by a cover of aluminum foil, removed after the liquid had been dried).

The auxin used was IAA. Auxin treatment was started 3 days after the retardant was applied (term 1) or 2 weeks later (term 2). In both terms, the fruits were treated with IAA for 3 consecutive days: on the first and second days by immersion in a solution containing 50 mg/L IAA plus 0.1 percent Tween 20, and on the third day by injecting a 0.1-mL 50 mg/L of IAA solution without Tween 20 into the core of the fruit.

The control trees were not treated with growth regulators. Two other treatments were introduced in both years: some trees were treated only with retardant and others only with auxin.

In 1988, two additional treatments were introduced. In the first, the apices of current-year shoots including the youngest leaves were pinched on the day when the retardants were applied in other combinations. This was the treatment in which the optimum suppression of growth was expected. In the second treatment, the current-year shoots on the branches were sprayed once with 500 mg/L GA_3 (Gibrescol), using glicerol (5 mg/L) as a surfactant. Gibberellin was applied to stimulate growth to obtain the greatest contrast with the effect of the retardant.

Five uniform branches with approximately 100 clusters of flowers were selected for each combination of treatments. Just after petal fall, the fruitlets were hand thinned in such a way that only one central fruitlet remained in every cluster. The treatments with retardant only and with retardant plus auxin in two terms were carried out on different branches of the same tree. One branch comprised one repetition.

The influence of the treatments on the uptake and distribution of ^{45}Ca was investigated. Plant material for a radioactive tracer study was taken 2 days after the second treatment with auxin (approximately 4 weeks after bloom). The fragments of branches with a single short shoot and with a fruit and a single long shoot were excised and their cut ends immersed in water. Five such branches were taken after every treatment. After being transported to the laboratory, the excised shoots were immersed in Hoagland's solution[37] labeled with 20 MBq of $^{45}CaCl_2$ (O.P.I.D.I., Swierk, Poland, specific activity 230 GBq/g Ca) per 100 mL of solution. Then they were placed in the greenhouse in conditions of high air humidity. After 3 days, the branches were removed from the solution and divided into short shoots, fruits, and long shoots. Each part was separately weighed and ashed. The resulting ash was suspended in a scintillation cocktail (7 g/L butyl-PBD in dioxane) and radioactivity was determined on a liquid scintillation counter (Beckman LS-1701, Beckman Instruments, Inc., Fullerton, California).

The results of radioactivity measurements of the analyzed plant parts, expressed in the fresh mass, were worked out statistically after logarithmic transformation using the analysis of variance. The differences between means were evaluated using Duncan's multiple range test at a 5 percent level of significance.

The result gave information on calcium uptake by fruits and short shoots growing in the vicinity of the fruits (i.e., organs that could compete with the fruits for calcium). In the papers of Benson and Stahly,[14] Marcelle et al.,[47] and Banuelos and Bangerth,[3] only the influence of growth regulators on calcium uptake by fruits was investigated.

Their results suggest that exogenously applied growth regulators may change the activity of particular sinks in the plant in relation to calcium.

In both years of the 2-year experiment, daminozide had no distinct influence on the uptake of radioactive calcium by whole plant segments (Tables 15.2 and 15.3). However, in 1988 this retardant caused a change in distribution of ^{45}Ca among particular parts of the plants (Table 15.3). Thus, treatments with daminozide, regardless of the method of application, caused a significant drop of ^{45}Ca in long shoots while concomitantly evoking an increase of the amount of isotope in the fruit (Table 15.3). Some increase in ^{45}Ca content was also visible in short shoots; however, the differences with control were insignificant. In 1987, calcium distribution among analyzed plant parts was similar in the control and daminozide-treated branches (Table 15.2).

Differences in calcium uptake by whole analyzed segments in both 1987 and 1988 are not surprising in the light of the results of other authors. According to Faust and Miller[26] and Tromp[72], the influence of daminozide on calcium uptake depends on numerous factors, for example, the relationship between shoots and fruit growth. Such factors change from year to year. The increased accumulation of ^{45}Ca in fruits in 1988 probably resulted from more efficient suppression of the long-shoot competition by daminozide.

Auxin diminished radioactivity in whole-plant segments in 1988 when it was applied 4 weeks after blooming (i.e., 3 days before the start of radioisotope feeding) (Table 15.3). However, there was no evident influence of IAA treatment on the distribution of ^{45}Ca in particular plant parts. Nevertheless, in 1987 the exogenous auxin applied 2 weeks after blooming changed the distribution of calcium in investigated plant segments (Table 15.2).

Translocation and uptake of calcium, and transport of IAA are closely independent.[2,29] Our results coincide well with those of Banuelos and Bangerth.[3] In their experiment, treatments of fruits with auxin caused increased translocation of calcium to the fruits. Increased ^{45}Ca uptake in short shoots in our experiment may also be interpreted as the effect of IAA transported from the fruits in a basipetal direction.

A lower percentage of ^{45}Ca in the long shoots probably resulted from increased sink activity of the fruits due to the application of IAA.

Treatment	Time of Treatment (weeks after blooming)	Radioactivity in the Whole Plant Segment (10^3 cpm)	The Distribution of Radioactivity (%)			
			Fruits	Short Shoots	Long Shoots
Control untreated	—	287.7 bc	1.17 a	12.56 a	86.27 b
IAA	2	137.6 ab	5.82 b	38.52 b	55.06 a
IAA	4	232.1 bc	0.94 a	13.21 a	85.85 b
Daminozide applied on shoots and fruits	2	386.8 c	0.70 a	18.11 ab	81.19 b
Daminozide as above plus IAA	2	92.2 a	5.51 b	42.95 b	51.54 a
Daminozide as above plus IAA	4	327.0 c	1.63 a	12.91 a	85.46 b

Means followed by the same letters do not differ at the 5 percent level of significance: Duncan's multiple range test.

TABLE 15.2 Influence of Retardant and Auxin on Uptake and Distribution of ^{45}Ca in 1987

TABLE 15.3 Influence of Retardant and Auxin on Uptake and Distribution of ^{45}Ca in 1988

Treatment	Time of Treatment (weeks after blooming)	Radioactivity in the Whole Plant Segment (10^3 cpm)	Fruits	Short Shoots	Long Shoots
Control untreated	—	210.8 a	3.41 a	35.96 ab	60.63 bc
GA$_3$	2	122.4 a	6.29 bc	38.13 abc	55.58 abc
IAA	2	183.8 b	4.14 abc	32.81 a	63.05 c
IAA	4	99.6 a	5.08 abc	47.65 bc	47.27 ab
Daminozide applied on shoots and fruits	2	200.2 b	6.52 bc	47.64 bc	45.84 a
Daminozide as above plus IAA	2	185.2 b	6.49 bc	45.42 abc	48.09 ab
Daminozide as above plus IAA	4	224.6 b	3.20 a	41.75 abc	55.05 abc
Daminozide applied on shoots only	2	204.8 b	6.81 c	49.70 bc	43.49 a
Daminozide as above plus IAA	2	240.2 b	4.84 abc	45.40 abc	49.76 ab
Daminozide as above plus IAA	4	234.5 b	2.83 a	44.76 abc	52.41 abc
Pinching of long shoots	2	353.7 c	3.91 ab	50.52 c	45.57 a

Means followed by the same letters do not differ at the 5 percent level of significance: Duncan's multiple range test.

Therefore, part of the calcium that would be normally destined for the long shoots is now being directed toward the fruits.

The long period between application of IAA and time of radioactive calcium application makes it possible that the auxin effect consists of auxin stimulation of the vascular connection between the fruit and the branch that sustains it.[54] There are examples in literature that auxin treatment increases the amount of calcium in the fruit.[42,47] The failure of the auxin application to substantially increase the Ca content of normal, pollinated fruits sometimes also happens.[2] This is probably due to the presence of seeds, which may produce sufficient auxin for this purpose. Then the exogenous auxin supply would not be effective.

The effect of auxin in daminozide-treated branches in 1987 was similar to that in branches not treated with this retardant. IAA applied in the earlier term decreased ^{45}Ca uptake by whole segments (Table 15.2). Also, ^{45}Ca uptake by long shoots was relatively lower but that was taken up by the fruits increased. The effect of auxin in the second term in these branches (daminozide treated) was insignificant, as in the branches not treated with retardant.

In 1988, auxin applied in the second term on daminozide-treated branches diminished calcium uptake by fruits, regardless of the method of treatment (Table 15.3). IAA treatment in the first term was without effect. There was no IAA influence on ^{45}Ca content in short or long shoots.

The results suggest that exogenous auxin applied 2 weeks after blooming may increase the accumulation of calcium in fruits but only when the retardant is not very effective. When the retardant augments calcium accumulation in fruits, the effect of auxin given on the same branch 2 weeks after blooming is not significant.

It is well known that it is difficult to predict the calcium content in apples after retardant treatment. This group of chemicals may influence Ca uptake by direct or indirect ways that may act independently, and their effects are combined.[2] Therefore, it does not seem that exogenous auxin can be a useful tool for improving calcium nutrition of fruits on retardant-treated trees.

Gibberellin A_3 applied on long shoots only significantly decreased the amount of ^{45}Ca taken up by whole investigated fragments of a plant. However, the percentage of calcium supplied to the fruits increased, whereas that in long shoots and in short shoots remained similar as in the controls (Table 15.3).

It seems surprising that gibberellin, which stimulated the growth of long shoots (Basak, results unpublished), augmenting its needs for Ca, concomitantly caused an increase of calcium accumulation in fruits. A possible explanation for this may lie in the known stimulatory effect of gibberellin on the formation of vascular systems.[22,54] Because GA is also transported basipetally, it can stimulate the development of vascular tissues in the short shoot to which the fruit is attached. Then the fruit can take advantage of this better vascular connection of the short shoot that sustains it.

Plants treated in such a manner showed the greatest uptake of the radioisotope by whole-plant segments (Table 15.3). Pinching had no influence on the uptake of calcium by fruits but significantly increased the amount of ^{45}Ca per unit of fresh mass in short shoots. At the same time, the uptake of ^{45}Ca by long shoots diminished markedly.

The great increase of calcium uptake by a whole segment may be tentatively explained, taking into account that about 4 weeks passed from the time when the long shoots were pinched to the time when the radioactive calcium was applied. During that time, two or three lateral buds started to develop and probably substituted the lack of terminal apex with their own auxin. The high supply of calcium in these segments caused a large amount of ^{45}Ca to accumulate (possibly transitory) in the burse. Although the fruits did not have a great relative share in the total calcium that was taken up, their share was great in absolute units.

In interpreting the results of this chapter, it should be taken into account that the experiment consisted in feeding with ^{45}Ca fragments of branches by their cut lower end. Faust and Shear[28] have shown great differences in Ca transport in cut branches and in noninjured seedlings.

In comparison with the research of other authors, this work introduces more factors that could influence calcium uptake. For instance, more growth regulators were applied. Besides retardant, auxin and gibberellin and combinations of auxin with retardant were also used. Instead of fruits alone, the whole system composed of a fruit, a short shoot, and long shoots was investigated. From the point of view of methods the retardant was applied to the whole system as well as to shoots only (excluding the fruit). This might demonstrate whether the influence of retardant on the fruit is direct or indirect.

The calcium uptake in such a set of combinations confirms the opinion of Faust and Miller[26] and Tromp[72] that calcium uptake by the apple depends on many factors.

Growth regulators may influence the calcium content of fruits, but the effect is not always repeatable, depending on environmental as well as physiological factors.

15.4 The Effect of Fruitlet Thinning Using Bioregulators on Mineral Content and Storage Quality of Fruits

It is well known that the storage ability of fruits depends on conditions prevailing during their growth and storage. One of the main measures that modifies fruit growth is flower or fruitlet thinning. The effect of thinning on apple size and their attractiveness is well documented.[8–10,20,38,44,75,76] Much less information exists concerning the influence of fruitlet thinning on the storage ability of apples.

Here, we present the results of experiments in which the storage ability of apple cultivars is compared after hand thinning or chemical thinning of fruitlets with NAA, CPPU, BA, and urea or with a mixture of urea and NAA.

The experiments were carried as thinning was achieved with the following preparations:

- Urea (in the form normally used as fertilizer), containing 45 percent a. i. (CH_4N_2O)
- α-Naphthylacetic acid (NAA), as the preparation Pomonit R-10 containing 10 percent a. i. (potassium salt of NAA, Organika, Poland)
- Mixture of potassium salt of NAA [(80 g /1 L) with urea (30 g/1 L)] in a form of the preparation Pommit Ekstra 110 SL (Varichem, Zabia Wola, Poland)
- Forchlorfenuron (CPPU) as the preparation Sitofex EC 1 percent (AlzChem Trostberg Gmbh, CHEMIEPARK TROSTBERG, Trostberg, Germany)
- Benzyladenine (BA) as the preparation Paturyl 100 SL containing 10 percent a. i. (BA) (Reanal Finechemical Co., Budapest, Hungary)

Experiment 1 was done with 9-year-old trees Gala cultivar grafted onto M26 rootstock. Thinning was done with urea in the concentrations from 0.75 percent up to 6 percent on two occasions: during petal fall and just after flowering, or with the mixtures of urea (in concentrations of 0.45 percent to 1.5 percent) plus NAA (in concentrations of 15 to 25 mg/L) just after flowering. Part of the trees was sprayed with NAA alone in concentrations of 25 or 40 mg/L also just after flowering. The other trees were sprayed with CPPU at 20 mg/L on two occasions: just after flowering or 2 weeks later.

Experiment 2 was set up on 8-year-old apple trees of Lobo cultivar grafted onto M26. For thinning, urea was used in concentrations of 2 and 4 percent, respectively, during petal fall or just after flowering. The mixtures of urea (0.45 and 1.5 percent) with NAA (15, 20, or 25 mg/L) were applied just after flowering but NAA alone (40 mg/L) and CPPU at 2.5, 5, or 10 mg/L were used at the end of petal fall and 2 weeks after blooming.

Experiments 3 to 6 were done on the following trees, respectively: 12-year-old trees cultivar Gala/M26, 9-year-old trees cultivar Lobo/M26, and 12-year-old trees cultivar Gloster/M26. The trees were sprayed just after flowering with NAA at 25 to 40 mg/L or with BA at 50 to 100 mg/L when the fruitlets attained diameters of 10 to 12 mm.

In all experiments, unthinned and hand-thinned trees (just after June drop, as is commonly done in commercial orchards) were the controls. Results relating the influence of thinning on crop load and on the size and appearance of the apples during harvest are found elsewhere. This chapter contains only data concerning the internal quality of apples, their firmness, soluble solids, and sap starch content during harvest and after storage, as well as the susceptibility to rotting, shriveling, and physiological disorders.

The apples were stored in common cold storage at 2 to 3°C and about 90 percent relative humidity during the period similar as for commercially treated apples.

Each treatment was done for six to eight similar trees (one tree was a plot). Liberal spraying was done with a lance sprayer, until the droplets started to drip.

Analysis of variance was applied with Duncan's t test at $P = .05$ to separate the means.

In the first experiment, on Gala apples (data not shown in Table 15.4), only urea at 6 percent applied at the beginning of petal fall caused reduction of the crop load. Urea also caused a distinct improvement in apple size. However, this compound, regardless of its concentration

| Measurements | Cultivar Gala/M26 12-Year-Old Trees ||||
	Control	Hand Thinning	NAA (40 mg/L)	BA (100 mg/L)
Percent fruit set	81 cd	42 a	48 ab	44a
Mean fruit weight (g)	107 a	132 d–g	143 f–i	139 e–h
Yield kg/tree	72 d	51 ab	66 a–d	53 abc
At Harvest				
Firmness (lb)	180 abc	18.6 c	17.9 abc	18.2 abc
Soluble solids (%)	12.7 cd	13.5 f	12.3 abc	13.3 def
Starch index	7.8 a–d	7.4 a	8.8 hi	7.9 d–e
Streif's index	28 a–f	34 h	25 ab	31 efg
After Storage				
Firmness (lb)	14.8 a	14.9 a	14.0 a	14.9 a
Soluble solids (%)	12.5 c	12.2 bc	11.1 a	12.2 bc
Physiological Disorders (%)				
Bitter pit	8.1 ab	21.1 abc	3.9 a	45.4 c
Rotten apples	11.3c	1.3 a	2.2 ab	1.5 a
Shriveled apples	11.2 bc	1.3 a	2.2 ab	1.5 a

Means in columns followed by the same letters do not differ at 5 percent level of significance: t Duncan's multiple range test.

TABLE 15.4 The Quality of Apples after Thinning Using NAA or BA in 1996

and date of application, caused a reduction of Ca content, especially when used at low concentrations. Urea also caused a decrease of K content, with the exception of the lowest concentration (0.75 percent), which caused the reverse effect, an increase in the K level. Urea treatment caused, regardless of its concentration, an increase of Mg content and an increase in P levels at higher concentrations (3 to 6 percent). A higher ratio of K and Mg content to Ca content indicates a lower storage ability of apples thinned with urea. Only urea applied at 3 percent (recommended in practice), applied after termination of blooming, did not change the proportion of K and Mg to Ca in comparison with the unthinned trees (control).

Treatment with urea usually did not change the firmness and soluble solids of apples in relation to apples from unthinned trees. Only treatment with urea at 6 percent resulted in apples that were firmer and contained more sugars.

Treatment with urea at 1.5 to 3 percent after the end of blooming had no visible effect on rotting and shriveling of apples and on the incidence of physiological disorders during storage.

NAA alone at 40 mg/L, applied just after blooming, caused a marked reduction of crop load and a marked increase in apple size without influencing Ca content. Nevertheless, this treatment caused K, Mg, and P content to increase. NAA at 40 mg/L did not influence firmness of the apples but raised the soluble solids of the sap and limited fruit shriveling during storage. Nevertheless, twice as many apples showed internal breakdown. This indicated that these apples ripened earlier than the controls.

Lower concentrations of NAA (25 mg/L) caused a reduction in Ca levels and lowered fruit firmness at harvest. This was probably connected with higher crop load.

CPPU at 20 mg/L caused a significant increase in fruit size as well as a reduction of Ca level in the fruit, regardless of the date of application. The influence of CPPU on the content of other elements depended on the date of treatment. Consequently, a less desirable ratio of K and Mg to Ca and related lowering of the storage ability was observed only when CPPU was applied 2 weeks after bloom. Despite the fact that these apples showed higher firmness than those from unthinned trees, the apples showed more internal breakdown (although the differences were statistically not significant). Bitter pit was found in some fruits.

Hand thinning done about 4 to 6 weeks after blooming did not change Ca content in fruit but did cause an increase in K and Mg levels and an increase of the ratio of K and Mg to Ca, which indicates a lowering of storage ability of the apples. Nevertheless, apples from hand-thinned trees showed the highest firmness ratings during storage.

In the second experiment on the Lobo cultivar, no single urea treatment caused a significant crop reduction, but urea applied at 4 percent (highest rate) at the beginning of petal fall caused a significant increase in size (similar to that found in hand-thinned apples) (Table 15.5).

Treatment	Yield (kg/tree)	Mean Fruit Weight (g)	Ca	K	Mg	At Harvest	After Storage	Soluble Solids (% at Harvest)	Percent of Rotted Apples After Storage
Control unthinned	57.9 c–f	136 a	0.025 a	0.87 c	0.049 b	9.0 d	9.2 b	10.50 ab	18.2 ab
Hand thinned	51.0 a–f	162 d–h	0.032 c	0.87 c	0.047 a	9.2 e	8.4 b	10.10 ab	8.2 ab
Urea 2% BPF*	56.5 c–f	145 a–d	0.038 ef	0.99 cd	0.052 d	9.0 de	8.8 b	9.9 ab	7.4 ab
Urea 2% EPF*	67.6 ef	136 a	0.035 ef	0.83 ab	0.056 e	—	8.9 b	—	5.6 ab
Urea 4% BPF	41.3 abc	158 c–h	0.031 b	0.95 c	0.055 e	8.4 bcd	—	9.6 a	7.7 ab
Urea 4% EPF	45.6 abc	146 abc	0.028 b	0.80 a	0.042 a	—	—	—	23.4 ab
Mixture: urea 1.5%+ NAA 25 mg/L EPF without adjuvant	51.5 a–f	155 b–g	0.035 ef	0.99 cd	0.061 g	8.8 cde	8.7 b	9.9 ab	37.1 b
Mixture urea 1.5%+ NAA 15 mg/L EPF + adjuvant	53.3 b–f	165 e–h	0.031 bcd	0.96 c	0,047 b	8.3 bc	8.8 b	10.3 ab	24.9 ab
Mixture: urea 0.45%+ NAA 20 mg/L EPF + adjuvant	39.1 abc	172 ghi	0.034 cde	1.03 d	0.060 f	8.1 ab	8.6 b	11.0 b	22.5 ab[†]

TABLE 15.5 The Quality of Apples (Cultivar Lobo) after Thinning in 1993

Treatment	Yield (kg/tree)	Mean Fruit Weight (g)	Mineral Constituents of Apples (% dry matter) Ca	K	Mg	Firmness (lb) At Harvest	Firmness (lb) After Storage	Soluble Solids (% at Harvest)	Percent of Rotted Apples After Storage
NAA 40 mg/L EPF	35.8 ab	169 f-i	0.034 def	1.03 d	0.052 d	7.6 a	7.7 a	9.7 a	34.0 ab[†]
CPPU 2,5 mg/L EPF	63.0 def	134 a	0.030 b	1.15 f	0.063 h	9.0 de	8.7 b	9.9 ab	19.7 ab
2 weeks APF*	69.2 f	142 abc	0.034 d	0.83 b	0.052 d	9.0 de	8.7 b	10.2 ab	12.2 ab[†]
CPPU 5 mg/L EPF	53.8 b-f	133 a	0.037 e	1.00 d	0.060 g	9.0 de	9.2 b	10.9 b	5.6 ab
2 weeks APF	53.0 b-f	139 ab	0.030 bc	1.06 e	0.058 f	8.5 b-e	9.0 b	9.7 ab	23.1 ab[†]
CPPU 10 mg/L EPF	57.8 c-f	131 a	0.035 d	0.77 a	0.051 c	8.6 b-e	9.0 b	10.0 ab	23.1 ab[†]
2 weeks APF	68.6 ef	150 a-f	0.037 e	1.00 d	0.055 e	8.5 bcde	9.0 b	9.9 ab	4.6 a[†]

*BPF = beginning of petal fall, EPF = end of petal fall, 2 weeks APF = 2 weeks after petal fall. Means in columns with the same letter do not differ significantly at $P = .05$. Duncan's t test was used.
[†]Single apples with symptoms of a brown core were noticed.

TABLE 15.5 The Quality of Apples (Cultivar Lobo) after Thinning in 1993 *(Continued)*

The influence of urea on the mineral content of the Lobo cultivar was different than in Gala. In Lobo, the ratio of K and Mg to Ca content was reduced, which may indicate the improvement of their storage ability. Treating with urea, especially at 2 percent at the beginning of petal fall, caused an increase in Ca content but also the augmentation of Mg and P levels jointly with the reduction of K levels. Surprisingly, urea treatment had no influence on the nitrogen content in the fruit. Urea also had no influence on the firmness and soluble solids of fruit sap, both at harvest and during storage. Fruit rotting was lower, as in case of hand thinning. In this experiment, Lobo apples, as well those treated with urea and hand thinned or unthinned, showed no symptoms of physiological disorders.

Urea in low concentrations (0.45 to 1.5 percent) applied as an adjuvant to improve the action of NAA (which was used also in low concentrations) caused an increase in apple firmness at harvest and during storage and also increased the sugar level (as compared with apples treated with NAA alone).

Thinning with NAA alone caused a rise in K and P levels (and insignificantly of Mg level) in comparison with apples from unthinned trees, similar to Lobo and Gala cultivars. Apples treated with NAA alone showed the lowest firmness at harvest and during stored. NAA-treated apples also showed a high tendency for rotting during storage. These apples, however, were not prone to bitter pit.

Spraying Lobo trees with CPPU, regardless of its concentration and date of treatment, had no distinct effect on fruit size or crop load. Nevertheless, CPPU caused a significant increase in Ca content and often in K, Mg, and P contents and the reduction of the ratio of K and Mg to Ca. In spite of this, bitter pit appeared in these apples, especially in those that were sprayed with CPPU 2 weeks after bloom. Bitter pit was not observed on the apples from nonthinned trees.

In the case of the Lobo cultivar, no single CPPU treatment visibly influenced the content of firmness soluble solids. Some treatments with CPPU limited rotting during storage.

Hand thinning of fruitlets on Lobo trees, in spite of the late date of administering this treatment, caused an augmentation of Ca content with no effect on K and Mg levels, and sometimes even lowering the level of these elements. These apples showed no bitter pit symptoms in storage and rotted less than apples from unthinned trees.

The results of four other experiments (3 to 6) indicated that hand thinning performed after the end of the June drop usually caused a marked increase in apple size and sometimes diminished rotting and shriveling of fruits during storage (e.g., Gala cultivar) (Table 15.4).

However, hand thinning promoted the tendency for bitter pit (e.g., in Gala). Hand thinning sometimes resulted in firmer fruits in apples from 11-year-old Lobo trees at harvest and during storage. Hand thinning influenced soluble solids by increasing it in Gala apples at harvest and in "Gloster" apples at harvest and subsequent storage. Hand thinning also caused an increase in starch content (in 9-year-old Lobo trees) and sometimes resulted in earlier ripening as indicated by a higher Streif's index.

NAA applied just after flowering, 4 to 5 weeks earlier than hand thinning, caused similar (in Gala and Gloster cultivars greater (in Lobo) or sometimes smaller increases in apple size than hand thinning. This may have been connected to the varying sensitivity of cultivars to NAA. However, apples from trees treated with NAA differed from those thinned by hand; they contained fewer sugars and starches (especially in "Gala" and "Gloster"), but they also showed smaller incidences of bitter pit. Apples from these trees rotted less and showed less shriveling than apples from unthinned trees.

Treatment with BA in general caused an increase in apple size compared with hand thinning or NAA treatment. Treatment with BA also limited fruit shriveling during storage. The influence of BA on the susceptibility of fruit to rotting was not uniform in different cultivars (in the Gala cultivar, BA treatment reduced rotting, whereas in Gloster and Lobo, BA evoked the tendency for more rotting). Firmness and soluble solids content at harvest and during storage in BA-treated apples were more similar to those in hand-thinned apples than to those in NAA-thinned ones.

These results indicate that thinning—one of the most important agricultural measures in modern apple orchards—may modify several fruit characteristics influencing the storage ability of apples. Thinning affects such fruit characteristics as firmness during harvest and after storage, soluble solids content of the sap, starch content, mineral composition, incidence of physiological disorders during storage, and susceptibility to rotting. Of course, the effect of thinning on storage ability depends on conditions prevailing in the orchard during the growing season, methods of thinning, and cultivar.

Summarizing the results, we may conclude the following.

Hand thinning, done after the end of June drop, in spite of the late date of its execution, resulted in bigger fruit, an increase in sugar content, and sometimes caused higher fruit firmness (especially after storage) as well as less rotting. Nevertheless, such late hand thinning may increase the susceptibility to bitter pit. The influence of hand thinning on the mineral composition of apples was not uniform: in Lobo apples it caused the augmentation of Ca content without influencing potassium, whereas in Gala apples the level of Ca was not changed and the content of K and Mg increased. Other authors indicate that storage ability of apples and their quality at harvest depend on timing of hand thinning during the season and on the localization of fruits on the tree.[73]

Urea applied during flowering influenced the storage ability of apples differently depending on concentration, date of spraying, and on the cultivar. A low concentration (0.75 percent, commonly applied in orchard for fertilization), caused a decrease in Ca level, an increase in K level, and a decrease in storage ability. Higher concentrations (4 to 6 percent), especially when applied at the end of flowering, did not change (e.g., Gala) or even improve the storage ability of apples (e.g., Lobo). Increasing the concentration of urea and delaying the date of its application resulted in improving the mineral composition of fruit, higher firmness, and higher sugar content.

Urea, applied just after flowering at a low concentration, as an adjuvant for low concentrations of NAA, caused an increase in firmness and sometimes increased the sugar content as compared with apples treated only with NAA.

NAA applied just after flowering mainly limited the occurrence of bitter pit and shriveling during storage. NAA may, however, diminish firmness and soluble solids content of the sap and may accelerate ripening. Martin et al.[50,51] reported that NAA applied just after flowering diminished the incidence of bitter pit and internal breakdown of apples. However, Link[40] reported that NAA increased the incidence of bitter pit. It is probable that the storage ability of apples thinned with NAA depends on the rate of crop load reduction and on the attained fruit size.

CPPU may augment the mineral content of apples, but this phenomenon depended on its concentration and date of application. In spite of this, CPPU treatment may intensify the occurrence of bitter pit and internal breakdown, especially when it was applied relatively late (2 weeks after flowering). CPPU may also increase the firmness and sugar content of apples.[9,10]

BA used for thinning, about 2 weeks after flowering, influenced apple quality similar to hand thinning but better than NAA. In spite of the late date of application, BA caused an increase in sugar content and sometimes improved apple firmness.[32] Apples treated with BA were firmer and contained more sugars than those treated with NAA.[8] In some conditions, BA may intensify the occurrence of bitter pit and internal breakdown.

15.5 The Influence of Thinning and Treatment with Calcium Chloride on the Quality of McIntosh Apples Treated with SADH

SADH applied during the summer on McIntosh apple trees effectively prevents fruit drop during harvest in Poland.[65] However, this treatment sometimes reduces apple size. That disadvantageous phenomenon usually occurs if SADH is applied too early in the season. To improve apple size, thinning is the method most commonly used by fruit growers. In these experiments, trials were conducted to investigate the influence of fruit thinning on size and other characteristics of apples treated with SADH.

SADH applied against a preharvest fruit drop stimulates development of core flush of apples during storage. McIntosh apples, the cultivar commonly cultivated in Poland, are particularly sensitive to this disorder.[65] A similar problem was reported in the United Kingdom[61] for the Cox Orange cultivar. To improve the storage ability of apples, it is recommended to spray or dip the fruits with calcium chloride. Calcium was reported to prevent apples from several storage disorders, among them the core flush.[15,49,50] One of the points of this chapter was to investigate the influence of calcium chloride on the quality of apples sprayed with SADH, with particular attention to the occurrence of core flush during storage.

Experiments were carried out from 1983 to 1984 at two experimental stations of the Institute of Pomology and Floriculture—Prusy and Sinoleka—on mature McIntosh trees.

Alar 85 (SADH, succinic acid-2-2-dimethylhydrazide) was applied at a concentration of 0.1 percent in the middle of July (about 60 days before harvest from untreated trees). Half of the trees had additionally been sprayed 5 times with 0.8 percent calcium chloride ($CaCl_2 \times 6\ H_2O$). The first application was done at the beginning of July, three others every 2 weeks, and the last one a week before the harvest of apples from untreated trees. Untreated trees and trees sprayed only with calcium chloride were used as controls. Half of the trees in each treatment were sprayed with Pomonit R-10 (NAA, 1-naphtaleneacetic acid) for apple thinning. It was applied at a concentration of 0.04 percent 2 days after bloom.

Observations were made to establish the influence of the treatments on fruit quality during harvest and, in the case of unthinned trees, also on fruit quality during storage. Additionally, the storage ability of apples sprayed with SADH and dipped immediately after harvest in 4 percent solution of calcium chloride were investigated. Apples were kept in cold storage at temperature of 3°C and relative humidity of 85 to 92 percent. In the second part of December and at the end of January and the beginning of February apples were examined for their firmness and the occurrence of physiological disorders.

Experiments were also carried out to determine the influence of treatments with SADH and calcium chloride on the content of calcium, potassium, and magnesium in apples during the last month of storage. The level of the components was investigated using the absorption spectrophotometer method.

Apples from trees sprayed with SADH were harvested 2 weeks later than those from other treatments. An investigation of each treatment was done on apples collected from seven uniform trees selected according to their growth and bearing. Results of the experiments were calculated statistically with analysis of variance, and the significance was evaluated using Duncan's t test at $P = .05$.

15.5.1 Apple Size

SADH applied to trees in the middle of July did not reduce the average size of the apples (Table 15.6) and sometimes increased the number of apples by 3.9 to 19.1 percent. Apple thinning and treatment with SADH have resulted in an increase of average apple size and increased the number of large-size fruits in total yield. These results were similar to those recorded from trees not treated with SADH. After thinning, apples treated with SADH were bigger in size than those from control unsprayed trees. Thus, SADH applied in the middle of July to McIntosh trees did not affect the beneficial effect of thinning on fruit size, which is what may happen if SADH is applied early in the season.[56]

The biggest quantity of extra apples has often been collected from trees thinned and subsequently treated with SADH and calcium chloride. These trees gave the lowest yields simultaneously.

Calcium chloride applied alone reduced apple size in both orchards but only in the first year of the experiments. In either case, calcium chloride induced some leaf injury. The reduction in fruit size was particularly evident on unthinned trees and trees not treated with SADH. These apples had been harvested earlier than those treated with SADH, and perhaps this and the leaf injury were reasons for the differences in size of the apples.

	Orchard 1		Orchard 2	
Treatment	**1983**	**1984**	**1983**	**1984**
Unthinned				
Control	12.8 ab	7.8 bc	9.2 a	10.7 bc
SADH	12.5 a	8.4 c	10.2 a	10.5 bc
SADH + CaCl$_2$ spray	12.5 a	8.1 bc	10.0 a	10.1 ab
CaCl$_2$ spray	13.1 abc	6.7 a	10.3 a	9.1 a
Thinned				
Control	14.2 abc	8.1 bc	9.2 a	11.7 cde
SADH	14.8 bc	8.6 c	10.9 a	12.2 de
SADH + CaCl$_2$ spray	15.0 c	7.7 abc	11.5 a	12.7 e
CaCl$_2$ spray	13.1 abc	7.1 ab	10.1 a	11.0 bcd

Orchard 1: Experimental Station Prusy, Orchard 2: Experimental Station Sinoleka. Means in columns followed by the same letters do not differ at the 5 percent level of significance: Duncan's multiple range test was used.

TABLE 15.6 Effect of Treatments on Fruits Weight in kg/100 Fruits

15.5.2 Red Color of Apples

Apples sprayed with SADH were much better colored than unsprayed ones. In trees with oval crowns (the orchard in Prusy) SADH increased the number of apples with blush covering more than 50 percent of the apple surface by 19.0 and 28.1 percent, respectively, in years 1983 and 1984. Apple thinning on these trees improved the coloration of the fruit. The number of fruits with blush covering more than 50 percent of apple surface have increased by 45.6 percent and 51.3 percent as compared to fruits collected from trees unthinned and untreated by SADH. Thinning alone has also increased fruit coloration but only by 22.0 and 31.3 percent, respectively. The influence of thinning on fruit coloration induced by SADH was less evident on trees with flattened crowns (the orchard in Sinoleka). Spraying these trees with $CaCl_2$ sometimes increased the blush surface (differences not statistically significant).

15.5.3 Ground Color of Skin

Apples sprayed with SADH generally had a greener ground color of the skin than apples not treated with SADH. Thinning itself has sometimes induced a yellowish skin ground color in apples both treated and untreated with SADH. Of all the treatments, the yellowest ground color of the skin was found in apples from trees thinned and sprayed with calcium chloride.

15.5.4 Apple Firmness

In general, SADH had no influence on the firmness of apples during harvest, and sometimes firmness was higher. Treatment with calcium chloride or thinning applied separately, in general, had no influence on the firmness of apples sprayed or unsprayed with SADH. In one of the experiments, the softest apples appeared to be those harvested from trees thinned and sprayed with SADH and calcium chloride.

During storage, the firmest apples appeared to be those treated with SADH alone, and often the apples had been dipped in calcium chloride. The lowest readings for firmness were obtained in these apples than for apples harvested from trees untreated by SADH or calcium chloride.

15.5.5 Physiological Disorders of Apples during Storage

From the experiments presented here, it is evident that only once did apples treated with SADH suffer more from core flush than did apples that did not undergo such treatment (Table 15.7). In experiments reported earlier, such phenomena occurred more often.[57] In such cases, application of calcium chloride reduced the amount of affected fruits. The smallest number of fruits with symptoms of the disease was found in apples that had been treated with both SADH

	Orchard 1			Orchard 2	
Treatment	2 Feb. 1984	27 Dec. 1984	30 Jan. 1985	9 Feb. 1984	30 Jan. 1985
Control	5.1 a (53.0 ab)	0 (43.0 cd)	59.2 b (73.2 bc)	77.5 c (97.6 c)	76.2 b (100.0 a)
SADH	40.8 b (59.1 b)	0 (21.4 bc)	24.6 a (67.5 b)	4.3 a (67.3 a)	46.7 ab (90.2 a)
SADH + CaCl$_2$ spray	18.3 ab (38. a)	0 (1.2 a)	39.0 a (45.7 a)	7.4 a (65.7 a)	29.7 a (77.9 a)
SADH + CaCl$_2$ dipp	13.5 ab (59.7 b)	0 (4.3 ab)	27.4 a (60.0 b)	14.5 a (80.6 ab)	55.2 ab (93.3 a)
CaCl$_2$ spray	31.2 b (55.6 ab)	0 (56.7 d)	69.3 b (82.1 c)	59.5 b (90.2 bc)	50.0 ab (98.9 a)

Orchard 1: Experimental Station Prusy, Orchard 2: Experimental Station Sinołeka.
Data in parentheses were obtained from apples taken out from cold storage and kept for 5 days at room temperature.
Means in columns followed by the same letters do not differ at the 5 percent level of significance: Duncan's multiple range *t* test was used.

TABLE 15.7 Effect of Treatments on Percent of Apples with Core Flush

and calcium chloride. Calcium chloride applied in the form of a spray showed a much better influence on the reduction of core flush than when it was used for dipping. When apples were kept for 5 days at room temperature (18 to 20°C), only samples treated with SADH and sprayed with calcium chloride differed from apples of other treatments in their susceptibility to core flush.

In other experiments, it was found that SADH, in general, reduced the number of fruits with core flush. In this respect, calcium chloride applied alone sometimes showed a similar influence as did SADH alone, but the number of affected fruits was greater. Then, apples treated with SADH and calcium chloride suffered less from core flush than fruits treated with calcium chloride alone.

Calcium chloride showed a pronounced influence on the incidents of other disorders of apples sprayed with SADH. Among them is a reduced occurrence of bitter pit, which is a confirmation of earlier reports.[63] Disorders such as bitter pit, internal breakdown, rotting, and shriveling of apples rarely occurred in our experiments and was of minor importance in comparison with core flush.

In most cases, SADH reduced superficial scald occurring on apples treated with calcium chloride. This disorder appeared more frequently on apples sprayed with calcium chloride than on apples treated with SADH alone.

15.5.6 Chemical Composition of Apples

SADH applied in the middle of July had no influence on the chemical composition of apples. In both experiments, in apples treated with SADH and calcium chloride, the ratio of K and Mg to Ca was the lowest one. This indicates that fruits from those treatments possessed the best storage ability.[46]

References

1. Bangerth, F. K. 1974. Calcium related physiological fruit disorders. *Proceedings of the 19th International Horticultural Congress* Warsaw, Poland 19:124.
2. Bangerth, F. 1976. A role for auxin and auxin transport inhibitors on the Ca content of artificially induced parthenocarpic fruits. *Physiology of Plants* 37:191–194.
3. Banuelos, G. and Bangerth, F. 1986. Interrelationship between IAA- and Ca-transport in respect to Ca nutrition of fruits. *Acta Horticulturae* 179:817–818.
4. Banuelos, G. S., Bangerth, F., and Marschner, H. 1987. Relationship between polar basipetal auxin transport and acropetal transport into tomato fruit. *Physiology of Plants* 71:321–327.
5. Basak, A. 1984/1985. Stosowanie etefonu łącznie z NAA i SADH w celu przyspieszenia zbioru oraz poprawy wybarwienia jabłek odmiany McIntosh. *Prace Instytutu Sadownictwa i Kwiaciarstwa* A(25):85–97.
6. Basak, A. 1992/1993. Wpływ preparatu Promalin na jakość jabłek odmiany Golden Delicious i Starkrimson. *Prace Insytutu Sadownictwa i Kwiaciarstwa* A(31):67–68.
7. Basak, A. 1993. Influence of retardant and auxin treatment of apple shoots and fruits on calcium uptake and distribution. *Journal of Fruit and Ornamental Plant Research* 1(4):103–114.
8. Basak, A. 1996. Benzyladenine as an apple fruitlets thinning agent—Preliminary results. *Horticultural Science* (Kerteszeti Tudomani) 28:54–57.
9. Basak, A. 1997. Storage quality of apples after fruitlets thinning. *Acta Horticulturae* 485:47–53.
10. Basak, A. 1998. Effect of forchlorfenuron (CPPU) combined with NAA and carbaryl on fruit set and fruit characteristics in two apple cultivars. *Acta Horticulturae* 463: 287–294.
11. Basak, A. 1999. Wpływ egzogennych bioregulatorów na skład mineralny owoców i ich zdolności przechowalnicze. *Postępy Nauk Rolniczych* 1/99: 59–70.
12. Basak, A., Soczek, Z., and Niezborała, B. 1986. The influence of thinning and treatments with calcium chloride on the quality McIntosh apples treated with SADH. *Acta Horticulturae* 179:761–765.
13. Ben, J. 1992. Wpływ paclobutrazolu na zawartość niektórych składników chemicznych w jabłkach odmiany Jonagold i Gloster oraz ich zdolnść przechowalniczą. *Prace Insytutu Sadownictwa i Kwiaciarstwa* C(3–4):134–135.
14. Benson, N. R. and Stahly, E. A. 1972. Restriction of 45 calcium translocation into apple fruit by 2,3,5-triiodobenzoic acid. *HortScience* 7/2:172–173.
15. Blanpied, G. D. 1981. The relationship between water loss and storage beakdown of McIntosh apples. *HortScience* 16/4:525–526.
16. Blatt, M. R. and Thiel, G. 1993. Hormonal control of ion channel gating. *Annual Review of Plant Physiology and Plant Molecular Biology* 44:543–567.
17. Bramlage, W. J., Drake, M., and Lord, W. J. 1980. The influence of mineral nutrition on the quality and storage performance of pome fruit in North America. In *Mineral Nutrition of Fruit Trees*, eds. D. Atkinson, J. E. Jackson, R. O. Sharples, and W. M. Waller. London, UK: Butterworths, 29–30.
18. Choi, J. S. and Lee, J. C. 1994. Seasonal trends of calcium accumulation into fruit in five apple cultivars (*Malus domestica* Borkh). *Journal of the Korean Society of Horticultural Science* 33(22):156–160.

19. Clijsters, H. 1971. Influence des regulateurs de croissance sur la maturation, la conservation et la qualite des fruits. *Fruit Belge* 354:69–77.
20. Curry, E. A. and Greene, D. W. 1993. CPPU influences fruit quality, fruit set, return bloom, and preharvest drop of apples. *HortScience* 28:115–119.
21. Curry, E. A. and Williams, M. W. 1986. Effect of paclobutrazol on fruit quality: Apple, pear and cherry. *Acta Horticulturae* 179:743–754.
22. Digby, J. and Wareing, P. E. 1966. The effect of applied growth hormones on cambial divisions and the differentiation of cambial derivatives. *Annals of Botany* 30:539–548.
23. Dilley, D. R. 1969. Hormonal control of fruit ripening. *HortScience* 4:111–114.
24. Elfving, D. C., Chu, C. L., Lougheed, E. C., and Cline, R. A. 1987. Effects of daminozide and paclobutrazol treatments on fruit ripening and storage behavior of "McIntosh" apple. *Journal of the American Society of Horticultural Science* 112(6):910–915.
25. Faust, M. 1975. The role of calcium in the respiratory mechanism and senescence of apples. *Colloque C.N.R.S.* 238:87–92.
26. Faust, M. and Miller, A. N. 1989. Effect of bioregulators and herbicides on calcium and phosphorus content of fruit. *Acta Horticulturae* 239:409–416.
27. Faust, M. and Shear, C. B. 1972. The effect of calcium on respiration of apples. *Journal of the American Society of Horticultural Science* 97:437–439.
28. Faust, M. and Shear, C. B. 1973. Calcium translocation patterns in apples. *Prace Instytutu Sadownictwa E* 3:423–436.
29. de la Fuente, R. K. and Leopold, A. C. 1973. A role for calcium in auxin transport. *Plant Physiology* 51:845–847.
30. Greene, D. W. 1986. Effect of paclobutrazol and analogs on growth, yield and fruit quality and storage potential of Delicious apples. *Journal of the American Society of Horticultural Science* 111:328–332.
31. Greene, D. W. 1991. Reduced rates and multiple sprays of paclobutrazol control growth and improve fruit quality of Delicious apples. *Journal of the American Society of Horticultural Science* 116(5):807–812.
32. Greene, D. W. 1993. A review of the use of benzyladenine (BA) as a chemical thinner for apples. *Acta Horticulturae* 329:231–236.
33. Greene, D. W., Lord, W. J., and Bramlage, W. J. 1974. Effects of low etephon concentrations on quality of McIntosh apples. *Journal of the American Society of Horticultural Science* 99(3):239–242.
34. Greene, D. W., Lord, W. J., Bramlage, W. J. 1977. Mid-summer application of etephon and daminozide on apples. I. Effect on McIntosh. *Journal of the American Society of Horticultural Science* 102(4):491–494.
35. Greene, D. W., Lord, W. J., and Bramlage, W. J. 1982. Effects of giberellins A_{4+7} bottom index and 6-benzylamino purine on fruit set, fruit characteristics, seed content, and storage quality of McIntosh apples. *HortScience* 17:653–654.
36. Griggs, W. H., Martin, G. C., and Iwakari, B. T. 1970. The effect of seedless versus seeded fruit development on flower bud formation in pear. *Journal of the American Society of Horticultural Science* 95:243–248.
37. Hoagland, D. R. and Arnon, D. J. 1938. The water culture method for growing plants without soil. University of California Agricultural Experiment Station Circular 347.
38. Jones, K. M., Graham, B., Bound, S. A., and Oakford, M. J. 1993. Preliminary trials to examine the effects of ethephon as a thinner of 'Gala' and 'Jonagold' apples. *Journal of Horticultural Science* 68:139–147.
39. Kallai, T., Buban, T., Fargo, M., and Szucs, E. 1987. Influence of modified source-sink relations on Jonathan apple fruit nutrition and quality. *Journal of Plant Nutrition* 10:1563–1571.
40. Link, H. 1973. Effect of fruit thinning on some components of fruit quality in apples. *Acta Horticulturae* 34:445–448.
41. Looney, N. E. 1973. Control of fruit maturation and ripening with growth regulators. *Acta Horticulturae* 34:397–406.
42. Looney, N. E. 1977. A four year study of single calcium chloride and growth regulator tree sprays to control storage breakdown of "Spartan" apples. *Journal of the American Society of Horticultural Science* 102:85–88.

43. Looney, N. E. 1979. Some effects of gibberellins A$_{4+7}$ bottom index plus benzyladenine on fruit weight, shape, quality, Ca content and storage behavior of Spartan apple. *Journal of the American Society of Horticultural Science* 104:389–391.
44. Looney, N. E. 1986. Chemical thinning of apple: Some new strategies and important refinements to old procedure. *Acta Horticulturae* 179:597–604.
45. Ludders, P. and Fischer-Bolukbasi, T. 1980. Einfluß von Alar und TIBA auf den Mineralstoffgehalt der Fruchte bei unterschiedlichem Fruchtbehang. *Gartenbauwissenschaft* 45(5):235–240.
46. Marcelle, R. D. 1984. Mineral analysis and storage properties of fruit. *Proceedings of the VIth International Colloquium for the Optimization of Plant Nutrition* 2:365–371.
47. Marcelle, R., Simon, P., and Lennes, G. 1981. The effect of IAA and TIBA on calcium transport and accumulation in fruits. *Acta Horticulturae* 120:193–197.
48. Martin, D., Lewis, T. L., Cerny, J., and Grassia, A. 1969. Effect of some chemical treatments on the incidence of bitter pit and breakdown in Cox apples. *Field Station Records Division Plant Industry CSIRO* (Australia) 8:57–76.
49. Martin, D., Lewis, T. L., Cerny, J., and Ratkovsky, D. A. 1976. The effect of tree sprays of Ca, B, Zn and naphtalenacetic acid, alone and in all combinations, on the incidence of storage disorders in Merton apples. *Australian Journal of Agricultural Research* 27:391–398.
50. Martin, D., Lewis, T. L., Cerny, J., and Ratkovsky, D. A. 1976. Der Einfluss erhohter Eisens-Bor und Calcium-Versorgung auf die Fruchtzusammensetzung und das Auftreten physiologischer Storungen bei Jonathan. *Erwerbstobstbau* 18/5/:70–74.
51. Miller, S. S. and Świetlik, D. 1986. Growth and fruiting response of deciduous fruit trees treated with paclobutrazol. *Acta Horticulturae* 179:563–566.
52. Naumann, W. D. 1971. Calcium distribution in apple trees and development of bitter as influenced by growth retardants. *Gartenbauwissenschaft* 36:63–69.
53. Pfammater, W. and Dessimoz, M. 1974. Influence des regulateurs de croissance sur la qualite de la variete de pommes Gravestein. *Revue Suisse de Viticulture, d'Arboriculture et d'Horticulture* 6:87–88.
54. Pieniążek, J. and Jankiewicz, L. S. 1967. Initiation of cambial activity during dormancy due to naphthaleneacetic acid, 6-benzylaminopurine and gibberellin in the apple shoot. *Wiss. Zeit. Univ. Rostock Math-Naturwiss* 16:651–652.
55. Robinson, J. B. D. 1975. The influence of some growth-regulating compounds on the uptake, translocation and concentration of mineral nutrients in plants. *Horticultural Abstracts* 45(10):611–618.
56. Rogers, B. L. and Thompson, A. H. 1983. Effects of dilute and concentrated sprays of NAA and cabaryl in combination with daminozide on fruit size and return bloom of Starkrimson Delicious apple. *HortScience* 18/1/:61–63.
57. Romaniuk, J., Soczek, Z., and Machnik, J. 1981. The influence of Alar (succinic acid 2,2-dumethyl hydrazide), nitrogen fertilization and the picking date on the incidence of storage diseases and the climacteric of McIntosh apples. *Fruit Science Reports* VIII/4/:163–172.
58. Sacher, J. A. 1973. Senescence and postharvest physiology. *Annual Review of Plant Physiology* 24:197–224.
59. Schumacher, R., Fankhauser, F., and Stadler W. 1980. Influence of shoot growth, average fruit weight and daminozide on bitter pit. In *Mineral Nutrition of Fruit Trees*, eds. D. Atkinson, J. E. Jackson, R. O. Sharples, and W. M. Waller. London, UK: Butterworth, 83–91.
60. Scott, K. J., Wills, R. B. H. 1976. Core flush of apples: I. Effect of absorption of carbon dioxide, ethylene and water from the storage atmosphere. *Journal of Horticultural Sciene* 51:55–58.
61. Sharples, R. O. 1972. *East Malling Research Station Report* 1971:73–74. Naumann, W. D. 1971. Calciumverteilung in Apfelbaumen und Entwicklung von Stippigkeit unter dem Einfluss von Wachstumregulatoren. *Gartenbauwissenschaft* 36:63–69.

62. Sharples, R. O. 1980. Influence of orchard nutrition on the storage quality of apples and pears grown in the United Kingdom. In *Mineral Nutrition of Fruit Trees*, eds. D. Atkinson, J. E. Jackson, R. O. Sharples, and W. M. Walker. London, UK: Butterworth 17–28.
63. Sharples, R. O. and Johnson, D. S. 1976. Post-harvest chemical treatments for the control of storage disorders of apples. *Annals of Applied Biology* 83:157–167.
64. Shear, C. B. 1975. Calcium-related disorders of fruits and vegetables. *HortScience* 10(4):361–365.
65. Soczek, Z. 1977. The growth and bearing of apple trees as affected by SADH and CCC applied for several years. *Acta Horticulturae* 80:437–466.
66. Stahly, E. A. and Benson, N. R. 1976. Calcium levels of Golden Delicious apples as influenced by calcium sprays 2,3,5-triiodobenzoic acid, and other growth regulator sprays. *Journal of the American Society of Horticultural Science* 101:120–122.
67. Stahly, E.A. and Benson, N. R. 1982. Seasonal accumulation of calcium and potassium in the cortex of Golden Delicious apple fruit sprayed with 2,3,5-triiodobenzoic acid. *HortScience* 17(5):781–783.
68. Steffens, G. L., Wang, S. Y., Faust, M., and Byun, J. K. 1985. Growth, carbohydrate and mineral element status of shoot and spur leaves and fruit of Spartan apple trees treated with paclobutrazol. *Journal of the American Society of Horticultural Science* 110:850–855.
69. van Stuivenberg, J. H. M. and Pouwer, A. 1950. Ondrzoek over de bestrijding van stip bij Notarisappols. *Meded. Dir. Tuinb.* 13:201–211.
70. Terblanche, J. H., Gurgen, K. H., Hesebeck, I. 1980. In *Mineral Nutrition of Fruit Trees*, eds. W. D. Atkinson, J. E. Jackson, R. O. Sharples, and W. M. Waller. London, UK: Butterworth, 71–82.
71. Tretyn, A. 1992. Elektrofizjologiczne badania nad rolą jonów wapnia w mechaniźmie działania auksyny w komórkach włośnikowych korzeni gorczycy (*Sinapis alba* L.) *Materiały Zjazdu*, Mechanizmy regulacji morfogenezy roślin. Rogów, Polska, June 15–16, 1992:95–98.
72. Tromp, J. 1989. Interrelationships between plant growth regulators and fruit tree mineral nutrition. *Acta Horticulturae* 239:409–416.
73. Voltz, R. K., Ferguson, I. B., Hewett, E. W., and Woolley, D. J. 1994. Wood age and leaf area influence fruit size and mineral composition of apple fruit. *Journal of Horticultural Science* 69(2):385–395.
74. Wertheim, S. J. 1972. De invloed van groeiregulatoren op de houdbaarheid van appelen peer. *Fruitteelt* 62:930–931.
75. Wertheim, S. J. 1997. Chemical thinning of deciduous fruit trees. *Acta Horticulturae* 463:445–462.
76. Williams, M. W. and Edgerton, L. J. 1989. Fruit thinning of apples and pears with chemicals. *U.S.D.A. Agricultural Information Bulletin 289*.
77. Wills, R. B. H. and Scott, K. J. 1974. Effect of phorone and other growth regulators on the incidence of storage breakdown in apples. *Journal of Horticultural Science* 49:199–202.

CHAPTER 16
Supercritical Fluid Extraction Applications for Biosystems Engineering

Terry H. Walker
Biosystems Engineering
Clemson University, Clemson, South Carolina

Meidui Dong
Abengoa Bioenergy, St. Louis, Missouri

Paresh Patel
Arkansas State University, Jonesboro, Arkansas

16.1 Introduction

There are growing concerns of health and environment problems associated with the use of traditional solvent extraction techniques employed in consumer products. Carbon dioxide is a potential extraction solvent alternative to the traditional organic solvents. Supercritical CO_2 has advantages of low cost, nontoxicity, high diffusivities with appreciable solubility, and low viscosity. Separating the solvent from the extract is easily accomplished by reducing pressure and returning CO_2 to a gaseous state. The solvent can be recycled for further use.

Supercritical fluid extraction (SFE) processes use fluids that has appreciable solubility near the sub- or supercritical point to extract selected components by regulating the temperature, pressure, or flow rate. Supercritical CO_2 is the most commonly used fluid due to its low critical properties, 31.1°C, 7.38 MPa, 0.468 g/cm^3 and other advantages. Physical properties of solvents considered for supercritical fluid extraction and other applications are shown in Table 16.1. Because supercritical CO_2 extraction has low solvent temperatures compared to conventional organic solvent extraction or steam distillation extraction, degradation of some thermolabile active compounds may be avoided. SFE has become an acceptable extraction technique used in many areas, far beyond the well-known decaffeination of tea and coffee and extraction of hops for beer flavoring. SFE of active natural products from herbal or more generally, from plant materials has become an important application area (McHugh and Krukonis 1994). Additionally, SFE technology using carbon dioxide as a solvent is generally recognized as safe, or "green," which is accepted both in Europe and the United States. Thermodynamic

Fluid	Molecular Weight (g/mole)	Boiling Point (°C)	Critical Constants Pressure (MPa)	Temperature (°C)	Density (g/cm^3)
Ammonia	17.03	−33.4	11.28	132.5	0.240
Butane	58.12	−0.49	3.80	152.0	0.230
Carbon dioxide	44.01	−78.5	7.38	31.1	0.468
Ethane	30.07	−88.6	4.88	32.2	0.207
Ethylene	28.05	−103.7	5.04	9.3	0.200
Ethanol	46.07	78.4	6.38	243.0	0.284
Propane	44.10	−44.5	4.25	96.7	0.220
Propylene	42.08	−47.7	4.62	91.9	0.230
Methane	16.04	−161.5	4.59	−82.6	0.163
Methanol	32.04	64.5	8.10	239.5	0.280
Nitrous oxide	44.01	−88.5	7.10	36.5	0.457
Toluene	92.1	110.0	4.11	318.6	0.290
Trichlorofluoromethane	137.40	23.8	4.41	196.6	0.554
Water	18.01	100.0	22.05	374.2	0.272

Source: Klesper (1980).

TABLE 16.1 Physical Properties of Common Supercritical Solvents

Supercritical Fluid Extraction Applications 449

Carbon Dioxide: Temperature-Pressure Diagram

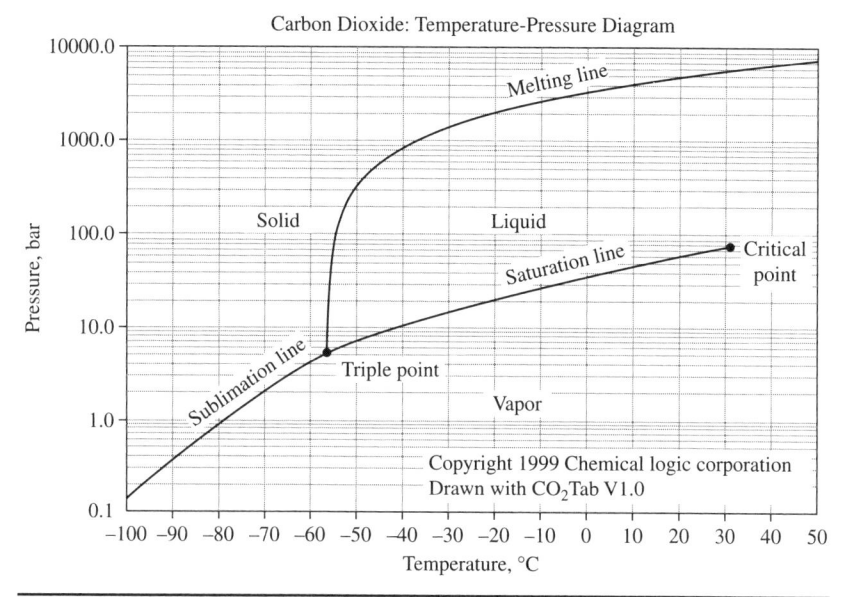

FIGURE 16.1 PT diagram of carbon dioxide. The supercritical region exists at all pressures and temperatures above the critical point. (*source*: http://www.chemicalogic.com/download/co2_phase_diagram.pdf)

properties of carbon dioxide are shown in the pressure-temperature phase diagram (Fig. 16.1).

Supercritical CO_2 extraction of vegetable oil has been studied from a processing point of view, and a wide range of seed species has been explored, including cottonseed, soybean, peanut, and rapeseed (Reverchon et al. 2000). Fattori et al. (1988) evaluated the feasibility and merits of supercritical CO_2 extraction as an alternative to hexane extraction in the canola oilseed industry. The experiment was conducted at temperatures ranging from 25 to 90°C and pressures ranging from 10 to 36 MPa. The oil solubility in CO_2 was found to be strongly dependent on pressure and weakly dependent on temperature. The highest observed oil solubility was 11 mg/g CO_2 and occurred at 36 MPa and 55°C. The amount of oil extractable from flaked and cooked seeds was comparable to that obtained by conventional hexane extraction. Przybylski et al. (1998) analyzed the composition and oxidative stability of oil fractions collected during the process of supercritical CO_2 extraction at 40°C and 41.4 MPa. The results showed that linolenic acid composition decreased, and the amounts of phospholipid increased as the CO_2 volume increased. The fraction obtained at the end of extraction contained higher amounts of unsaponifiables and phospholipids and showed better oxidative stability.

Supercritical CO_2 extraction of lipids from lipid-bearing fungal or algal biomass has also been studied (Walker et al. 1999; Badal and

Walker 2002; Andrich et al. 2005; Cantrell and Walker 2006). Badal and Walker (2002) conducted the SFE of raw and fermented rice bran for value-added oil. This study focused on determining the effect of particle size (>48 mesh size and 16 to 48 mesh size) and biotreatment on the yield and quality of rice bran oil. The results showed that eicosapentaenoicacid (EPA) and arachidonic acid (ARA) produced during biotreatment were extracted by SFE at 40°C and 4000 psi.

16.2 Supercritical Mass-Transfer Mechanisms

The SFE process involves removing solutes from porous matrices into the supercritical CO_2 via internal and external mass-transfer mechanisms. Generally, the extraction process consists of three periods. The first, the constant extraction rate period or the solubility-controlled period, is governed by the solubility equilibrium between the CO_2 solvent and the extract. The transition period starts when diffusion-controlled mass transfer occurs and extraction rate decreases. Finally, diffusion completely controls the mass-transfer process, and the extraction is said to be in a diffusion-controlled period. Diffusion of the extracts through the bulk material becomes a more integral part of the extraction process; over time, product accumulation approaches zero (Reverchon et al. 2000; Sovova 2005). An extensive discussion on modeling techniques may be found in McHugh and Krukonis (1994) and Walker et al. (2007).

The mass-transfer rate of high initial concentration of extracts (such as oilseeds) from a fixed bed containing natural material typically remains constant and then declines. Supercritical fluid extraction involves controlling the solubility by manipulating temperature and pressure. Natural materials contain multiple components making solubility and extractability difficult to predict (Lira 1996). Various mathematical aspects related to SFE of lipids were described by King and List (1996) that include solubility, phase equilibria, mass transfer, fractionation, and modeling.

The Goto et al. (1993) model for extraction of essential oil from peppermint leaves has been successfully applied to many biological materials. Solute is assumed to be extracted after being desorbed from a porous biological solid substrate. During the process, diffusion occurs inside its pores and the film surrounding the particle offers mass-transfer resistance. This model was also applied to fatty acids and carotenoid extractions from microalgae *Spirulina maxima* by Canela et al. (2002). In their application, the mass of extract at the bed outlet was described by the following equation:

$$m(t) = \left[\frac{\varepsilon_p}{K} + (1-\varepsilon_p)\right]\frac{\varepsilon_p}{K} + (1-\varepsilon_p)X_0\rho_s Q_{CO_2} A\tau \left\{\frac{1}{a_1}\left[\exp\left(a_1\frac{t}{\tau}\right) - 1\right]\right.$$

$$\left. + \frac{1}{a_2}\left[1 - \exp\left(a_2\frac{t}{\tau}\right)\right]\right\} \quad (16.1)$$

where A is a constant defined by the following equation:

$$A = \frac{\phi}{[\varepsilon_p + (1-\varepsilon_p)K](a_1 - a_2)}\left(\frac{1-\varepsilon_1}{\varepsilon_1}\right) \quad (16.2)$$

where

$$a_1 = \frac{1}{2}(-b + \sqrt{b^2 - 4c}) \quad (16.3)$$

$$a_1 = \frac{1}{2}(-b - \sqrt{b^2 - 4c}) \quad (16.4)$$

$$a_1 = \frac{1}{2}(-b - \sqrt{b^2 - 4c})$$

$$b = \frac{\phi}{\varepsilon_p + (1-\varepsilon_p)K} + \frac{1}{\varepsilon_1} + \frac{\phi(1-\varepsilon_1)}{\varepsilon_1} \quad (16.5)$$

$$c = \frac{\phi}{[\varepsilon_p + (1-\varepsilon_p)K]\varepsilon_1} \quad (16.6)$$

and

$$\phi = K_p A_p \tau \quad (16.7)$$

where ε_1 = interstitial bed porosity
ε_p = particle porosity
K_p = combined mass-transfer coefficient
A_p = specific surface area
K = particle coefficient of the solvent in the solute
X_0 = initial solute mass ratio in the solid phase
ρ_s = solid true density
Q_{CO_2} = volumetric solvent flow rate
t = extraction time
τ = CO_2 residence time

16.3 Carbon Dioxide Explosion Processes

High-pressure CO_2 explosion is a process of rapid release of pressurized CO_2 when an exit valve is quickly opened. Different names of similar processes have been used in the literature, such as explosive depressurization, rapid depressurization, explosive decompression, flash decompression, flash discharge, blowdown, and the like. This process was developed for chemical and biological applications to treat a sample in a high-pressure vessel. For example, this process was investigated for sterilization (Spilimbergo et al. 2002; White et al. 2006; Zhang et al. 2006), activation, and denaturation of enzymes

(Dunford and Temelli 1996; Bauer et al. 2000), disruption of microbial cells to aid extraction (Castor and Hong 1995), and pretreatment of lignocellulosic substrates (Zheng et al. 1998; Kim and Hong 2001). The mechanisms for these applications involve mechanical cell rupture, chemical and biochemical modification of structure of the treated sample and their synergistic effects, giving several advantages of high-pressure carbon dioxide (see Fig. 16.2).

The mechanism of cell disruption by pressure gradient for the application of explosion process was proposed and supported by x-ray and nuclear magnetic resonance examination (Zheng et al. 1998) and by scanning electron microscopy (Gaspar et al. 2001).

To extend the solubility-controlled period and improve the later diffusion-controlled period, some technologies were developed to improve the extraction by disrupting the cell wall (e.g., high-pressure homogenizer), agitating the biomass with glass beads, and even ultrasonification. A high-pressure CO_2 explosion could take some physical and chemical advantages of supercritical CO_2, comparing it to the high operation temperature of a steam explosion. Pretreatment using supercritical CO_2 explosion on pure cellulose and industrially processed

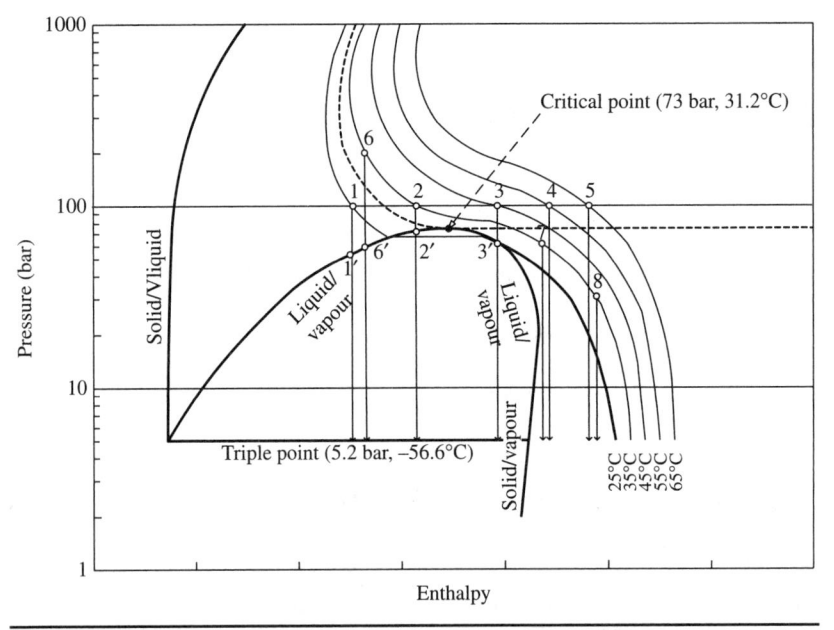

FIGURE 16.2 Schematic diagram of the pressure-enthalpy of carbon dioxide (referenced from http://www.chemicalogic.com/download/phase_diagram.html, last updated 1999). Points corresponding to each initial pressure and temperature are shown by the open circle. A thin solid line indicates the temperature. The critical point is shown as a filled black circle. The region above the dashed line indicates the supercritical fluid. Thick solid lines indicated the solid/liquid, solid/gas, liquid/gas, and triple point, respectively.

materials improved glucose yield by as much as 50 percent (Zheng et al. 1998). Explosion effects on pine and aspen resulted in a glucose yield 12 to 14 percent beyond untreated materials (Kim and Hong 2001). Gaspar et al. (2001) studied the disruption of essential oils glandular trichomes by contact with compressed CO_2 followed by rapid decompression at isothermal conditions, 37°C. The efficiency of the disruption process was deduced from the results of subsequent extraction tests using compressed CO_2 under standard conditions. They found that pre- and postexpansion pressures, exposure time, and rate of decompression have a significant effect on the efficiency of the disruption process. Gaspar et al. (2003) proposed a theoretical model for evaluating the disruption efficiency of essential oils glandular trichomes (glands) with compressed CO_2. The glands were described as closed structures slightly permeable to CO_2. The gas slowly penetrated the glands and dissolves in the intraglandular oil until the solubility limit was reached. During fast decompression of the bed, the dissolved gas was desorbed from the oil phase and discharged to the bulk solvent. The inability of the glands to discharge gas generated a pressure gradient across the glands that might lead to its rupture. In the model, the excess pressure was described by an equation similar to Hagen–Poiseuilles formula for viscous flow due to a pressure gradient. Cantrell and Walker (2006) investigated the integration of current SFE technology with an explosion process and applied these techniques to microbial metabolite recovery. These two downstream processing techniques could work consecutively without the need to transfer the materials from one unit to another. The effect of explosion pretreatment on the biomass was evaluated by the extractability of polyunsaturated fatty acids (PUFAs). The results showed the pretreatment with a 60-min saturation time did result in a statistically significant increase in extraction yield over the control. The explosion pretreatment at 10.3 MPa and 60 min provided the highest oil extraction yield of the four treatments, whereas treatment at 10.3 MPa and 20 min showed no statistical enhancement of SFE over the control.

16.4 Biological Systems Applications

An initial high capital cost of supercritical extraction equipment often results in uses limited to high-value food or pharmaceutical components where savings in operating costs are pursued and capital costs are offset by the value of the product. Using microbial fermentation pathways to yield high-value products has shown great promise in the past century with a remarkable list of bioactive compounds produced efficiently from fermentation mechanisms. Supercritical carbon dioxide is directly suitable for extraction and fractionation of edible oil compounds. The commercial production of fungal oil containing high ARA is an example of a successful product suited for this extraction technique.

16.4.1 Fungal Production of Extractable Oils

Single-cell oil production was commercially explored in the 1980s to produce cocoa butter substitutes, as cocoa butter was in short supply (Gunstone 2001). Few microbial oils have been commercialized because of the high production cost as compared with the low cost of conventional edible oil extracted from natural plant sources. With research on physiological functions of lipids, the production of microbial lipids has once again become attractive. Among the known microbial lipids, PUFAs have attracted great interest because of their valuable functions. ARA exists widely in the animal kingdom and has been isolated from lipids extracted from the adrenal gland and the liver of animals. However, isolation from these organs with small amounts is insufficient to meet the demand. Various microorganisms capable of producing ARA were investigated as summarized by Ward and Singh (2005). Among them, microorganisms belonging to the *Mortierella* genus have been extensively studied by many groups, particularly, the *alpina* species, which show a high potential for ARA production. *Mortierella alpina* is an attractive source for production of ARA because 60 percent of ARA in the total lipids predominates the content of long-chain PUFAs, with only traces of EPA and no DHA (docosahexaenoic acid) being produced. This strain is currently regarded as the most effective industrial producer of ARA (Ward and Singh 2005; Barclay 2007). Amano et al. (1992) analyzed the fatty acid composition of 50 *Mortierella* subgenus isolates. They found that ARA composition of *M. alpina* was higher than those of any other species. ARA was present in all major lipid structures, and a high concentration was detected in phosphatidylcholine (PC) and phosphatidylethanolamine (PE). However, triacylglycerols (TAGs) are the main lipid structure in the fungal intracellular lipid and their content in total lipids was 60 to 90 percent (Certik and Shimizu 2000).

Extensive research and patents on the production of ARA by the fungus *M. alpina* were carried out over the past decade (Willis et al. 1998; Ward and Singh 2005; Barclay 2007). Research on microbial PUFA production was basically aimed at improving the economic competitiveness of microbial lipids compared to plant- and animal-derived lipids. Emphasis was placed on screening for more efficient strains, increasing product value, optimizing culture conditions, using inexpensive substrates, and reducing the processing steps necessary for lipid recovery from the cells (Certik and Shimizu 2000; Zhu et al. 2003; Ratledge 2005; Streekstra 2005).

Effect of Carbon and Nitrogen Sources

To produce ARA commercially, a high biomass concentration is required for high productivity because ARA is stored primarily as an intracellular product. Moreover, ARA formation requires adequate oxygen because the enzymatic desaturation process requires oxygen

(Ratledge 1992). In cultures of *Mortierella* for ARA production, glucose and starch are most frequently used as the primary carbon source. In terms of nitrogen sources, yeast extract (YE) and soybean meal (SM) are widely used. Bajpai et al. (1991) studied the effect of carbon sources in the medium on the growth of fungi, *M. alpina* ATCC 32222. This medium was prepared according to the procedure of Hansson and Dostalek (1988). Glycerol, glucose, fructose, xylose, maltose, sucrose, starch, olive oil, and linseed oil were compared. Fungal growth was poor in the medium containing xylose or sucrose. With other carbon sources, biomass production ranged from 16.9 g/L for dextrin to 28.2 g/L for linseed oil. However, ARA content of biomass and content of lipids in biomass was very low when linseed oil was the carbon source. ARA content of biomass produced with fructose, maltose, glycerol, and glucose ranged from 9.6 to 10.8 percent (w/w) with ARA accounting for 44.7 to 54.1 percent of the total fatty acids. Jang et al. (2005) also investigated various carbon sources, including soluble starch, glucose, glycerol, galactose, and maltose, as well as nitrogen sources including KNO_3, $NaNO_3$, NH_4NO_3, $(NH_4)_2SO_4$, urea, YE, and their combinations, and found that soluble starch at 10 percent and the mixture of KNO_3 and yeast extract 2:1 (w/w) was the best carbon and nitrogen combination for ARA and total PUFA production. Totani et al. (1992) reported the effect of glucose concentration on ARA production, and showed that more than 20 percent glucose inhibited the growth of *M. alpina*. Totani et al. (2000) also reported that when peptone and wheat bran were used as nitrogen sources, a high ARA productivity of 11 g/L was obtained. Only a small amount of biomass was obtained in the medium comprising NH_4NO_3, $NaNO_3$, KNO_3, or $(NH_4)_2SO_4$ as nitrogen sources, suggesting that *Mortierella* fungi hardly assimilate inorganic nitrogen source, and require an amino acid or a protein to obtain a certain amount of biomass. Singh and Ward (1997) also reported when $NaNO_3$ was replaced with corn steep liquor, ARA yield was significantly improved.

Lipid production was one of a few biotechnological processes that proceeded quite well in a simple batch culture. In this system, the carbon-to-nitrogen (C/N) ratio was an important parameter (Streekstra 2005). Jang et al. (2005) conducted an elemental analysis of *M. alpina* ATCC 32222, with a C/N ratio of 4.9, and reported that a C/N ratio of 9.0 gave a high cell yield, whereas a C/N ratio of 5.2 gave maximal ARA and total PUFAs production. However, an optimal C/N ratio of 15 to 20 for ARA production was reported by Koike et al. (2001). Sajbidor et al. (1990) reported that *Mortierella* sp. had high ARA production at a C/N ratio of 20. The difference might result from different strains or culture conditions.

The ratio of C/N also affects the morphology of fungi, which affects the whole fermentation process, including the productivity of targeted metabolites and product recovery/purification in downstream processes (Gibbs et al. 2000). Koike et al. (2001) reported the effect of C/N on ARA production and mycelial morphology in the culture of *M. alpina*. The optimal C/N ratio of the medium was 15:20 for ARA production with a balance between carbon and nitrogen sources. When an enriched medium was used at a fixed C/N ratio of 20, the cellular and ARA concentration were proportional to the total concentration of carbon and nitrogen sources. The whole pellet size did not change with increasing C/N ratio when the ratio was below 20. When the C/N ratio was higher than 20, pellet size increased in proportion to the C/N ratio.

Generally, morphology of the fungi will affect the fermentation (mycelia growth and metabolite formation) by changing the viscosity (or rheology) of broth. Filamentous growth of fungi results in highly viscous broth with a non-Newtonian, pseudoplastic flow behavior (Papagianni 2004). Pelleted growth exhibited low viscosity and approached Newtonian flow behavior, thus pellet morphology allowed easier mixing and better mass transfer to culture broth (Hamanaka et al. 2001). Although pellet-fermentation broths might be Newtonian and have low viscosity, problems could arise with the transport of nutrients inside the pellets, thus reducing productivity. Therefore, dispersed forms of the pellets predominate in most industrial fermentations (Riley et al. 2000). The morphology of fungi was found to affect the productivity of ARA, with small pellets (1 to 2 mm in diameter) showing the highest productivity, indicating that the pellet form was more suitable, although a filamentous morphology would be more suitable for ARA production from the viewpoint of oxygen and mass transfer (Higashiyama et al. 1999b). From the cultures in a shaker and jar fermenter, featherlike morphology with high concentration of small-size pellets was found to be the most suitable for ARA production (Park et al. 1999).

The change of the rheology of broth will directly or indirectly affect other factors that synergistically determine the fermentation process, for example, mass- and energy-transfer properties, biomass concentration, mixing and aeration, nutrition and oxygen consumption, temperature, pH, cell differentiation and growth, metabolite formation, and specific growth rate. In turn, morphology of fungi will be affected by the viscosity of broth, mutually together with other factors mentioned above. Gibbs et al. (2000) reviewed the problems associated with mycelial morphology of fungal fermentation, including formation of heterogeneous and stagnant zones, nutrients and oxygen gradient and limitation, and inaccurate temperature control and locally excessive heat. These problems may result in cellular stress, causing strain degradation or mutation, especially on a larger scale.

The effect on nutrient and oxygen consumption is not only affected by rheology, mixing, or aeration of the media but also by the different morphology of the biomass. Oxygen is probably the most important factor physiologically, because oxygen limitation may result in the suppression or even total inhibition of metabolite production of lipids in aerobic fungi. PUFAs are formed through elongation and desaturation. Desaturation involves an aerobic reaction by oxygenation. Therefore, oxygen availability is important to synthesize ARA, an n-6 PUFA with four double bonds, in *M. alpina*. Cellular growth of fungi was restrained due to substrate limitation in the region of the dense pellet core when the pellet radius exceeded a critical value. The lipid was produced on the edge of the pellet, where mycelial density was high (Hamanaka et al. 2001). Higashiyama et al. (1999a) studied the effects of dissolved oxygen (DO) on the morphology of *M. alpina* in a 50-L fermentor. When DO concentration was maintained at 20 to 50 ppm using the oxygen-enrichment method, the morphology changed from filaments to pellets, and ARA yield decreased drastically because of stress due to the limited mass transfer through the pellet wall. Studies of fungal morphology and metabolite production in submerged mycelial processes over the past 50 years were reviewed by Papagianni (2004). The content included growth mechanisms of filamentous fungi, dynamics of mycelial aggregation, influence of process parameters on the morphology, productivity and rheology, and some related models.

Optimum morphology for production usually varies among fungi and products. There is still no obvious or unequivocal conclusion to tell which is better for ARA production by *M. alpina*, the filamentous or the pellet form. ARA is an intracellular primary metabolite that is accumulated highly in active and mature cells between 3 to 7 days. Small and compacted pellets consist of oil-rich cells. Furthermore, small pellets reduce nutrients and oxygen limitation and autolysis of cells in the core part. Small pellets also have better rheological properties than filamentous mycelia or large pellets, which may reduce the problems associated with filamentous growth. Also, small pellet formation has a relatively suitable fluid property for transportation in downstream processing, such as biomass centrifuge recovery (preventing compacted cake forming), spraying, drying, and extraction.

Effect of Oil Supplement to the Culture Medium

Various studies have indicated that vegetable oil addition was beneficial to enhance ARA yield (Shinmen et al. 1989; Singh and Ward 1997; Jang et al. 2005) in *M. alpina* culture. The addition of corn, soybean, peanut, and canola oils (1 percent) to the medium stimulated biomass and lipid production. Cultures supplemented with corn and canola oil exhibited the maximum ARA yield and was 50 percent higher than that without any supplementation (Singh and Ward 1997). Jang et al. (2005) also investigated the effect of added oil. Supplementation with 1.0 percent of linseed oil gave the highest biomass, followed by sunflower oil and soybean oil. Linseed oil gave a high ARA production, followed by soybean oil and sunflower oil. The effect of the supplemented oil concentration was also studied using soybean oil. Biomass increased with the concentration of soybean oil and had a plateau at 1.0 percent. ARA yield also increased and had a maximal value of 1.0 percent of soybean oil.

Linseed oil was often supplemented for EPA production because of its high percentage of α-linolenic acid. Shimizu et al. (1989) investigated the conversion of α-linolenic acid to EPA and suggested two metabolic routes (n-3 and n-6 routes) for synthesizing PUFA. In *Mortierella* species, C18:2 was desaturated to form C18:3 and then was elongated forming C20:3, and finally desaturated to ARA. Adding vegetable oil to the basal media significantly enhanced ARA yield, where some of the fatty acid supplement (C18:1, C18:2, or C18:3 as the major fatty acid) was used as a precursor for ARA synthesis. Bajpai et al. (1991) investigated the direct use of olive oil and linseed oil as carbon sources for ARA production. Both linseed oil and olive oil had a higher biomass production and lipid content of the biomass than other carbon sources (e.g., glucose, starch, and glycerol). However, linseed oil had a low ARA content of biomass and total ARA yield. This result is different from that when linseed oil was used as a supplement (Jang et al. 2005). Olive oil had a comparable ARA content of biomass and total ARA yield to glucose and ARA yield higher than starch. EPA content of biomass and yield was not reported in this research. High percentage of α-linolenic acid of linseed oil might promote EPA formation via the n-3 pathway, but this may compete with ARA formation because of the shared enzyme system for both ARA and EPA formation routes.

Waste Material as a Fermentation Substrate

The demand of reducing processing costs, coupled with environmental pressure caused by the release of agricultural/industrial by-products and waste, has resulted in the need for using by-product/waste as fermentation feedstock. Bioprocessing of these materials shows a promising potential for effectively using agricultural and food-processing wastes and by-products.

Zhu and Walker (2001) investigated the feasibility of usingrice bran as a fermentation substrate for *Pythiumirregulare* to produce EPA and ARA. This study showed the feasibility of rice bran use in submerged fermentation to produce lipids containing target PUFAs. Rice bran media was competitive with glucose–yeast extract media for EPA and ARA production at low cost as a potential feedstock. In the 5 percent (w/v) rice bran medium, which contained about 2 percent carbohydrates with an estimated C/N ratio of 10, after 7 days incubation, lipid yield increased nearly 10 percent compared with that in the initial rice bran. Composition of the fatty acid fraction was improved with the addition of EPA and ARA synthesized by the fungus. Maximum EPA and ARA yields reached 207 mg/L and 70 mg/L, respectively, after 7 days' incubation. Bioconversion percentages from rice bran to EPA and ARA were 0.41 percent and 0.14 percent of total rice bran, respectively, where the bioconversion percentage was defined as fatty acid yield (g/L) divided by rice bran yield in media (50 g/L). Lindberg and Hansson (1991) examined the use of rapeseed meal and beet molasses as substrates on γ-linolenic acid production. Chaudhuri et al. (1998) determined that 0.5 percent deoiled mustard meal was a good substrate for producing ARA-rich oil. Cheng et al. (2006) investigated the fungal production of ARA and EPA using industrial by-product streams including sucrose waste stream (SWS) and soymeal waste stream (SMW), and crude soybean oil (SBO). The strains *Pythium irregulare* and *M. elonga* were used for comparison. The greatest EPA yield was 1.4 g/l in the medium composed of 4 percent SBO and 1 percent SMW at 12°C and that of ARA was 2 g/L in the media composed of 4 percent SBO by *P. irregulare*. Zhu et al. (2003) studied the production of ARA in the fungus *M. alpina* using an inexpensive medium. Glucose derived from maize starch hydrolysate was the sole carbon, and defatted soybean meal and sodium nitrate were the nitrogen sources. The results showed that a mixture of soybean alkali–extract protein and sodium nitrate was an excellent nitrogen source for fungal growth, lipid accumulation, and ARA production. A maximum yield of 1.87 g/L ARA was obtained with a soybean protein concentration of 4.6 g/L and a sodium nitrate concentration of 2.3 g/L.

Genetic Engineering Technology in ARA Production

Several strains have been extensively studied for the practical production of ARA. These strains were bred using the conventional mutagenesis method, especially for creating desaturase and elongase mutants with unique pathways (Shimizu et al. 1989). In addition to studying desaturase and elongase mutants, research was conducted in the area of cloning specific desaturase genes from various *M. alpina* strains and expressing them in a number of other organisms. Some promising results showed that cloned desaturase genes could be used for transformation to other organisms, which were potential PUFA producers. These include cloning the desaturase gene from *M. alpina* and expressing it in microorganisms and plants (Knutzon et al. 1998; Michaelson et al. 1998; Chen et al. 2006).

Cloning Δ5 desaturase from *M. alpina* was expressed in *Saccharomy cescerevisiae*. The transformed yeast was able to accumulate ARA (Michaelson et al. 1998). Knutzon et al. (1998) reported the expression of the cloning of Δ5 desaturase from *M. alpina* in *S. cerevisiae* revealed that the recombinant product had Δ5-desaturase activity. Expression of the *M. alpina* Δ5-desaturase cDNA in transgenic canola seeds resulted in the production of taxoleic acid (Δ5, 9-18:2) and pinolenic acid (Δ5, 9, 12-18:3), which were the Δ5-desaturation products of oleic and linoleic acids, respectively. Chen et al. (2006) reported the production of ARA soybean seed by the specific expression of genes encoding Δ6 desaturase, fatty acid elongase, and Δ5 desaturase from *M. alpina*, as well as the down regulation of an endogenous gene of the soybean, the Δ15 desaturase gene. These results indicated that metabolic engineering of the fatty acid biosynthetic pathway to produce ARA in oilseed crops was possible.

16.4.2 Extracting Oil from Rice Bran

Rice bran, which includes the pericarp, the aleurone and the subaleurone layers and parts of the germ, embryo, and small portions of the starchy endosperm is a valuable by-product in the rice-milling industry. Bran, almost 10 percent the weight of rough rice, is rich in oil content ranging from 15 to 22 percent depending on the milling procedure used and the variety of the rice (Houston 1972; Randall et al, 1985; Martin 1994). Crude rice bran oil contains 68 to 71 percent triglycerides, 2 to 3 percent diglycerides, 5 to 6 percent monoglycerides, 2 to 3 percent free fatty acids, 2 to 3 percent waxes, 5 to 7 percent glycolipids, and 3 to 4 percent phospholipids with 4 percent of the phospholipids unsaponifiable (Saunders 1990; McCaskill and Zhang 1999). Rice bran oil contains oleic acid (38.4 percent), linoleic acid (34.4 percent), and linolenic acid (2.2 percent) as unsaturated fatty acids, and palmetic (21.5 percent) and stearic acids (2.9 percent) as saturated fatty acids (Rukmini and Raghuram 1991; Xu 1998). Rice bran is also a source of linoleic acid that is essential to human health (Ramezanzadeh et al. 2000).

The growing interest in rice bran oil is due, in part to its high unsaponifiable level (4.2 percent) compared to other vegetable oils (Orthpoefer 1996; Lloyd et al. 2000; Dunford 2001). Rice bran is a rich source of vitamin E (~300 mg/kg or 0.1 to 0.14 percent) and has high concentration of oryzanols (~3000 mg/kg or 0.9 to 2.9 percent) (Xu and Godber 1999; Lloyd et al. 2000). Vitamin E consists of components of tocopherols (α, β, γ, and δ) and tocotrienols (α, β, and γ). Oryazanols, tocopherols, and tocotrienols are the antioxidant compounds that improve stability and frying quality in rice bran oil (Sonntag 1979; Xu 1998; Lloyd et al. 2000). Fat-soluble antioxidant compounds of vitamin E protect cell membranes by blocking the oxidation of the unsaturated fatty acids and acting as a scavenger of free radicals (Komiyama et al. 1992; Nesartetnam et al. 1998). Gamma oryzanol is reported to reduce cholesterol absorption (Hu 1995; Rong et al. 1997). Oryzanols also have hypolipidemic effects; promote growth, gonadotrophoic action, and hypothalamus stimulation effects (Sugano and Tsuji 1997).

Oryzanols are the mixture of ferulate (4-hydroxo-3-methoxycinnamic acid) esters of sterols (campesterol, stigmasterol, and β-stigmasterol) and triterpene alcohols (cycloartenol, cyccloartenol,

24-methylenecycloartanol, cyclobranol). Major portions of γ-oryzanol are cycloartenylferulate, 24–methylenecycloartanylferulate and campesterylferulate. γ-Oryzanol makes up 1.5 to 2.9 percent of rice bran oil and is a white or yellowish tasteless powder with little or no odor (Hu 1995; Xu 1998; Xu and Godber 1999). Vitamin E, which is a mixture of tocopherols and tocotrienols, is a pale, yellow, and viscous oil (Budavari et al. 1989; Hu 1995). Tocopherols and tocotrienols differ in the number and positions of methyl groups on the fused chromonol ring, and the absence and presence of three double bonds in the isoprenoid side chain (Hua 2000). Major forms of tocopherols in rice bran oil are 5,7,8-trimethyltocol (α-tocopherol), 7,8-dimethyltocol (γ-tocopherol), and 8-methyltocol (δ-tocopherol). Similarly major tocotrienol forms are 5,7,8-trimethyltocotrienols (α-tocotrienol), 7,8-dimethyltocotrienol (γ-tocotrienol), and 8-methyltocotrienol (δ-tocotrienol) (Hu 1995; Xu 1998; Xu and Godber 1999; Hua 2000).

Solvent extraction is the conventional method for lipid recovery from rice bran that uses toxic and flammable solvents like hexanes, petroleum ether, and isopropanol, among others. Proper disposal, toxic residue in final food products and environmental regulations represent key problems associated with the use of these solvents. These issues have prompted scientists to search for alternative, non-hazardous extraction techniques. SFE is a prominent alternative technique, which promises to meet a growing demand for natural, green and organic extracts from food and biological materials. McHugh and Krukonis (1994) have given a detailed historical perspective of the developments related to supercritical fluids. SFE, apart from overcoming problems associated with conventional solvent extraction, also offers additional advantages of the selective extraction and fractionation capability for high-value components in the extract at optimized extraction conditions. Carbon dioxide is an appropriate SFE solvent for biological materials like rice bran because of the possible thermal degradation of important minor components at higher temperatures (Rizvi 1994; Williams and Clifford 2000).

Supercritical Extraction of Rice Bran Oil

SFE in food, pharmaceutical, nutraceutical, and other natural and biological products has received significant attention. Several recent detailed studies, reviews, and books have reported that SFE of various biological, food, and natural products (Canela et al. 2002; Rozzi and Singh 2002; Prieto et al. 2003). Lipids are an important part of the food system. Some of the recent studies related to application of SFE of lipids from natural products include extraction from lavender (Reverchon and Porta 1995), ginger oil (Roy et al. 1996), Turkish mint plant leaves (Ozer et al. 1996), corn (Moreau et al. 1996) soybeans (Montanari et al. 1996), spearmint oil from mint plant leaves

(Ozer et al. 1996), cloudberry seed oil (Manninen et al. 1997), sunflower oil (Perrut et al. 1997), almond oil (Marrone et al. 1998), pistachio nut lipids (Palazoglu and Balaban 1998), rapeseed, sunflower and soybean (Bjergegaard et al. 1999), corn bran (Taylor and King 2000), grapeseed oil (Lee et al. 2000), hiprose seed oil (Reverchon et al. 2000), lavender essential oils and waxes (Akgun et al. 2000), and Romanian mentha hybrids oil (Eugenia and Danielle 2001).

Rice bran lipid extraction with supercritical fluid is also reported. Figure 16.3 shows a pilot-scale SFE extraction system. Zhao et al. (1987) studied fractional extraction of rice bran oil at 40°C and obtained an oil yield in the range of 18.6 to 22.0 percent with pressure variations ranging from 14.7 to 34.3 MPa. Fractions obtained with SFE contained 8.8 percent free fatty acids compared to 11.9 percent for hexane-extracted oil. Ramsey et al. (1991) compared rice bran oil extraction with hexane, SFE (29.99 MPa, 35°C, 5 h, 20.5 g/min) and SFE with cosolvent (29.99 MPa, 35°C, 5 h, 20.5 g/min, 5 percent ethanol) extraction processes. He was able to recover 20.2, 18.0, and 18.2 percent oil yield; sterol yields were 9.4, 7.3, and 8.3 mg/g of rice bran for hexane, SFE, and SFE–cosolvent extractions, respectively. Garcia et al. (1996) recovered 16 to 60 percent of solvent extractable oil from rice bran at 28 MPa and 70°C. Kuk and Dowd (1998) reported 19.2 to 20.4 percent RBO yield in SFE extraction (48.26 to 62.05 MPa) compared to 20.5 percent in hexane extraction. Xu and Godber (1999) compared solvent extraction (50 percent hexane and 50 percent isopropanol v/v) of rice bran with supercritical carbon dioxide extraction (50°C and 68.9 MPa) for γ-oryzanol fractionation. Their study

Figure 16.3 Pilot-scale SFE system for extraction of rice bran oil (Patel 2005).

suggested that SFE extracts up to 4 times greater γ-oryzanol (5.39 mg/g of rice bran) compared to solvent extraction.

Dunford and King (2000) studied enrichment of rice bran oil (20.5 to 32.0 MPa and 45 to 80°C) to reduce free fatty acids and minimize the loss of phytosterols and found that low-pressure high-temperature combinations are better for reducing loss of triglycerides and phytosterols during removal of free fatty acids from crude rice bran oil. Badal and Walker (2002) in their rice bran lipid study with supercritical carbon dioxide (40°C, 27.58 MPa) found that oil yield was a function of particle size during SFE. Badal and Walker's yield was 51.5 percent of the total ether extractable oil in 2 h from small particles (16 to 48 mesh) compared to 41.2 percent extracted from larger particle size (>48 mesh) rice bran.

16.5 Summary and Future Trends

Supercritical fluids were discovered in the 1800s, but have only been applied since the 1960s as a means for separations and analytical techniques. Applications toward extraction of caffeine from coffee and hops extraction were initial commercial products. Since then, applications ranging from high-value oil extractions, fractionation of pharmaceuticals, sterilization of medical equipments, explosion pretreatment of cellulosic and oil-based sources, production of biodiesel with supercritical methanol, and rapid expansion of supercritical fluid solutions (RESS) for bioactive component delivery mechanisms have been realized. The technology takes advantage of the broad thermodynamic nature of fluids to give unique process characteristics that allow for greater functionality or versatility than many conventional methods, thus, extending the potential operational capabilities inherent to biological systems processing. With greater emphasis on reduction of toxic waste streams, environmental impacts, carbon reduction, and use of GRAS solvents, supercritical fluids have a bright future for many new biological system applications.

References

Akgun M., Akgun N. A., and Dincer S. 2000. Extraction and modeling of lavender flower essential oil using supercritical carbon dioxide. *Ind. Eng. Chem. Res.* 39(2):473–477.

Amano, N., Shinmen, Y., Akinoto, K., Kawashima, H., Amachi, T., Shimizu, S., and Yamada, H. 1992. Chemotaxonomic significance of fatty acid composition in the genus *Mortierella*. *Mycotaxon* 94:257–265.

Andrich, G., Nesti, U., Venturi, F. Zinnai, A., and Fiorentini, R. 2005. Supercritical fluid extraction of bioactive lipids from the microalga *Nannochloropsis* sp. *European Journal of Lipid Science and Technology* 107(6):381–386.

Badal, R. and Walker, H. T. 2002. Supercritical carbon dioxide extraction of lipids from raw and bioconverted rice bran. M.S. thesis, Louisiana State University, Baton Rouge, LA.

Bajpai, P. K., Bajpai, P., and Ward, O. P. 1991. Production of arachidonic acid by *Mortierellaalpina* ATCC32222. *Journal of Industrial Microbiology* 8:179–186.

Barclay, R. W. 2007. Method for production of archidonic acid. U.S. patent no. 7,195,791.

Bauer, C., Steinberger, D., Schlauer, G., Gamse, T., and Marr, R. 2000. Activation and denaturation of hydrolases in dry and humid supercritical carbon dioxide. *Journal of Supercritical Fluids* 19:79–86.

Bjergegaard C., Buskov S., Sorensen H., Sorensen J. C., and Sorensen S. 1999. Supercritical fluid techniques as preparative and analytical tools in the analytical gray area between volatiles and hydrophilic compounds. *Proceedings of 10th international rapeseed congress*, Canberra Australia.

Budavari, S., O'Neil, M. J., Smith, A., and Heckelman, P. E. eds. 1989. *The Merck Index*, 11th ed. Rahway, NJ: Merck & Co. Inc.

Canela, A. P. R. F., Rosu, P. T. V., Maraques, M. O. M., and Meireles, M. A. A. 2002. Supercritical extraction of fatty acids and cartenoids from microalgae sporulina maxima. *Ind. Eng. Chem. Res.* 41:3012–3018.

Cantrell, K. B. and Walker, H. T. 2006. Integrated bioprocessing of *Pythiumirregulare* to obtain PUFA-rich oil. Ph.D. dissertation, Clemson University, Clemson, SC.

Castor, T. P. and Hong, G. T. 1995. Supercritical fluid disruption of and extraction from microbial cell. U.S.patent. 5,380,826, filed Sept. 29, 1992, issued Jan. 10, 1995.

Certik, M. and Shimizu, S. 2000. Kinetic analysis of oil biosynthesis by and arachidonic acid-producing fungus, *Mortierellaalpina* 1S–4. *Applied Microbiology and Biotechnology* 54:224–230.

Chaudhuri, S., Ghosh, S., Bhattacharyya, D. K., and Bandyopadhyay, S. 1998. Effect of mustard meal on the production by *Mortierellaalpina* isolate. *Journal of the American Oil Chemists'Society* 74:569–578.

Chen, R., Matsui, K. M., Oe, M. M. Ochiai, Kawashima, H. Sakuradani, E., et al. 2006. Expression of Δ6, Δ5 desaturase and GLELO elongase genes from *Mortierellaalpina* for production of arachidonic acidin soybean [*Glycine max*(L.) Merrill] seeds. *Plant Science* 170:399–406.

Dunford, N. T. and Temelli, F. 1996. Effect of supercritical CO_2 on myrosinase activity and glucosinolate degradation in canola. *Journal of Agricultural and Food Chemistry* 44:2372–2376.

Dunford, N. T. 2001. Health benefits and processing of lipid based nutritional. *Food Technology* 55(11):38–43.

Dunford, N. T. and King, J. W. 2000. Phytosterol enrichment of rice bran oil by a supercritical carbon dioxide fractionation technique. *J. Food Sci.* 65(8): 1395–99.

Eugenia, G. P. and Danielle, B. 2001. Supercritical fluid extraction of Z-sabinene hydrate rich essential oil from Romanian Mentha hybrids. *Pure Appl. Chem.* 73(8):1287–91.

Fattori, M., Bulley, N. R., and Meisen, A. 1988. Carbon dioxide extraction of canola seed: oil solubility and effect of seed treatment. *Journal of the American Oil Chemistry Society* 65(6):968–974.

Garcia, A., de Lucas, A., Rincón J., Alvarez, A., Gracia, I., and García M. A. 1996. Supercritical carbon dioxide extraction of fatty and waxy material from rice bran. *Journal of the American Oil Chemists' Society*. Springer, Berlin. 73(9):1127–1131.

Gaspar, F., Leeke, G. A., Al–Duri, B., and Santos, R. 2003. Modeling the disruption of essential oils glandular trichomes with compressed CO_2. *Journal of Supercritical Fluids* 25:233–245.

Gaspar, F., Santos, R., and King, M. B. 2001. Disruption of glandular trichomes with compressed CO_2: Alternative matrix pre–treatment for CO_2 extraction of essential oils. *Journal of Supercritical Fluids* 21:11–12.

Gibbs, P. A., Seviour, R. J., and Schmid, F. 2000. Growth of filamentous fungi in submerged culture: Problems and possible solutions. *Critical Reviews in Biotechnology* 20:17–48.

Goto, M., Sato, M., and Hirose, T. 1993. Extraction of peppermint oil by supercritical carbon dioxide. Journal of Chemical Engineering of Japan26, 401–407.

Gunstone, F. D. 2001. *Structure and Modified Lipids*. New York, NY: Marcel Dekker.

Hamanaka, T., Higashiyama, K., Fujikawa, S., and Park, E. Y. 2001. Mycelial pellet intrastucture and visualization of mycelia and intracellular lipid in a culture of *Mortierelaalpina*. *Applied Microbiology and Biotechnology* 56:233–238.

Hansson, L. and Dostalek, M. 1988. Effect of culture conditions on mycelial growth and production of linolenic acid. *Applied Microbiology and Biotechnology* 28: 240–246.

Higashiyama, K., Fujikawa, S., Park, E.Y., and Okabe, M. 1999a. Image analysis of morphological change during arachidonic acid production by *Mortierellaalpina* 1S–4. *Journal of Bioscience and Bioengineering* 87(4):489–494.

Higashiyama, K., Murakami, K., Tsujimura, H., Matsumoto, N., and Fujikawa, S. 1999b. Effects of dissolved oxygen on the morphology of an arachidonic acid production by *Mortierellaalpina* 1S–4. *Biotechnology and Bioengineering* 63(4):432–448.

Houston, D. F. 1972. Rice bran and polish. *In* Rice Chemistry and Technology, Houston D. F. ed. American Association of Cereal Chemists, St. Paul, Minnesota.

Hu, W. 1995. Production of an enriched tocopherol, tocotrienol, oryzanol fraction from extrusion stabilized rice bran. M.S. Thesis, Louisiana State University.

Jang, H., Lin, Y., and Yang, S. 2000. Polyunsaturated fatty acid production with Mortierellaalpina by solid substrate fermentation. *Botanical Bulletin of the Academia Sinica* 41:41–48.

Jang, H., Lin, Y., and Yang, S. 2005. Effect of culture media and conditions on polyunsaturated fatty acids production by *Mortierellaalpina*. *Bioresource Technology* 96(15):1633–1644.

Kim, K. H. and Hong, J. 2001. Supercritical CO_2 pretreatment of lignocellulose enhances enzymatic cellulose hydrolysis. *Bioresource Technology* 77:139–144.

Klesper, E. 1980. Chromatography with supercritical fluids. Extraction with supercritical gases. Schneider, G., Stahl, E., and Wile, G., ed., VerlagChemie, Deerfield, FL.

Knutzon, D. S., Thurmond, J. M., Huang, Y., Chaudhary, S., Bobik, E. G., and Chan, G. M. 1998. Identification of Δ5 desaturase from *Mortierellaalpina* by heterologous expression in baker's yeast and canola. *Journal of Biological Chemistry* 273:29360–29366.

Koike, Y., Cai, H., Higashiyama, K., Fujikawa, S., and Park, E. Y. 2001. Effect of consumed carbon to nitrogen ratio of mycelial morphology and arachidonic acid production in cultures of *Mortierellaalpina*. *Journal of Bioscience and Bioengineering* 91(4):382–389.

Komiyama, K., Hayashi, M., Cha, S., and Yamaoka, M. 1992. Antitumor and antioxidant activity of tocotrienols. In *Lipid Soluble Antioxidants: Biochemistry and Clinical Applications*. Ong ASH and Packer L. eds. Basel, Switzerland: BirkhauserVerlag, pp. 152–159.

Kuk, M. S. and Dowd, M. K. 1998. Supercritical extraction of rice bran. *J. Am. Oil. Chem. Soc.* 75(5):623–628.

Lindberg, A. and Hansson, L. 1991. Production γ–linolenic acid by the fungus *Mucorrouxii* on cheap nitrogen and carbon sources. *Applied Microbiology and Biotechnology* 36:26–28.

Lira, C. T. 1996. Molecular Thermodynamics of Supercritical Fluids with respect to Lipid-Containing Systems, in *Supercritical Fluid Technology in Oil and Lipid Chemistry*, J.W. King, G.R. List, eds., Urbana, IL: AOCS Monograph, AOCS Press.

Lee, W. Y., Cho, Y. J., Oh, S. L., Park, J. H., Cha, W. S., Jung, J. Y., and Choi, Y. H. 2000. Extraction of grape seed oil by supercritical CO_2 and ethanol modifier. *Food Sci. Biotechnol.* 9(3):174–178.

Lloyd, B. J., Sibernmorgen, T. J., and Beers, K. W. 2000. Effect of commercial processing on antioxidants in rice bran. *Cereal Chem.* 77(5):551–555.

Martin, D. E. 1994. Extrusion Stabilization and near infrared analysis of rice bran. M.S. Thesis, Louisiana State University.

McCaskill, D. R. and Zhang, F. 1999. Use of rice bran oil in foods. *Food Technology* 53(2):50

McHugh, M. A. and Krukonis, V. J. 1994. *Supercritical Fluid Extraction*, 2nd ed. Boston, MA: Butterworth–Heinemann.

Michaelson, L. V., Lazarus, C. M., Griffiths, G., Napier, J. A., and Stobart, A. K. 1998. Isolation of Δ5 fatty acid desaturase gene from *Mortierellaalpina*. *Journal of Biological Chemistry* 273:19055–19059.

Manninen, P., Pakarinen, J., and Heikki, K. 1997. Large-scale supercritical carbon dioxide extraction and supercritical carbon dioxide counter current extraction of cloudberry seed oil. *J. Agric. Food Chem.* 45:2533–2538.

Marrone, C., Poletto, M., Reverchon, E., and Stassi, A. 1998. Almond oil extraction by supercritical CO_2: experiments and modeling. *Chem. Eng. Sci.* 53(21): 3711–3718.

Montanari, L., King, J. W., List, G. R., and Rennick, K. A. 1996. Selective extraction of phospholipid mixtures by supercritical carbon dioxide and co solvents. *J. Food Sci.* 61(6):1230–1233.

Moreau, R. A., Powell, M. J., and Hicks, K. B. 1996. Extraction and quantitative analysis of oil from corn fiber. *J Agric Food Chem.* 44:2149–2154.

Nesartetnam, K., Stephen, R., Dils, R., and Darbe, P. 1998. Tocotrienols inhibit the growth of human breast cancer cells irrespective of estrogen receptor status. *Lipids*.33:461–69.

Orthpoefer, F. T. 1996. Rice Bran Oil: Healthy lipid Source. *Food Technology.* December: 62–64.

Ozer, E. O., Platin, S., Akman, U., and Hortacsu, O. 1996. Supercritical carbon dioxide extraction of spearmint oil from mint plant leaves. *The Canadian J Chem Eng.* 74:920–928.

Palazoglu, T. K., and Balaban, M. O. 1998. Supercritical CO_2 Extraction of lipids from roasted pistachio nuts. *Trans ASAE* 44(3):670–84.

Papagianni, M. 2004. Fungal morphology and metabolite production in submerged mycelial processes. *Biotechnology Advances* 22(3):89–259.

Park, E.Y., Koike, Y., Higashiyama, K., Fujikawa, S., and Okabe, M. 1999. Effect of nitrogen source on mycelial morphology and arachidonic acid production in cultures of *Mortierellaalpina*. *Journal of Bioscience and Bioengineering* 88:61–67.

Perrut, M., Calvier, J. Y., Poletto, M., and Reverchon, E. 1997. Mathematical modeling of sunflower seed extraction by supercritical CO_2. *Ind. Eng. Chem. Res.* 36(2):430–435.

Patel, P. 2005. Supercritical fluid extraction of rice bran with adsorption on rice hull ash. Ph.D. dissertation, Louisiana State University, Baton Rouge, LA.

Prieto, M. S. G., Caja, M. M., Herraiz, M., and Maria, G. S. 2003. Supercritical fluid extraction of all-trans-lycopene from tomato. *J Agric Food. Chem*.51:3–7.

Przybylski, R., Lee, Y., and Kim, I. 1998. Oxidative stability of canola oils extracted with supercritical carbon dioxide. *Lebensmittel-Wissenschaft und-Technologie* 31:687–693.

Ramezanzadeh, F. M., Rao, R. M., Prinyawiwatkul, W., Marshall, W. E., and Windhauser, M. 2000. Effects of microwave heat packaging and storage temperature on fatty acid and proximate compositions in rice bran. *J. Agric. Food Chem.* 48(2):464–467.

Randall, J. M., Sayre, R. N., Schultz, W. G., Fong, R. Y., Mossan, A. P., Tribelhorn, R. E., and Saunders, R. M. 1985. Rice bran stabilization by extrusion cooking for extraction of edible oil. *J Food Sci.* 50(2): 361–364, 368.

Ramsey, M. E., Hsu, J. T., Novak, R. A., and Reghtler, W. J. 1991. Processing rice bran by supercritical fluid extraction. *Food Technology.* November: 98–104.

Ratledge, C. 1992. Microbial lipids: Commercial realities of academic curiosities. In *Industrial Application of Single Cell Oils*, eds. D. J. Kyle and C. Ratledge. Champaign, IL:American Oil Chemists' Society Press. 1–15.

Ratledge, C. 2005. Single cell oils for the 21st century. In *Industrial Application of Single Cell Oils*., eds. Z. Cohen and C. Ratledge. Champaign, IL: American Oil Chemists' Society Press. 1–20.

Reverchon, E., Kaziunas, A., and Marrone, C. 2000. Supercritical CO_2 extraction of hiprose seed oil: Experiments and mathematical modelling. *Chemical Engineering Science* 55(12):2195–2201.

Reverchon, E. and Porta, G. D. 1995. Supercritical extraction and fractionation of lavender essential oil and waxes. *J. Agric. Food Chem.* 43:1654–1658.

Riley, G. L., Tucker, K. G., Paul, G. C., and Thomas, C. R. 2000. Effect of biomass concentration and mycelial morphology on fermentation broth rheology. *Biotechnology and Bioengineering* 68(2):160–172.
Rizvi, S. S. H. ed. 1994. *Supercritical Fluid Processing of Food and Biomaterials.* Glasgow, New Zealand: Blackie Academic and Professional.
Rong, N., Ausman, L. M., and Nicolosi, R. J. 1997. Oryzanol decreases cholesterol absorption and aortic fatty streaks in hamsters. *Lipids* 32:303–309.
Roy, B. C., Goto, M., and Hirose, T. 1996. Extraction of ginger oil with supercritical carbon dioxide: Experiments and modeling. *Ind. Eng. Chem Res.* 35:607–612.
Rozzi, N. L. and Singh, R. K. 2002. Supercritical fluids and food industry. *Comprehensive Reviews in Food Science and Food Safety*. 1:33–44.
Rukmini, C. and Raghuram, T. C. 1991. Nutritional and biochemical aspects of the hypolipidemic action of rice bran oil: A review. *J Am. Col. Nutr.* 10:593.
Sajbidor, J., Dobronova, S., and Certik, M. 1990. Arachidonic acid production by Mortierella sp. S-17: influence of C/N ratio. *Biotech Lett.* 12:455–456.
Saunders, R. M. 1990. The properties of rice bran as a food stuff. *Cereal Foods World.* 35 (7):632–636.
Shimizu, S., Kawashima, H., Akimoto, K., Shinmen, Y., and Yamada, H. 1989. Microbial conversion of oil containing α–linolenic acid to oil containing eicosapentaenoic acid. *Journal of American Oil Chemists' Society* 66(3):342–347.
Shinmen, Y., Shimizu, S., Akimoto, K., Kawasashima, H., and Yamada, H. 1989. Production of arachidonic acid by *Mortierella* fungi. *Applied Microbiology and Biotechnology* 31:11–16.
Singh, A. and Ward, O. P. 1997. Production of high yield of arachidonic acid in a fed-batch system by *Mortierellaalpina* ATCC 32222. *Applied Microbiology and Biotechnology* 48:1–5
Sonntag, N. O. V. 1979. Composition and characteristics of individual Fats and oils. In *Bailey's Industrial Oil and Fat Products,* Vol. 1, D. Swern, ed. New York, NY: Wiley, p. 122.
Sovova, H. 2005. Mathematical model for supercritical fluid extraction of natural products and extraction curve evaluation. *Journal of Supercritical Fluids* 33:35–52.
Spilimbergo, S., Elvassore, N., and Bertucco, A. 2002. Microbial inactivation by high–pressure, *Journal of Supercritical Fluids* 22:55–63.
Streekstra, H. 2005. Arachidonic acid: fermentative production by *Mortierella* fungi. In *Single Cell Oil*, eds. Z. Cohenand C. Ratledge. Champaign, IL:American Oil Chemists' Society Press. 73–85.
Sugano, M. and Tsuji. 1997. Rice bran oil and cholesterol metabolism. *The Journal of Nutrition.* 127(3):521S-524S.
Taylor, S. L. and King, J. W. 2000. Optimization of the extraction and fractionation of corn bran oil using analytical supercritical fluid instrumentation. *J of chromatographic science.* 38(3):91–94.
Totani, N., Hyodo, K., and Ueda, T. 2000. A study on new nitrogen source for cultivation of genus *Mortierella. Journal of Japan Oil Chemists' Society* 49:479–485.
Totani, N., Someya, K., and Oba, K. 1992. Industrial production of arachidonic acid by *Mortierella*. In *Industrial Applications of Single Cell Oils,* ed. C. Ratledge. Champaign, IL: American Oil Chemists' Society Press, pp. 52–60.
Walker, T. H., Cochran, H. D., and Hulbert, G. J. 1999. SFE of lipids from *Pythiumirregulare. Journal of American Oil Chemists' Society* 76:596–602.
Walker, T. H., Patel, P., and Cantrell, K. B. 2007. Supercritical fluid extraction and other technologies for extraction of high-value food processing co-products. In *Waste Management and Co-Product Recovery in Food Processing*, ed. K.Waldron. London:Woodhead Publishing Ltd.
Ward, O. P. and Singh, A. 2005. Omega–3/6 fatty acids: Alternative sources of production. *Process Biochemistry* 40(12):3627–3652.
White, A., Burns, D., and Christensen, T. W. 2006. Effective terminal sterilization using supercritical carbon dioxide, *Journal of Biotechnology* 123:504–515.
Williams, J. R. and Clifford, A. A. (ed.) 2000. Supercritical fluid methods and protocols. Totowa, NJ: Humana Press.

Willis, M. W., Lencki, W. R., and Marangoni, G. A. 1998. Lipid modification strategies in the production of nutritionally functional fats and oils. *Critical Reviews in Food Science and Nutrition* 38(8):639–674.

Xu, Z. and Godber, J. S. 1999. Purification and identification of components of Υ-oryzanol in rice bran oil. *J. Agric. Food Chem.* 47(7):2724–2728.

Xu, Z. 1998. Purification and antioxidant properties of rice bran γ-oryzanol components. Louisiana State University. PhD dissertation.

Zhang, J., Davis, T.A., Matthews, M.A., Drews, M.J., LaBerge, M., and An, Y.H. 2006. Sterilization using high–pressure carbon dioxide–a review. *Journal of Supercritical Fluids* 38(3):354–372.

Zhao, W., Shishikura, A., Fujimoto, K., Arai, K., and Saito, S. 1987. Fractional extraction of rice bran oil with super critical carbon dioxide. *Agric. Biol. Chem.* 51(7): 1773–1777.

Zheng, Y., Lin, H.M., and Tsao, G.T. 1998. Pretreatment for cellulose hydrolysis by carbon dioxide explosion. *Biotechnology Progress* 17:890–896.

Zhu, H. and Walker, H.T. 2001.Utilization of rice bran by *Pythiumirregulare* for lipid production. M.S. thesis, Louisiana State University, Baton Rouge, LA.

Zhu, M., Yu. L., and Wu, Y. 2003. An inexpensive medium for production of arachidonic acid by *Mortierellaalpina*. *Journal of Industrial Microbiology and Biotechnology* 35(1):75–79.

CHAPTER 17

Agriculture Management

Wouter Saeys,* Bijoy Chandra Ghosh,[†]
Erich Vareed Thomas,[†] and Ahindra Nag[‡]

17.1 Introduction

In prehistoric times, humans used to gather food from forests. Around 10,000 years ago, humans discovered that by planting seeds, they could grow plants that gave them food; this activity gradually became more and more systematic, known as *cultivation*. The practice of cultivating land for growing crops is known as *agriculture* (*agriculture* is the English adaptation of the Latin word *agricultūra*, from *ager*, "a field," and *cultūra*, "cultivation," in the strict sense of "tillage of the soil"). Thus, an agricultural land is a human-managed ecosystem, which is managed scientifically to achieve maximum sustained productivity.

Due to the increase in population, the demand for food grains has increased tremendously. To be able to feed all these extra people, food production should grow at least as fast as the world population. However, in 1995/1996 there were already 839 million undernourished people. Thus, if we want to halve the number of undernourished by 2015, as is the target of the World Food Summit, we should make sure that crop production increases significantly faster. To maintain a steady supply of food, farmers have several systematic activities for managing the soil—water, pest control, fertilizer applications, use of high-yielding seeds, and the like—spread over a period of time.

---

*Biosystems Department, K.U. Leuven, Heverlee, Belgium.
[†]*Department of Food and Agriculture Engineering, Indian Institute of Technology, Karagpur, India.*
[‡]*Natural Product Laboratory, Department of Chemistry, Indian Institute of Technology, Karagpur, India.*

Because agricultural land is limited, modern agrotechniques and ways to manage crops are now practiced to increase productivity.

Farmers have accumulated adequate knowledge to grow healthy crops. This requires a balanced water system, fertilizer, protecting crops against diseases, and proper storage of harvested crops. Rice production has increased from 25 million tons in 1954 and 1955 to about 78 million tons in 2000 and 2001. Efforts are now being made all over the world to increase production using biotechnological methods.

Because different crops have different seasons to grow, crops are categorized as Kharif and Rabi crops. Paddy rice and maize are main Kharif (autumn harvest) crops that grow during the months of June to October that are dependent on western monsoons, whereas barley and wheat are the main Rabi crops (spring harvest) that are grown from April to November and that are not dependent on monsoon.

17.2 Modern Agriculture Systems

Modern agriculture system is involved with the following components for increasing food productivity:

1. Land-use planning and irrigation management
2. Preparing and testing the soil
3. Seed testing
4. Agricultural mechanizing
5. Manure and vermicomposting
6. Biologically controlling pests and integrated pest management
7. Harvesting

17.2.1 Land-Use Planning and Irrigation Management

This has been thoroughly discussed in Chaps. 3, 5, and 6.

17.2.2 Preparing and Testing the Soil

Soil testing is a useful tool for recommending which fertilizers should be applied to crops. The major objectives of soil testing are as follows:

1. Grouping soil into classes relative to their levels of nutrients for suggesting fertilizer practices.
2. Helping to evaluate soil productivity.
3. Determining specific soil conditions such as alkalinity, salinity, or acidity, conditions that limit crop growth.

Soil Sampling

The success or failure of soil analysis is an aid to fertilizer guidance or subsequent handling operations.[1] General principles of soil sampling are as follows.

1. A series of cores, taken according to a systematic grid layout of the area of equal diameter and comparable depth (volume), should be collected.

2. Separate soil cores should be analyzed, or replicate sets of composites made to determine the statistical significance of results on the final composite.

3. Contamination from soil surface materials (crop residues, manures, fertilizers, etc.) should be avoided.

Method of Sampling Farm Fields The farm or field to be sampled is given a general inspection, and a diagram is prepared showing the different fields, the drainage pattern, and the main types of soil, such as upland or lowland. Separate samples are taken to represent each distinct kind of topography, soil texture, soil organic content (light colored as separate from dark colored), fertility status (as indicated by crop growth), and management unit. Abnormal conditions not to be included in the samples are soil near buildings, gates field margins, highways, and soil from fertilized or manure points.

The operator traverses each area to be sampled separately, taking a core or slice of the plow layer at intervals of 15 to 20 steps and places them together in a water-resistant bag.

Handling the Soil Sample in the Laboratory

Drying: The soil sample is usually partially air dried at a temperature of about 25 to 35°C and relative humidity of 20 to 60 percent. The common weight basis for expressing the results is 100 to 110°C oven dry weight of soil material.

Sieving: The bulk soil sample, in its natural field-moist conditions, is passed through a 6-mm (4 meshes/in) sieve, usually by rubbing it with the fingers. While the soil is sampled in the field, stones and large gravels are ordinarily ignored, and a few stones and gravel particles remaining on the 6-mm screen are discarded. When the air is dry, the soil should be worked through a 2-mm screen again, discarding small gravel but not soil crumbs.

Grinding: Soil aggregates are broken up by lightly grinding them with a roller and a rubber pestle in a motorized grinder. Crushing the primary sand and gravel particles is generally avoided.

Partitioning the sample (quartering method): The mixed soil material is coned in the center of the mixing sheet with care to make it symmetrical with respect to the fine and the coarse soil material. The cone is flattened and divided through the center with a flat metal spatula or metal sheet, one half of the cone being moved to the side quantitatively. Then each half is further divided into half, and the four quarters are placed as separate piles. For small samples, a paper-quartering technique is used. Two diagonally opposite quarters are discarded quantitatively. The other two are mixed by rolling if the entire half is to be retained or resieved with finer sieves if further partitioning of sample is to be made.

Weighing: The sample is weighed with an analytical balance.

Determining Soil Properties

The properties of soil are measured as follows.

Determining the pH of the Soil   Soil reaction or pH is meant to express the acidity or alkalinity of the soil. This is very important because it determines the soil's capacity for plant growth, availability of nutrients, bacterial activity, and physical conditions.

Take exactly 10 g of soil and add 25 mL of distilled water. Shake the mixture occasionally, and, after 30 min, dip both the glass and calomel electrode in the soil suspension. Connect the electrode to the pH meter, which has been already checked with a standard buffer of known pH. Now read the pH of the soil from the pH meter scale.

Determining Organic Carbon   Soil is digested with chromic and sulphuric acids. The excess of chromic acid, not reduced by organic matter of the soil, is determined by titrating with a standard ferrous ammonium sulphate solution.

Take 2 to 10 g of soil in a 500-mL conical flask. Add 10 mL of 1-N potassium dichromate solution and 20 mL of concentrated sulphuric acid. Shake well for a minute or two and allow it to stand on an asbestos mat for about 30 min. Add 200 mL of water, 10 mL of phosphoric acid, and 1 mL of diphenyl amine indicator solution. A deep violet color will appear. Titrate it with 0.5-N ferrous ammonium sulphate solution until the violet color changes to purple.

Add 0.5 mL of potassium dichromate solution accurately with a burette and titrate with ferrous ammonium sulphate solution until the color changes to green. In the same way, carry out a blank determination:

Weight of soil taken = W g

Volume of 0.5-N ferrous ammonium sulphate required for reducing 10-mL $K_2Cr_2O_7$ solution (blank reading) = X mL

Volume of 0.5-N ferrous ammonium sulphate required for reducing excess dichromate (experimental reading) = Y mL

Difference = $X - Y$ mL

1 mL of 1-N $K_2Cr_2O_7$ = 0.003 g carbon

% of carbon in the soil = $\dfrac{(X-Y) \times 0.003 \times 100}{2 \times W}$

There is incomplete oxidation of organic matter in this procedure. The organic carbon is multiplied by 1.3 on the assumption that there is a 77 percent chance of recovery.

Actual organic carbon (%) = organic carbon estimated × 1.3 percent of organic matter in soil = 2 × 1.724.

Determining Total Nitrogen in the Soil Ammonia gas that is formed due to the reaction of soil with excess of alkaline $KMnO_4$ is absorbed in a known volume of standard acid. Excess acid is titrated with standard alkali using methyl red as an indicator.

Place a 20 g of soil sample in a distillation flask and add 20 mL of water. Then add 1 mL of liquid paraffin and a few glass beads to avoid frothing and bumping, respectively, during distillation.

Now add 100 mL of 0.32 percent $KMnO_4$ solution and 100 mL of 2.5 percent NaOH solution and immediately place it in the distillation apparatus. Pipette out 20 mL of boric acid solution (with a mixed indicator) in a conical flask and dip the end of the delivery tube in it. Distill ammonia gas from the distillation flask and collect about 100 mL of the filtrate in about 30 min. With the absorption of ammonia, the pink color of the boric acid solution turns green. Titrate the contents with 0.02-N H_2SO_4 to the original shade (pink). A blank correction (without soil) is to be made for final calculations:

Weight of soil taken = 20 g
Volume of 0.02-N H_2SO_4 required for titration = R mL
1 mL 1-N H_2SO_4 = 14 mg N
Available N in 20-g soil = R × 0.02 ×14 mg
Available N kg/ha = R × 31.36

Determining the Available Phosphorus To determine available phosphorus in soils, use the Olsen method for neutral–alkaline soils, whereas the Bray–Kurtz P1 method[1] is used for acidic soils.

Olsen's Method This method is based on the fact that sodium bicarbonate solution extracts some exchangeable or surface-adsorbed Al-P, Fe-P, calcium phosphates, and other phosphates.

Reagents of this method are prepared as follows:

1. Take sodium bicarbonate (Olsen's reagent) 0.5-M $NaHCO_3$, pH 8.5. Dissolve 84 g of $NaHCO_3$ in up to 2 liters of water. Adjust pH 8.5 with 1-M NaOH (4 g/100 mL) solution. Store in a glass or polyethylene bottle.

2. Reagent A: Dissolve 12 g of ammonium molybdate $(NH_4)_6 Mo_7O_{24} 4H_2O$ in 250 mL of distilled water. Dissolve 0.2908 g of antimony potassium tartrate [K (SbO) $C_4H_4O_6 \frac{1}{2}H_2O$] in 100-mL of water. Add these two solutions to 1000 mL of 2.5-M H_2SO_4, mix thoroughly; make up to 2000 mL. Store in a Pyrex glass bottle in a dark and cool place.

3. Reagent B: Dissolve 1.056 g of ascorbic acid ($C_6H_8O_6$) in 200 mL of reagent A and mix. This does not keep for more than 24 h at room temperature. Prepare daily as required.

4. Sulphuric acid (2.5 M): Dilute 140 mL of concentrated H_2SO_4 to 1 liter of water.

5. Standard stock potassium solution: Dissolve exactly 0.439 g of potassium dihydrogen orthophosphate (KH_2PO_4) AR grade (dried in the oven at 60°C for 1 h and cooled in a desiccator) in ½ L of distilled water. Add 25 mL of 7-N H_2SO_4 and make up to 1 L with distilled water. This gives us a 100-ppm potassium standard stock solution. Diluting this 50 times gives a 2-ppm solution.

To prepare the standard curve, 1, 2, 3, 4, 5, and 10 mL of 2-ppm potassium solution are placed in 25-mL volumetric flasks. To these flasks, add 5 mL of the extracting solution (Olsen's or Bray–Kurtz P1). When Olsen's extractant is used, acidify a 5-mL aliquot with 2.5-M H_2SO_4 to pH 5.0. Add distilled water to increase the volume to 20 mL and then add 4 mL of reagent B. Prepare a blank with $NaHCO_3$ solution, distilled water, and 4 mL of reagent B. After 10 min, read the intensity of the blue color in a photoelectric colorimeter using a Whatman No. 730 to 840 nm filter on a spectrophotometer at 730 nm. Weigh 2.5 g of 2 mm air-dry soil into a 150-mL Erlenmeyer flask and add a little Darco G-60 or equivalent-grade potassium-free activated charcoal. Then add 50 mL of Olsen's reagent and shake the reciprocating shaker for 30 min. Similarly, run a blank without soil. Filter through a 42 filter paper into a clean and dry beaker. Shake the 5-mL aliquot of extract in a 25 mL volumetric flask and acidify with 2.5 M H_2SO_4 to pH 5.0. Add distilled water to acidify to 20 mL and then add 4 mL of reagent B. After 10 min, read the color intensity on a spectrophotometer as described for the standard curve:

$$\text{Available P (kg/ha)} = \frac{R \times \text{volume of extract } 2.24 \times 106}{\text{volume of aliquot wt (g)} \times 10^6}$$

where R = µg of P in the aliquot (obtained from the standard curve).

$$\frac{R \times 50 \times 2024}{5 \times 2.5} = \mu g\ P \times 8.96$$

Bray–Kurtz P1 Method   The dilute acid–fluoride extractant easily removes acid-soluble potassium from phosphates bound to Al, Fe, and Ca. Phosphates in the extract are determined calorimetrically as phosphomolybdenum blue with ascorbic acid as a reducing agent and antimony added to the solution to give a stable Mo-P-Sb compound.

Reagents of this method are prepared as follows:

1. *Bray–Kurtz P1 extracting solution:* Dissolve 22.2 g of NH_4F and 41.6 mL of concentrated HCl making 20 L. This gives us a solution of 0.03-M NH_4F in 0.025-M HCl. This should be stored in a polyethylene bottle.
2. *Reagent A:* As described in Olsen's method.
3. *Reagent B:* As mentioned earlier.
4. *Sulphuric acid 2.5 M.*
5. *Standard stock potassium solution:* As mentioned in Olsen's method. Prepare a standard curve as described under Olsen's method. Weigh 2.5 g of 2-mm air-dried soil into a 150-mL Erlenmeyer flask. Add 25 mL of extracting solution. Place a stopper and shake the suspension for 5 min on a reciprocating shaker. Immediately filter through a Whatman No. 42 filter paper. If the filtrate is turbid, quickly pour the filtrate back through the same filter. To avoid the interference of fluoride, add 7.5 mL of 0.8-M boric acid (50 g of H_3BO_3/L) to 5 mL of extract, if necessary. Place a 5-mL aliquot of the extract in a 25 mL volumetric flask. Add distilled water to 20 mL and add 4 mL reagent B to the volume. After 10 min, read the intensity of blue color as described under Olsen's method:

$$\text{Bray's P (kg/ha)} = \frac{R \times 25 \times 2.24}{5 \times 2.5} \times \mu gP \times 4.48$$

where $R = \mu g$ of P in the aliquot.

Determining Exchangeable Potassium   The term "available potassium" (K) conventionally refers to exchangeable and water-soluble K. Exchangeable K constitutes the major portion of available K except in saline or saline sodium soil. Available K or exchangeable K, along with Ca and Mg, are usually determined in neutral normal ammonium acetate (in NH_4OAC) extract. K is estimated using a flame photometer and Ca and Mg either by EDTA titration or by using an atomic absorption spectrophotometer.

To determine potassium, reagents are prepared as follows:

1. *Ammonium acetate (1 N, pH 7.0):* Mix 700 mL of distilled water and 57 mL of 99.5 percent glacial acetic acid and 69 mL of concentrate ammonium hydroxide. Dilute to a volume of 900 mL

and adjust pH to 7.0 by adding 3-N ammonium hydroxide or 3-N acetic acid; make up to 1 L. Store in a Pyrex or polyethylene bottle.

2. *Potassium chloride standard solution:* Make a stock solution of 1000-ppm K by dissolving 1.908 g of AR-grade potassium chloride (dried at 60°C for 1 h in distilled water and diluting up to 1 L). Prepare a 100-ppm standard solution by diluting 100 mL of 1000-ppm stock solution to one with the extracting solution.

Pipette 0, 5, 10, 15, and 20 mL of the 10-ppm solution into a 100-mL volumetric flask and bring the volume to the mark with the extracting solution. Place 5 g of soil in a 150-mL Erlenmeyer flask and pour in 25 mL of neutral 1-N ammonium acetate. Shake in a reciprocating shaker for 5 min and immediately filter through a Whatman No. 1 filter paper. Determine K in the extract in a flame photometer using a K filter at the necessary setting and calibrating the instrument. Obtain the standard curve by plotting the reading against the different concentrations of K:

$$\text{Available K (9 kg/ha)} = \frac{R \times \text{volume of extract} \times 2.24 \times 106}{\text{wt of soil taken} \times 10^6}$$

$$= \text{ppm of K} \times 11.2$$

where R = ppm of K in the extract (obtained from the standard curve).

Determining Available Zinc, Copper, Manganese, and Iron Chelating agents combine with free metal atoms in solution to form complexes. DTPA (diethylene triamine pentaacetic acid) offers the most favorable combination of stability constants for the simultaneous complexing of Zn, Cu, Mn, and Fe.

We used the following reagents:

DTPA 0.005 M

$CaCl_2$, $2H_2O$ 0.01 M

TEA 0.1 (triethanol amine)

The reagents were prepared as follows.

Extracting Solution Dissolve 13.1-mL reagent-grade TEA, 1.967 g of DTPA (AR grade) and 1 g of $CaCl_2$ in 100 mL of glass-distilled water. Allow some time for the DTPA to dissolve and dilute to approximately 900 mL. Adjust the pH to 7.3 with 1:1 HCl while stirring and dilute to 1 L.

Standard Solutions

Zinc standard solution: Dissolve 0.439 g of AR-grade $ZnSO_4$, $7H_2O$ in 200-mL (middle) distilled water in a beaker. Add 5 mL of 1:5 H_2SO_4.

Transfer to a liter-measuring flask and fill the flask until you have a 100-ppm standard solution. Transfer 10 mL of this solution to a 100-mL volumetric flask and dilute to the mark with DTPA extracting solution to have a 10-ppm stock solution. To prepare working standards, transfer 1, 2, 4, and 6 mL of stock solution (10 ppm) to a series of clean 100-mL of volumetric flasks and dilute each to the mark with DTPA extracting solution.

Volume of stock Zn solution taken: 0, 1, 2, 4, and 6 mL

Concentration of Zn now in solution: 0, 0.1, 0.2, 0.4, and 0.6 ppm

Iron standard solution: Dissolve 0.702 g of AR-grade ammonium ferrous sulphate in 300 mL of distilled water in a beaker. Add 5 mL 1:5 H_2SO_4. Transfer to a liter-measuring flask and fill volume to the mark. This is a standard 100-ppm solution. To prepare a working standard, transfer 1, 2, 4, and 6 mL of stock solution and dilute each to the mark with DTPA extracting solution volume of a stock Fe solution.

Volume of stock Fe solution taken: 0, 1, 2, 4, and 6 mL

Concentration of Fe in solution: 0, 1, 2, 4, and 6 ppm

Manganese standard solution: Dissolve 0.288 g potassium permanganate in 300 mL of distilled water in a beaker. Add 20 mL of concentrated H_2SO_4, warm to about 60°C, and add oxalic acid solution dropwise to render the solution colorless. Cool and transfer to a liter-measuring flask and make volume to the work. This is 100-ppm Mn solution. To prepare working standards, transfer 1, 2, 4, 6, and 8 mL of the standard solution to a series of 100-mL volumetric flasks and dilute each to the mark with DTPA extracting solution.

Volume of stock Mn solution taken: 0, 1, 2, 4, 6, and 8 mL

Concentration of Mn in solution: 0, 1, 2, 4, 6, and 8 ppm

Copper standard solution: Dissolve 0.392 g of copper sulphate in 400 mL of distilled water in a beaker. Transfer to a liter-measuring flask and fill up to the mark with distilled water. This is a standard solution containing 100-ppm Cu. To prepare working standards, transfer 1, 2, 4, and 6 mL of stock solution to a series of clean 100-mL volumetric flasks and dilute each to the mark with DTPA extracting solution.

Volume of stock Cu solution taken: 0, 1, 2, 4, and 6 mL

Concentration of Cu in solution: 0, 1, 2, 4, and 6 ppm

At least three to four standards and a blank of each micronutrient cation are used to draw a calibration curve. The blank solution 0 (ppm) is used to determine the zero level of the atomic absorption spectrophotometer. The standards are then analyzed with the lowest concentration first and a blank run between standards to ensure that the line (zero point) has not changed. A graph of absorbance versus concentration of standard solution is plotted on graph paper. After setting the instrument, aspirate the sample, read the absorbance, and determine the concentration against the absorbance from the curve.

Weigh 10 g of air-dried soil in a 125-mL conical flask. Then add 20 mL of DTPA extracting solution. Cork the flask and shake for 2 h at 120 cycles/min. Filtrate in polypropylene bottles to be analyzed for Zn, Cu, Mn, and Fe with an atomic absorption spectrophotometer. When samples need to be diluted before measuring, DTPA solution should be used to maintain a constant matrix.

Determining Molybdenum Molybdate absorbed on soil colloids and clay is presumably replaced by oxalate ions. This exchange is made irreversible by the formation of strong Mo–oxalic acid complexes to make a single extraction effective.

Reagents used to determine molybdenum are as follows:

1. Potassium iodide solution (50 percent aqueous).
2. Ascorbic acid solution (50 percent aqueous).
3. Tartaric acid solution (10 percent aqueous).
4. Thiourea solution (10 percent aqueous).
5. Dithiol solution (toluene-3, 4-dithiol): 1 g of melted dithiol is added in 100 mL of 10 percent NaOH solution. The content is warmed up to 51°C with frequent stirring for 15 min on a hot plate. Then 1.8 mL of thioglycolic acid is added and stored in a refrigerator.
6. Isoamyl acetate solution.
7. Ferrous ammonium sulfate solution: Dissolve 63 g of ferrous ammonium sulphate in distilled water until you have 1 L.
8. Extracting solution: Dissolve 24.9 g of ammonium oxalate and 12.6 g of oxalic acid per liter of solution. Adjust pH to 3.3.

Prepare a 100-ppm molybdenum solution by dissolving 0.150 g of AR-grade Mo_3O_4 in 100 mL of 0.1-N NaOH solution, rendering it slightly acidic with HCl; and make the volume to 1 L. Take 10 mL of this stock solution in a liter-volumetric flask until you get a working standard of 1 ppm. To make working standards, transfer 0.5, 1.0, 2.0, 3.0, and 4.0 mL of the 1-ppm solution to series of clean 125-mL separator funnels. Fifty milliliters of each working standard along with a Mo-free blank is taken in a separator funnel (250-mL capacity). Then 0.25 mL of ferrous ammonium sulphate solution is added in 20 mL of distilled water. Ascorbic acid is added dropwise in the solution to remove liberated iodine coming from the excess of KI, which was added to the solution. Then, add 1 mL of tartaric acid, 2 mL of thiourea, five drops of dithol, and 10 mL of amyl acetate to the solution. Shake the contents of the separator funnel thoroughly. Absorbance of green color complex is then measured at 680 nm in a spectrophotometer. A standard curve is then drawn by plotting concentration on the x axis and absorbance on the y axis using semilog graph paper. This graph is used to determine the concentration of the unknown sample.

Extracting and Determining Molybdenum Twenty–five grams of soil are placed into a 500-mL conical flask. To this, 250 mL of ammonium oxalate of pH 3.3 is added. The flask is shaken in a horizontal shaker for 10 h and is then filtered through Whatman filter paper No. 50. Two hundred milliliters of the filtrate is placed into a 250-mL beaker and is evaporated to dryness on a water bath. The contents of the beaker are heated at a high temperature (500°C) in a furnace for 5 h to destroy organic matter and oxalates. The contents are then digested with 5 mL diacid (NHO_3-$HCIO_4$ 4:1) after keeping it overnight, then with 10 mL of 4-N H_2SO_4 and to dryness. Add 10 mL of 0.1-N HCl to the beaker and filter paper wash with another 10 mL of 0.1-N HCl and redistill until the filtrate volume obtained is 100 mL. The filtrate is stored in corked plastic reagent bottles to estimate the available Mo by the procedure described under standard curve.

Concentration of Mo against standard reading (80) from the standard curve = 0.1/µg (say) dilution factor = 2.

Thus, Mo in 20 g of soil = 0.2/µg Mo in soil = 0.01/µg/g or ppm.

Estimating Boron Boron is absorbed in soil as soil colloids, and its determination is done using the Berger–Truog (1939) method. It is found as ppm level in soil.

The following reagents are used for estimating boron:

1. *Buffer solution:* Dissolve 250 g of ammonium acetate and 15 g of EDTA disodium salt in 400 mL of distilled water and slowly add in 125 mL of acetic acid. The mixture is shaken for some time for thorough mixing.
2. *Azomethine–H reagent:* Dissolve 0.45 g of azomethine H in 100 mL of 1 percent L–ascorbic acid solution. Fresh reagent should be prepared each week and stored in a refrigerator.

To prepare the standard stock solution, dissolve 0.570 g of boric acid (H_3BO_3) in 1 L of distilled water. Take 5 mL of this stock solution in a 100-mL volumetric flask and dilute to this mark. This solution contains 5-ppm boron. To a series of 25-mL volumetric flasks, add 0.25, 0.5, 1.0, 2.0, and 4.0 mL of 5-ppm boron solution, respectively, and increase this final volume to 25 mL. To each volumetric flask containing a standard working solution, add 2 mL of buffer solution and 2 mL of azomethine–H reagent, mix thoroughly, and allow to stand at room temperature for 30 min. Make the volume to 30 mL with distilled water and measure the absorbance at 420 nm with a spectrophotometer.

Plot the strength of concentration versus absorbance on semilog paper. Take distilled water as the standard solution.

Extracting and Determining Boron Soil sample (25 g), 50 mL of water, and 0.5 g of activated charcoal are boiled for 5 min in a quartz flask and filtered immediately through a Whatman No. 42 paper. Pour 5 mL of extract in 25-mL volumetric flask and add 4 mL of buffer solution, 4 mL of azomethine–H reagent solution, and make the volume 25 mL with distilled water. Allow the color to develop for 1 h. The color intensity is measured spectrophotometrically at 420 nm. By comparing the measured absorbance to that for the standard curve, we can calculate the unknown concentration.

Calculation Boron (ppm) in soil sample = concentration of boron in analyzed sample = reading from the standard curve × 10.

17.2.3 Seed Testing

Seed development is one of the most important steps in agricultural development. Seed technology includes the development of superior crop plant varieties, seed production and processing, seed storage and testing, seed certification, quality control, and seed marketing and distribution. Routine seed-testing steps are shown in Fig. 17.1.

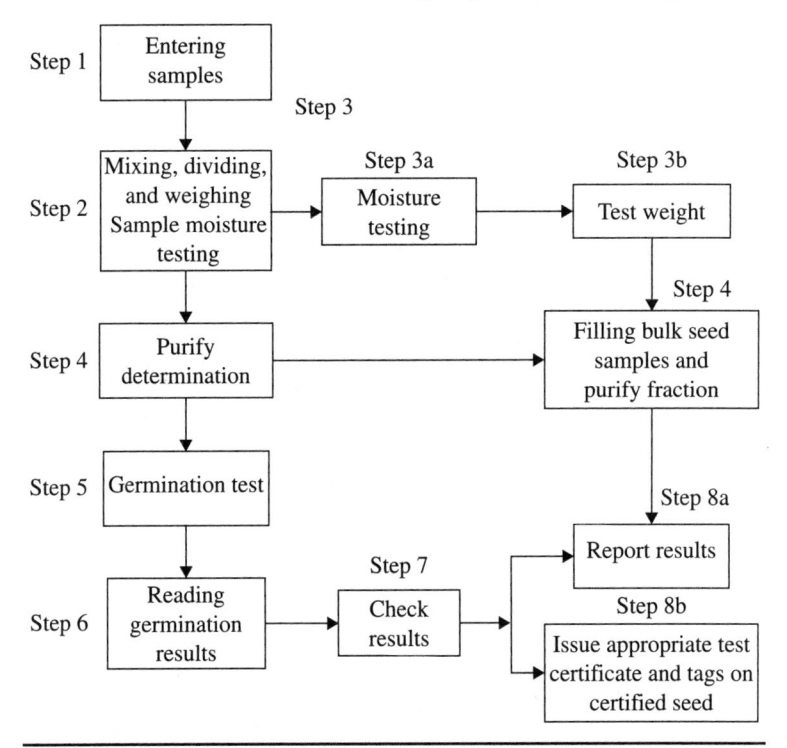

Figure 17.1 Routine seed testing.

Rapid Tests for Seed Viability

Rapid seed viability tests are required when large quantities of seed have to be procured in the shortest possible time (within 24 h). Although a germination test is still the most reliable and satisfactory, it is a time consuming and slow process. Rapid tests are also useful when a seed is dormant or a seed has been injured due to heat, frost, or unsuitable storage. Rapid seed tests are based on enzyme activity as measured and have proved to be the most useful because the enzyme activity is closely bound up with the life of the seed. Enzymes are involved in oxidation/reduction reactions with certain organic compounds, some of which give a different color in their oxidized state from that in the reduced state. Thus, a simple color change can be used to demonstrate the presence or absence of dehydrogenase enzymes.

The chemical used for the test is a cream or light yellow soluble powder 2,3,5-triphenyl tetrazolium chloride (or bromide). The colorless solution is reduced in the presence of certain enzymes to an insoluble formazan. When this occurs in living tissues, the tissue is stained red. The test reaction should be performed in the dark.

The tetrazolium solution must come in contact with the seed embryo. Since the coats of many seeds are relatively impermeable to the chemical, it is often necessary to prepare the seed. A period of preconditioning in which the seed is soaked in water at 30°C is recommended. This allows the seed coat to soften and swell and ensure a bright strain. The preconditioned seed may then

1. Be placed directly in the tetrazolium solution (peas and beans).
2. Be bisected longitudinally (cereal and grasses).
3. Be pricked below the embryo or bisected laterally (small grass seed).
4. Have the seed coat removed (melon, cotton).
5. Be scratched above the embryo (lettuce).
6. Have a piece cut off the embryo (*Pinus*).

Some seeds such as small clovers and *Brassica* sp. may be placed directly in the tetrazolium solution without preconditioning. After a period in the solution, the length of which varies with the species and with the temperature of the test conditions, the seed is removed from the tetrazolium solution, washed in water, and classified according to the staining pattern. The time that it takes to complete the test varies with the species, with the number of species (preferably between 6 and 8), with the strength of the solution (0. 1, 0.5, and 1.0 percent are used), and with the temperature. It may be difficult to interpret the results, and the analyst must have a proper understanding of the importance of the parts of the embryo in order to interpret the staining pattern correctly.

Advantages of tetrazolium test are

1. Results are available within a few hours.
2. Very little equipment is necessary to carry out the test.
3. Close examination of individual seeds will often reveal the reason for poor germination.

Germination Method Using Paper

Top of Papers (TP)   The seeds are germinated on top of one or more layers of paper, which are placed in any of the following ways:

1. Enclosed in transparent Petri dishes
2. Put directly on germination trays in a cabinet or in a room-type germinator

In the first method, water is supplied to the paper at regular intervals, and germination takes place under ambient temperature. In the second method, the relative humidity in the germinator must be maintained close to saturation level. Moistened porous paper or absorbent cotton can be used as a base for the paper or even as an immediate substratum.

Between Paper (BP)   The seeds are germinated between two layers of germination paper that are placed in any of the following ways:

1. Directly on germination trays in a cabinet or a room-type germinator
2. In metal, plastic, or glass boxes

In this method, the relative humidity will be maintained close to saturation level. The paper can be folded or rolled and placed in a flat or upright position. Metal, glass, or plastic frames can be inserted between papers to ensure ventilation. Moistened porous paper or cotton can be used as a base for paper or even as an immediate substratum.

Germination Method Using Sand

In a sand culture, the seeds are placed in a uniform layer of moist sand and covered to a depth of 1 to 3 cm with sand, which is left loose. The amount of water to be added to the sand will depend on its characteristics and size of seed to be tested. Cereals except maize may be germinated in sand moistened to 50 percent of its water-holding capacity. For large-seeded legumes and maize the sand should be moistened to 60 percent of its water-holding capacity. Proper spacing of seeds should be maintained to reduce the contact of the seedlings with each other during germination to a minimum level. This is especially important because large seeds, diseased seeds, and seedlings, which may affect healthy seeds, may be removed before final count.

Germination Method Using Soil

Soil may be considered as the substrate, which is sometimes used to confirm the evaluation of seedlings in doubtful cases and for testing samples that produce seedlings with phytotoxic symptoms when germinated on paper or sand.

Water should be added until the consistency of the soil is such that the ball formed by squeezing it in the palm of the hand is easily broken when pressed between two fingers. After the soil is wet, it should be rubbed through a sieve and put into a container without packing.

17.2.4 Agricultural Mechanization

Agricultural mechanization is a process used to increase farm output. In the modern era, powered machinery has replaced many tedious and time-consuming farm operations carried out by humans or animals. The first pervasive mechanization of agriculture came with the introduction of the plough, usually powered by animals. Invention of engines brought a major change in mechanization. Current mechanized agriculture practices include the use of airplanes, helicopters, trucks, tractors, and combine harvesters among other vehicles. Almost all such mechanization has led to the following broad conclusions:

1. Farm mechanization led to an increase in input on account of higher average cropping intensity and larger area and increased productivity of farm labor.
2. Farm mechanization increased agricultural production and profitability on account of timely operation, better quality of work done, and more efficient utilization of inputs.
3. Farm mechanization has led to a serious reduction of the number of people employed in agriculture. The cultivated area per farm has increased while the number of farms and the number of people per farm has decreased.
4. Farm mechanization displaced animal power to the extent of 50 to 100 percent from place to place. The time spent per ha has been reduced, but thanks to mechanization farms have grown.

Farm implements and machines include all tools, implements, and machines used in agriculture to produce crops. Some important and generally used farm implements and machines for tilling the soil will be described in what follows.

Tillage Implements

Tillage is cultivation carried out for nurturing crops. Tilling helps to create a better environment for seed germination, root growth, weed control, soil erosion, and moisture control.[2] It can be divided in two categories: primary tillage and secondary tillage.

Primary Tillage Primary tillage is used so that the soil can attain a reasonable depth (10 to 15 cm) with varying clod sizes, bury trashes by inversion, and mix different tilled layers. The action is rather more aggressive, the operation being relatively heavy, leaving a rough surface. Primary tillage helps soil aeration and water accumulation. Depending on the soil type and the plough used, the soil will normally be inverted aerating the deep layers and trapping water during a rainfall. During primary tillage crop residues are also chopped and incorporated. Common implements used for this purpose are moldboard, one-way disc, offset disc, and tine implement. The differences between these tools are given in Table 17.1.

Secondary Tillage Secondary tillage implements are used to create refined soil conditions, control weeds, incorporate fertilizers, puddling, and leveling soil surface. Many types of machines are available for this type of tillage. The implements used for secondary tillage are the peg tooth harrow, disc harrow, tined cultivator, rotary tiller, and

	Moldboard	**One-Way Disc**	**Offset Disc**	**Tine Implement**
System	Animal, 2-wheel, 4-wheel tractor	2-wheel, 4-wheel tractor	4-wheel tractor	Animal, 2-wheel, 4-wheel tractor
Width	1–3 shares	2–4 disc	9–21 discs	1–15 tine
Soil Disturbance	High	Medium-high	High	Low
Plough Action	Total inversion	Inversion	Inversion/ cutting	Cutting
Soil Penetration	Share design	Machine weight	Machine weight	Weight/design
Weed control	Bury	Bury	Cut/bury	Cut
Handle Obstacles	Poor	Good	Good	Medium-good
Power Requirement	Heavy	Medium-heavy	Medium-heavy	Low-medium

Source: en.wikipedia.org/wiki/Plough.

TABLE 17.1 Different Types of Secondary Tillage

interrow cultivators. With improvements in planting equipment, it becomes less necessary to do secondary tillage after primary tillage. Crop residue can sometimes create a problem with secondary tillage if there is not a good residue flow through the tillage machine.

Sowing and Planting Equipment

The basic objective of sowing is to put the seeds in the soil at a desired depth and seed-to-seed spacing, to cover the seeds with soil, and to provide proper compaction over the seed. The recommended row-to-row spacing, seed rate, seed-to-seed spacing, and depth of seed placement vary from crop to crop and for different agroclimatic conditions to achieve optimum yields.[3] Different crops have different methods of sowing, depending on the size of the seed and other crop factors.

Seed Drills Seed drills are generally used for sowing small-size seeds at a depth where seeds can get proper moisture.

Planters Planters deposit seeds at equal distances in the row at specified depth and cover the seeds with soil. Unlike with the seed drill, plant-to-plant distance is maintained in the planter. They are used for row crops such as maize, cotton, beans, and potatoes, which require accurate placement of seeds for obtaining good yields, and to permit intercultivation operations.

Transplanters have been used to plant seedlings. Transplanters need considerably less time and less labor than manual transplanting. They are used with rice and some vegetables (e.g., cabbages and lettuce).

Intercultivation and Plant-Protection Equipment

Intercultivation operations like weeding and hoeing, and plant-protection measures such as spraying have to be carried out to maintain a healthy crop.

Intercultivation Operations Intercultivation operations are carried out to nourish the plant. They provide favorable conditions to the plants for promoting plant growth. These operations include weeding, harrowing, and hoeing.

Plant Protection Plant protection includes using chemicals as sprayers and dusters. The main function of a sprayer is to break suspension (pesticides in solvent) into droplets of effective size and distribute them uniformly over the surface to be protected. Another function is to regulate the amount of herbicide or pesticide to avoid excessive application that might prove harmful or wasteful.[3]

17.2.5 Manure and Vermicomposting

Recently, there has been a great increase in food production due to the use of fertilizers to maintain soil fertility and control pests. Chemical fertilizers no doubt have increased food production, but they also have the following harmful effects:

1. Chemical fertilizers are more costly than biological fertilizers.

2. Chemical fertilizers are manufactured from depleted coal and petroleum as sources of energy

3. Chemicals used as fertilizers are often released in the environment causing air, water, and soil pollution.

4. Chemical fertilizers can persist in the soil for a long time and may damage soil fertility and ecofriendly microbes.

Application of Manure

Although it is recognized that manure can increase crop yields by improving soil structure and adding nutrients, most farmers are not eager to replace the larger part of their artificial fertilizer use with manure. Even when farmers apply large amounts of manure, as is the case in Flanders, they are in doubt about the precise amounts of manure and artificial fertilizer that they have to apply. Therefore, farmers typically add a safety margin and provide more nutrients to the soil than what the crop needs. The main reason for this is that farmers fear productivity losses resulting from lack of certain nutrients due to the unknown composition of the manure. Matching fertilizer application to the soil reserve, measured by soil sampling and crop needs, found in the literature (or measured), can be easily done with artificial fertilizer, as the composition is known and guaranteed. Manure can only become a viable alternative if the composition is known.

There is a large variation in manure composition resulting from variations in the most important influencing factors: animal species and type (e.g., meat or dairy), ration, type of barn used, presence of straw, length of storage, sedimentation, gas emissions, mixing, spilling, and/or rain water. All these factors are highly farm and management dependent.

Because no two farms or farmers are the same, manure composition will also be different.

Based on this variation between farms, one could decide that it might be sufficient to perform one analysis per farm. Although this would already result in a considerably better nutrient management, it would still be rather inaccurate, because several influencing factors vary with the time of the year and evolve over the years. For example, in spring manure will be typically more concentrated and more digested due to longer storage during the winter, whereas in the fall manure will contain more cleaning water and is typically less digested.

Liquid manure is a suspension of solid particles, containing most of the dry organic matter, phosphorus, calcium, and magnesium in a liquid matrix.

In pig and poultry manure, the solids sink, leaving the liquid on top, whereas in cattle manure the solids float on top of the liquid.

Manure Composition Measurement Sampling of liquid manure is a crucial step in a complete manure composition analysis procedure.[4] As mentioned above, manure composition varies with depth in the storage facility due to gravitational separation effects. Therefore, a representative sample should contain equal amounts of manure at different depths in the pit. This is accomplished by lowering a tube with constant diameter vertically to the bottom of the pit. Once the bottom of the pit is reached, the downside of the tube is closed with a rubber stop and the filled tube is pulled out of the pit. Apart from the vertical variation, composition in the manure pit also varies in the horizontal direction (e.g., places where more spilled water or feedstuffs enter the pit). Therefore, it is also necessary to take several subsamples in the horizontal plane. Eight to ten subsamples are sufficient to obtain a representative sample for an unmixed manure storage facility. Instead of sampling manure storage, it is also possible to sample at haulout before transport. In this case, a sampler is mounted on the tank to take several subsamples throughout the filling process. The obtained sample will then be representative for the manure in the tank but cannot be applied to the rest of the manure in storage. In the framework of a mineral declaration system, farmers in the Netherlands are obliged to sample each manure transport in this way so that the government can monitor the movement of nutrients from livestock to crop-production regions.

Once a representative sample has been taken, manure is stored in a closed container and cooled until it is chemically analyzed. This cool and airtight storage is done to avoid compositional changes by microbial action and gaseous losses (e.g., NH_3, CO_2, CH_4). In the laboratory, the concentration of different components is analyzed chemically. Parameters and methods vary throughout the world, but they typically give information about the nitrogen, phosphorus, and potassium content of the manure, as these are the most important plant nutrients. Often, there are also measures of the dry and organic matter content. In Flanders, for example, manure samples are typically analyzed for pH, dry matter content (DM), organic matter content (OM), total nitrogen by Kjeldahl method, phosphorus (P), potassium (K), sodium (Na), calcium (Ca), and magnesium (Mg).

The sequence of sampling, transport, wet chemical analysis, and communication of the results induces a time delay of 5 to 7 days between the time of sampling and the moment that the farmer knows the results. The cost for the farmer of one full compositional analysis lies in the range of 50 to 80 euros. The time delay induced by wet chemical analysis makes good manure nutrient management very difficult, because the land application may be completed before the sample results are known. Therefore, several on-farm analysis techniques have been developed to rapidly estimate the nutrient (primarily N, but also P) content of manures. Most of these quick tests are portable, simple to use, relatively inexpensive, and take less than 10 min to run.

Van Kessel et al. give an overview of several commercially available quick tests for rapid on-farm analysis of manure nutrients.[5] One of the simplest tests is based on the idea that a sample with a higher dry matter content will also contain more nitrogen and phosphorus.[5] Although a higher dry matter content indeed indicates more concentrated manure, and thus higher nutrient contents, the correlations are rather weak. Therefore, hydrometer measurement of the specific gravity of manure only allows a very rough estimation of the total nitrogen and phosphorus contents. Another quick test is based on measuring the electrical conductivity of the manure, which is related to the concentrations of cations and anions that perform electron transport. Measuring electrical conductivity has been successfully used to estimate the NH_4^+–N concentration in manure but not the phosphorus content.[6] However, conductivity is not specific for NH_4^+–N, but it measures the flow of electrons due to all cations and anions in the solution. Therefore, large shifts in the other ions (e.g., K^+, Na^+) could alter the conductivity and give a false indication of a change in NH_4^+–N concentration.

Ammonia and dissolved phosphorus concentrations can also be measured colorimetrically with test strips. The color intensity on the test strip can then be quantified using a small reflectometer. With this combination of colorimetric reaction and reflectometer, rather accurate estimations of ammonia and dissolved phosphorus concentrations have been obtained.[6] When hypochlorite is added to the manure, the NH_4^+ is oxidized and N_2 gas is produced. By measuring the amount of N_2 that is released from a manure sample, the ammonia concentration in the manure can be estimated. Both the Agros N Meter and the Quantofix-N-Volumeter were found to give rather accurate predictions of the NH_4^+–N in liquid manure but were not successful for drier manures.[7]

Discrete sampling and analysis of manure is time consuming and increases the time it takes to spread manure. To be able to perform good manure nutrient management Scotford et al.[8] developed an in-line nutrient–sensing system for liquid manures based on the measurement of several physical and chemical properties of the manure. They reported useful results for NH_4–N but were less successful for K and total Scotford et al.[8] also had some difficulties with the transferability of their calibrations and the robustness of their sensors.

Although several of the above-mentioned quick tests are quite capable of measuring the mineral nitrogen (ammonia) content, this is not sufficient for good nutrient management. A large part of the total nitrogen content in manure is bound to organic molecules and will therefore not be measured by the test. However, this nitrogen will become available to the plant after application of the manure when the organic molecules are decomposed by microorganisms. Moreover, the correlation between the mineral and total nitrogen content is too small to estimate the total nitrogen content based on the ammonia content. This makes it very difficult to estimate the nitrogen fertilizer value of the manure based only on the ammonia content.

During the last decades, visible and near-infrared (VIS/NIR) spectroscopy has proven to be a valuable, nondestructive technique for rapid analysis of quality and composition of agricultural products.[9] The measurement of suspensions like hog manure on an NIR spectrophotometer is, however, rather difficult due to the presence of particulate matter. For clear liquids with very few particles, transmission would be dominant, making transmission measurement the best choice. However, transmission measurement is not possible for a liquid sample with high concentration of particles because the suspended particles will prevent the light from going through the sample, even for a small path length. Therefore, when a so-called infinite thickness, which could be due to the scattering caused by sufficient particles in suspension, is achieved, the reflectance mode can be applied. On the other hand, if attenuation of the light is not that hard, sample presentation in transflectance mode can be considered. For transflectance measurement, a reflector made of a diffusely scattering material is placed at the opposite side of the light source. In this presentation mode, the signal collected by the detector consists of two parts. Part of the incident light could transmit through the sample to reach the reflector and then be reflected back into the sample. Some of the energy can pass through the depth of the sample again and get sensed by the detector at the same side of the light source. The second part of the signal comes from the light scattered back by the suspended particles without hitting the reflector.

Several researchers have shown the feasibility of NIR spectroscopy for the rapid analysis of manure. Most of them presented the manure in reflectance mode to a scanning monochromator or a portable diode array instrument[10,11] and obtained promising results. However, the clearly better results obtained by Malley et al.[12] using a monochromator with a sample cell in transflectance mode suggest that the presentation mode may play an important role in predicting the nutrient contents of hog manure. Therefore, Saeys et al.[13] compared the performance of transflectance and reflectance sample presentation modes for the rapid analysis of hog manure (slurry) using VIS/NIR spectroscopy and concluded from the cross-validation results that measuring in transflectance mode gave a more accurate composition analysis than measurement in reflectance mode. Saeys et al.[13] reported that very good quantitative predictions for total N; good quantitative predictions for K, DM, OM; approximate predictions for P, Mg, and NH_4–N; very approximate predictions for Ca; and discrimination between high and low values for Na can be obtained.

However, in the case of online application of the transflectance mode, the small pathway of 1 mm between the measurement glass and the reflector acts as a filter for the larger particles in the manure and is easily blocked by them, so that it would be very difficult to build a robust online sensor with it. Online application of reflectance measurement is far easier because the manure does not have to flow through a small slit. Therefore, the reflectance mode is preferential for this type of application because it combines reasonable accuracy with rather easy online implementation.

Another important aspect of the measurement setup with respect to accuracy is the spectroscopic measurement principle and the wavelength range. To be able to select the "best" instrument for on-site and online manure analysis, Mouazen et al.[14] compared the performance of four visible and near-infrared (VIS/NIR) spectrophotometers with different measurement principles and wavelength ranges. Mouazen et al.[14] concluded that a scanning monochromator (400 to 2500 nm) and a Fourier transform spectrophotometer (1000 to 2500 nm) performed quite well. However, these are typical laboratory instruments that are very stable in the laboratory but are too big and too vulnerable to be used on-site or online. The investigated diode array (306 to 1710 nm) and diode array plus scanning monochromator (350 to 2500 nm) instruments were designed to be used outdoors and are easily portable so that they both can be used for laboratory and online field measurements. Results showed, however, that the diode array plus scanning monochromator performed the worst of the four due to its fixed and very short integration times in the NIR. On the other hand, the diode array spectrophotometer was equally or more accurate than the other three instruments for the different components considered. Moreover, the latter is robust, cost effective (is half the price of the other instruments), and can be used in the field for online measurement. The integration time of this type of instrument can also be modified, which allows optimizing it for the measurement of highly absorbing materials such as dark brown manure samples. Because the only instrument that stopped at 1700 nm performed best, it was also concluded that the wavelengths beyond 1700 nm are not critical for manure analysis.

Saeys et al.[15] calibrated a VIS/NIR diode array instrument in reflectance mode for a compositional analysis in cross validation on a set of 420 pigs' manure samples and validated this sensor on an independent test set of 164 pigs' manure samples. The prediction results on the independent test set were evaluated based on the ratio of prediction to deviation (RPD) and the determination coefficient r^2 values as excellent for the dry matter (DM), good for organic matter (OM) and total nitrogen (N), and approximate for ammonium nitrogen (NH_4–N), phosphorus (P) and magnesium (Mg). The calibrations for potassium (K) and calcium (Ca) were found to be only able to discriminate between high and low values. Saeys et al.[15] also compared the accuracy of the prediction to the standard error of laboratory analysis and the standard deviation of the manure composition with the depth in an unmixed storage facility. Laboratory analysis was found to have a 3 to 8 times smaller error on the compositional measurement of one sample. However, the short measurement time and the low extra cost per measurement of the online VIS/NIR spectroscopic sensor promoted a higher sampling rate, which would reduce the prediction error on the estimated mean composition. Therefore, Saeys et al.[15] concluded that an on-site composition measurement with the VIS/NIR spectroscopic sensor for each depth region in an unmixed storage facility would be two to four times more reliable than the mean composition obtained by sampling and laboratory analysis.

Emissions during Application

Agriculture is considered to be one of the most important sources of ammonia in the atmosphere with a share of 50 percent in total ammonia emission, more than 70 percent in regions with intensive livestock farming, such as Europe,[16] and even 93 to 96 percent in Flanders. Because ammonia forms the major component of acid rain, together with sulphur oxide SO_2, and different nitrous oxides NO_x, its gaseous emission by agriculture leads to acidification and eutrophication of the environment and should therefore be minimized. Without emission reduction initiatives, 50 percent of the ammonia emission would come from manure application, 36 percent from barns and manure storage facilities, and 14 percent from pasturing. Therefore, reducing ammonia emission during manure application is the most efficient way to reduce agriculture's share in total emission.

Ammonia emission during manure application ranges from 0 to 60 percent of the total amount of ammonia nitrogen in manure and depends on manure composition, soil properties, weather, and application technique. Ammonia emission is a function of the transfer of ammonia from the air phase just above the manure to the atmosphere. This is performed by convection and diffusion, which are functions of weather conditions (temperature, precipitation, etc.), contact time, and contact surface. Therefore, this transfer can be influenced by the method of application.[17]

The ammonia concentration just above the manure is in equilibrium with the ammonia in solution in the manure. This equilibrium is mainly influenced by chemical and physical conditions at the manure surface. Ammonia NH_3 in solution is in equilibrium with ammonium NH_4^+ in solution. Most of the ammonium in manure originates from hydrolysis of the urea in urine, but about 10 percent comes from the organic nitrogen components in feces. Because the hydrolysis of urea in manure is very fast, and manure is stored for several weeks, urea hydrolysis will have been completed at the moment of application. The equilibrium between NH_4^+ and NH_3 depends highly on the pH and the ionic strength of the manure. This means that the pH of the manure or the manure/soil mixture has a major effect on ammonia emission. Ammonia emission lowers the pH, whereas CO_2 emission, by decay of organic matter, increases the pH. Because the solubility of CO_2 in water is 200 times lower than that of NH_3, the CO_2 will be emitted at a higher rate directly after application of the manure. This increases the pH of the manure, which will then promote ammonia emission and results in a very high ammonia emission in the first hour after application.[17] After a period of 24 to 48 h, the pH in the soil surface will decrease again by the buffering action of the soil and manure. Once the manure is in contact with the soil, the NH_4^+ will be captured on cation exchange places of the soil.

The rate of ammonia emission is thus largest directly after application, due to the high initial concentration of ammonia and the increase in pH by the emission of CO_2. After this initial emission, the rate decreases rapidly because the ammonia concentration decreases by emission, infiltration, and nitrification. Typically, emission already reaches 50 percent of its cumulative maximum within 12 h after application. However, sometimes emission can go on for longer periods when the meteorological conditions limit evaporation, and there is a very low speed of infiltration. It can, therefore, be concluded that the contact surface and time between manure and air should be minimized. To reach this goal, several manure application strategies (band application, incorporation, injection) have been introduced.

When weather conditions minimize ammonia emission, there can still be important emissions originating from manure in the form of surface runoff. As long as the manure is lying on top of the soil, runoff of excess precipitation can transport the nutrients away from the field into surface waters. In the context of the manure problem, this is directly attributed to the nutrient concentrations in the surface waters and should therefore also be avoided. This can be brought about by putting the manure into the soil as soon as possible so that the nutrients will not be brought into solution and transported away by the water flow.

Apart from the negative impact of these emissions on the environment, there is another reason why we want to minimize them. It would be very difficult to do precise application of manure when nutrient losses by gaseous emission of ammonia and surface runoff reduce the nutrient content available to the crop.

Manure application techniques that bring the manure directly into the soil reduce both ammonia emission and surface runoff. The deeper the manure is placed in the soil, the higher the emission reduction, but energy consumption and crop damage will also be higher. Therefore, there exists an optimal injection depth. This injection depth is traditionally controlled with depth wheels or a constant pressure system, which are highly sensitive to variations in the soil penetration resistance. These systems work deeper on soft, moist soil and energy consumption and crop damage increase, whereas emissions increase when the injector is pushed out of the ground on hard, dry soil. Therefore, Saeys et al.[18,19] developed an automatic depth control system for shallow slurry injection which maintains the injection depth within a narrow range around the desired value independent of the soil penetration resistance.

Vermicomposting

Vermicomposting is the process of recycling organic matter into nutrient-rich compost using worms. There are several advantages of vermicomposting:

1. Vermicomposting is a natural and ecofriendly process.
2. Organic wastes generated from agricultural and forest activities, food processing, household activity, and natural vegetation can be recycled into a vermicompost.
3. A vermicomposting facility can be designed and operated to minimize environmental impacts by controlling odors and bioaerosols.
4. A mature vermicompost is a stable, usable product.
5. Vermicompost can be considered as a balanced food to plants.

Recyclable Organic Waste   Crop and animal wastes from agricultural activities are directly or indirectly the major source of recyclable materials. The different waste materials differ in their composition, biochemical nature [carbon/nitrogen (C/N) ratio], and stage of maturity. Waksman and Cordon[20] indicated that plant residue decomposition is influenced by material preparation. According to these authors, the most important factors are (1) nature of the composting feedstock, especially the proportion of nitrogenous compounds to carbohydrates, (2) decomposition temperature, and (3) microbial population of the compost. The latter two factors are very interactive. The most important factors affecting composting include oxygen, C/N ratio, temperature, and moisture.

The process is carried out generally by aerobic conditions to produce organic manure in a shorter time and to get rid off the bad odor of the degrading organic waste (see Fig. 17.2).

Earthworms belong to the class Oligochaeta. Earthworms are bilaterally symmetrical and externally segmented, with a corresponding internal segmentation. Three broad categories of earthworm are as follows:

Epigeic (manure worm)	Lives in and consumes decomposable organic matter, is small in size, and is uniformly pigmented, suitable for vermicomposting Species: *Eisenia foetida*
Endogeic (subsoil dweller)	Small to large in size, weekly pigmented, lives in burrows in organic mineral complex of soil, consumes soil Species: *Apporrectodea caliginosa, Octolasion cyaneum*
Anecic (topsoil dweller)	Large in size, has brown pigment anteriorly and dorsally, lives in a deep vertical burrow in soil, casting on surface Species: *Lumbricus terrestris, Allolobophora longa*

In a tropical climate, endogeic and anecic earthworms dominate the epigeic species. The species most commonly used for vermicomposting is *Eisenia foetida*. It can thrive well under a wide range of temperatures, 4 to 30°C and moisture of about 60 to 30 percent, and can live in organic matter in a wider pH range, from 6 to 8.

Earthworms feed on any organic waste; however, the acceptability and palatability of the waste depends on the texture as well as the chemical nature of the food material. Earthworms swallow the food material, which passes through the alimentary canal equipped with a minigrinder gizzard. After consumption, the organic matter must have undergone physical and chemical breakdown through the activity of the muscular gizzard, which grinds the material to a particle size of 1 to 2 μm. Usually, under ideal conditions, earthworms consume

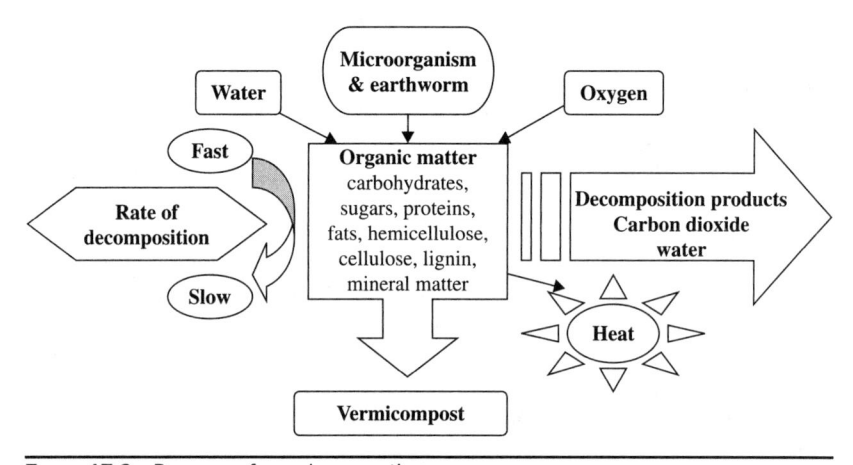

FIGURE 17.2 Process of vermicomposting.

organic waste as per their body weight in 24 h. Therefore, on average, a 1-kg earthworm can consume 1 kg of organic waste per day. Of the total consumption, 5 to 10 percent feedstock used for their growth and any remaining comes out as vermicast. Numerous enzymes have been reported to be present in earthworm gut, which includes mainly protease, lipase, amylase, chitinase, and cellulase. It is not clear, however, which of these enzymes are directly secreted by the gut and which are produced by the diverse symbiotic flora and fauna of microorganisms that are commonly present in earthworms. Thus, the body of an earthworm is considered to be a natural bioreactor. The excreted vermicast, therefore, is a rich source of macro- and micronutrients, vitamins, enzymes, antibiotics, growth hormones, and immobilized microflora.

A wide range of organisms that have been isolated from compost include actinomycetes, fungi, and bacteria. Organisms associated with composting fall into the two classes: mesophiles (25 to 45°C) and thermophiles (>45°C); some microbes may even grow in compost heaps at temperatures exceeding 75°C. In composting, the primary interest is in the organisms that decompose organic matter at the most efficient temperature and other conditions. Major agricultural wastes constitute a decomposition-resistant organic compound known as lignocellulose, which increases composting. Attempts have been made to accelerate the composting process by optimizing the growth of certain lignin decomposers such as *Phanerochaete chrysosporium*, *Trichoderma viridae*, and *Paeciliomyces fusisporus*. The white rot fungus, *P. chrysosporium*, is known to biodegrade lignin-related aromatic compounds such as benzoic, p-OH–benzoic, vanillic, veratric, syringic, and p-coumeric acids. The use of cellulolytic and lignolytic microorganisms as inoculants is known to speed up the process of vermicomposting. Inoculation with mesophyllic cellulolytic fungi is known to reduce the time needed for completion of the process and improve the quality of the final product. For regular availability, multiplication of pure culture of microorganisms in appropriate growth media under aseptic laboratory conditions is important. Microorganisms multiplied in the laboratory are mixed with the organic wastes thoroughly before placing the microorganisms in the vermibed to hasten the decomposition process.

Humidification is an ultimate stage of the vermicomposting process (Fig. 17.3). It involves decomposition of high- and low-molecular-weight plant, animal, and microbial cell constituents and products synthesized by microorganisms into complex amorphous colloids containing phenolic materials. However, humification is also accelerated by the passage of organic material through the gut of earthworms when they feed on decomposed organic matter together with mineral soil. Probably some of the final stages of humification are due to the intestinal microflora in the earthworm's gut because most of the evidence indicates that the chemical processes of humification are caused more by the microflora than by the fauna.[21]

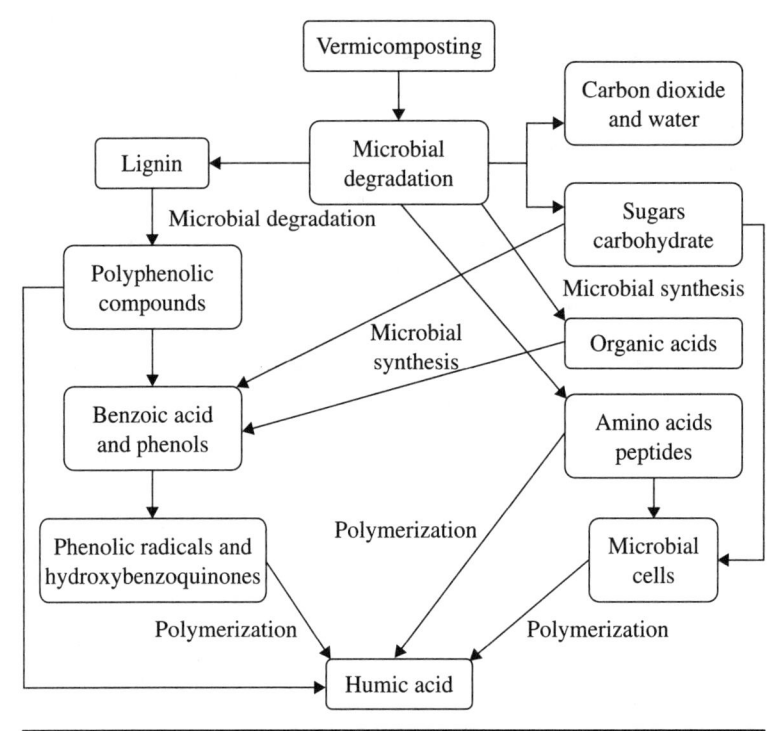

FIGURE 17.3 Potential pathway for the formation of humic acid during vermicomposting.

Technology of Vermicomposting

Bed Preparation and Loading Vermicomposting can be practiced indoors on a small scale or outdoors commercially. The compost infrastructure should be developed in shade in a covered area to avoid adverse climatic conditions like heavy rainfall, direct sunlight, and low temperature. The area also should be protected from attack by ants, rodents, and insectivorous birds. Vermicompost can be done in earthen pits, concrete tanks, plastic/wooden crates, or tin containers.

To establish a small unit at home or in a courtyard, a regular flow of organic wastes like kitchen waste, crop waste, animal waste, etc. will be available. Generally, a container/box/bin made of wood/tin/plastics, or two- or three-chambered containers can be used for vermicomposting, depending on the amount of wastes generated in a family. Usually, in a multichambered bin, composting is done alternatively following the "by filling one and making vacant other" policy. Wooden bins are preferred because they provide better insulation against adverse temperatures. The size and number of containers depends on the quantity of waste material generated per day. The height of the bin should not be more than 15 to 18 in. Holes ¼ in in diameter made in the body of the container will ensure better aeration and drainage of excess water. The container should be covered by a gunny bag or a lid to conserve moisture and avoid light for better worm activity because worms function against sunlight and prefer dark conditions.

Before putting household waste in the container, a layer of coconut coir/leaf mold/peat moss/banana peels biodegradable in nature should be placed as the first layer 3 to 4 in thick as the vermibed. Thereafter, household waste, particularly kitchen waste, can be added continuously in layers. Earthworms can consume all kind of household wastes like yard waste, teabags, vegetable and fruit waste, pulverized egg shells, and the like. Waste like garlic and onion scales, citrus foods, bones, meats, dairy products, and other household chemicals like metals, plastic, glass, soap, and insecticide should be avoided. Composting efficiency can be improved by reducing particle size and increasing surface area of the waste. Therefore, waste material may be dumped after grinding or blending. A good mixture, preferably of equal proportions, of brown (dry and dead organic waste like straw, weeds, etc.) and greens (fresh plants or kitchen waste materials) would be best for a balanced nutrition for the earthworm. On average, a container 2 ft wide, 3 ft long, and 12 in deep can accommodate 1 kg of earthworms (approximately 2000 adults), which can recycle a minimum of half a kilogram of organic waste in a day (Fig. 17.4).

Maturity and Harvesting When the vermicompost starts maturing, it shows a definite change in the physical appearance of the compost material. At maturity, the material will be soft, spongy, dark brown, and have an earthy odor (no foul smell). At this point, watering can be stopped. The mature vermicompost can also be judged by its C/N ratio, BOD, nitrate–N, VFA (volatile fatty acids) and ratio of reducing sugars.

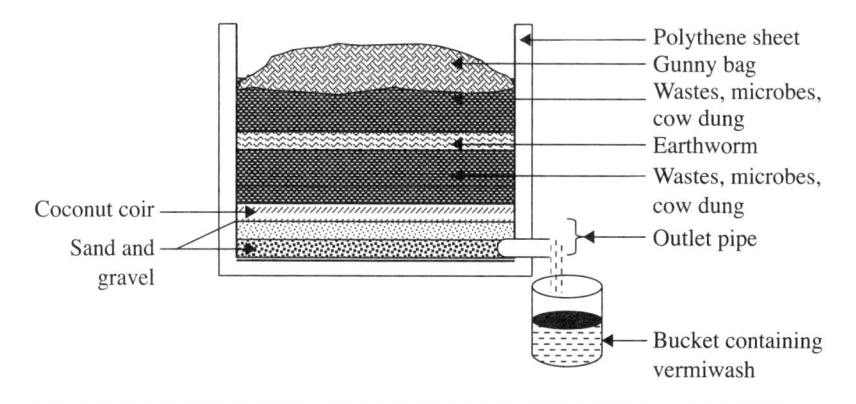

FIGURE 17.4 Schematic structure of a compost unit.

After the vermicompost is harvested, earthworms can be separated either manually or by screening through a net 2 mm in size. However, the latter process of earthworm separation is more convenient and economical than the former. After separation, the cocoons and the baby earthworms may remain with the vermicompost. After earthworm separation, the moisture is brought to a 20 to 25 percent level before bagging. During the dry season, vermicompost gets dried in the open sun, whereas, during the rainy season, drying in the open becomes difficult.

The harvested vermicompost is spread on a screen where the earthworm gets separated and remains above the screen while the vermicompost moves downward through a series of plates. During that movement, hot air blows with an air temperature of about 50°C to bring down the moisture content of the compost to a desired level. Finally, the dried compost falls on a conveyor belt, which is bagged in the same unit. The capacity of the machine is 100 kg/h.

Vermiwash

Besides the production of solid vermicompost, another product (i.e., liquid manure or vermiwash) is obtained in the composting process. During watering of compost beds, excess water percolates through the compost material and is collected in a tank. All the soluble nutrients, enzymes, humic acids, and hormones are washed through the body of the earthworm and vermicompost, which is a rich fertilizer material used as a liquid manure. A minimum 100 L of vermiwash can be harvested per year from each vermibed (see Table 17.2). Vermiwash is generally used as a foliar spray to the crops and also in bodies of water in fish farming to increase production of phytoplankton and zooplankton.

Parameters	Vermicompost	Vermiwash*
pH	6.9	6.9
N, %	1.60	0.0046
P, %	0.98	0.0025
K %	1.10	0.063
O, C, %	14.1	—
Cu, ppm	38.0	0.117
Zn, ppm	180	0.132
Ca, ppm	2760	786
Mg, ppm	4100	328
Fe, ppm	11200	0.151
Mn, ppm	1290	213

*From Basak and Ghosh.[22]

TABLE 17.2 Composition of Vermicompost and Vermiwash

17.2.6 Biological Control

Chemical pesticides produce harmful effects to the environment as well as to human and animal health. Safer alternative methods to control these pests are urgently required by biological control as well as suppressing the insect population by the action of their native or introduced natural enemies.[23]

Integrated pest management is important for environmental safety (Fig. 17.5). This process involves the selection, integration, and implementation of pest control based on predicted economic, ecological, and sociological consequences. It is based on different physical, chemical, biological, and mechanical control, use of pesticide at the most susceptible stage of the pest, use of resistant varieties, mixed cropping, and crop rotation.

In the United States, integrated pest management (IPM) was formulated into national policy in February 1972 by President Richard Nixon. In 1979, President Jimmy Carter established an interagency IPM Coordinating Committee to ensure development and implementation of IPM practices.

Biopesticides

Biopesticides are made by directly using the biological organism as microbial pesticides or by using organismal products as biochemical pesticides or as plant-incorporated protectants. Based on the target organism, microbial pesticides can be broadly classified as follows.

Bioinsecticides Biopesticides should completely control the specific target and be biodegradable and cost effective. Bioinsecticides are based on mutant strains of the bacterium *Bacillus thuringiensis* (B.t.), the most widely used and successful microbial pesticide. There have been over 30 subspecies of B.t. classified, containing over 140 described crystalline toxins. These toxins have not only shown activity against lepidoptera, diptera, and coleoptera, but recent isolates

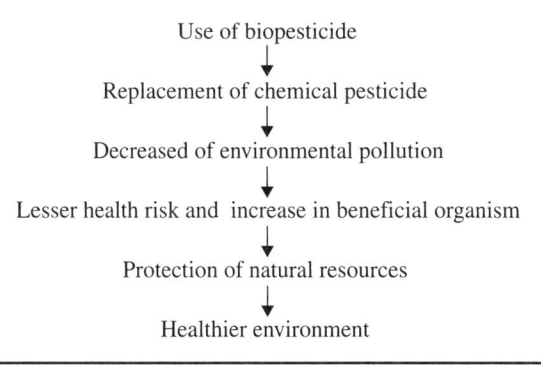

Use of biopesticide
↓
Replacement of chemical pesticide
↓
Decreased of environmental pollution
↓
Lesser health risk and increase in beneficial organism
↓
Protection of natural resources
↓
Healthier environment

FIGURE 17.5 Gradual replacement of chemical pesticides for environmental safety.

have been discovered that show activity against nematodes, mites, lice, aphids, and ants (Lacey and Goettel 1995). There are two other species of bacteria currently registered[24] as microbial insecticides: *Bacillus sphaericus* and *Serratia entomophila*. *Bacillus sphaericus* is strictly registered for mosquito control.

Entomopathogenic viruses have been isolated from a number of viral families, but a few from the family Baculoviridae have been registered as viral insecticides. Commercially available baculoviruses are mostly targeted for lepidopteran hosts. The majority of baculoviruses used as biological control agents are in the genus *Nucleopolyhedrovirus*.

The most widely used fungal insecticide is *Beauveria bassiana* to control the maize corn borer. Other strains of *B. bassiana* are used to control white flies, thrips, aphids, and mealybugs. *Metarhizium anisopliae* spores are used to control the wheat cockchafer, and *Metarhizium flavoviride* controls grasshoppers and locusts. *Paecilomyces fumosoroseus* controls white flies, thrips, aphids, and spider mites; *Lagenidium giganteum* is currently used for mosquito control. These fungi generally penetrate the host with enzyme action of enzymes; death is caused by mycelial colonization, causing starvation, or by the toxin released.

Bioherbicides The first bioherbicide (de vine) was used in 1981, which was a mycoherbicide based on the fungus *Phytophthora palmivora*. It controls the growth of milkweed vines (*Morrenia odorata*) in citrus orchards.

Templeton et al. (1986)[25] suggest that bioherbicides have greatest potential for controlling (1) weeds infesting small specialized areas where chemical herbicide development is costly, (2) weeds that have been intransigent to chemical control, (3) crop mimics, and (4) parasitic weeds. Weed species may be annual (e.g., velvetleaf, *Abutilon theophrastii*), or perennial (e.g., field bindweed, *Convolvulus arvensis*, and dandelion, *Taraxacum officinale*).

Biofungicides *Trichoderma harzianum* is a naturally occurring fungus, which is used as a biofungicide to protect crops from root rot and wilting diseases caused by other harmful fungi (*Fusarium, Rhizoctonia, Pythium, Botrytis*, etc.). *Gliocladium virens* acts against *Rhizoctonia solani* and *Pythium* spp., whereas *Ampelomyces quisqualis* protects plants from powdery mildew disease.

Insect Hormones Insect hormones such as pheromones and juvenile, ecdysone have been found useful in controlling insects. Pheromones that attract males of the appropriate species are secreted by the female. If the traps are made using synthetic analogs of female pheromones, the males are attracted and eliminated. Similarly, molting hormone ecdysone and juvenile hormones have been found to control insects. Juvenile hormone should be present in the early stages to prevent early maturation, but if this hormone is given at a later stage, the insects become transformed into giant larvae (immature adult) that die quickly.

Natural Insecticides Natural insecticides are extracted from plants and sometimes from microorganisms. Alkaloids like nicotine from tobacco (*Nicotiana* spp.) and pyrethrum from pyrethrum inflorescence (*Chrysanthemum cinerarifolium*) are also used. Neem (*Azadirachita indica*) is considered to be one of the most useful natural insecticides.

17.2.7 Harvesting

It is the process of collecting mature crops from the agricultural fields. The cutting of grain for harvest is typically done by using a scythe, a sickle, or a reaper. After harvesting postharvest handling begins including cooling, sorting, cleaning, packing up for further on-farm processing, or shipping to wholesale or consumer markets.

It has been suggested that an earlier harvest date may avoid damaging conditions but might result in poorer yield and quality. Delaying a harvest may result in a better harvest, but it increases the risk of weather problems.

Harvesting involves various operations and varies from crop to crop. It includes cutting, digging, picking, laying, gathering, curing, transporting, and staking of crops. Cereals crops, such as wheat, rice, and jawar are earheads containing grains, whereas most of the oilseed and pulse crops have branches, which creates problems in harvesting by manual or mechanical means (Pranav and Pandey 2008).[26]

Traditional Harvesting Methods
Crop harvesting is traditionally done with manual methods. Harvesting major cereals, pulses and oilseed crops is carried out using sickles, whereas tuber crops are harvested using country ploughs or spades.

Mechanical Harvesting
Timeliness of the harvest is of prime importance. Rains and storms occur often during the harvesting season causing considerable damage to the standing crop. Rapid harvesting also facilitates extra days for land preparation and earlier planting of the next crop. The use of machines can help to harvest at the proper stage of crop maturity and reduce drudgery and operating time requirements. Considering these factors, improved harvesting tools, equipment, and combines are being adopted by farmers to use as a reaper. The invention of two successful reaping machines—both independently by Obed Hussey in Cincinnati, Ohio, who obtained the first patent in 1834. The first reapers cut the standing grain and, with a revolving reel, swept it onto a platform from which it was raked off into piles by a person walking alongside (Fig. 17.6).

FIGURE 17.6 Agostini Meccanica Reaper Binder. (*Source: www.ferraritractors.com/selfpropelled.htm*)

The Agostini Meccanica Reaper Binder uses a walking tractor to power the reaper binder for grain harvesting. The harvester consists of a platform carrying the reaper binder mechanism, the driver's seat, and the rear steering wheel. The walking tractor is installed temporarily to supply the ground drive and the power cutting and tying mechanism.

The reaper was eventually replaced by the self-propelled combine harvester, operated by one person, which cuts gathers, threshes, and cleans the grain mechanically.

On smaller farms with minimal mechanization, harvesting is the most labor-intensive activity of the growing season. On large, mechanized farms, harvesting uses the most expensive and sophisticated farm machinery which reaches ever higher capacities (up to 5 ha or 50 tons per hour) and operator comfort thanks to the incorporation of automatic monitoring and control systems.[27-34]

References

1. Nag, A. 2006. *Analytical Techniques in Agriculture, Biotechnology and Environmental Engineering*. New Delhi, India: Preienticel of India.
2. Srivastava, A. K., Goering, C. E., Rohrbach, R. P, and Buckmaster, D. R. 2006. Engineering principles of agricultural machines. *Agricultural Mechanization and Some Methods of Study*, 2nd ed. St. Joseph, MI: American Society of Agricultural and Biological Engineers. 1–14.
3. Sahay, J. 2006. Tillage implements. *Elements of Agricultural Engineering*, 4th ed., Delhi, India: Standard Publishers Distributors. 224–276.
4. Ndegwa, P. M. and Zhu, J. 2003. Sampling procedures for piggery slurry in deep pits for estimation of nutrient content. *Biosystems Engineering* 85(2):239–248.
5. Van Kessel, J. S., Thompson, R. B., and Reeves, J. B. III. 1999. Rapid on-farm analysis of manure nutrients using Quick Tests. *J. Prod. Agric.*,12:215–224.
6. Lugo-Ospina, A., Dao, T. H., Van Kessel, J. A., and Reeves, J. B. III 2005. Evaluation of quick tests for phosphorus determination in dairy manures. *Environmental Pollution* 135:155–162.

7. Van Kessel, J. S., and Reeves, J. B. III 2000. On-farm quick tests for estimating nitrogen in dairy manure. *Journal of Dairy Science* 83:1873–1844.
8. Scotford, I. M., Cumby, T. R., Richards, P. A., Keppel, D., and Lenehan, J. J. 1999. Development of an in-line nutrient sensing system for livestock slurries. *Journal of Agricultural Engineering Research* 74:313–316.
9. Shenk, J. S., Workman, Jr., J. J., and Westerhaus, M. O. 1992. Application of NIR spectroscopy to agricultural products. *Handbook of Near-Infrared Analysis*, eds. D. A. Burns and E. W. Ciurczak. New York, NY: Marcel Dekker. 681.
10. Millmier, A., Lorimor, J., Hurburgh, Jr.C., Fulhage, C., Hattey, J., and Zhang, H. 2000. Near-infrared sensing of manure nutrients. *Transactions of the ASAE* 43(4):903–908.
11. Dolud, M., Andree, H., Hügle, T. (2005). Rapid analysis of liquid hog manure using near-infrared spectroscopy in flowing condition. *Precision Livestock Farming '05 (Cox S. ed)*, Wageningen Academic Publishers, Wageningen, The Netherlands, 115–122.
12. Malley, D. F., Yesmin, L., and Eilers, R. G. 2002. Rapid analysis of hog manure and manure-amended soils using near-infrared spectroscopy. *Journal of the American Soil Science Society* 66:1677–1686.
13. Saeys, W., Xing, J., De Baerdemaeker, J., and Ramon, H. 2005. Comparison of transflectance and reflectance to analyse hog manures. Journal of Near Infrared Spectroscopy 13:99–108.
14. Mouazen, A. M., Dumont, K., Maertens, K., and Ramon, H. 2003. Two-dimensional prediction of spatial variation in topsoil compaction of a sandy loam field based on measured horizontal force of compaction sensor, cutting depth and moisture content. *Soil Tillage Research* 74(1):91–102.
15. Saeys, W., Mouazen, A. M., and Ramon, H. 2005. Potential for on-site and on-line Analysis of Pig Manure using visual and near-infrared Reflectance Spectroscopy. Biosystems Engineering 91(4):393–402.
16. Schlesinger, W. H. and Hartley, A. E. 1992. A global budget for atmospheric NH_3. *Biogeochemistry* 15:191–211.
17. Sommer, S. G. and Olesen, J. E. 1991. Effect of dry matter content and temperature on NH_3 losses from surface-applied cattle slurry. *Journal of Environmental Quality* 20:679–683.
18. Saeys, W., Engelen, K., Ramon, H., and Anthonis, J. 2007. An automatic depthcontrol system for shallow manure injection, part 1: Modelling of the depth control system. *Biosystems Engineering* 98(2):146–154.
19. Saeys, W., Wallays, C., Engelen, K., Ramon, H., and Anthonis, J. 2008. An automatic depth control system for shallow slurry injection, part 2: Control design and field validation. *Biosystems Engineering* 99(2):161–170.
20. Waksman, S. A. and Cordon, T. C. 1939. Thermophillic decomposition of plant residues in composts by pure and mixed cultures of microorganisms. *Soil Science* 47(3):217–225.
21. Edwards, C. A. and Bohlen, P. J. 1977. *Biology and Ecology of Earthworms*, 2nd ed. London: Chapman and Hall. 159–160.
22. Basak, S. and Ghosh, B. C. 2004. Application of fertilizers and pesticides through fertigation system and its effect on yield, nutrient content and quality of tea. M. Tech. thesis, Indian Institute of Technology, Karagpur, India.
23. Nag, A. and Vizaykumar, K. 2007. *Environmental Education and Solid Waste Management*. New Delhi, India:New Age Publishers.
24. Nag, A. 2008. *Text Book of Agriculture Biotechnology*. New Delhi, India: Preienticel of India.
25. Templeton, D. M., Dean, P. A, and Cherrian, M. G. 1986. The reaction of metallothionein with mercuribenzoate. A dialysis and 113Cd-n.m.r. study. *Journal of Biochemistry* pp. 234, 685–689.
26. Pranav, P. K. and Pandey, K. P. 2008. Computer simulation of ballast management of agricultural tractors *Journal of Terramechanics* 45(6):185–192.
27. Saeys, W., Deblander, J., Ramon, H., and Anthonis, J. 2008. High-performance flow control for site-specific application of liquid manure. *Biosystems Engineering* 99(1):22–34.

28. Wallays, C., Saeys, W., Missotten, B., and De Baerdemaeker J. 2009. Hyperspectral waveband selection for on-line measurement of grain cleanness. *Biosystems Engineering*. In Press
29. Saeys, W., Lenaerts, B., Craessaerts, G., and De Baerdemaeker, J. 2009. Estimation of the crop density of small grains using LIDAR sensors. *Biosystems Engineering*, 102:22–30.
30. Craessaerts, G., Saeys, W., Missotten, B., and De Baerdemaeker, J. 2008. Identification of the cleaning process on combine harvesters. Part I: A fuzzy model for prediction of the material other than grain (MOG) content in the grain bin. *Biosystems Engineering* 101(1):42–49.
31. Coen, T., Vanrenterghem, A., Saeys, W., and De Baerdemaeker, J. 2008. Autopilot for a combine harvester. *Computers and Electronics in Agriculture* 63(1): 57–64.
32. Coen, T., Saeys, W., Missotten, B., and De Baerdemaeker, J. 2008. Cruise control on a combine harvester using model-based predictive control. *Biosystems Engineering* 99:47–55.
33. Craessaerts, G., Saeys, W., Missotten, B., and De Baerdemaeker, J. 2007. A genetic input selection methodology for identification of the cleaning process on a combine harvester, Part II: Selection of relevant input variables for identification of material other than grain (MOG) content in the grain bin. *Biosystems Engineering* 98:297–303.
34. Craessaerts, G., Saeys, W., Missotten, B., and De Baerdemaeker J. 2007. A genetic input selection methodology for identification of the cleaning process on a combine harvester, Part I: Selection of relevant input variables for identification of the sieve losses. *Biosystems Engineering* 98:166–175.

Index

Note: Page numbers referencing figures are followed by an "*f*;" page numbers referencing tables are followed by a "*t*."

A

actuator
  hydraulic, subsystems, 41–42, 42*f*
  mass-spring-damper system, 43
  rate limit, 72
  saturation, 45, 45*f*
adaptive double self-organizing map (ADSOM), 20
adaptive resonance theory, 19
ADSOM. *See* adaptive double self-organizing map
Affymetric data preprocessing, 17
Affymetrix GeneChip, 15–16, 15*f*, 16*f*, 31
AFM. *See* atomic force microscopy
AGNPS. *See* Agricultural Nonpoint Source Pollution Model
agricultural mechanization, 483–485, 484*t*
Agricultural Nonpoint Source Pollution Model (AGNPS), 108, 171, 176–177. *See also* Annualized AGNPS
agricultural practices
  soil/water conservation and, 91
  support practice factor, 116–117
Agricultural Research Service (ARS), 184
Agricultural Runoff Management (ARM), 178
AGWA, 187–188, 189*f*
alar, 85, 438
albedo, 126
algorithms, 11, 12, 149. *See also* genetic algorithms
  classification, 21
  clustering, 18–19
  induction, 3
ALOP. *See* appropriate level of protection

aluminum packaging, 334. *See also* metal packaging
AM. *See* available moisture
ammonia, 489, 491–492
amplification, 59, 79–80, 79*f*
amplitude margin, 69
anhydrous alcohol, 269
Animal Plant Health Inspection Agency (APHIS), 351
AnnAGPS. *See* Annualized AGNPS
Annualized AGNPS (AnnAGPS), 171, 176–177
antioxidants
  activities, measuring, 385–389, 388*f*
  functions, 382–383
  mechanism of, 380–382
  oil stabilization, 384, 385*f*
  olive oil, 398–399
  primary, 383
  stability of, 379–385
  synergetic, 383–385
  types, 383–385
APHIS. *See* Animal Plant Health Inspection Agency
appropriate level of protection (ALOP), 314
ARA. *See* arachidonic acid
arachidonic acid (ARA), 450
  carbon and production of, 454–457
  fungal production of, 454–460
    oil supplement, effect on culture medium and, 457–458
    waste material as fermentation substrate and, 458–459
  genetic engineering and production of, 459–460
  nitrogen and production of, 454–457

505

ArcGIS, 190
ARM. *See* Agricultural Runoff Management
ARS. *See* Agricultural Research Service
ash, 215, 217
atomic force microscopy (AFM), 361–362, 362f
auxins
  calcium in fruit and, 423–429
  fruit, 414
  fruit mineral status and, 406–408, 421
  radioactivity in plants and, 425
available moisture (AM), 198. *See also* moisture
azeotropic mixture, 372

B

BA. *See* benzyladenine
backlash, 44
back-propagation algorithm, 11
bacteria
  cell disruption and, 288
  drug-resistant strains of, 311
  genes, transgenic trees and, 352
  resilient, risk analysis and, 317
bandwidth, 71–72
BASINS. *See* Better Assessment Science Integrating Point and Nonpoint Sources
Bayesian information criterion (BIC), 19–20
Bayesian networks, 21, 25
benzyladenine (BA), 409, 412–413
  fruit thinning with, 430, 437
best management practices (BMP), 117–118
  commonly used, 118–120
  MUSLE, 175
  watershed models, 117–118, 118f
Better Assessment Science Integrating Point and Nonpoint Sources (BASINS), 178, 186–187, 187f
BIC. *See* Bayesian information criterion
bio-based economy, 240
bio-based materials, 345
biodegradable packaging, 345
biodiesel, 234
  consumption, promoting, 254–255
  drawbacks, 256
  emissions, 263t
    CO, 262, 264f
    engine, 253–278
    hydrocarbon, 262, 264f
    load effect on, 261–262
    NO_x, 261–262, 262f
  exhaust gas temperatures, 265–266, 265f
  load effect on performance of, 260–261, 261f

biodiesel (*Cont.*):
  manufacturing costs, 254
  performance of, in engines, 253–278
  production of, 253–254
  tax exemption, 255
  vegetable oil processed into, 254, 255, 257
bioenergy, 253. *See also* energy
biofungicides, 500
bioherbicides, 500
bioinsecticides, 499–500
biological control, 499–501, 499f
biological systems
  input/output and, 46
  machine learning methods and, 3–4
  SFE applications for, 453–463
biomass
  cellulose and, 215, 216f, 218–219
  composition, 215–217
  devolatilization, 222, 222t
  economy, model of, 214, 214f
  hemicellulose and, 219–220
  lignin and, 220
  particle, cylindrical cells and, 217, 217f
  primary thermal decomposition reactions and, 217–218
  processing, 215
  second-generation transportation fuels converted from, 215
  single-particle models, 220–222
  transportation fuels converted from, 215
biomass pyrolysis, 213–242. *See also* pyrolysis
  reactions of, 217–220
bio-oils
  aging, 231
  aqueous phase catalytic processes, 239
  commercial use of, 233
  DTG curves, 226, 226f
  furans in, 236–237
  gasification synthesis, 238
  hydroprocessing, 239
  hydrotreatment of whole, 240
  liquid, 242
  multiphase structure of, 229–230, 230f
  phenolic groups and, 236
  phenols in, 235–237
  producing chemicals from, 234–238, 235f
  properties of, 241
  pyrolytic sugars and, 238–239
  refineries, 213–242
  steam reforming studies, 238
  storage/handling of, 230–231
  transportation fuels production and, 238–240
  whole, 234–235, 240

bio-oils, crude, 226–240
  chemical composition of, 226–231, 226f
  classification of, 232
  combustion of, 232
  commercialization of, 242
  diesel engine use of, 232–234
  families of, 227–228
  fuel applications of, 231–234
  fuel properties of, 231–232
  gas turbine use of, 232–233
  low-molecular-weight organic molecules and, 231–232
  octane numbers for, 233
  petroleum and, 228, 228t
  physicochemical properties of, 228–229, 229t
  upgrading, 233–234
biopesticides, 499–501
biorefineries, 240
bioregulators (BRs)
  calcium in fruit and, 406
  exogenous, fruit and, 405–442
  fruitlet thinning using, 429–437
  retardants as, 409–410
bioseparation
  cell disruption in, 284–288
    liquid homogenization for, 287
    mechanical methods and, 284–287
    ultrasonic vibration, 287
  centrifugal separation, 288–290
  chromatographic techniques, 300–304
  crystallization, 297–307
  drying, 297–307
  ELM, 306
  evaporation, 296–297
  filtration, 292–296, 292t
  flocculation, 291–292
  hollow fiber-supported liquid membrane, 307, 308f
  liquid membrane, 306
  membrane technology, 305
  thickening, 290–291, 291f
  thin supported liquid membrane, 307
  ultrafiltration membrane, 305–306, 306f
  unit operations in, 284–307
bioseparation processes, 281–308
  chemical engineering unit operations in, 283
  chemical separation processes, 281–282
  disadvantages, 282
  fermentation, 283
  methods, 283, 283t
  necessity of, 281
  properties, 283
  stages of, 282–283

biosystem
  analysis, 33–38, 47–57
  engineering, 195, 447–463
  identification, 57–66
  modeling, 78–83
  nonlinear optimization example, 75–88
  optimization, 33–38
    nonlinear, 75–88
  representation, 34–47
  step input and, 47–48, 47f
  transfer function, 48–51
Biot number, 221–222
biotechnology, 350–351, 400–401, 400t
block diagrams
  dynamic system, 34–35
  example, combine harvester, 35, 35f
  inputs, 35
  mathematical representation, 36–38
    Laplace transform, 38–44
  outputs, 35
  representation, 34–36
  subsystems, 35–36, 35f
blow molding, 330–332, 330f, 331f
BMP. *See* best management practices
Bode diagram
  constant term (gain term), 52–53, 53f
  derivative term (zero at the origin), 54
  integral term, 53–54, 53f
  quadratic lag, 55f, 56–57, 58f
  simple lag/lead, 54–55, 55f
  transfer function, 51–57
Bode plot
  empirical, 60–62, 61f, 66, 66f
  feedback control system, 72
  feedback controller design, 73–74, 74f
  model parameters, 58
  simulated, 66, 66f
Boolean logic. *See* conventional two-valued logic
Boolean networks, 21–23
bootstrapping methods, 12–13
boron, 479–480
bovine spongiform encephalopathy (BSE), 309
Bowen ratio, 102
brake power
  CO emissions and, 275–276, 275f
  diesel fuel consumption and, 272–273, 272f
  diesel substitution and, 273–274, 273f
  ethanol fumigation and, 273–278, 273f
  exhaust gas temperature and, 277–278, 277f
  NO_x emission and, 274–275, 274f
  smoke number and, 276–277, 276f

Index   507

brake specific fuel consumption (BSFC), 261, 272–273, 272f
brake thermal efficiency, 265, 266f, 271–272, 271f
brake-specific diesel fuel consumption (BSDFC), 272–273, 272f
Bray-KurtzP1 method, 475
BRs. *See* bioregulators
BSDFC. *See* brake-specific diesel fuel consumption
BSE. *See* bovine spongiform encephalopathy
BSFC. *See* brake specific fuel consumption
bubbling fluidized beds, 224–225, 225f

C

cablegation system, 203–204
CAC. *See* CODEX Alimentarius Commission
calcium
  daminozide and uptake of, 425, 426t, 427t
  fruit, 405, 415
    auxin and, 423–429
    BRs and, 406
    deficiency, 410
    distribution of, 423–429
    gibberellins and, 408–409, 428–429
    growth regulators and, 429
    leaves and, 417, 418f
    quality and, 413–414
    retardants and, 410–411, 415, 416f, 420, 423–429
    seeds and, 407, 415–417
    uptake of, 423–429
  plant uptake of, retardants and, 412–413
calcium chloride, 438–441, 439t, 441t
  apple treatment with, 437–442
Campylobacter, 310
cans, 335–337, 335f, 336f. *See also* metal packaging
capacitor, 37
capillary forces, 196–197, 196f
carbon
  ARA production and, 454–457
  neutral, 213
  soil, determining, 472–473
carbon dioxide
  explosion processes, 451–453
  high-pressure, 452, 452f
  SFE, 448–450
  thermodynamic properties of, 449, 449f
carbon monoxide (CO) emissions
  biodiesel, 262, 264f
  brake power, 275–276, 275f
  ethanol fumigation, 275–276, 275f

carbon/nitrogen (C/N) ratio, 455–456. *See also* nitrogen
carbonyl groups, bio-oils and, 236
carboxylic acids, bio-oils and, 235
CART. *See* classification and regression trees
catalytic processes, aqueous phase, 239
CCA pressure treatment. *See* copper-chrome-arsenate pressure treatment
cDNA, 13–15
cDNA microarrays, 13–15, 31
  background correction, 16
  high-level analysis, 18–26
    classification and, 20–21
    clustering and, 18–20
    genetic network modeling, 21–26
  image quantification, 16
  low-level analysis, 16–18
  normalization, 16–17
  quality control, 16
  variability, 17
cell(s)
  filtration resistance and pretreating, 295–296
  microbial, 284
cell disruption
  bacterial cell membrane, 288
  bioseparation and, 284–288
  colloid mill, 284–285, 284f
  efficiency of, 453
  explosion process and, 452
  freeze/thaw method, 288
  liquid homogenization, 287
  mechanical methods of, 284–287
  pan crusher, 285, 286f
  ultrasonic vibration, 287
cell rupture, 284
  breakage by impact/shear and, 284–285
  by pressure, 285–287
cell walls
  AFM visualization of, 362, 362f
  cellulose, 355–358
  components, 355
  density, 353
  hemicellulose, 355–358
  lignin, 355–357
  transgenic tree, 353
  wood, 347
    chemical composition, 355–358
    stiffness, 354, 355f
cellulose, 215, 216f, 357–358
  active form, 218, 218f
  biomass and, 215, 216f, 218–219
  cell wall, wood, 355–358
  deposition, 357–358
  ethanol production from, 268
  fibril networks, 357

cellulose (*Cont.*):
  hemicellulose and, 357
  primary reactions, mechanisms of, 218–219
  synthesis, 357–358
  thermal degradation reactions, 218–219, 219*f*
  transgenic tree content of, 353
centrifuge, 288–290, 289*f*
cetane number (CN), 256–257, 259
channel erosion, 114
charcoal, 213–214, 223
chemical dyes, microarray, 14
chemical engineering unit operations, 283
chlorophyll, 375
chromatin structure, microarray technology and, 1
chromatography, 388
  affinity, 303–304, 304*t*
  column, 301–302, 301*f*
  gel filtration, 303, 303*f*
  ion-exchange, 302–303
  paper, 302, 302*f*
  techniques, 300–304
circulating fluid beds, 225
classification and regression trees (CART), 20
climate, soil erosion and, 112
closed-loop control system, 73–75, 75*f*
closed-loop transfer, 67
cluster validity index, 20
clustering, 18–20, 31
  fuzzy logic-based model, 23
  gene expression and, 19, 23–24, 23*f*
  genes, coregulated, 21
CN. *See* cetane number
C/N ratio. *See* carbon/nitrogen ratio
CO. *See* carbon monoxide emissions
CODEX Alimentarius Commission (CAC), 311–312
  international food trade and, 321
  PC, 314
colloid mills, 284–285, 284*f*
combine harvester, cruise control, 75–88
  amplification and, 79–80, 79*f*
  block diagrams, 35, 35*f*
  control objective, 77–78
  discretization, 81–83
  disturbances, 83
  dynamic model, 80–81
  engine speed, 85
  equality constraints, 84–85
  experimental setup, 86
  inequality constraints, 84–85
  linearization, 81–83
  motivation for, 76
  MPC, nonlinear, 83–86

combine harvester, cruise control (*Cont.*):
  operator comfort in, 76, 86–87
  pump-setting, 86
  results, 86–88, 87*f*
  speed and, 77–79, 78*f*, 87–88, 88*f*
  static model, 78–80, 78*f*
  system description for, 77, 77*f*
  system modeling, 78–83
condensation, 94
condenser, 37
conservation, 116–117. *See also* water conservation
  soil, 91–122
constant volume combustion apparatus (CVCA), 259
constituent entrapment, 119
construction site impact reduction, 119
control spillover, 71
conventional two-valued (Boolean) logic, 7
copper, 476–478
copper-chrome-arsenate (CCA) pressure treatment, 358
cottonseed oil, 376
cover management factor, 116
CPPU, fruit thinning with, 430, 437
crop
  cooling, irrigation and, 201–202
  management techniques, soil and, 140–141
  water demand, 200–201
crossover frequency, 71–72. *See also* frequency
crude soybean oil (SBO), 459
crystal, formation of, 299
crystallization, 297–307
crystallizers, 299–300
CVCA. *See* constant volume combustion apparatus
CWA. *See* U.S. Clean Water Act
cytokines, synthetic, 412–413

D

Dalton's law, 102
daminozide, 410. *See also* retardants
  calcium uptake and, 425, 426*t*, 427*t*
  shoot growth and, 414
dead time, nonlinear system model, 46–47
dead zone, 44, 45*f*
defuzzification, 31
degumming, vegetable oil, 256–257
demethoxylation, lignin, 220
demethyloleuropein, 395–396
detention basins/ponds, 120
devolatilization, biomass, 222, 222*t*
diaminozide, fruit growth and, 420

510 Index

diesel, 253, 267t. See also biodiesel
  brake thermal efficiency, 265, 266f
  emissions, 263t
  engine, bio-oil use in, 232–234
  fuel consumption, brake power and, 272–273, 272f
  substitution, brake power and, 273–274, 273f
differentiation theorem, 39, 40
Diffuse Reflectance Absorbance Spectroscopy Taking in Chemometrics (DRASTIC), 170
dimensionality reduction, 18
disc filter, 293. See also filtration
discretization, 81–83
discriminant analysis, 20
disturbance attenuation, feedback control system, 75, 76f
disturbance rejection, 70–72
DNA microarray, 25, 31
Dorr thickener, 290, 291f
double self-organizing map (DSOM), 20
double-ring infiltrometers, 106
Dounce homogenizer, 287
drainage, systems, 207–209
DRASTIC. See Diffuse Reflectance Absorbance Spectroscopy Taking in Chemometrics
drop structures, 209
drum filter, 293. See also filtration
drying
  bioseparation, 297–307
  conditions, 298
  dryer classification, 297–298
  liquid flow, internal mechanism of, 298
  mechanism, 298
  removing liquid and, methods, 297
dry/wet swales, 120
DSOM. See double self-organizing map
DTG curves. See thermogravimetry curves
Dust Bowl, 91
dynamic behavior, 36
dynamic system, 33–34
  actuator saturation, 45
  block diagrams and, 34–35
  modeling, 44
  output, controlling, 67, 67f

E

earthworms, 494–495
EC. See electrical conductivity
edible oils
  bleachable, 376
  bleaching, 375–378
    energy/enthalpy/entropy and, 376–377, 377f
    loss of oil in, 378, 378f

edible oils (Cont.):
  deodorization, 378–379
  extraction, 367–389
    solvent, 372, 372t
  refining, 367, 374–375
  stabilization, 367–389
  unbleachable, 376
effluent disposal, irrigation and, 202
eicosapentaenoic acid (EPA), 450
electric cable heating, 153–154
  analysis of, 154–159
  cable depth, 156
  energy consumption, 159
  power demand, 158, 158f
  spacing, 156
  temperature, operating of, 156–157, 157f
electrical conductivity (EC), 210–211
ELM. See emulsion liquid membrane
emissions
  biodiesel, 253–278, 263t
  CO, 262, 264f
  engine, 253–278
  hydrocarbon, 262, 264f
  load effect on, 261–262
  NO_x, 261–262, 262f
  CO, 262, 264f
  brake power and, 275–276, 275f
  diesel, 263t
  ethanol, 253–278
  hydrocarbon, 262, 264f
  manure application, 491–493
  NO_x, 261–262, 262f
  vegetable oil, 260
empirical, nonparametric, transfer function estimate (ETFE), 60, 61f
emulsion liquid membrane (ELM), 306
energy
  balance, 101–102, 101f, 145
  cable heating consumption of, 159
  edible oil bleaching, 376–377, 377f
  efficiency, 147, 222
  evaporation and, 93–94
  heating, 147, 159
  substrate, 147, 156–157
ENergy and WATer BALance (ENWATBAL), 144. See also water
engine
  biodiesel, emissions, 253–278
  diesel
    bio-oil, crude use in, 232–234
    Perkins P4, 270–271
  speed, 78
  vegetable oil use in, 256
enthalpy, edible oil bleaching, 376–377, 377f
entropy, edible oil bleaching, 376–377, 377f

Environmental Protection Agency
  (EPA), 350–351
ENWATBAL. *See* ENergy and WATer
  BALance
EPA. *See* eicosapentaenoic acid;
  Environmental Protection Agency
EPIC. *See* Erosion Productivity Impact
  Calculator
equation of motion, 40
erosion. *See also* Kinematic Erosion
    Model; rainfall-runoff erosivity
    factor
  channel, 114
  control, 119
  gully, 114
  hydraulic, 111
  interrill, 114
  rainsplash, 111
  rill, 113–114
  soil, 110–117
    factors affecting, 112–113
    surface drainage and, 209
    types of, 113–117
Erosion Productivity Impact
  Calculator (EPIC), 108
error function, 11–12
Escheria coli, 310
EST. *See* expressed sequence tag
ET. *See* evapotranspiration
ETFE. *See* empirical, nonparametric,
  transfer function estimate
ethanol, 347–348
  as alternative fuel, 266–278
  cellulose material, 268
  corn, 348
  cost, 269
  emissions of, in engines, 253–278
  fumigation, 269–270
    brake power and, 273–278, 273f
    BSDFC and, 273
    emissions and, 274–276, 274f, 275f
    exhaust gas temperature and,
      277–278, 277f
    smoke number and, 276–277, 276f
  hydrocarbon gases, 268–271
  molasses, 267
  performance of, in engines,
    253–278
  production of, 267–271, 348
  properties, 267t
  starch, 267–268
  sugar cane, 267
  sulfite waste liquor, paper
    manufacturing and, 268
  tapioca material, 268
  wood, 268
ethylene, 268
European Food Safety Authority,
  312–313

evaporation, 93–94, 197
  actual, 94
  bioseparation, 296–297
  Dalton's law and computing, 102
  energy and, 93–94
  energy balance method for
    estimating, 101–102
  evaporator performance and, 296
  latent heat flux and, 102
  pan, 103
  potential, 94, 103
  rainfall measurement and,
    99–100
  water and, 93–94, 100
evapotranspiration (ET), 94, 99–100,
    103–104, 197
  estimating, 104, 104t
  potential, 103
  reference crop, 104
  soil moisture and, 177
exfiltration, 94. *See also* filtration
exhaust gas temperature. *See also*
    temperature
  biodiesel, 265–266, 265f
  brake power and, 277–278, 277f
  ethanol fumigation and, 277–278,
    277f
expressed sequence tag (EST), 13
extractives, 215, 217
extrusion, 327–330, 330f

F

FAO. *See* Food and Agriculture
  Organization
farming machinery
  nonlinear optimization example,
    75–88
  control objective, 77–78
  experimental setup, 86
  motivation for, 76
  MPC, nonlinear and, 83–86
  results, 86–88, 87f
  system description for,
    77, 77f
  system modeling, 78–83
  operator comfort in, 76, 86–87
fast Fourier transform (FFT), 149
fats, 367. *See also* saturated fatty acids;
    unsaturated fatty acids
  deterioration, 379–380
  hydrolysis, 379
  oxidation stability of, measuring,
    387
  polymerization, 380
  rancidity, 379
  reversion, 380
fatty acids, 388, 389. *See also* saturated
    fatty acids; unsaturated fatty acids

FC. *See* field capacity
feedback control system, 72
   disturbance attenuation, 75, 76f
   high-frequency models and, 71
   unmodeled dynamic immunity of, 69
feedback controller design, 67–75
   Bode plot, 73–74
   depth control for slurry injection example, 73–75
   example, 73–75, 75f
   loop-shaping, 69–73
      design procedure for, 72–73
   stability, 68–69
   structure, 67–68, 67f
   transfer function, 69–70
feedforward neural networks, 4–5, 4f, 31
FEM. *See* finite-element method
fermentation
   ARA production and waste material as, substrate, 458–459
   bioseparation processes, 283
   pyrolytic sugar, 238–239
FFT. *See* fast Fourier transform
field capacity (FC), 197
figure of merit (FOM), 19
filtration, 292–296, 292t
   bioseparation, 292–296, 292t
   cake resistance, 294
   at constant pressure/rate, 294–295
   cycle, 295
   fluid flow and, 293
   gel, chromatography and, 303, 303f
   rate, 293–294
   resistance, pretreating cells to alleviate, 295–296
   streptomyces, 296
   theory, 293–294
   time, 295
   washing rate, 295
   water, irrigation and, 206
finite-element method (FEM), 143–144
   substrate heating, 152–153
flocculation, 291–292
flooding, wild, 203. *See also* National Flood Frequency
FOM. *See* figure of merit
Food and Agriculture Organization (FAO), 311
food contamination, 309
food hygiene, risk analysis, 316
food packaging
   engineering of, 325–345
   importance of, 325
   materials, 326–328
food production chain, 314, 314f

food safety
   drug-resistant bacteria, 311
   European regulations on, 313–314
   management, 309–321
      evolution of, 321
      legislative requirement for, 320
      perspective, 320
      risk analysis and, 315–320
      zoonotic foodborne hazards and, 310–311
   measures, 311–314
   meat/meat products, 310
   risk analysis, 315–320, 315f
   targets, 314, 314f
   worldwide, 309–310
   trade and, 311–312
Food Safety Objectives (FSO), 314, 314f
food trade
   European, 312–314
   globalization of, 310
   international, 321
   risk analysis, 316
   world, 311–312
foodborne pathogens, 316
Fourier transform infrared (FT-IR)
   microimaging spectroscopy, 363
Fourier transformation, multisine split with, 59
French press, 285–286
frequency
   closed-loop control system, 72
   domain characteristics, 59
   domain model, 65–66
   excited, response at, 62, 63
   ground speed, 62–64, 64f
   nonexcited, response at, 62, 63
   nonlinearity and, 62
   response function, 66
   spectrum, 60
frequency-shift theorem, 39
frost protection, 201
fruit
   auxins and, 404–408, 414, 421, 423–429
   BA thinning of, 430, 437
   calcium chloride treatment of, 437–442
   calcium in, 405, 415
      auxins and, 423–429
      BRs and, 406
      deficiency of, 410
      distribution of, 423–429
      gibberellins and, 408–409, 428–429
      growth regulators and, 429
      leaves and, 417, 418f
      quality and, 413–414
      retardants and, 410–411, 415, 416f, 420, 423–429

fruit, calcium in (*Cont.*):
   seeds and, 407, 415–417
   uptake of, 423–429
  chemical composition, 442
  color, 440
  CPPU thinning of, 430, 437
  diaminozide and growth of, 420
  firmness, 440
  gibberellins and, 408–409
  ground color of skin, 440
  leaves, calcium in, 417, 418*f*
  magnesium in, 417–423
  mineral composition
   auxins and, 406–408, 421
   BR fruitlet thinning and, 429–437
   BRs, exogenous and, 405–442
   calcium chloride treatment and, 438
   quality and, 413
   retardants and effect on, 410–412, 421–423, 422*t*
   SADH treatment and, 438
   urea and, 435
  NAA thinning of, 430, 432, 435–437
  potassium in, 417, 419*f*, 421
  quality, 413–423, 433*t*
  SADH treatment of, 437–442
  seeds, calcium in, 407, 415–417
  shoot growth and, 410
   retardants and, 414
  size, 439, 439*t*
  storability
   BR fruitlet thinning and, 429–437
   BRs, exogenous and, 405–442
   nutrients and, 421, 422*t*
   physiological disorders of, 440–441, 441*t*
   retardants and, 421
  thinning of, 430–432, 431*t*, 437–442
   hand, 432, 435–436
   quality and, 432, 433*t*
   urea, 431–432, 436–437
FS. *See* fuzzy systems
FSO. *See* Food Safety Objectives
FT-IR microimaging spectroscopy. *See* Fourier transform infrared microimaging spectroscopy
fuels. *See specific fuels*
fumigation
  brake thermal efficiency and, 271, 271*f*
  ethanol, 269–270
   brake power and, 273–274, 273*f*
fungus
  ARA production, 454–460
  C/N ratio and morphology of, 455–456
  morphology, 456–457
furans, bio-oil, 236–237

fuzzification, 31
fuzzy c-means, 19, 20
fuzzy logic, 31
fuzzy logic-based model
  gene interaction, 21–23, 22*f*
  model structure selection, 10–11
  noise and, 24
fuzzy rule, 8
fuzzy set, 8, 8*f*, 9
fuzzy systems (FS), 3, 7–9, 8*f*
  operators, 11
fuzzy union operation (logic OR), 24
fuzzy values, 9

G

gain margin (GM), 69
Galerkin method, 144
gap statistic, 19
GAs. *See* genetic algorithms
gas turbines, crude bio-oil use in, 232–233
GATT. *See* General Agreement on Tariffs and Trade
Gauss-Newton method, 12
GC/MS analysis, 388
gene(s), 1–2
  applied time-lagged correlation studying, 24–25
  clustering coregulated, 21
  TFs and, targets, 25
gene expression, 31
  analysis, 2
  Bayesian networks, 25
  cDNA, 13–14
  clustering and, 19, 23–24, 23*f*
  data analysis, 25–26
  data matrix, 31
  fluorescence, 14
  genomic information integrated with, 25, 26
  levels, 14
  microarrays and, 1–2
  patterns, 19
  signal "noise" in, 15
  SOM and, 19
  transcriptionally regulated, 19
gene interactions, 21–23, 22*f*
General Agreement on Tariffs and Trade (GATT), 311–312
genetic algorithms (GAs), 3, 9–10, 24, 31
genetic engineering
  ARA production and, 459–460
  forest-tree research/improvement and, 364
  lignin, 356
  tree-breeding, 349
  wood, 359, 363

514 Index

genetic network, 21–26, 31
genetics, natural, 9
genome organization, microarray technology and, 1
genomewide location analysis, 25
genomic information, gene expression integrated with, 25, 26
geographic information system (GIS), 181–182. *See also* ArcGIS
   customized, models, 190
   data structure, 182
   GUI and, 190
   Internet, 191
   NPS pollution and, 183
   predictive model interfacing with, 185
   relational database structure of, 185
   software, 181–182
   time dimension within, 185
   watershed models and, 169–191
      data transfer between, 184–185
      future of, 190–191
      interfacing, 184–185
      state-of-the-art, recent, 186–190
Geographically Referenced Information System, 181
Geomatics, 181
Geoscience Information System, 181
germination, 125–126, 482–483
gibberellins
   fruit and, 408–409, 428–429
   retardants reducing, 409
ginger, 384–385
GIS. *See* geographic information system
glass packaging, 332–334
GM. *See* gain margin
graphical user interface (GUI), 186, 190
Green-Ampt model, 107
greenhouse
   solar radiation, 127–128
   substrate heating, 146–148, 151–152
Gridded Surface Subsurface Hydrologic Analysis (GSSHA), 189
ground speed, frequency, 62–64, 64*f*
groundwater, 95, 179–180. *See also* water
growth regulators, calcium in fruit and, 429
GSSHA. *See* Gridded Surface Subsurface Hydrologic Analysis
GUI. *See* graphical user interface
gully erosion, 114

H

HACCP. *See* Hazard Analysis Critical Control Points
harvesting, 501–502, 502*f*

Hazard Analysis Critical Control Points (HACCP), 312
HDPE. *See* high-density polyethylene
heat
   accumulators, 151
   capacity, 100, 128, 135
   exchange process, temperature and, 154
   flow
      soil, 140, 142, 149
      substrate heating, 154–156, 155*f*
   flux, 142–143
      evaporation and latent, 102
      plates, 133–134
   latent, 100, 102
   packaging material, 340–341
   pulse method, 133
   sensible, 100
   soil, 128, 135, 155
      conduction of, 143–144
      flow, 140, 142, 149
      flux of, 142–143
      transfer of, 125–128, 145
   transfer, 152, 221–222
      models, 125–159, 141
      soil, 125–128, 145
heating, 149. *See also* electric cable heating; substrate heating
   elements, 146
   energy, 147, 156–157, 159
Hebbian learning, 6
HEC model. *See* Hydrologic Engineering Center model
hemicellulose, 215–216, 216*f*, 357–358
   biomass and, 219–220
   cell wall, wood, 355–358
   cellulose and, 357
   content, 358
   primary reactions, mechanisms of, 219–220
   sugars, 358
   thermal degradation, 219–220
high-density polyethylene (HDPE), 329
holdout methods, 12–13
hollow fiber-supported liquid membrane, 307, 308*f*
Horton model, 106
HRUs. *See* hydrologic response units
HSPF. *See* Hydrologic Simulation Program—Fortran
Hughes press, 285–286
humidification, 495, 496*f*
HU/WQ. *See* Hydrologic Unit Water Quality Model
hydraulic erosion, 111
hydrocarbon
   biodiesel emission of, 262, 264*f*
   ethanol production from, 268–271

Hydrologic Engineering Center (HEC) model, 188
hydrologic models, 189–190
hydrologic response units (HRUs), 173
Hydrologic Simulation Program-Fortran (HSPF), 171, 177–178. *See also* WinHSPF
Hydrologic Unit Water Quality Model (HU/WQ), 184
hydrological cycle, 92–97, 93f. *See also* Gridded Surface Subsurface Hydrologic Analysis
  components, 93
  condensation, 94
  ET, 94
  evaporation in, 93–94
  exfiltration, 94
  groundwater, 95
  interception, 94
  land phase, 173, 174f
    MIKE SHE and, 179–180
  percolation, 94–95
  precipitation, 94, 96–97
  processes, 93
  routing phase, 173, 174f
  runoff, 95
  snow melt, 95–96, 95f
  sun in, 92–93
  transpiration, 94
hydrolysis, 379
hydroxyacetaldehyde, 237
hydroxytyrosol, 398
HYDRUS-1D code, 145

I

ILM. *See* immobilized liquid membrane
immobilized liquid membrane (ILM), 306
Impervious Land Segment Module (IMPLND), 178
IMPLND. *See* Impervious Land Segment Module
infiltration, 104–107, 105f. *See also* filtration
  capacity, 94, 105
  cumulative, 105
  excess runoff, 108
  models, 106–107
  rate, 105, 106
  runoff and, 108
input-output model, 59
insect hormones, 500–501
insecticides, natural, 501. *See also* bioinsecticides
integrated pest management, 499. *See also* pesticides

integration theorem, 39
interception, 94
interfacing, 184–185. *See also* graphical user interface
interference, 31
International Organization for Standardization (ISO), 259
Internet GIS, 191
interrill erosion, 114
ion-exchange, 302–303
iron, 476–478
irrigation
  crop cooling, 201–202
  drop tubes, 204
  efficiency, 202
  effluent disposal and, 202
  frost protection and, 201
  management, 470
  need for, 200–202
  salinization, minimizing and, 201
  sprinklers, 204, 204f, 205f
  systems, 202–206
    center-pivot sprinkler, 204, 204f, 205f
    drip emitter and, 206
    drop, 205–206
    furrow, 203
    sprinkler, 204–205
    subsurface, 206, 206f
    surface, 203–204, 203f
  water
    application efficiencies and, 205–206
    excess, 210
    filtration and, 206
    quality, 195
    saline, 210–211
ISO. *See* International Organization for Standardization
isohyetal method, 98, 99, 99f

K

Kelly filter, 293. *See also* filtration
Kersten function, 138
Kinematic Erosion Model (KINEROS), 171, 188, 189f
KINEROS. *See* Kinematic Erosion Model
KINEROS-2, 178–179
Kohonen learning rules, 6
Kostiakov model, 106
Kraft pulping experiments, 363

L

land
  forming, 208
  use, 112–113, 470

Index

land cover, 112–113
  cover management factor, 116
  soil solar radiation and, 127
Land Information System, 181
land mines, detecting, 145
Laplace domain, 39
  differential equation, 43
  nonlinearities, 44–47
  subsystems, transfer functions of, 43–44
  transfer function in, 40, 41
Laplace transform
  block diagram, 38–44
  properties, 39–40
Laplace variable, 38–39
LDPE. *See* low-density polyethylene
leaf color analysis, 200
leaf water potential, 199–200
learning vector quantization (LVQ), 6
leave-one-out method, 12–13
Levenberg-Marquardt learning algorithm, 12
LID. *See* low-impact designs
lignin, 215–216, 216*f*
  biomass, 220
  cell wall, wood, 355–357
  demethoxylation, 220
  genetic engineering, 356
  primary reactions, mechanisms of, 220
  pyrolytic, 239
  structure, 356
  thermal degradation, 220
lignocellulosic materials, pyrolysis of, 218
ligustroside, 395–396
linear system, 44, 62
linearity, 39, 81–83
linolenic acid, 368
linseed oil, ARA production by fungi and supplementation with, 458
lipids
  extraction, rice bran, 462, 462*f*
  oxygen, oxidation, 381
  production, 455
liquid homogenization, 287
liquid membrane, 306–307, 308*f*. *See also* immobilized liquid membrane
liquid removal, methods, 297
Listeria monocytogenes, 310
logic OR. *See* fuzzy union operation
loop-shaping controller design, 69–73
  loop gain and, 70–71
low-density polyethylene (LDPE), 329
low-impact designs (LID), 117
low-molecular-weight organic molecules, 231–232
LVQ. *See* learning vector quantization

M

machine automation, 76
machine learning methods, 3–13, 31. *See also specific methods*
  algorithms, 21
  applications, 4
  biological systems and, 3–4
  data quality and, 26–27
  induction algorithms, 3
  learning from data and, 11
  microarray data analysis using, 1–32
  modeling, 12
  steps in developing, 10–13
magnesium, fruit, 417–423
manganese, 476–478
manure, 485–498
  ammonia, 491–492
  analysis, 488
  application of, 486–491
  composition measurement, 487–491
  emissions, 491–493
  sampling, 488
  VIS/NIR spectroscopy, 489–490
mass-spring-damper system, 37–38, 37*f*
  actuator force, 43
  equation of motion, 40, 42–43
  input/output, 40–41
  second mass, 41–43, 42*f*
  mathematical representation, block diagram, 36–38
  Laplace transform, 38–44
MBEI. *See* model-based expression index
membrane technology, 305
metal packaging, 334–337
  cans, 335–337, 335*f*, 336*f*
  manufacturing, 335–337
  welding of, 335
MFA. *See* microfibril angle
microarray(s), 1–2. *See also* cDNA microarrays
  analysis, computational methods for, 26
  chemical dyes, 14
  DNA, 25, 31
  gene splicing variants and, 1–2
  oligonucleotide, 15–16
  predesigned chips, 13
  scanned, 14–15, 14*f*
  spotted, 13
microarray data analysis
  analysis methods, 3
  data, 2–3, 17–18
  dimensionality reduction, 18
  using machine learning methods, 1–32
MicroArray Suite, 17

microarray technology, 13–16
  applications, 1–2
  gene expression and, 1–2
  high-level analysis, 18–26
    classification, 20–21
    clustering and, 18–20
    genetic network modeling, 21–26
  low-level analysis, 16–18
  oligonucleotide arrays, high-density, 15–16
microbiological risk assessment (MRA), 318
microfibril angle (MFA), 354–355, 354f
MIKE SHE, 179–180
MIKE-11, 180
MIMO system. *See* multiple-input multiple-output system
minerals
  apple uptake of, 414
  BRs and plant, composition, 405–406
  fruit
    auxins and, 406–408, 421
    BR fruitlet thinning and, 429–437
    BRs, exogenous and, 405–442
    calcium chloride treatment and, 438
    quality and, 413
    retardants and effect on, 410–412, 421, 422t
    SADH treatment and, 438
    urea and, 435
  materials, 215, 217
  uptake of, 413
mismatch (MM) probe, 15
MISO system. *See* multiple-input single-output system
MLP network. *See* multilayer perceptron network
MM probe. *See* mismatch probe
model evaluation, 12–13
model parameters, 57–59
model structure
  selection, 10–11
  system, 57
model-based expression index (MBEI), 17
model-based predictive control (MPC), 75–76, 80–81, 83–84
  discretization, 81–83
  linearization, 81–83
  nonlinear, 83–86
Modified Rational Method (MODRAT), 188
modified universal soil loss equation (MUSLE), 175
MODRAT. *See* Modified Rational Method

moisture, soil, 129, 157, 177
  critical water content and, 136
  predicting, 140
  PWP and, 136
  submodels, 136–137
molecular classification, 21
molybdenum, 478–479
MPC. *See* model-based predictive control
MRA. *See* microbiological risk assessment
mRNA, 13–14
multilayer perceptron (MLP) network, 5, 11
multiple-input multiple-output (MIMO) system, 35
multiple-input single-output (MISO) system, 35, 35f
multiscale systems, 34
multisines, 59, 60f, 62, 63
MUSLE. *See* modified universal soil loss equation

N

NAA, fruit thinning with, 430, 432, 435–437
nanoindentation, 360–361
National Flood Frequency (NFF), 188
Natural Resource Conservation Service (NRCS), 184
natural selection, 9
near-infrared spectroscopy (NIR), 362–363
neural networks, 4–6, 31
  architecture, 10
  feedforward, 4–5, 4f, 31
  soil temperature model, 145–146
neural networks (NNs), 3
neuron, 5, 5f
Newton's method, 12
NEXRAD. *See* Next Generation Radar
Next Generation Radar (NEXRAD), 98
NFF. *See* National Flood Frequency
NIR. *See* near-infrared spectroscopy
nitrogen
  ARA production and, 454–457
  soil, determining, 473
nitrogen oxide (NO_x)
  biodiesel emission of, 261–262, 262f
  brake power and emission of, 274–275, 274f
  ethanol fumigation and, 274–275, 274f
NMR. *See* nuclear magnetic resonance
NNs. *See* neural networks
noise
  filtering, 70, 71
  fuzzy logic-based, 24
  multisines and, 63
  potential causes for, 24

Index

nonlinear optimization example, 75–88
  control objective, 77–78
  experimental setup, 86
  motivation for, 76
  MPC, nonlinear and, 83–86
  results, 86–88, 87f
  system description for, 77, 77f
  system modeling, 78–83
nonlinearity
  estimating, 62–64
  frequency and, 62
  Laplace domain, 44–47
  MPC, 83–86
  system model, 44–47
nonpoint source (NPS) pollution, 117.
  See also Agricultural Nonpoint
    Source Pollution Model; Better
    Assessment Science Integrating
    Point and Nonpoint Sources
  GIS and, 183
  watershed models, 170–171
normalization, 16–17
NO_x. *See* nitrogen oxide
NPS pollution. *See* nonpoint source
  pollution
NRCS. *See* Natural Resource
  Conservation Service
nuclear magnetic resonance (NMR),
  258, 258f
nutraceutical properties,
  398–399, 399t
nutrients, watershed model and
  movement of, 173–175
Nyquist stability, 68

O

observation spillover, 70
oils, 369. *See also specific oils*
  ARA production by fungi and
    supplementation with, 457–458
  extraction of, seed, 369–374, 371f
  properties, 367
  recovery, 372, 372t
  SFE, 373–374
  stabilization, antioxidant, 384, 385f
oleuropein, 395–397
oligonucleotide arrays, high-density,
  15–16
olive(s), 392–394
  biotechnological properties of,
    400–401
  chemical composition of, 392, 392t
  mill products related to, 392–394
    biotechnological properties of,
      400–401, 400t
    phenolic substances from,
      391–401, 395f
  oleuropein and, 395–397

olive(s) (*Cont.*):
  phenolic substances, 391–401, 395f,
    400t
  phenols in, 397
  polyphenols, 394, 395, 395f, 400, 400t
    formulas, 395, 396f, 397f
    nature of, 394–398
    nutraceutical properties of, 399,
      399t
olive oil
  antioxidant properties, 398–399
  ARA production by fungi and
    supplementation with, 458
  benefits of eating, 391
  extraction, 393–394, 393f
    process, 398
  extra-virgin, 393
  nutraceutical properties, 398–399
  phenols, 392
  virgin, 393
Olsen's method, 473–474
oriented strandboard (OSB), 358
oryazanols, 460–461
OSB. *See* oriented strandboard
oxidation
  fat, 387
  fatty acid, 388
  lipid oxygen, 381

P

Pacific Northwest National Laboratory
  (PNNL), 239–240
packaging materials, 326–338
  biodegradable, 345
  food, 325
  glass, 332–334
  mass transport, properties of,
    342–343
  mechanical properties, 339–340
  metal, 334–337
  optical properties, 341–342
  paper, 337–338
  physical properties of, 338–343
  polymer, 326–332, 327t
  recent advances in, 345
  state/phase transition temperature,
    341
  tear strength, 339–340
  thermal properties, 340–341
paclobutrazol, 411, 414. *See also*
  retardants
pan crusher, 285, 286f
paper packaging, 337–338
Parameter Estimation (PEST) tool,
  64–66, 187
particle swarm optimization (PSO),
  3–4, 9–10, 32
PC. *See* performance criteria

Index **519**

PCA. *See* principal component analysis
peak flow rates, rational method for estimating, 109
peak response, 48
percolation, 94–95, 177
perfect match (PM) probe, 15
performance criteria (PC), 314
Perkins P4 diesel engine, 270–271
PERLND. *See* Pervious Land Segment Module
permanent wilting point (PWP), 136, 197–198
peroxide value method, 386
Pervious Land Segment Module (PERLND), 178
pest management, integrated, 499
PEST tool. *See* Parameter Estimation
pesticides, 173–175. *See also* biopesticides
PET. *See* polyethylene terephthalate
petrol, properties, 267*t*
petroleum, 228, 228*t*, 253
phase margin (PM), 69, 72
phase shift, 59
  integral term, 53
  quadratic lag, 56–57
phenol(s)
  bio-oils and, 235–237
  complex, 398
  olive, ripe, 397
  olive oil, 392
  simple, 398
phenolic groups, bio-oils and, 236
phenolic substances
  olive, 391–401, 395*f*, 400*t*
  olive mill products, 391–401, 395*f*, 400*t*
  vegetable oil, 384
Philip model, 106–107
phosphorus, in soil, 473–475
phytoremediation, 349
plants. *See also* fruit; tree(s)
  growth/development, 125–126
  soil, water and, 195
  water, relations, 197
  water needs of, 198–200
plastics. *See* polymers
plate and frame filter, 293, 293*f*. *See also* filtration
PM. *See* phase margin
PM probe. *See* perfect match probe
PNNL. *See* Pacific Northwest National Laboratory
Poiseuille's law, 293
pollution, 349. *See also* Agricultural Nonpoint Source Pollution Model; nonpoint source pollution
polyethylene terephthalate (PET), 329

polymer packaging
  copolymer, 327
  homopolymer, 327
  packaging, 326–332, 327*t*
  processing, 327–332
    blow molding, 330–332
    extrusion, 327–329, 328*f*
    injection molding and, 329, 329*f*
    thermoforming, 332
  synthetic, 345
  thermoplastic, 326–329
polymerization, 380
polymers, permeability of, 343, 344*t*
polyphenols, olive, 394, 395, 395*f*, 400, 400*t*
  formulas, 395, 396*f*, 397*f*
  nature of, 394–398
  nutraceutical properties of, 399, 399*t*
polypropylene (PP), 329
polystyrene (PS), 329
polyunsaturated fatty acids (PUFAs), 453
pomace, 393
position vectors, 20
potassium
  exchangeable, determining, 475–476
  fruit, 417, 419*f*, 421
Potter-Elvehjem homogenizer, 287
power
  cable heating demand of, 158, 158*f*
  centrifuge machine, 289
PP. *See* polypropylene
precipitation, 94, 96–97. *See also* rainfall
primary thermal decomposition reactions, 217–218
principal component analysis (PCA), 18, 32
PS. *See* polystyrene
PSO. *See* particle swarm optimization
PUFAs. *See* polyunsaturated fatty acids
PWP. *See* permanent wilting point
py-MBMS. *See* pyrolysis molecular beam mass spectrometry
pyrolysis
  ablative, 225
  bubbling fluidized beds, 224–225, 225*f*
  energy efficiency, 222
  fast, 223–226
  lignocellulosic material, 218
  reactors, 224, 224*t*, 225
    designing, 220–221
  slow, 222–223
  technologies, 222–226
  vacuum, 225–226
pyrolysis molecular beam mass spectrometry (py-MBMS), 363
pyrolytic lignin, 239
pyrolytic sugars, 238–239

520 Index

Q

quadratic lag, 55f, 56–57, 58f
Quintek Measurement Systems, 359

R

radial basis function (RBF) network, 6
radioactivity, auxins and, 425
rain gauges, 97–100, 98f, 99f
rainfall, 97
    mean areal, depth, 98–99, 98f, 99f
    measurement, 97–100
    runoff to, ratio, 109
rainfall-runoff erosivity factor, 115
rainsplash erosion, 111
RAM. *See* readily available moisture
ramp input, 59, 60f
rancidity, 379
rancimat method, 387
rational method, 109
RBF network. *See* radial basis function network
RCHRES. *See* Reach/Reservoir Routing Module
Reach/Reservoir Routing Module (RCHRES), 178
readily available moisture (RAM), 198. *See also* moisture
real pole first-order system. *See* simple lag
real zero first-order system. *See* simple lead
recyclable organic waste, 493–495
reference trucking, 70
Regulation EC, 313
resistor, 36
retardants, 409–410
    calcium in fruit and, 410–411, 415, 416f, 420, 423–429
    calcium uptake and, 412–413
    fruit mineral composition and, 410–412, 421–423, 422t
    gibberellins reduced by, 409
    orchard use of, 409
reversion, 380
revised universal soil loss equation (RUSLE), 114–115. *See also* universal soil loss equation
rice bran
    ARA production and, 458
    lipid extraction, 462, 462f
rice bran oil, extraction, 460–463
rill erosion, 113–114
riparian buffers, 119–120
rise time, 48
risk analysis, 315, 317
    food safety, 315–320, 315f
    hazard identification and, 318
    industrial processes, 316

risk analysis (*Cont.*):
    methodologies, 319–320
    qualitative, 317, 319
    quantitative, 320
    risk characterization and, 319
    stages, 318–319, 318f
RMA. *See* robust multiarray analysis
robust multiarray analysis (RMA), 17
ROTO. *See* routing outputs to outlets
routing outputs to outlets (ROTO), 175
runoff, 95, 104, 107–108. *See also* Agricultural Runoff Management; rainfall-runoff erosivity factor
    rainfall to, ratio, 109
    soil moisture and, 177
    urban, SWMM and, 180–181
Runoff Block, 180
run-on, 107
RUSLE. *See* revised universal soil loss equation

S

SADH
    apple treatment with, 437–442
    fruit and, 438–441, 439t, 441t
saline parent material, 210
salinity control, 209–211
salinization, 129, 195–196
    controlling, methods for, 210–211
    minimizing, irrigation and, 201
Salmonella, 310, 314
salt. *See* salinization
SAM. *See* standard additive model
sanitary and phytosanitary (SPS) agreements, 311
saturated fatty acids, 367–368
saturation vapor pressure, 102
SBO. *See* crude soybean oil
Scientific Geographic Information System (SGIS), 183–184
SCS. *See* Soil Conservation Service
secoiridoids, 395–396
sediment, 110
    delivery rate, 111
    transport, 110–117
    watershed model and movement of, 173–175
    yield, 111–113, 112f
seeds
    fruit, calcium in, 407, 415–417
    germination method, 482–483
    oil content in, 369–374
    oil extraction in, 369–374
        methods, 370, 371f
        solvent, 371–372
    testing, 480–483, 480f
    viability, rapid tests for, 481–482

Index

self-organizing maps (SOM), 6, 12, 19. *See also* adaptive double self-organizing map; double self-organizing map
SFE. *See* supercritical fluid extraction
SGIS. *See* Scientific Geographic Information System
Sharples Super Centrifuge, 288. *See also* centrifuge
SilviScan, 359, 360*f*
SIMO system. *See* single-input multiple-output system
simple lag (real pole first-order system), 54–55, 55*f*
simple lead (real zero first-order system), 55, 55*f*
SIMPLEX algorithm, 149
simulator for water resources in rural basins (SWRRB), 175. *See also* water
simulator for water resources in rural basins-water quality (SWRRBWQ), 175. *See also* water
single-input multiple-output (SIMO) system, 35
single-input single-output (SISO) system, 35
SISO design toolbox, 73
SISO system. *See* single-input single-output system
slope length/steepness factor, 116
smoke number, brake power and, 276–277, 276*f*
SMW. *See* soymeal waste stream
snow melt, 95–96, 95*f*
soil
  BMP, 117–118
  boron in, estimating, 479–480
  carbon in, determining organic, 472–473
  composition, 112
  conditions, 138
  conservation, 91–122
  copper in, determining available, 476–478
  crop management techniques and, 140–141
  deposition/detachment, 110–111
  dry, 136, 207
  energy balance in, 145
  erodibility factor, 115
  erosion, 110–117
    factors affecting, 112–113
    surface drainage and, 209
    types of, 113–117
  frozen, 138
  heat in, 155
    capacity of, 128, 135
    conduction of, 143–144

soil, heat in (*Cont.*):
    flow of, 140, 142, 149
    flux of, 142–143
    transfer of, 125–128, 145
  heated/unheated, 157
  heterogeneity, 140
  iron, determining available, 476–478
  manganese, determining available, 476–478
  medium-textured, 137
  moisture, 129, 157
    factors affecting, 177
    predicting, 140
    submodels, 136–137
  molybdenum in, 478–479
  nitrogen in, determining, 473
  nonuniform, thermal properties of, 131–132
  packing, degree of, 129
  pH, determining, 472
  phosphorus in, determining, 473–475
  physical properties of, 196–197, 196*f*
  plant water needs assessed by, 198–199
  porosity, 129
  preparation, 470–480
  properties, 134–135, 472–480
  root zone, 158
  runoff water infiltration of, 107
  salt concentration, 129
  sampling, 471–472
  saturated, 137
  solarized, 141
  structure, 132
  surface color/cover, 127
  temperature, 125, 137, 157–159, 159*f*
    abnormally high, 141
    estimation, 125–126
    predicting, 139–146, 139*f*
    solar radiation and, 126–127
    surface, 149
    thermal conductivity and, 129
  testing, 470–480
  thermal conductivity, 128–130, 138
    effective, determining, 135–136
  thermal diffusivity, 130, 131
  thermal properties, 128–138
    determining, 130–138, 131*f*
  thermal regime, 125–126, 140
  THM behavior in, 144
  uniform, thermal properties of, 132
  unsaturated, 137
  volcanic, 141
  water, 140

522   Index

soil (*Cont.*):
  water in
    content of, 127
    flow of, 142
    movement of, 105, 196
    plants and, 195
    retention of, 134–135, 196–197, 196*f*
  waterlogged, 207
  zinc, determining available, 476–478
soil and water assessment tool (SWAT), 108, 171, 175–176, 188, 189*f*. *See also* water
Soil Conservation Service (SCS), 91, 175
  AnnAGNPS and, 176–177
  curve number method, 108–109
solar radiation, 101, 126–128
solenoid, 36
solvent extraction techniques, 447
  mass transfer mechanisms and, 450–451
  oil, 371–372, 372*t*
  SFE, 448, 448*t*
SOM. *See* self-organizing maps
source reduction, 119
source scale conservation, 92
soymeal waste stream (SMW), 459
Spatial Information System, 181
specific heat capacity, 100
SPS agreements. *See* sanitary and phytosanitary agreements
standard additive model (SAM), 24
steady state, 48, 76
  transfer relation, 50
  value, 75
steel packaging, 334. *See also* metal packaging
steepest gradient method, improving performance of, 12
Stefan-Boltzmann law, 101
step input, biosystem, 47–48, 47*f*
Storm Water Management Model (SWMM), 175–176, 180–181. *See also* water
stormflows, subsurface, 104
stretomyces, filtration, 296
sublimation, 95, 100
substrate heating
  design, 151
  energy, 147, 156–157
  FEM, 152–153
  greenhouse, 146–148, 151–152
  heat flows, 154–156, 155*f*
  temperature, 148–154, 158–159
  thermal performance of, model, 150–151
  thermal properties of, 149
  warm-water buried pipes, 151

subsystems, 34
  block diagrams, 35–36, 35*f*
  hydraulic actuator, 41–42, 42*f*
  Laplace domain, transfer functions, 43–44
  in parallel, 37–38
  in series, 41–44
sucrose waste stream (SWS), 459
sulfite waste liquor, paper manufacturing, 268
sun, 92–93. *See also* solar radiation
supercritical fluid extraction (SFE), 373
  biological systems applications of, 453–463
  biosystem engineering applications of, 447–463
  carbon dioxide, 448–450
  rice bran, 461–463, 462*f*
  solvent, 448, 448*t*
supercritical mass transfer mechanisms, 450–451
supervised learning, 5, 32
support vector machines (SVMs), 3–4, 7, 7*f*, 11, 26, 32
sustainable water management systems. *See also* water
  design of, 195–211
  drainage systems and, 207–209
  irrigation systems and, 202–206
  salinity control and, 209–211
  water quality issues and, 211
SVMs. *See* support vector machines
swarm intelligence, 10
SWAT. *See* soil and water assessment tool
Sweetland filter, 293. *See also* filtration
SWMM. *See* Storm Water Management Model
SWRRB. *See* simulator for water resources in rural basins
SWRRBWQ. *See* simulator for water resources in rural basins-water quality
SWS. *See* sucrose waste stream
system identification
  biosystem, 57–66
  excitation experiments, 59
  experimental data and, evaluating, 60–62
  nonlinearities, estimating and, 62–64
  PEST tool, 64–66
systems analysis, 33
  application of, 34
  biosystem, 34, 47–57

Index

T

TDR. *See* time-domain reflectometry
temperature
  electric cable heating and, 156–157, 157*f*
  heat exchange process and, 154
  packaging materials and state/phase transition, 341
  plant growth/development and, 125–126
  soil, 125, 137, 157–159, 159*f*
    abnormally high, 141
    estimation, 125–126
    predicting, 139–146, 139*f*
    solar radiation and, 126–127
    surface, 149
    thermal conductivity and, 129
  substrate heating, 148–154, 158–159
  transition, state/phase, 341
  variations, ambient, 132–133
  water conductivity and, 137
terpenoids, bio-oils and, 236
terraces, 120
TFs. *See* transcription factors
thermal conductivity, 127
  factors, 129
  packaging material, 341
  soil, 128–130
    conditions and, 138
    effective, determining, 135–136
    water and, 141
thermal degradation
  cellulose, 218–219, 219*f*
  hemicelluloses, 219–220
  lignin, 220
  xylan, 219
thermal diffusivity, 130, 131, 143
thermal efficiency, brake, 265, 266*f*, 271–272, 271*f*
thermal expansion, coefficient, 341
thermal insulation, 146, 151
thermal properties
  packaging material, 340–341
  soil, 128–138
    determination of, 130–138, 131*f*
    uniform, 132
  substrate heating, 149
thermal regime, soil, 125–126, 140
thermal-hydraulic-mechanical (THM) behavior, 144
thermogravimetry (DTG) curves, 226, 226*f*
thickening, 290–292, 291*f*
Thiessen polygon method, 98–99, 98*f*
thin supported liquid membrane, 307
thiobarbituric acid test, 386–387
THM behavior. *See* thermal-hydraulic-mechanical behavior
time delay, 39, 46–47, 46*f*

time dimension, GIS, 185
time-domain reflectometry (TDR), 198
time-lagged correlation, 24–25
time-scaling theorem, 39
time-series experiments, DNA microarray, 25
TMDLs. *See* total maximum daily loads
tocopherols, 461
tocotrienols, 461
topography, 113, 116
total maximum daily loads (TMDLs), 169–170, 186
transcription factors (TFs), 25
transesterification, vegetable oil, 256–257
  analyzing products of, 257–258
  property measurement of, 258–260
TRANSFAC, 26
transfer function, 36, 51*f*
  biosystem, 48–51
  Bode diagram, 51–57
  controller, 69–70
  ETFE, 60, 61*f*
  feedback controller design, 69–70
  Laplace domain, 40, 41, 43–44
  magnitude, 48–51
  parameters and, 65–66, 65*f*
  phase, 48–51
transfer relation, steady-state, 50
transgenic trees, 348–349
  bacterial genes and, 352
  cell walls, 353
  cellulose content, 353
  escape, impact of, 350
  evaluation of, 347–364
    challenges for, 350–351
  growth rate, 351–352
  herbicide-resistant, 349
  mechanical properties of, 352–355
  MFA, 354–355, 354*f*
  physical properties of, 352–355
  regulation of, 350–351
  safety of, 350–351
  sexually sterile, rendering, 350
transient response, 49
transpiration, 94, 99–100, 197. *See also* evapotranspiration
transport scale conservation, 92
transportation fuels. *See also specific fuels*
  bio-oil production of, 238–240
  second-generation, biomass converted into, 215
tree(s), 349. *See also* transgenic trees
tree-breeding programs, 348–349, 351
tree-ring machine, 359
tung oil, 368–369

524 Index

Tustin discretization rule, 81. *See also* discretization
2DSOIL, 144

U

ultrafiltration membrane, 305–306, 306f. *See also* filtration
ultrasonic vibration, 287
universal soil loss equation (USLE), 114–115. *See also* soil
unsaturated fatty acids, 368–369
unsupervised learning, 6, 32
urea
    fruit mineral content and, 435
    fruit thinning with, 431–432, 436–437
    NAA and, 435
U.S. Clean Water Act (CWA), 169–170. *See also* water
U.S. Department of Agriculture (USDA), 350–351
U.S. Environmental Protection Agency (USEPA), 169
USDA. *See* U.S. Department of Agriculture
USEPA. *See* U.S. Environmental Protection Agency
USLE. *See* universal soil loss equation

V

Vallez filter, 293. *See also* filtration
vaporization, 100. *See also* saturation vapor pressure
variable source area, 108
vegetable oil
    ARA production by fungi and supplementation with, 457–458
    biodiesel processed from, 254
    calorific value, 259
    CN, 256–257, 259
    degumming, 256–257
    emission, measurement of, 260
    engine use of, 256
    long-term use of, drawbacks to, 256
    NMR, 258, 258f
    nonedible, 255
    performance, measurement of, 260
    phenolics in, 384
    products of, analyzing, 257–258
    properties of, 259, 260t
    transesterification of, 256–257
        analyzing products, 257–258
        property measurement and, 258–260, 260t
    transesterified, 265–266
        load effect on performance of, 260–261, 261f
    viscosity, 256, 259
vegetation cover, 140
vermicomposting, 493–495
    bed preparation, 496–497, 497f
    harvesting, 497–498
    humidification and, 495, 496f
    loading, 496–497
    maturity, 497–498
    process of, 493, 494f
    recyclable organic waste and, 493–495
    technology of, 496–498
    vermiwash, 498, 498t
visible and near infrared (VIS/NIR) spectroscopy, 489–490
VIS/NIR spectroscopy. *See* visible and near infrared spectroscopy
vitamin E, 461

W

wastewater, olive oil extraction, 394
water. *See also* hydrological cycle
    balance, 173
    BMP, 117–118
    conductivity, 137
    content, critical, 136
    crop, demand, 200–201
    drainage, 207–209
    evaporation, 93–94, 100
    irrigation
        application efficiencies, 205–206
        excess, 210
        filtration of, 206
        quality of, 195
        saline, 210–211
    leaf, potential, 199–200
    management, 197–200
    system, 195–196
    plant
        needs and, 198–200
        relations and, 197
    runoff, 107–108
    soil, 140
        content, 127
        flow of, 142
        movement of, 105, 196
        plants and, 195
        retention of, 134–135, 196–197, 196f
        temperature and conductivity of, 137
        thermal conductivity and, 141
    volume management, 119
water conservation, 91–122
    agricultural practices and, 91
    hydrological cycle and, 92–97, 93f
    stages, 92
water quality
    irrigation, 195
    issues, dealing with, 211
    NPS pollution, 170
    treatment, 119

Index   525

Watershed Modeling System (WMS), 178, 188–190
watershed models, 117–118, 118f
  AGNPS, 176–177
  AnnAGNPS, 176–177
  application, 182–183
  BMP, 117–118, 118f
  characterization, 171–172
  commonly used, examples of, 175–181
  components, 172–173
  continuous, 172
  deterministic, 171
  empirical, 171
  event, 172
  GIS and, 169–191
    data transfer between, 184–185
    future of, 190–191
    interfacing, 184–185
    state-of-the-art, recent, 186–190
  HSPF, 177–178
  input data for, 183
  KINEROS-2, 178–179
  linear, 171, 172
  mechanistic, 171
  MIKE SHE, 179–180
  need for, 170–171
  nonlinear, 171, 172
  NPS pollution, 170–171
  nutrient movement and, 173–175
  origin, 171
  pesticides movement and, 173–175
  sediment movement and, 173–175
  stochastic, 171
  subdivision, 173
  SWAT, 175–176
  SWMM, 180–181
Watershed Science Institute (WSI), 118
waterways, vegetated, 120
Waterways Experiment Station (WES), 189
weather, plant water needs assessed by, 199
WES. See Waterways Experiment Station
wetlands, 120, 121t
WHO. See World Health Organization
Wiener-Hammerstein structure, 80, 81f
WinHSPF, 187
WMS. See Watershed Modeling System

wood
  CCA pressure treatment, 358
  cell walls, 347
    chemical composition, 355–358
    stiffness, 354, 355f
  density, 351–353, 352f, 359, 360f
  durability, genetic modification of, 359
  ethanol production from, 268
  genetic modification, 363
  growth rate, 351–352
  processing, 363
  productivity, 347–364
  quality, 347–364
    density of, 352–353
  quantity, 351
  SilviScan analysis of, 359, 360f
  species, 348, 349
  transgenic
    advanced analysis tools, 359–363
    AFM, 361–362
    durability, 358–359
    evaluation of, 347–364
    mechanical properties of, 352–355
    nanoindentation, 360–361
    NIR, 362–363
    physical properties of, 352–355
    process and, 363
    productivity of, 351–363
    quality of, 351–363
    SilviScan, 359
    tree-ring machine, 359
    utilization of, 363
World Health Organization (WHO), 309
World Trade Organization (WTO), 311, 314
WSI. See Watershed Science Institute
WTO. See World Trade Organization

X
xylan, thermal degradation, 219

Y
Yersinia enterocolitica, 310

Z
zinc, 476–478
zoonotic foodborne hazards, 310–311